ERRATA

Hormone Action in the Whole Life of Plants

Page	column	line	
4	2	15	should read "cell contents and that of the soil solution, modified by the resistance to entry"
15	diagram		arrow should point from Dk to Pr
105	figure 5-7		add to legend "The asterisks locate the quiescent zones."
137	1	17	10° should be 15°
157		18	"crosses" should be "circles"
159	1	17	"large squares" should be "crosses"
203	figure 7-19		add to legend "Left, control; center, EL 531; right, EL 531 + GA_3"
232	2	38	"indoleacetoneni-" should be "indoleacetoni-"
234	2	21	"fig. 8-13" should be "fig. 8-14"
252	2	17	[up from the bottom] "syringic" should be "sinapic"
277	2	19	"thin" should be "thick"
		25	"dotted" should be "dashed"
279	figure 9-10		[line 3 of legend] "dotted" should be "thin"
300	1	11	*Codium* should be *Caulerpa*
329	1	12	"11" should be "14"
337	2	26	delete "(fig. 11-5)"
		37	should read "ments, e.g., figure 11-4 (below) and figure 11-5."
363	1	20	"protein" should be "chlorophyll"
		21	"chlorophyll" should be "protein"
378	figure 12-15		legend should read "Senescence of one primary leaf of *Phaseolus* brought on by treating the other primary leaf with benzyladenine."
387	1	23	"fig. 13-4" should be "fig. 13-3"
413	figure 14-11		add to legend after ". . . time zero": "17 x 10^{-6} added 2 hours before time zero. Below, IAA added *at* time zero. Data from several experiments, shown by different symbols."

Hormone Action in the Whole Life of Plants

Hormone Action in the Whole Life of Plants

KENNETH V. THIMANN

THE UNIVERSITY OF MASSACHUSETTS PRESS AMHERST 1977

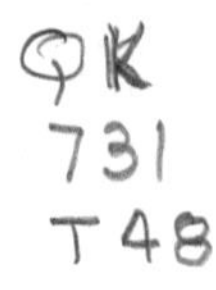

Library of Congress Catalog Card Number 76-26641
ISBN 0-87023-224-X
Printed in the United States of America
Designed by Mary Mendell

Support of the author's researches over many years by the National Science Foundation is gratefully acknowledged.

Library of Congress Cataloging in Publication Data
Thimann, Kenneth Vivian, 1904-
Hormone action in the whole life of plants.
1. Hormones (Plants) 2. Growth (Plants) I. Title.
QK731.T48 581.1'927 76-26641
ISBN 0-87023-224-X

Foreword

When Seymour Shapiro, our present dean of the Faculty of Natural Sciences and Mathematics, was head of the Department of Botany, University of Massachusetts, Amherst, he had the foresight to realize that no department, however large, was likely to embrace completely all of the pertinent subdisciplines of botany, but that this deficiency could be minimized by the annual appointment of a visiting professor. This concept has served us well: to fill whatever instructional lacunae existed at the time of appointment, and, simultaneously, to bring distinguished scientists into our midst for brief periods of time. As a result, over the past several years we have been privileged to have had a succession of outstanding botanists share their rich experiences and knowledge with us: Professors John Harper and Jeffrey G. Duckett of the University College of North Wales; Professor Hans Mohr of the University of Freiburg, Germany; University Professor Friedrich Ehrendorfer of the Botanical Institute, and Director of the Botanic Gardens, University of Vienna; Professor John Heslop-Harrison, formerly Director of Kew Gardens, England, and now Royal Society (London) Research Professor at the University of Wales, Aberystwyth.

Kenneth Vivian Thimann was a member of this distinguished company. A luminary among other notables, Professor Thimann was born and educated in England, receiving his Ph.D. from the Department of Biochemistry and Plant Physiology, Imperial College, the University of London. After five years at the California Institute of Technology where he was associated with Professor Fritz Went, he joined the faculty of Harvard University where he was eventually elevated to the Higgins Professorship of Biology. During this period, marked by a number of significant, original publications in hormonal research, he was honored by election to every important scientific society in this country and to honorary membership in a number of foreign academies and societies. He has been awarded honorary degrees at Harvard, and in Switzerland and France; has held office in a number of scientific societies and served on the editorial boards of many of their publications; has been a Fulbright Scholar and twice, a

Guggenheim Fellow; and has been a recipient of the Stephen Hales Prize of, and is a Charles Reed Barnes Life Member in, the American Society of Plant Physiologists. Among his peers, Professor Thimann is recognized as a distinguished senior statesman of the scientific community in this country and abroad. Most important, he is also a botanist's botanist.

Retiring from Harvard University in 1965, Professor Thimann began a new career as Dean, and then Provost, of Crown College, University of California, Santa Cruz. The naming of the Kenneth V. Thimann Biological Laboratories at Crown College is indicative of the esteem in which he is held at Santa Cruz. A second retirement in 1975, this time to a part-time teaching status, permitted Professor Thimann to join us here for the Fall semester of that year as a visiting professor. His lecture course, *The Role of Hormones in the Whole Life of Higher Plants,* was an immensely rewarding experience for the campus community, students and faculty alike; it led to the present volume, an important historical document since his investigative career has covered most of the period of hormonal research in the higher plants.

I would be remiss if I did not also point out that Professor Thimann's tenure as visiting professor was made all the more delightful by the presence of his lovely wife, Ann. And, if I may add a very personal note, it was a distinct pleasure for me to have had him as a colleague, however temporarily, since Professor Thimann was a member of my doctoral oral committee, more years in the past than either of us wishes to recall.

C. P. Swanson
Amherst, May 1977

Contents

Author's Preface

This book is based on a series of lectures which were delivered at the University of Massachusetts in the fall term of 1974. Since they were addressed primarily to graduate students in botany, it was assumed that the audience had a basic knowledge of the form and function of the flowering plants. As a result, the explanations offered have been somewhat more fundamental in biochemical matters than in basic botany. In order to make the lectures more easily assimilated at first hearing, those specific points which are often slightingly referred to as "technicalities" also were deliberately kept to a minimum, and this practice has been followed in the revised version presented here. The reader should not get the impression that precise technical details are not important, either in the description or the evaluation of the work; they are, and critically so. It is simply that their introduction would generally impede the flow of ideas. Then too, since lectures are a verbal communication between lecturer and audience, and are therefore subject to spontaneous modifications in accord with the evident signs of audience comprehension, they develop something of a conversational nature. That special property is difficult to convey in the written form, though here and there an attempt has been made at it. Whether any of this "spoken" character comes through in the text will of course be for the reader to judge.

The aim of the book is to try to visualize and integrate the multitudinous ways in which hormones initiate and control the growth and development of the higher plants. We follow the plant from birth to death, that is, from seed germination through all growth stages till the new seed has again freshly fallen on the ground. To my knowledge this is the first time that such a lifetime sequence has been attempted. While it is necessarily still full of gaps, and many new discoveries will doubtless illuminate and complicate it, still we have enough solid information now to assemble a composite picture of the life of a plant as influenced by its hormones. I believe it is somewhat different, and perhaps more lively, than the usual textbook picture. In many ways, plant physiologists can be rather proud of the extent to which they have painstakingly elucidated what a recent pseudoscientific writer has called "the secret life of plants." This is particularly justified because the number of researchers in this field is far smaller than those working on the hormonal control of growth and development in animals.

The viewpoint on the material presented is frankly a personal one. Historical details of developments in which the writer has been concerned have often been included. For the same reason, no apology is offered for drawing widely on my own work and on that of my students and colleagues, more especially for the figures and other detailed data. In a span of over forty years of research on plant hormones, illustrative materials in several areas of the general subject have naturally been accumulated. In many cases the data obtained by others could have about equally well been used, but our own materials were at hand. The work of others has not been slighted, of course, since plant physiology is no exception to the globally integrated character of science. However, truly encyclopedic coverage was the last thing at which the lectures aimed, since integration and interrelation constituted the prime targets. Like experimental details, the marshalling of too many facts would obscure the main lines of thought. Instead, I have selected those studies that seem to contribute most to the integration, and have skimmed more lightly over those that either introduce secondary complications or that would take us away from the hormonal concepts being developed.

On this account I have made no attempt to cite in detail the immense number of individuals whose ultimate results are presented. The references given at the end of each chapter are merely those needed to acknowledge sources of the figures and tables. Occasionally I have added a few of the most recent contributions, especially those that appeared after the lectures were transcribed, and bore closely on the matters dealt with. In *The Natural Plant Hormones*, volume VI B of the Academic Press treatise on plant physiology (edited by F. C. Steward) my co-authors and I have listed over 1300 selected references to the original literature, and it seemed unnecessary to duplicate most of that list here. For those who would appreciate a guide to general supplementary reading, I have given at the end a small group of books, symposia, and recent review articles which should provide a useful entry into the original literature.

I am grateful to Professors Seymour Shapiro and Otto Stein for the invitation and sponsorship of the lectures, and for the important exercise in rethinking, encompassing, and re-ordering of the field which was thus forced upon me. I must also thank the several transcribers of the recorded lectures, and especially Leone Stein and the staff of the University of Massachusetts Press for their help in the many details involved in the preparation of the book.

Santa Cruz, California
May 27, 1976

Hormone Action in the Whole Life of Plants

Historical and General Introduction

1

It is sometimes worthwhile to remind ourselves that plants form the background of our civilization—both the plants that we grow and those that favor us with their spontaneous presence. It was our ancestors' discovery that certain wild plants could be grown from seed, harvested, and eaten, that led them to abandon their wanderings and settle down in what we should now call a farming community—the first village. Furthermore it was the productivity of those plants that enabled some of the men or women to turn their attention away from the endless getting of food and the making of tools and weapons for it, to the development of cultural pursuits. So whatever we can learn about plants and their growth is closely related to our fundamental needs for food and fiber, to the history of the human race, and to our plans for the future. The discovery of plant hormones has shed such a flood of light on the growth and behavior of plants that it now constitutes a major part of plant biology, and therefore carries an importance to us all which needs no further justification.

However, the concept that all the delicate adjustments of a complex organism like a flowering plant are made possible by the flow of diffusible chemicals is in some ways an elusive one. It has a curiously *accidental* quality and does not at first sight seem to be capable of sufficient precision. Perhaps it is for this reason that cranks are constantly coming up with claims that plants can appreciate music, be affected by prayer, or distinguish base motives in their watchers from charitable ones—powers that could only be exerted through the possession of a developed nervous system. While we can unhesitatingly dismiss such wishful thinking, it remains true that a nervous system gives a much more appealing impression of both delicacy and immediacy of control than does the secretion and flow of a group of chemical substances.

Indeed, the concept of hormonal control in animals, which preceded the corresponding concept in plants by half a century, was itself slow to be accepted. It was back in 1849 that Berthold made the first experiments which clearly demonstrated the existence of a hormone, and one with multiple effects at that, yet the generality of the idea was not widely accepted until well into the twentieth century. Control by the nervous system seemed much more directly in the chain of events between the environment, the senses, and the bodily reaction. Gowland Hopkins, the biochemist, once wrote: "Up to near the end of the last century, nearly every expert looked to the influence of the nervous system alone as concerned with the coordination of functions in the body; the conception of chemical regulation and coordination had achieved no place in the minds of the majority."

Let me tell you first about Berthold's experiments. He had been observing the behavior of young chickens, relating the characteristic behavior of each sex to its characteristic internal organs. Among other experiments, he removed the testes from two young cockerels; after their recovery from the operation he found that their behavior was completely changed: their combs and wattles remained small and pale, they no longer crowed, and they did not fight as cocks normally do. In other similar birds he removed the testes as before but re-inserted one testis into the abdomen of each. His surgical technique must have been excellent, for they survived, and now they developed into typical males: they crowed, they fought, and their combs and wattles enlarged and darkened in color. (It is interesting that about 100 years later Butenandt used the growth and development of the combs as an assay for testosterone, the male hormone, which he purified.) The climax of Berthold's experiment was that he subsequently killed the cockerels and could locate the inserted testes, which had established no nervous connections but seemed to be growing and were secreting semen. In a strikingly botanical simile, he compared them to grafts in plants, where the scions "bear fruit not of the parent stock but characteristic of themselves." Their action on behavior, he decided, must be exerted via the blood and thus on the whole animal. Thus was formulated, well over a century ago, a clear concept of the secretion of a controlling substance into the circulating blood stream. It was a further century before the actual substance was isolated, but meanwhile, in 1902, there were the classical experiments of Bayliss and Starling which gave the world the name *hormone*.

As is well known, their work began from the observation that if HCl is injected into the duodenum or the small intestine, the pancreas at once begins to secrete its characteristic juice. This happens even if the nervous connections to the pancreas are severed. So they scraped the mucous membrance from a loop of the small intestine of a dog, treated it with HCl, then filtered off the extract and injected it into the blood stream; the pancreas began at once to secrete its juice. The active substance liberated by the HCl was called *secretin,* since it stimulated secretion, and for the class of substances whose action they considered to be that of chemical messengers or stimulants they coined the term *hormone*. Again a further half century has elapsed before work began on the isolation of secretin. Indeed it is a peculiarity of hormone work that a very long time elapses between the experiments that show that a hormone exists and those that elucidate its actual chemical nature.

The discovery of the plant hormone *auxin* is too well known for me to repeat it here.

Most people have read of the Darwins' demonstration in 1880 that when seedlings bend towards the light, the curvature involves the whole length of the shoot, yet the whole response can be prevented by just covering up the extreme apical 2 mm; thus the tip must detect the light and the basal part (as well as the tip) responds. The Darwins, father and son, deduced that some "influence" travels down from the tip to the base. In 1910–1913 Boysen Jensen showed that when the tip was cut off and stuck on again, the plant regained its ability to respond to the light. Seven years later Arpad Paál cut off the tip again but replaced it asymmetrically; the plants now showed a curvature similar to that caused by light but in *darkness*. Thus the asymmetry of the tip was imitating the asymmetry caused by light. Paál was the first to reason that the tip must be producing a growth-promoting substance which travels toward the base, and light must make it travel asymmetrically. Finally, in 1928, Frits Went placed the surviving tips on agar for an hour or two and showed that the agar now had received the growth-promoting property and could produce curvature when applied asymmetrically to other seedlings (see fig. 1-1).

1-1 Summary of the major developments in the early studies of auxin.

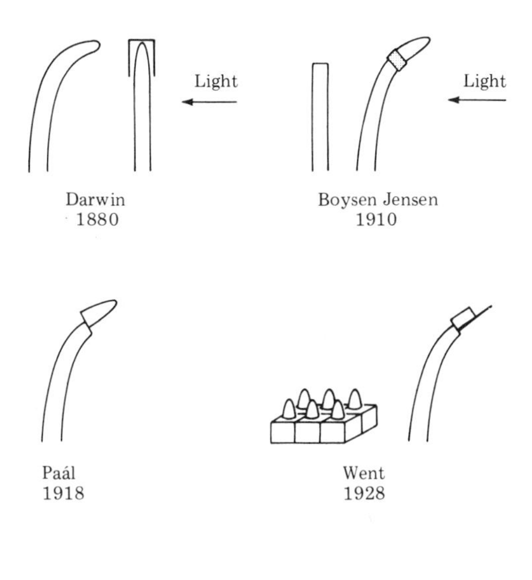

The term *hormone* had actually been applied to plants in 1909, when Hans Fitting had found that the pollen-bearing organs of orchids caused the swelling, withering, and abscission of the petals—the normal changes which follow pollination. The effect of pollination could be duplicated by an aqueous extract of the pollen. He postulated a "pollen hormone" as the causative agent. Strangely enough, the substance concerned turned out to be the same one as Paál had recognized, though its action is probably more complex.

The discovery of the *gibberellins* came close on the heels of that of the auxins. Kurosawa, a Japanese agronomic botanist working in Taiwan in 1926, studied a rice disease in which the plants became excessively tall and rather yellow, and identified the causative organism as *Gibberella fujikuroi*, the perfect form of a Fusarium, which was attacking the roots. When he grew the fungus in an artificial medium, he found that the medium itself had now acquired the ability to cause similar disease symptoms. Thus the disease was due to a special substance which the organism secretes. Work on the isolation of the "gibberellin" from cultures of Gibberella was begun in Japan by Yabuta and Sumiki, but because of the war it remained unknown to the Western world. It was only when an Imperial Chemical Industries correspondent in Japan called it to the attention of his colleagues in Britain that their chemists, especially Jacob MacMillan and his co-workers, with P. W. Brian as botanist, took up the work and finally isolated a whole series of closely related gibberellins. This group of substances now numbers around forty and new members are still being discovered and their structures determined. They are not just fungal products, but actually occur in higher plants and function as growth hormones in a wide variety of ways.

The cytokinins, ethylene (the gaseous hormone), and the several types of inhibiting substances, especially Abscisic acid (abbreviated AbA) will be discussed later, as they enter upon the scene. Their discovery is more recent; like the gibberellins and the auxins, they appear to exert a variety of controls on the development of plants. But at this point we need to take up the problem of the movement of substances in plants.

The concept of hormones in animals is totally bound up with their *movement*; they are defined as messengers, produced in one organ or tissue or gland and transported to another part of the body (the "target organ") to exert their effects. In vertebrates, with their vigorous blood circulation, this poses no problem. In insects the action of hormones shows considerably more localization. In higher plants movement is of three types, each somewhat special, and each justifying brief review, even if the facts are more or less familiar to most readers.

In the first type, water or soil solution enters through the root hairs, and moves across the several layers of root parenchyma cells in response to a gradient of Water Potential, or Water Entering Tendency, which is compounded of the difference between the osmotic potential of the cell contents and the resistance to entry exerted by the mechanical strength of the cell wall. The water or soil solution also passes not only *through* the cells but along the continuous, though thin, medium of the cell *walls*. When it reaches the vascular system and passes through the xylem walls it encounters (at least in fine weather) a negative pressure inside the vessels which draws the water in vigorously. Also in passing through the root parenchyma the water has picked up a small amount of soluble cell contents. Thus a very dilute solution, containing mineral salts from the soil and some organic compounds from the root cells, enters the xylem and is drawn up by the transpiration stream into the stem and leaves. The water goes out by transpiration through the stomata, and the solutes remain.

The second system begins in the leaves, where photosynthesis has produced glucose and fructose phosphates. These enter the various non-photosynthetic cells, including the companion cells, which are rich in phosphatase; here they are converted into monophosphates, synthesized to sucrose and to some smaller amounts of oligosaccharides, and secreted into the sieve elements. Here the concentrated solution exerts enough osmotic pressure to bring in water from adjacent cells, and the resulting hydrostatic pressure makes the sugar solution flow along the sieve tubes—out of the leaf and down the stem—to where the adjacent cells take in the sucrose and raffinose, etc., and with minimal energy expenditure synthesize it into cell-wall polysaccharide. Some also is oxidized to provide phosphorylating energy for other syntheses in these cells.

As a result of these two systems an upward flow of dilute, largely inorganic, solution occurs in the xylem, and a generally downward flow (except upward into young fruits and buds) of more concentrated solution, largely sugars, takes place at the same time in the adjacent phloem. Both are more rapid during the day than at night—in the xylem because the transpiration rate is higher in the warm day temperature and higher in the light, and in the phloem because the photosynthetic rate is, of course, higher in the light.

There is a third type of movement, much slower, but perhaps especially important in seedlings, whose vascular system is poorly developed. This is the movement of solutes from cell to cell. It is only in part a simple diffusion and is in many cases an active secretion. In those cases it is necessarily dependent on oxidations to supply the needed energy. Such movement can be against the gradient of solutes, and in at least some cases a solute can move against its own concentration gradient. This type of active secretion is the one which most frequently takes part in the movement of hormones, especially auxin. Accordingly it is governed by quite special rules, which differ from one transported compound to another; its temperature dependence is like that of enzyme systems and its peculiar qualities are still under active investigation. We shall discuss this process in some detail in a succeeding chapter. In the case of auxin the peculiarity of the transport is that it is morphologically *polar*, i.e., it moves from the apex towards the base of a plant and even on into the roots. This has most important results in the distribution of hormones in the whole plant, and therefore in the localization of various aspects of morphogenesis.

What we shall do in this volume is to take up all the steps or phases in the life of the flowering plant (with occasional excursions into the life of simpler forms) and show how hormones participate in the processes themselves or in their control. We shall start with the seed and go through the elongation of the seedling; its responses to light and to gravity, consisting of modifications in its elongation; the growth and development of individual organs; the interaction between organs, especially the phenomenon of apical dominance; the abscission of leaves; senescence; flowering; and finally the formation of fruits and seeds. Thus we shall touch on most aspects of the total life cycle, or rather those parts of it in which hormones can be shown to play a part—a view of plant growth, metabolism, and development in hormonal terms. It will be a first attempt at a much-needed synthesis.

Hormones and the Germination of Seeds 2

A. FERTILIZATION AND FORMATION OF THE SEED

Let us start at the beginning of the plant's life. Of the fertilization of the ovule by the two nuclei of the pollen tube we can say nothing because we know of no hormonal influences in this process. The pollen contains some auxin, and the elongation of the pollen tube down the style probably depends on the same nutrients as when it elongates on an agar medium, namely sugar, balanced mineral salts and a modicum of boric acid (of which pollen is more demanding than other parts of the plant). What directs the growing tip of the tube to go into the ovule through the micropyle? Evidently this is chemotaxis—chemically directed movement. Is there perhaps a gradient of some special substance coming out of the micropyle, like the gradient of cyclic-AMP which directs the amebae of a slime mold to come together? If there is such an "attraction hormone" we do not know of it. Then when well inside the ovule the tip of the pollen tube bursts. Why? Experiments with pollen germinating on agar show that it often bursts in an auxin solution, but there is no special evidence that I know of to show that the interior of the unfertilized ovule is specially rich in auxin. Some careful quantitative work is needed here.

Then the diploid nucleus begins to divide to produce an embryo with root and shoot initials. The triploid nucleus formed from union of the second pollen nucleus with the two central nuclei of the ovule also divides; at first it produces free-floating nuclei and only later forms cell walls and becomes a tightly packed, undifferentiated tissue, the endosperm. In monocotyledons this accumulates, while in some dicotyledons it becomes absorbed into the two cotyledons, which are diploid, and as a result they become thick and fleshy. Sometimes, especially in the Compositae and some Rosaceae, a thin layer of endosperm remains unabsorbed and surrounds the whole embryo. The morphologists in their wisdom call the first type *mono*cotyledons. But (if it is not blasphemous) I suggest that while this may be true for the lilies, the grasses (Gramineae) could equally be called *non*-cotyledons, for the endosperm is not a cotyledon physiologically, nor is the scutellum, though its origin is somewhat like that of the two cotyledons. Thus it is not clear that they have a cotyledon at all. In any case we have very limited hormonal insight here.

The most remarkable behavior of the embryo is that now, just when it is fully formed and provided with endosperm or cotyledons (a provision which takes generally only a few days or weeks), its growth *stops*. In some cases, like the cereals and the legumes, it seems to run out of water, and the seeds or fruits dry up. But in the succulent fruits—the apples, oranges, melons, and berries—the growth stops while the rest of the fruit is juicy and succulent. The seed, consisting of embryo, endosperm or cotyledons, and seed coat, dries out to a water content of 15% or less, while the ovary tissue may be 90% water. Nobody knows why or how this happens, and oddly enough nobody seems to care much, for I know of no great research programs aimed at the water relations of maturing seeds. In animals the fertilized egg immediately grows into an embryo and continues growth at an undiminished rate; the S-shaped growth curve of a human infant begins from the zygote, shows no great change in slope at the time of delivery, and goes smoothly on for 15 to 20 years. Only in insects is there a possible parallel to the behavior of the seed, for the pupa does represent a stopping place, and pupae can cease growing for years; the 17-year Cicada is an extreme and notable case. Indeed, though it is not often pointed out, the physiologies of higher plants and insects have a number of features in common.

Some microorganisms, of course, show long periods of dormancy. Spores of bacilli can be stored for years, and even frozen in liquid hydrogen, without losing the ability to germinate. Recently spores of two species of Thermoactinomyces found in the mud under Lake Windermere in Britain, and dated as 1330–1500 years old, were claimed to have germinated. Such phenomena are more extreme than the dormancy of most seeds.

B. CONDITIONS LEADING TO GERMINATION

The dried-out seed, which is referred to as "dormant" (literally, sleeping), falls to the ground, and after a while it germinates, i.e., it takes up water again and resumes its interrupted growth. Different genera of plants require different treatments to restart the growth in this way. There are at least five major sets of conditions to induce it:

1. *Simply add water*. This suffices for mature seeds (technically, *caryopses* or one-seeded dry fruits) of the cereal grains, and indeed in a rainy autumn in Britain or western Europe wheat or barley will sometimes germinate in the gathered sheaf. It also suffices for the edible peas and beans, though not for many other legumes.

2. *Stratify*, that is, bury them in moist soil and maintain them for many weeks or months at a moderately low temperature, e.g., 6°–10°C. Continued washing will sometimes have the same effect by washing away growth inhibiting substances. Seeds of forest trees and of some rosaceous fruits respond well to this type of treatment.

3. *Chill*, usually to a few degrees below freezing, for a few days, in the moist state. On rewarming, germination may begin.

Even if chilling is not needed for germination it may have other useful effects, especially in inducing early flowering, as in the *vernalization* of winter wheat.

4. *Weaken the seed coat*, usually with mechanical scraping or partial peeling. Many seeds have a tight, water-impermeable seed coat whose cells are mainly dead; its mechanical strength prevents swelling and therefore prevents the uptake of water. Acacia and many other leguminous trees produce such seeds. The seeds of the East Indian Lotus (*Nelumbo nucifera* or *Nelumbium nelumbo*) have been reported to remain ungerminated in the mud of old ponds for several *centuries*; then after part of the seed is filed off they germinate readily.

5. *Expose to light*. This requirement is the simplest to analyze and will be discussed further on. But first let us consider the first type, for, although it appears the simplest, the process is still complex and offers a nice example of hormonal function on a small scale.

C. THE GERMINATION OF BARLEY

The passion for beer in Britain and western Europe led to much study of the germination of barley, which produces malt. The formation of amylase in the germinating barley grain (fig. 2-1) led to its extraction and purification—the first extracted enzyme—by Payen and Persoz in 1833. Brown and Morris in 1890 made clear how the sugars produced by starch hydrolysis were used in the growth of the embryo, and observed that, if the embryo were removed before soaking, the starch was not hydrolyzed.

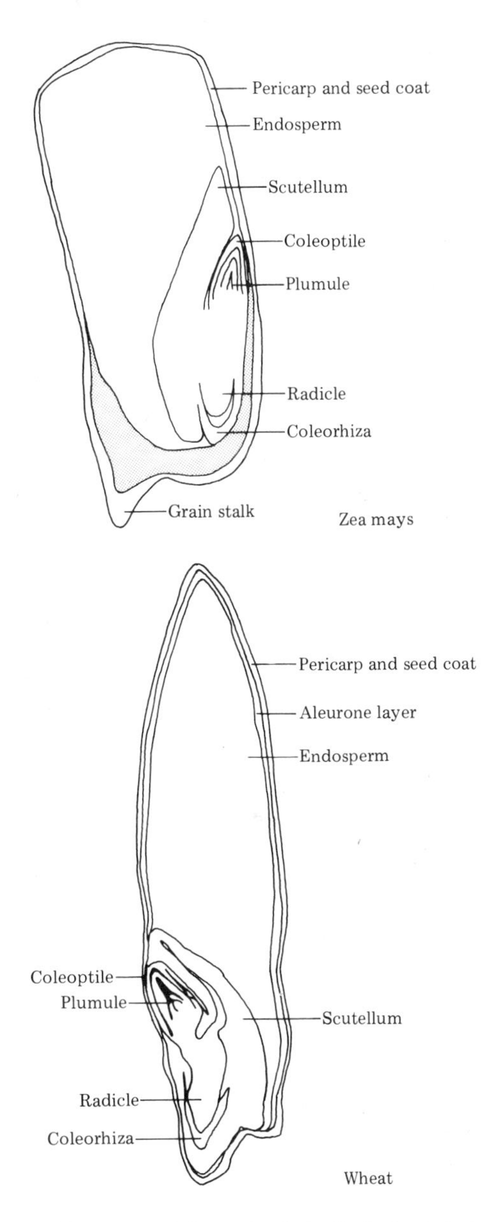

2-1 Structures of the grains (actually fruits, technically *caryopses*) of corn (above) and wheat (below). Oats and barley are very like wheat. From Mayer and Poljakoff-Mayber, 1975.

They placed isolated embryos on a layer of moist starch and noted that the starch became hydrolyzed in the region round the scutellum, much as it did in the endosperm. Thus the embryo appeared to be the source of the amylase. Shortly afterwards, Haberlandt (1900) noted that cell walls in the endosperm were also hydrolyzed, especially close to the aleurone layer. However, it was a long time before the true nature of the process was revealed.

This happened in the late 1950s when Yomo, in Japan, cultured isolated barley embryos on a sugary solution and applied the used culture medium to halved barley grains, using the half without the embryo. The solution activated starch hydrolysis in the half-seed, and Yomo deduced that the medium contained an "amylase-activating substance," AAS. On purifying this material (1960) he found it to have an Rf in chromatography and an infrared spectrum like those of gibberellic acid, GA_3. At this time Leslie Paleg, in Australia, found that, though gibberellic acid had no effect itself in an amylase assay, it did greatly increase the production of maltose and glucose in the embryoless "half-seed." It had no such effect if the half-seed had been briefly heated to 100°, therefore it must be operating *by way of* an enzyme. He also noted that hydrolysis involved not only starch but also protein. Thus Paleg concluded that his postulated "endosperm-mobilizing hormone" was gibberellic acid and was the same as Yomo's AAS.

However, there is another complexity. Endosperm is not the only tissue in the embryoless half-seed, for Varner in 1964 pointed out that the only *living* cells present were the *aleurone*, which consists of a layer one to three cells thick around the endosperm proper (fig. 2-1). Endosperm dissected out by itself does not respond to the GA, but the aleurone does, and its amylase diffuses out into the endosperm. Untreated, it shows no amylase activity. (In wheat, which is otherwise very similar, the untreated aleurone does have some amylase activity, though GA greatly increases this.) Careful study of the enzyme by Briggs in England (1965) showed that the amylase is accompanied by cellulase, protease, ribonuclease, and peroxidase. All in all, it seems clear that in the whole grain the embryo produces gibberellic acid which diffuses to the aleurone and there stimulates the production of a whole group of hydrolytic enzymes; these diffuse out into the endosperm, which consists of dead cells packed with hydrolyzable polymers. The products of hydrolysis then diffuse back into the scutellum to furnish growth material to the embryo and growing axis. The scutellum has an elaborate system of single sieve tubes which facilitates rapid transport to the axis (fig. 2-2).

An almost immediate application of these findings was to the malting process in the brewery. Addition of a modest concentration of GA_3 to the barley on the malting floor hastens saccharification by about half a day—a time saving of some 16%—and costs almost nothing.

D. POSSIBLE MECHANISMS OF GA ACTION AND RELEASE

The above description raises two questions: how does the embryo release GA, and how does the GA release the enzymes? The second is the easier one and we will take it up first.

Since the amylase and other enzymes are not present in untreated barley aleurone they must be either synthesized de novo or released from a bound form. Briggs in England and Varner in USA both found that if phenylalanine labeled with C^{14} was applied to the aleurone, the amylase set free contained C^{14}, hence some of it at least must have been newly synthesized. Varner supplied H_2O containing O^{18} and found that the amylase was increased in density by about 1%; this represents almost complete labelling of the oxygen atoms, which means that virtually all of the enzyme must have been newly synthesized.

Experiments with inhibitors lead to the same conclusion. Parafluorophenylalanine, chloramphenicol and cycloheximide, all of which prevent protein synthesis, all inhibit the appearance of amylase. Abscisic acid does too, and, surprisingly, ethylene partly reverses that inhibition. Actinomycin D, which binds to the guanosine of DNA and thus prevents it from forming messenger RNA, inhibits amylase appearance only if it is applied within 6 to 8 hours of the GA; thereafter it has no effect, indicating that the messenger for amylase has already been formed. On the other hand cyclo-

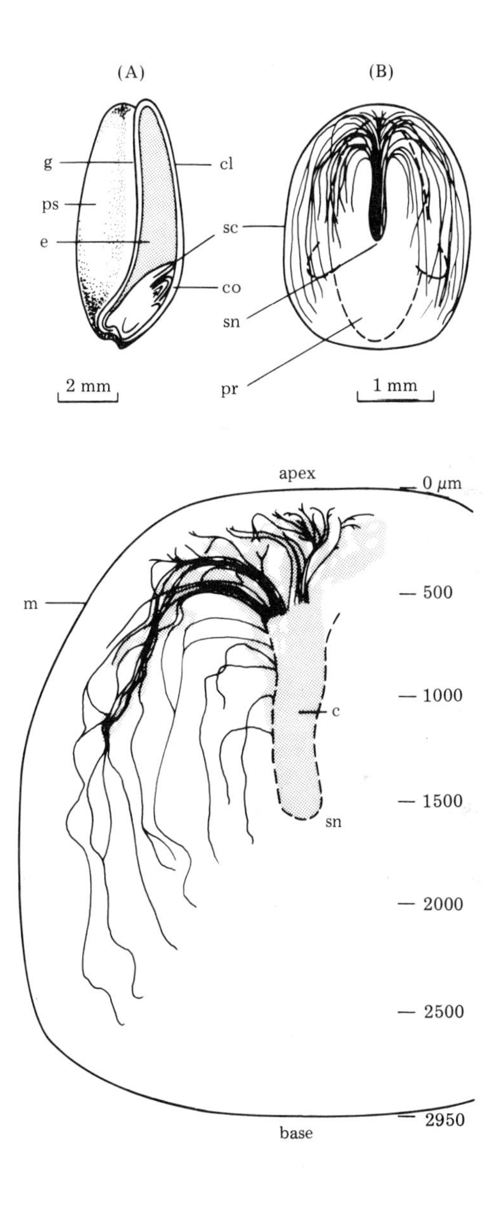

2-2 Above, (A) the wheat grain and (B) its scutellum, as viewed from the endosperm face, the coleoptile and radicle being on the side away from the drawing. *co*, coleoptile; *e*, endosperm; *g*, groove; *pr*, primary root; *ps*, pericarp surface; *sc*, scutellum; *sn*, scutellar node. Below, detail of the sieve tubes of part of the scutellum. *c*, central bundle; *sn*, scutellar node, where the central bundle joins that of the embryo. From Swift and O'Brien, 1970.

heximide, which by preventing attachment of amino acids to the ribosomes directly prevents protein synthesis, inhibits amylase appearance if applied at any time. These two inhibitors thus give us a skeleton timetable for the synthetic process. Gibberellin must somehow activate the DNA to make it start forming the messengers; the 6 hours needed for this step includes the time needed for the GA to penetrate into the aleurone cells and reach its threshold concentration therein.

The question of how the embryo produces or releases the GA is harder to answer. Both Paleg in Australia and Radley in Britain found that the dry embryo, or the embryo directly after soaking in water, yielded no gibberellin on extraction, but 12 to 24 hours after the start of water intake it yielded 0.5 to 1.0 nanogram of GA. The GA seems to come first from the scutellum (supporting our statement above that the scutellum is not a cotyledon), and an isolated scutellum can produce GA for a time. Later it may come from the embryo. Yomo found that after 60 hours the composition of the mixture of gibberellins changes, and suggests that up to this time the substance was released from an inactive bound form and thereafter is newly synthesized. But this has not been proved.

E. SOME COMPARABLE CASES

The sprouting of buds on tubers presents a parallel to the germination of seeds, since, at least in potato, the tuber cells are packed with starch. Most varieties contain a small amount of free sugar, and will sprout slowly without GA treatment. Presumably once the buds have started growth they produce some GA, causing starch hydrolysis. In the Dutch variety Bintje, the buds will not grow (at least early in the season) unless some GA is applied. The slow disappearance of the dormant state has been ascribed to the presence of growth inhibitors which are slowly metabolized away, and indeed Hemberg has shown that an inhibitor of Avena coleoptile elongation is present in the dormant tuber and gradually disappears on storage. But it may also be that this is a coincidence and that the control is exerted by the failure to form gibberellin in the dormant state and the slow development of a GA-synthesizing system on storage. Such a slow development might be the basis for the action of stratification ("method 2" above).

If potato or artichoke tubers are sliced and the slices treated with auxin under aerobic conditions, the cells will enlarge greatly, and the starch disappears within 4 to 6 days (at 25°C). This prompts the suggestion that perhaps auxin somehow brings about the release of gibberellin. So far this problem has not been tackled, but in the seed of blue lupines auxin actually inhibits amylase formation.

One might hope that the germination of dicotyledonous seeds would show behavior similar to that of the cereal grains, and indeed the light-sensitive seeds discussed below do have limited behavior in common with barley. But in the rather well-studied case of peas the control of germination is not at all clear. Isolated cotyledons in a nutrient solution do not hydrolyze their proteins and polysaccharides, and their mitochondria soon deteriorate. But gibberellin seems not to induce proteolytic activity, and indeed protease can be shown to be present from the start. Thus the embryo must exert on the cotyledons some enzyme-activating and mitochondrion-stimulating influence which is not replaceable by gibberellin. Perhaps, as seems to be true in some other seeds, a complex of hormones is required.

In beans (*Phaseolus vulgaris*) the half-cotyledons, freed from the embryo, do develop amylase, so that gibberellin from the embryo is not needed, even though both gibberellin and kinetin do promote amylase development. If the cotyledons are left attached to the plant, the amylase decreases when the embryo begins to grow, which suggests secretion of an inhibitor by the growing axis. Indeed, Abscisic acid does inhibit amylase development in these seeds, and, as reported for the barley seed, ethylene largely reverses this inhibition. In seeds of the peanut (*Arachis hypogaea*) germination is improved by some weeks of moist storage ("after-ripening") and at the same time the auxin and gibberellin contents increase. The gibberellin can hardly be determining, though, because in milkweed seeds which similarly respond to after-ripening, the gibberellin content does not increase at all. In the peanut, like the bean, AbA inhibits germination and ethylene again reverses this inhibition. Also the germination of very old peanut seeds is promoted by auxin, an effect so far not noted in any other seeds. In the seed of fenugreek, a legume much grown for flavoring, hydrolytic enzymes are secreted by the aleurone layer even when the embryo has been wholly removed. Either gibberellin is not needed or perhaps it is already present. Again, AbA inhibits the secretion.

In regard to the inhibition by AbA, Yomo considers that, in *Phaseolus* at least, AbA does not inhibit the enzyme formation directly, but prevents some other change "which must occur in the cotyledons" before amylase formation can begin. Thus in some seeds there may be an additional step, of which, as yet, we know nothing. It is interesting too that the fungal product Fusicoccin, which in many respects acts like an auxin (chap. 14) promotes germination strongly in several seeds, including wheat, and also reverses the AbA inhibition.

In contrast to all these complications, seeds of Avocado (*Persea americana*) present a picture basically like that of barley. Lesham et al. (1973) find that when the seeds are immature the endosperm is very soft. At this time GA released from the embryo causes hydrolysis to begin by acting on the thin layer of sheath cells around the endosperm. Later, when the seeds are mature, the starch has been absorbed into the cotyledons, which have become hard, and the sheath is now reduced to a papery double layer. The seed is now dormant. When germination begins it is this layer which is acted on by GA to cause starch hydrolysis. Thus GA applied to these seeds can release dormancy.

That an elaborate multi-hormone control system must exist within the few millimeters that encompass the average seed is a remarkable conclusion to come to, and makes us wonder about the still smaller system in the pollen grain. Evidently, too, the hormonal control system differs in different seed species. In view of the agricultural importance of understanding seed germination, more research on this topic is obviously needed.

F. GERMINATION PROMOTED BY LIGHT

Just as the morning light wakens the light sleeper, so some seeds which are more or less dormant in the dark are caused to germinate by light. Seeds of this type are found in many families; more than 323 species, in 53 families, have been listed, including most tobacco cultivars, dandelion, the black-seeded varieties of lettuce, the European beech, and several rush species. Most of these are "light sleepers" in that they will germinate to a small extent in the dark (often varying with the individual seed lot) and sometimes are stimulated by chilling or other mild treatments.

Nearly all the quantitative work has been done on the Grand Rapids cultivar of lettuce, a dark brown seeded form which shows some germination in the dark. Flint and McAlister at the U.S. Department of Agriculture were the first to study the light sensitivity. In testing the response to different wavelengths they found that red light of around 660 nm was by far the most active, while the so-called far-red, of 720–740 nm, actually had the opposite effect, inhibiting the germination of those seeds that would have germinated in the dark. Later, Borthwick and Hendricks, of the same institution, showed that if seeds stimulated by red light were exposed to the far-red within an hour or two, before germination had begun, their germination could be 100% inhibited. Furthermore, if

Table 2-1 Germination of Grand Rapids lettuce seeds after imbibition in the dark (16 hours) followed by brief illumination with red light (1 minute), R or far-red light of 740 nm (4 minutes), FR

Illumination	Percentage Germination
Dark	8.5
R	98
R, then FR	54
R, then FR, then R	100
R, then FR, then R, then FR	43
R, then FR, then R, then FR, then R	99
R, then FR, then R, then FR, then R, then FR	54
R, then FR, then R, then FR, then R, then FR, then R	98

From Borthwick et al., 1952.

Table 2-2 Germination of seeds of three tomato cultivars after imbibition in the dark (9 hours) then brief illumination with red or far-red light

	Percentage Germination		
Illumination	Ace	Porte	Glamour
Dark	85	88	71
15 seconds far-red	33	85	56
60 seconds far-red	10	22	28
10 minutes far-red	15	25	33
10 minutes far-red, then 15 seconds red	88	88	74

From Mancinelli et al., 1967.

then re-exposed to red light the inhibition was completely relieved again (table 2–1). Among many others, seeds of several tomato varieties behave in the same way (table 2–2). Thus the germination is controlled by a fully reversible system; it was deduced that a single substance was involved, existing in two forms, one absorbing in the red and the other in the far-red. The deduction was soon borne out by Warren Butler and colleagues who actually observed with a spectroscope that when maize coleoptiles were illuminated at 660 nm their light absorption at that wavelength *decreased*, and it *increased* correspondingly at 735 nm. Extraction and purification, over several years, finally yielded a substance giving a bright blue solution (hence absorbing in the red) and called phytochrome. One form of this, P_r, absorbs at 660 nm and is thereby converted to the other form, P_{fr}, which absorbs at 735 nm (fig. 2-3). From the measured absorption at either wavelength and the maximum change on illumination one can calculate the percentage of each form present at any time.

The matter is not as simple as it first appeared, however, for after exposure to red, if not immediately reversed, the total absorption (i.e. the sum of absorption at 660 and 735 nm) steadily decreases. This means that P_{fr} decomposes, and part of it reverts to the P_r form slowly in the dark. Although there are several intermediate steps in the system, P'_r, P'_{fr}, etc., the basic system can be written:

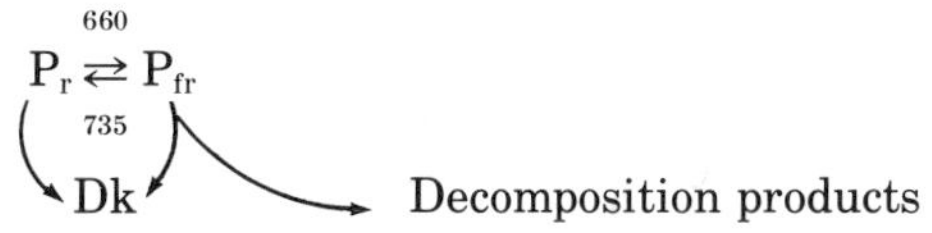

$$P_r \underset{735}{\overset{660}{\rightleftarrows}} P_{fr} \xrightarrow{} \text{Decomposition products}$$
$$P_{fr} \xrightarrow{Dk} P_r$$

Generally very low light dosages ("irradiances") are involved, exposures of less than a minute to a 60-watt bulb through filters usually sufficing for the full conversion. There are some effects of blue light on the system, but perhaps these may involve a quite different photoreceptor (chap. 6). Also there are plants like *Artemisia monosperma* whose seeds respond equally to all wavelengths, but they are rare exceptions.

So much for the stimulus; now for its interaction with the hormone system.

The primary break came from the observation of Khan, Goss, and Smith at Cornell (1957) that simple soaking in gibberellic acid, GA_3, had the same effect as red light. This surprising result was widely confirmed, and gave rise to the notion that "GA acts like light." As we shall see in chapter 11, GA acts on many plants like long days, and indeed the statement is better put in reverse, for there is good evidence that in some plants exposure to long days causes an increased formation of gibberellin. For the lettuce seed many questions at once arise: is the action of GA comparable, or related, to its effect in barley and other cereals? Is GA acting by simply promoting elongation of the embryo, just as it promotes elongation of stems and leaf sheaths in whole plants? Does the 660 nm red light produce a gibberellin in the seed? And lastly, can the two effects be made use of to analyze and understand the nature of germination? Some of these questions can be answered, some not.

First it must be stated that if the seeds are lightly crushed, or the seed coats removed, germination can take place, in the

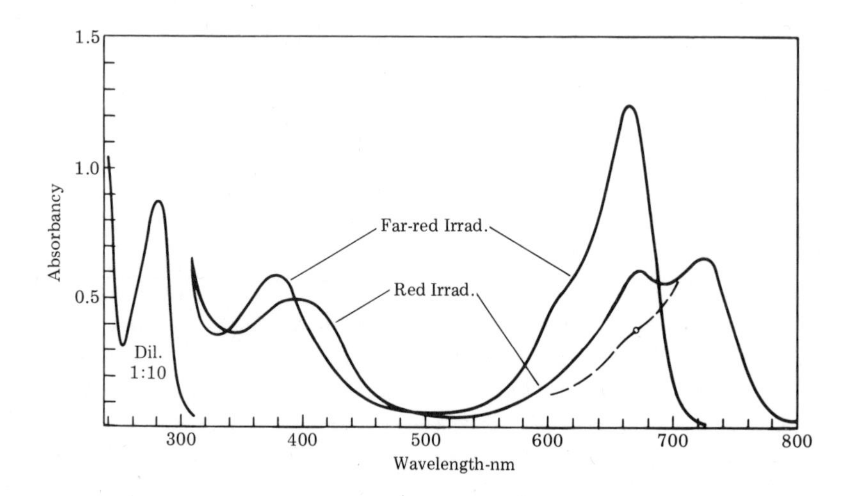

2-3 Absorption spectra of the two forms of phytochrome—P_r (after far-red irradiation) with peak near 660 nm, and P_{fr} (after red irradiation) with main peak about 735 nm, but some P_r still present. Dashed curve, absorbance calculated for pure P_{fr}. From Butler et al. 1959.

majority of seeds, in the dark.* This observation can be made use of to produce dark controls for certain light experiments and, as we shall see, it has important implications.

It would be logical to ascribe the action of GA to its ability to promote elongation. But the first stage in germination is protrusion of the radicle, and GA promotes elongation of shoots, not of roots. Besides, figure 2-4 shows that the embryos of GA-treated seeds do not elongate any faster than in decoated seeds, both being kept in the dark. Thus germination is not *simply* growth; they are two separable phenomena, although some growth participates in the germination process.

An important property of the system is the timing. Figure 2-5 shows that light sensitivity is not permanent. It is absent in the dry seed, of course, and begins to appear during the imbibition of water; the seed becomes maximally sensitive in about two hours (though imbibition is not yet complete) and then gradually loses its sensitivity again. The "decay curve" is very repeatable.

A second important property of the system is its sensitivity to temperature. Soaking at 1°C leads ultimately to complete germination in the dark, even though the germination is necessarily very slow at that temperature. This suggests a parallel with stratification (above). On the other hand, at 35° the seeds can hardly be induced to germinate if the red light is delayed till after 16 hours (fig. 2-6). If the sample of seeds gives partial germination in the dark (as most samples do) then that germination in

* It is worth noting that in all work of this sort, seeds behave as a *population* of varying sensitivity.

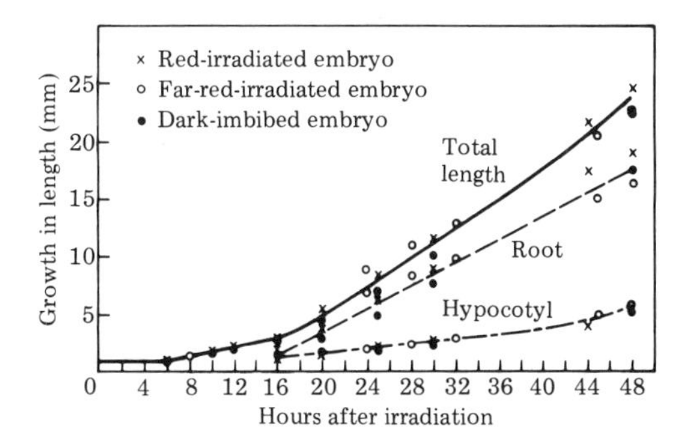

2-4 Elongation of root and hypocotyl of lettuce seeds irradiated with red, with far-red, and unirradiated, then all decoated. Note no effect of light on the elongation of either organ. From Ikuma and Thimann, 1963.

2-5 Germination of lettuce seeds given 1 minute of red light at different times after soaking. The ability to respond decreases steadily, to reach zero (in other experiments) at about 40 hours. For up to 8 hours the red irradiation can be "perceived" as well in nitrogen as in air. However, those few seeds that germinate in the dark are soon affected by the nitrogen. From Ikuma and Thimann, 1964.

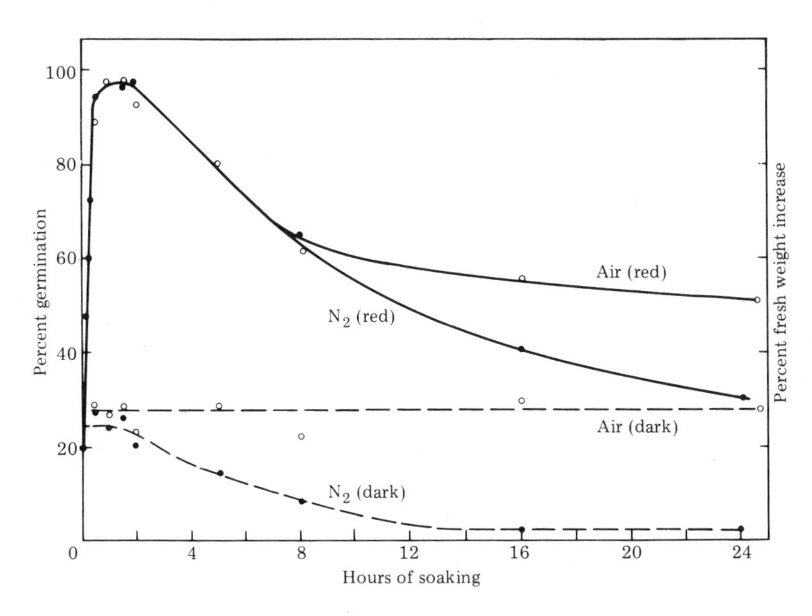

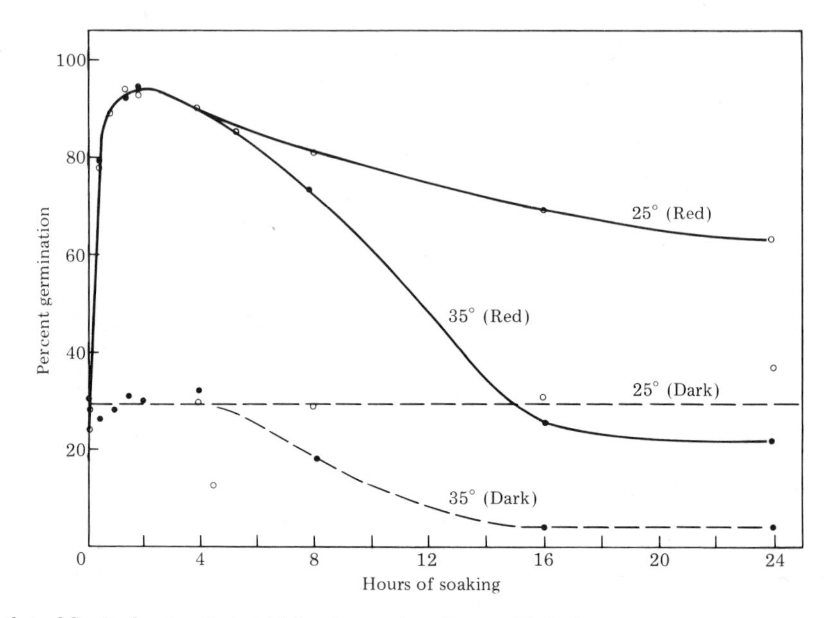

2-6 Experiment like that of fig. 5, showing that at higher temperature the sensitivity to light falls off more rapidly. All seeds in air. From Ikuma and Thimann, 1964.

the dark is also greatly decreased at 35°. The seeds have entered a stage of "harder dormancy" i.e., the progress to the end of the decay curve has been greatly accelerated by the high temperature, and this tells us that the decay of light sensitivity is a chemical or enzymatic process, with the expected large temperature coefficient, a Q_{10} of 2 or 3.

Since the first stages in the onset of light sensitivity go nearly as fast at 3° and 15° as at 25°, with a Q_{10} of not over 1.5, we can correspondingly deduce that a nonchemical process rules, and thus probably that it is the imbibition of water that allows light sensitivity to develop.

That this first process, the "pre-induction phase," is purely physical was borne out by the observation (fig. 2-5) that for up to 8 hours after imbibition it can proceed perfectly well in an atmosphere of nitrogen. However, the exposure to nitrogen cannot be continued for long after the red light has been administered, for then some other process, not growth itself, but preparative for growth, comes into play. Visible growth does not begin until 17 hours after the start of imbibition. Hence we can distinguish four phases in the germination of lettuce seed: pre-induction phase → light-sensitive phase → oxygen-requiring phase → growth. Growth itself also requires oxygen, as we shall see in the next chapter.

The problem as to whether the action of

red light was due to the production of GA was tackled by extracting large numbers of seeds which had been optimally illuminated after 2 hours' imbibition, partially purifying, and testing the concentrated extract on dwarf corn seedlings which are known to be very sensitive to gibberellins (chap. 3). The extracts assayed for no more gibberellin content than the similar extract of seeds which had imbibed in the dark.

The fact that the seeds are able to germinate in the dark—at least to a small degree—suggests that germination (at 25°C) is limited by some kind of a light-demanding system which is highly localized and thus could be easily destroyed. To test this we experimented with making cuts in specific places on the seed. To make such cuts accurately on seeds no more than 2-3 mm long in a very weak green light (since red or white would obviously cause germination in controls) is a difficult matter. I was fortunate to have as colleague a very dexterous and talented Japanese, Hiroshi Ikuma (now at Michigan). Figure 2-7 and table 2-3 show the cuts made and the results after 2 days in moist conditions in the dark.

If the cut was made near the "top" of the seed (the end remote from the embryo), germination was not much greater than in darkness, but the large majority germinated if given 5 minutes red light 1½ hours later. If the cut was made longitudinally, at C, then the seeds all germinated in the dark, but did so by the cotyledons emerging from the slit end; the radicle sometimes protruded in the usual way but more often it simply pushed the whole seed coat along with it (extreme right of the figure). If the cut was made at B, through the widest part of the seed, all the seeds germinated in the dark but again most radicles did not protrude; the halved seed coat was simply pushed off. In some seeds the radicle penetrated this half seed coat later.

Apparently, therefore, the seed's maintenance of the light requirement when cut at

Table 2-3 Effects of seed coats on photocontrolled germination of Grand Rapids lettuce seeds at 25°

	Percent germination					
	Dark control		Light treatment			
Treatment	Expt. 1	Expt. 2	R	FR	FR→R	R→FR
Normal seed	0	8	88	2	90	2
Cut at A	2	16+ (4)	86+ (2)	2+ (6)	98+ (2)	20+ (4)
Cut at B	(100)	(94)	(96)	(98)	(96)	—
Slit at C	(95)	(100)	(90)	(76)	(80)	(86)
Cut at A and Slit at C	(100)	—	—	—	—	—
Seeds with fruit coat removed	12	20	96	14	98	6
Cut at A	30	—	—	—	—	—
Cut at B	(100)	—	—	—	—	—

R: red light. FR: far-red light. ( ): atypical germination.
From Ikuma and Thimann, 1963.

A was due to the inability of the radicle to push the seed coat off, because of the constriction by the wider part of the seed below it. As soon as the cuts allow it, either the radicle or the cotyledons will emerge *without the need for light*. What light does, then, is to allow the radicle to *penetrate* the seed coat. This conclusion, which is critical, has as corollary that the light sensitivity of the seed depends upon the integrity of the seed coat, which imposes a *mechanical* restriction. The seeds of *Acacia*, *Nelumbo* and other highly resistant types, which will not germinate unless scratched deeply or otherwise freed from the tough seed coat, offer a classic example of *mechanical* prevention of germination, but this idea had not been carried over to light-sensitive seeds. Bierhorst has found, too, that *Gladiolus* corms show a similar mechanical restriction and need to be scraped before they can begin growth. All such types of dormancy have been referred to as "coat-imposed."

Another conclusion is that the common belief that germination of lettuce is prevented by a light-sensitive inhibiting substance can hardly be correct, for the cut at A should be large enough to let such a substance diffuse out. The later experiments with light-inhibited seeds (section G below) confirm this. As a matter of fact, Böhmer's classical work on lettuce seeds in 1928 had made the inhibitor theory unlikely, because he showed that such growth-inhibiting substances as could be washed out of seeds were highly nonspecific in their effects, and that similar materials could be extracted from the young leaves of a number of plants.

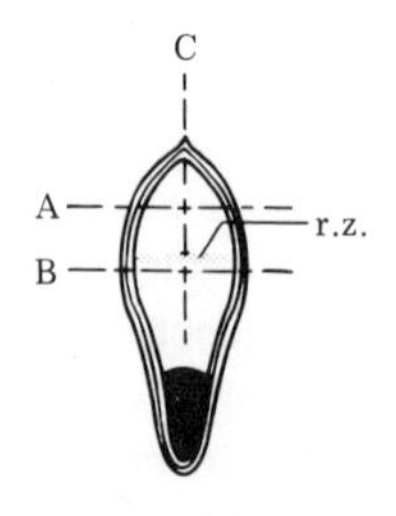

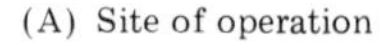
(A) Site of operation

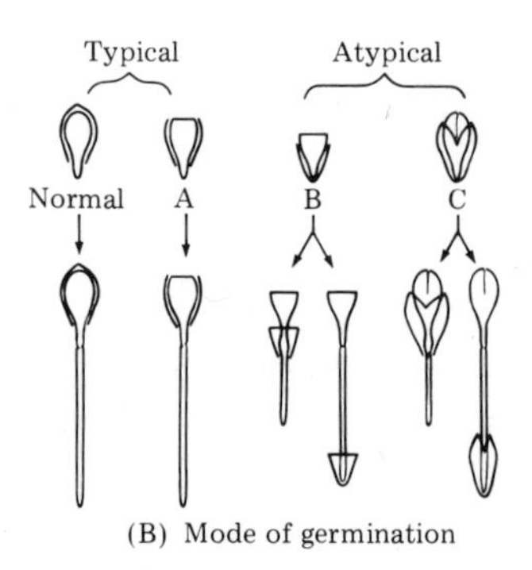

(B) Mode of germination

2-7 Cuts made on lettuce seeds and the modes of germination. Cuts made at A or B, slit made at C. r.z., reference zone; black area, embryonic axis. Above right, immediately after germination; below, two days later. From Ikuma and Thimann, 1963.

The seed coat, which is embryologically a thin layer of endosperm in these seeds, is indeed very tough. However, since like most plant materials it consists of polysaccharide, our next idea was to try the effect of injecting enzymes which hydrolyze such polysaccharides. Unfortunately the insertion of a micropipette and injection of water, as control, seems to weaken the mechanical restriction somewhat, so that a third to a half of the water controls germinated in darkness. But table 2-4 shows nevertheless that cellulase, pectinase, and pentosanase do strongly promote germination in the dark, and are nearly as effective as red light.* In each of the three experiments the boiled enzyme behaved essentially like water.†

Several workers have reported that kinetin exerts an effect on light-sensitive seeds of both lettuce and tobacco, either allowing some germination in the dark or increasing the action of red light. Table 2-5 demonstrates, with lettuce, a very strong synergism between kinetin and gibberellin, in the dark. Kinetin alone has only a small effect. The action is the more remarkable in that kinetin at 10 ppm *inhibits* the elongation of lettuce radicles. When the germinated seeds were dissected the effect was traced to the marked swelling of the cotyledons (table 2-6). The synergism between kinetin and GA is not shown at all in elongation of the parts, but is evident in the fresh weights of the cotyledons, whether

* Since this was written, the endosperm cell walls have been found to be largely a polymer of mannose. Hence the enzymes used were not the most active (Halmer et al. 1975).

† In this and in table 2-3 a very small number of seeds germinating atypically have been omitted.

Table 2-4 Effect of injected enzyme solutions on germination of Grand Rapids lettuce seeds in the dark; 50-100 seeds in each group. Injections made in weak green light.

Expt.	Solution injected	Percentage germination after 2 days at 25°
I	Water	46
	Cellulase (from Myrothecium)	81
	Boiled cellulase	51
II	Water	33
	Pectinase (Rohm & Haas)	86
	Boiled pectinase	49
III	Water	47
	Pentosanase (from bacteria)	91
	Boiled pentosanase	49

From Ikuma and Thimann, 1963.

Table 2-5 Synergism between kinetin and gibberellin on germination of Grand Rapids lettuce seeds in the dark. Data show percentage germination after 2 days.

GA_3 concentration ppm	Kinetin concentration 0	1	10 ppm
0	1	5	11%
1	8	13	35%
10	29	38	80%
20	51	84	100%

From Ikuma and Thimann, 1960.

the isolated cotyledons are soaked in the solutions or the seeds are left intact and then dissected after germination. Kinetin also swells the cotyledons of pumpkin seeds. Such behavior supports the idea of mechanical restriction very well, for the swelling of the cotyledons, like the injection of water in table 2-4, would weaken the strength of the seed coat and thus allow the radicle to penetrate better.

But in some seeds cytokinin may play a more determining part. For instance, the seeds of *Rumex obtusifolius* (whose leaves have been much used in senescence studies [chap. 13]) are also light-promoted and show a red far-red system like lettuce. In these, a brief exposure to red light causes a huge increase in the extractable cytokinin (fig. 2-8). After two more days of darkness or a few minutes of far-red light the cytokinin decreases again, along with the ability to germinate.

So far we have a picture of the germinating process and its control, but we have not explained the roles of red light and gibberellin. We have seen that in barley gibberellin acts to release hydrolytic enzymes, which convert polymers (starch, protein, RNA, etc.) to small molecules. These small molecules exert osmotic pressure. Thus we believe that the action of red light and GA is essentially *to increase the osmotic pressure* within the seed, and thus to put enough strain on the seed coat to allow penetration by the radicle. Earlier we considered the possibility that the action was to liberate cell wall hydrolyzing enzymes and to weaken the seed coat that way, but the osmotic effect, which is equally plausible in other respects, has the advantage that it explains certain experiments in the next section. Perhaps both phenomena participate.

G. GERMINATION INHIBITED BY LIGHT

In contrast to the large number of plants, the germination of whose seeds is promoted by light, there are at least 28 known species, scattered through a dozen families, in which seed germination is inhibited by light. To some extent tomato seeds show light inhibition, but a much more powerful effect is seen in the seeds of *Phacelia tanacetifolia*, a plant of the southwest deserts. A number of other desert plants be-

Table 2-6 Expansion of the cotyledons of lettuce seeds after treatment with light and hormones

	Percentage increase in fresh weight of cotyledons when seeds were:					
	Exposed to light after 2 hours' imbibition, then dissected and treated with solutions. Examined 50 hours later			Exposed to light after 2 hours' imbibition, then treated with solutions, then dissected. Examined 50 hours later		
Treatment	Dark	Red	Far-red	Dark	Red	Far-red
Water	41	39	39	52	82	53
Kinetin (10 ppm)	92	124	88	136	150	120
GA_3 (100 ppm)	72	73	76	83	109	86
Kinetin plus GA_3	184	204	195	130	166	129

From Ikuma and Thimann, 1963.

have similarly, and an ecological interpretation could be that inhibition by light would prevent the seed from germinating too close to the surface of the soil and thus drying out.

As with the light-promoted seeds of lettuce, scraping the seed coat ("scarification"), especially at the radicle end, allows germination in dark or light. In other words, scarification releases the seeds from the inhibition by light. Embryos dissected out from the seeds also germinate in light. Gibberellic acid acts similarly, as Rollin, in Paris, had reported earlier. Gibberellic acid also somewhat increases germination in the dark, so that it is not simply an antagonist of light but a promoter in its own right. Unlike the situation with lettuce, gibberellin here *acts like darkness*. Also as with the light-promoted seeds, light and temperature interact, and low temperatures (around 6-7°C) allow some germination in light. In the wild oat (*Avena fatua*), 3% CO_2 allows germination in light also.

With all these parallels in mind, we undertook to make cuts in specific locations as before, and these have clarified our understanding considerably. Figure 2-9 shows the operations and table 2-7 the results. Cuts across the top of the seed, at B, had little effect, though in some seeds the cotyledons emerged to give *atypical* germination as in lettuce. Cuts at a wider point, as at E, allowed nearly all of them to germinate atypically. Lateral cuts, as at C or D, had little or no effect. If germination were controlled by inhibiting substances, these cuts should have provided ample surface from which such substances could diffuse out. But there was always the possibility

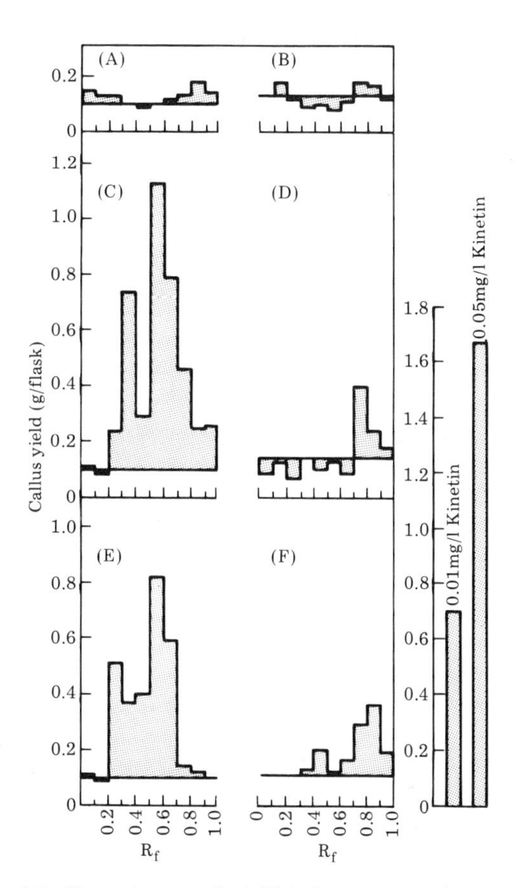

2-8 Chromatograms of cytokinin (by assay on soybean tissue cultures) in extracts of Rumex seed which have been: (A) two days in dark, (B) 4 days in dark, (C) 2 days in dark, then 10 minutes red light, (D) as C but after 2 more days in dark, (E) as C but after 5 minutes far-red, (F) same but 20 minutes far-red. From Van Staden and Wareing, 1972.

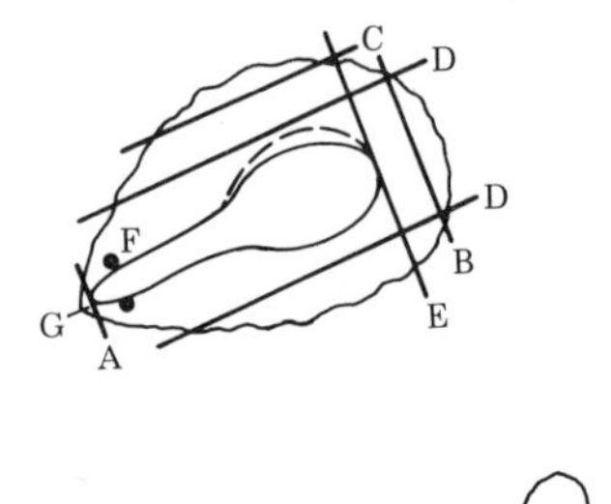

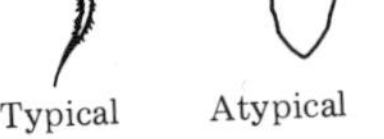

2-9 Cuts made on the seed of *Phacelia tanacetifolia* and the two types of germination resulting. Only the cut at A and the tiny slit at G gave normal (typical) germination. From Chen and Thimann, 1964.

Table 2-7 Germination of seeds of *Phacelia tanacetifolia* following cuts made as in fig. 2-8. Figures in parentheses indicate atypical germination.

Cut made at:	Percentage germination after 2 days in bright light
Intact control	7
A	90
B	5 + (15)
C	5
D	15
E	5 + (75)
F	0*
G	64

* These seeds subsequently germinated 90% on transfer to darkness.
From Chen and Thimann, 1964.

that the inhibitor might be localized close to the radicle. To test that, two fine holes were bored on either side of the radicle as at F. (All such operations are much easier with *Phacelia* than with lettuce because they can be done in white light). Seeds so treated did not germinate in light, but the radicles were evidently not damaged because on subsequent transfer to darkness they germinated 89%. In contrast to the ineffectiveness of this treatment, a very small cut just at the tip, either as A or as G, caused good germination in the light. This confirms that penetration of the radicle through the seed coat is the limiting factor. R. L. Jones of Berkeley has just recently confirmed this idea by direct microscopy, which shows that the thick cell walls of the endosperm or seed coat of the Grand Rapids lettuce seed are rapidly hydrolyzed and dissolved at the time of germination (fig. 2-10). Because the hydrolysis appears to be initiated from within, he concludes that even if the stimulus comes from the embryo, the actual enzyme is secreted within the endosperm cells themselves.

Another point against the role of an inhibitor is this: if an inhibitor were produced in light, it should still be present when the seed is transferred to darkness; yet when light-inhibited seeds are put in darkness they germinate at the same rate as do the unilluminated controls.

What then is the role of GA (or of Fusicoccin which often acts similarly)? It could be to activate cell-wall hydrolyzing enzymes or to produce osmotic pressure (by hydrolysis also) as discussed above. It is probably the latter. Scheibe and Lang, at Michigan, found that a certain concentra-

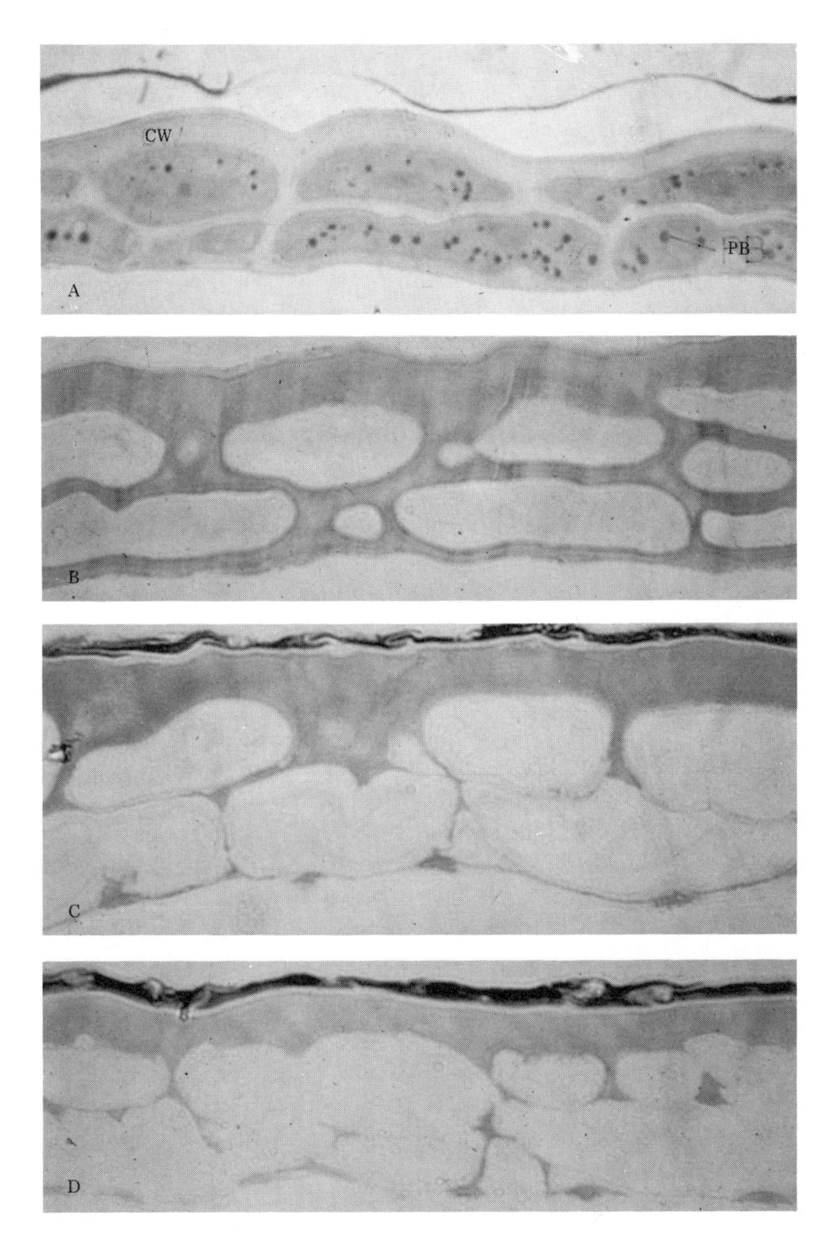

2-10 Sections through lettuce endosperm from seeds imbibed in light for (from top down) 1,3,15 and 24 hours respectively. CW, cell wall; PB, protein bodies. From Jones, 1974.

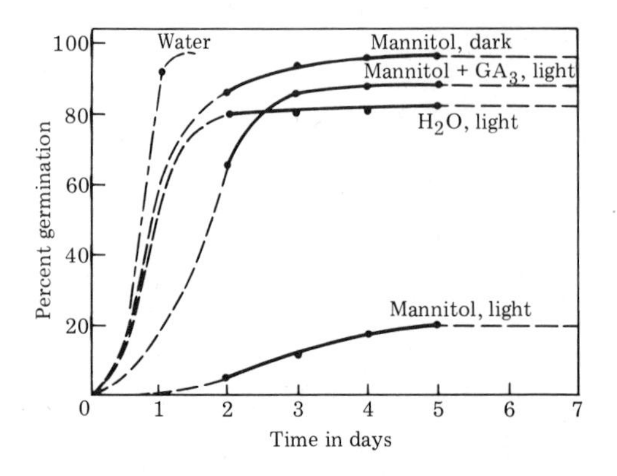

2-11 Restoration of light sensitivity in *Phacelia* seeds, after removal of the tip, by 0.3 M mannitol, and removal of the mannitol-induced light sensitivity by gibberellin. From Chen and Thimann, 1966.

tion of mannitol—0.46 molar for lettuce seed—would just balance the germinating force; red light or GA would just allow expansion against this, while far-red would prevent it. Similarly in *Phacelia*, we found that if the tip were removed as at A and then the seeds soaked in mannitol, their germination was now *inhibited by light* (fig. 2-11). Thus a reduced osmotic pressure reinstates the light sensitivity. GA must therefore cause an increased osmotic pressure. Since actinomycin D or cycloheximide, which prevent protein synthesis, prevent the germination, the needed hydrolytic enzymes must be synthesized de novo in the seed. Light evidently antagonizes this; so does AbA, in some seeds at least, and GA or Fusicoccin reverses that antagonism. As we shall see in the next chapter, light also antagonizes elongation. In total darkness, elongation is excessive, and very suggestively, treatment with gibberellin acts somewhat like darkness in causing excessive elongation.

Thus for barley and for light-promoted and light-inhibited seeds, the limiting factor in germination is the *activation of hydrolysis*. The normal role of hormones, the action of exogenous gibberellin, and the positive or negative action of light, receive a unified explanation in this way.

The changes in old age, whether in plants or in animals, are characterized by marked hydrolysis of polymers: leaves lose their starch and protein (chap. 12); mammals go into "negative nitrogen balance." It is a strange thought that seed germination, the *first* stage in a plant's life, is also made possible by active hydrolysis. Birth and death have their similarities.

REFERENCES*

Borthwick, H. A., Hendricks, S. B., Parker, M. W., Toole, E. H., and Toole, V. K. A reversible photo-reaction controlling seed germination. *Proc. Nat. Acad. Sci.* 38, 662–666, 1952.

Briggs, D. E. Biochemistry of barley germination. Action of GA on barley endosperm. *J. Inst. Brewing* 69, 13–19, 1965.

Butler, W. L., Norris, K. H., Siegelman, H. W., and Hendricks, S. B. Detection, assay, and preliminary purification of the pigment controlling photosensitive development of plants. *Proc. Nat. Acad. Sci.* 45, 1703–1708, 1959.

Chen, S. S. C., and Thimann, K. V. Studies on the germination of light-inhibited seeds of *Phacelia tanacetifolia*. *Israel J. Bot.* 13, 57–73, 1964.

* As stated in the Foreword, these are references only to papers specifically cited, sources of tables and figures, or to a few published since 1972.

———. Nature of seed dormancy in *Phacelia tanacetifolia*. *Science* 153, 1537-1539, Sept. 23, 1966.

Flint, L. H., and McAlister, E. D. Wavelengths of radiation in the visible spectrum inhibiting the germination of light-sensitive seed. *Smithsonian Misc. Coll.* 94, 1-11, 1935; also 96, 1-8, 1937.

Halmer, P., Bewley, J. D., and Thorpe, T. A. Enzyme to break down lettuce endosperm cell wall during germination. *Nature* 258, 716-718, 1975.

Ikuma, H., and Thimann, K. V. The role of the seedcoats in germination of photosensitive lettuce seed. *Plant and Cell Physiol.* 4, 169-185, 1963.

———. The action of kinetin on photosensitive lettuce seed germination as compared with that of GA. *Plant and Cell Physiol.* 4, 113-128, 1963.

———. Action of GA on lettuce seed germination. *Plant Physiol.* 35, 557-566, 1960.

———. Analysis of germination processes of lettuce seed by means of temperature and anaerobiosis. *Plant Physiol.* 39, 756-767, 1964.

Jones, R. L. The structure of the lettuce endosperm. *Planta* 121, 133-146, 1974.

Khan, A., Goss, J. A., and Smith, D. E. Effect of gibberellin on germination of lettuce seed. *Science* 125, 645-646, 1957.

Lesham, Y., Seifer, H., and Segal, N. On the function of seed-coats in gibberellin-induced hydrolysis of starch reserves during avocado (*Persea americana*) germination. *Ann. Bot.* 37, 383-388, 1973.

Mancinelli, A. L., Yaniv, Z., and Smith, P. Phytochrome and seed germination 1. Temperature dependence and relative P_{fr} levels in the germination of dark-germinating tomato seed. *Plant Physiol.* 42, 333-337, 1967.

Mayer, A. M., and Poljakoff-Mayber, A. *The germination of seeds*. 2nd. ed. Oxford and New York: Pergamon Press, 1975.

Paleg, L. Physiological effects of GA. 1. On carbohydrate metabolism and amylase activity of barley endosperm. *Plant Physiol.* 35, 293-299, 1960. 2. On starch hydrolyzing enzymes of barley endosperm. Ibid. 35, 902-906. 3. Observations on its mode of action on barley endosperm. Ibid. 36, 829-837, 1961.

Radley, M. Site of production of gibberellin-like substances in germinating barley embryos. *Planta* 75, 164-171, 1967.

———. The effect of the endosperm on the formation of gibberellin by barley embryos. *Planta* 86, 218-223, 1969.

Swift, J. G., and O'Brien, T. P. Vascularization of the scutellum of wheat. *Australian J. Bot.* 18, 45-53, 1970.

Van Staden, J., and Wareing, P. F. The effect of light on endogenous cytokinin levels in seeds of *Rumex obtusifolius*. *Planta* 104, 126-133, 1972.

Varner, J. E. GA-controlled synthesis of α-amylase in barley endosperm. *Plant Physiol.* 39, 413-415, 1964.

———, Ram Chandra, G., and Chrispeels, M. J. GA-controlled synthesis of α-amylase in barley endosperm. *J. Cellular Comp. Physiol.* 66, 55-68, 1965.

Yomo, H. Amylase Activating Substance. *Hakko Kyokaishi* 18, 494-496, 1960.

———, and Iinuma, H. Production of gibberellin-like substance in the embryo of barley during germination. *Planta* 71, 113-118, 1966.

Cell Enlargement and Growth 3

THE ACTION OF AUXINS

A. THE USE OF ISOLATED SEGMENTS OF SEEDLINGS

To consider the action of auxins in elongation seems like old stuff now, because it takes us back to the late 1930s and early 1940s when the action of auxin was identified. As we saw in chapter 1, Went's demonstration of the action of auxin was done by applying auxin asymmetrically to a decapitated (i.e., auxin-starved) oat coleoptile. It could have been done on other seedlings equally well: Söding, in fact, used the dicot seedling *Cephalaria*. The curvature that results is due to the fact that the side that receives the auxin elongates more than the other side. So the very earliest discovery of auxin rested on the fact that it promoted elongation. However, a system in which tiny blocklets of auxin in agar have to be applied to seedlings in semidarkness does not lend itself very well to extensive physiological study—the procedure was very good for bioassay, and very good for its day; but it was soon found in several laboratories that it was much simpler to take segments of the seedlings and simply float them on the auxin solution, when as a result they would elongate more than control segments. In detail: we take an oat coleoptile, remove the apex, which of course is the source of the natural auxin, remove the primary leaf, cut a fixed length of the central part of the seedling and place this in solution. We later found that it was necessary to have the material in contact with air, so either the segments were allowed to float directly or they were fixed to some support which would float; if they were submerged by as little as 1 mm the growth was depressed by about 50%. The segments are cut at standard lengths so that they can be easily measured thereafter with the eyepiece scale of a microscope. My colleague Charles Schneider soon found that in order to get a good response to auxin it was necessary to add sugar. After all, we were removing them from their source of carbohydrates in the endosperm, so it was only natural that they should respond to sugar; nevertheless it was some years before this was realized. Given a suitable concentration of sucrose—about 2% is adequate—the time relations would look like those of figure 3-1. In water, segments would grow for some 5 hours and then virtually stop; in sucrose or auxin alone

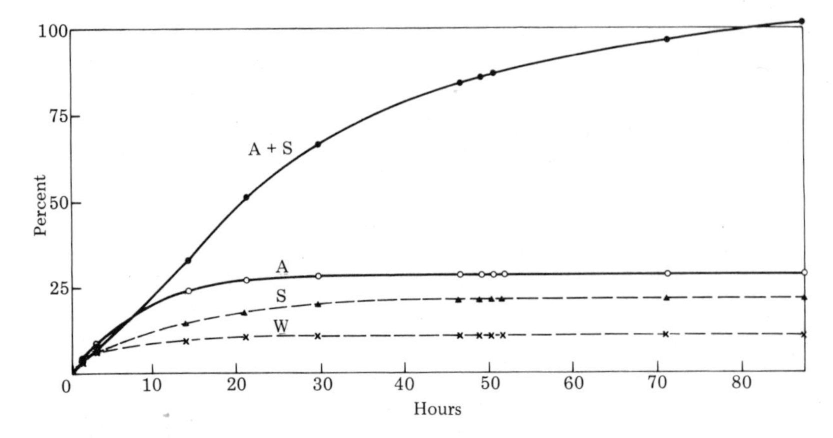

3-1 Elongation of *Avena* coleoptile segments in water (W), sucrose 1% (S), auxin (A), and auxin plus sucrose (A + S). From Schneider, 1938.

they would grow a little longer, while in auxin plus sucrose they grow much faster and continue for about 3 days, to reach double their length. In general they tend not much to exceed double their initial length: e.g., segments 10 mm long grow to 20 mm. One can measure these easily under the microscope to a tenth of a millimeter; hence, with an error of 1%, the floating segments provide a very satisfactory system for study of growth and growth rate. The growth is to some extent a function of the age of the seedling, for when the plants are smaller, the cut segments grow somewhat more. However, most workers have used a standard seedling length of around 25 mm, which gives 10 mm segments able to elongate 100–125% in optimum auxin plus sucrose. This procedure is not restricted to the classical oat coleoptile, for wheat coleoptiles respond equally well, and, as we shall see below, many dicotyledonous seedlings furnish segments that also elongate quite well.

In these seedling segments we have plant material which is not enlarging in width, but is only elongating, and furthermore it is material in which cell division is not important, so that one could simply study cell elongation pure and simple. In oat coleoptiles we have the very great advantage that when they are about 6 mm long, cell divisions are largely completed, and from then on to a final length of about 50 mm only elongation takes place. This was established by simply counting the number of cells in each layer: it was shown long ago by George Avery, then at Connecticut College, that cell division in the epidermal cells ceases at around 6 mm in length and in all other layers at about 10 mm; from then on there is only elongation.

This was historically more important than it seems now, for it was believed in the

1930s that auxin was the "cell enlargement hormone," and therefore one wanted to study uncomplicated cell enlargement. A few years afterwards, when Robin Snow showed that auxin also caused cambial activation in the sunflower seedling, he caused a sensation: auxin had been thought only to control cell enlargement, but now it was evident that it stimulated cell division also.

B. THE INTERACTION BETWEEN AUXIN AND CATIONS

The first findings in studies of this sort were that in presence of indoleacetic acid plus sugar the segments do not respond to added sources of nitrogen, but they do respond (rather surprisingly) to several cations. They are particularly sensitive to potassium, which will increase the elongation around 20%, whereas calcium has the opposite effect, decreasing the elongation. In later studies, James Bonner found that manganese had a promoting effect, and still later a very surprising finding was that cobalt also had a promoting effect.

Now manganese is known to be needed by plants, and of course potassium is a major requirement. But manganese is needed by plants in very low concentrations; in nutrient solutions for whole plants it is present at 2-3 $\times 10^{-6}$ M; nevertheless manganese at an optimum concentration of about 10^{-4} M promotes the growth of segments. The optimum concentration for cobalt is one to two orders of magnitude lower than for manganese. A curious fact is that cobalt is not needed in nutrient solutions, as far as we know. It is needed by legumes when their nitrogen is being supplied by fixation in the nodules, but that is not because the plant needs it but because the rhizobia in the nodules need it. Thus there is no well-established requirement for cobalt in whole plants. It is believed now that the effect of cobalt is to inhibit the production of ethylene (S. F. Yang, 1976), which is produced by most plant tissues when they are treated with auxin, and which inhibits elongation, as we shall see later.

As far as manganese is concerned, the low concentration which suffices for optimal growth when supplied to the roots is probably accumulated by the roots to an internal concentration 30 to 100 times the external. Such a ratio is not uncommon—potassium has often been found to reach internal concentrations up to 100 times the external, and values up to 1000 times the external concentration have been recorded for phosphate. Thus the optimum value actually needed internally in growing plants may be some 30 to 100 times the value supplied in nutrient solutions, and this would be the value needed by nonaccumulating stem tissue. Applying this reasoning inversely to cobalt we could deduce that the requirement for this ion in nutrient solutions might be one hundredth of that which is optimal for the segments, i.e. about 10^{-7} M. In that case it would be near the level of the impurities in even carefully recrystallized nutrient salts. Copper, for example, is required at about 3×10^{-8} M and copper deficiency is very difficult to obtain. Thus it is conceivable that there might be a microscopic need for cobalt for whole plants too.

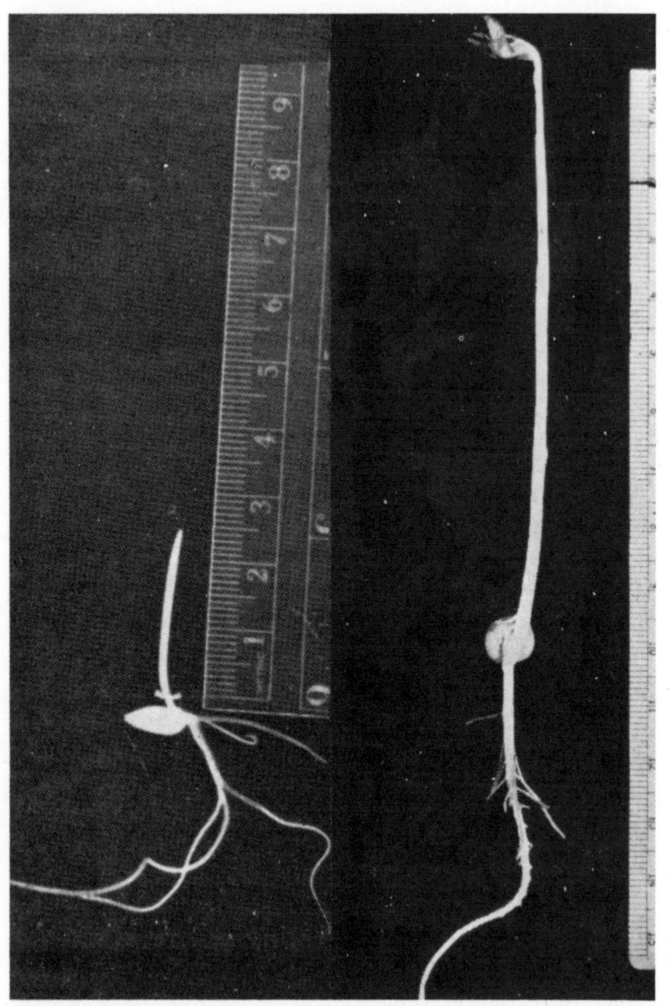

3-2 Etiolated seedlings of *Avena* and *Pisum* at 72 hours and 7 days (at 25°C) respectively, as used for most growth experiments.

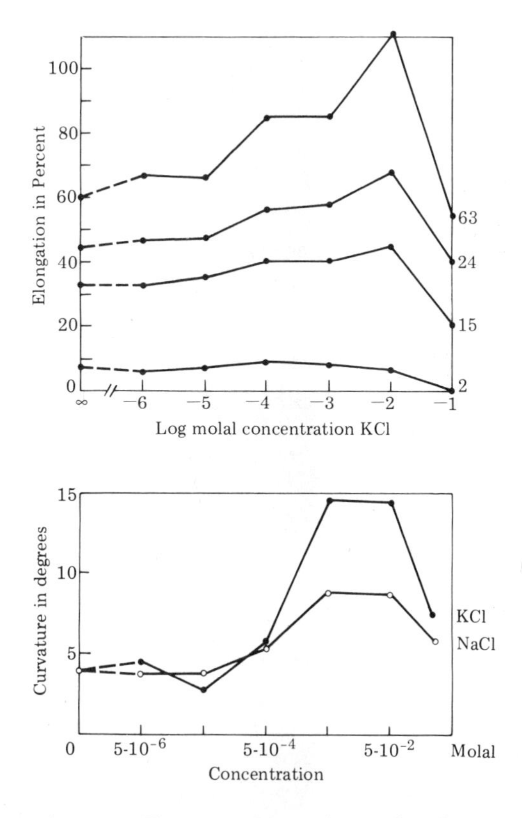

(above) **3-3** Elongation of 3-mm *Avena* coleoptile segments in IAA plus sucrose plus serial concentrations of KCl. Figures at right show the number of hours after cutting and floating on solutions. From Thimann and Schneider, 1938 a.
(below) **3-4** Curvatures produced by agar blocks soaked in IAA plus serial concentrations of KCl or NaCl. From Thimann and Schneider, 1938 a.

Not only coleoptiles and other monocotyledonous seedlings, but dicotyledonous seedlings are very suitable material for growth studies; pea seedlings offer epicotyl segments which are rapidly elongating and have been used by many workers. An etiolated pea seedling is shown in figure 3-2. It will have a node or two, depending on its age; most workers use the cultivar Alaska as a standard plant. Alaska is a variant of another pea called Telephone, which as its name implies, grows very tall in the light, and still taller in the dark. In these young epicotyls the internodes which have just elongated provide a material very similar to a coleoptile, in which cell divisions have been almost completed, and pure elongation is occurring. These segments respond to potassium and cobalt, just as do coleoptiles. Their response to added sucrose is very slight, however.

Figure 3-3 summarizes the influence of potassium in promoting elongation of coleoptile segments. (These segments were 3 mm long, but more of the work has been done with 10 mm sections.) The measurements taken at different times show how the potassium effect, which is barely detectable after two hours, becomes clear in 15 hours, and by 24 hours or longer has become quite large. At 24 hours plants without potassium have elongated 44% and plants with optimum potassium, 68%. Thus the potassium, at about 10^{-2} M, had increased the elongation by about 50%. Figure 3-4 shows that potassium even promotes the curvature in the agar-block bioassay of Went. Figure 3-5 shows the effects of cobalt and manganese on coleoptile segments. The elongation was measured at

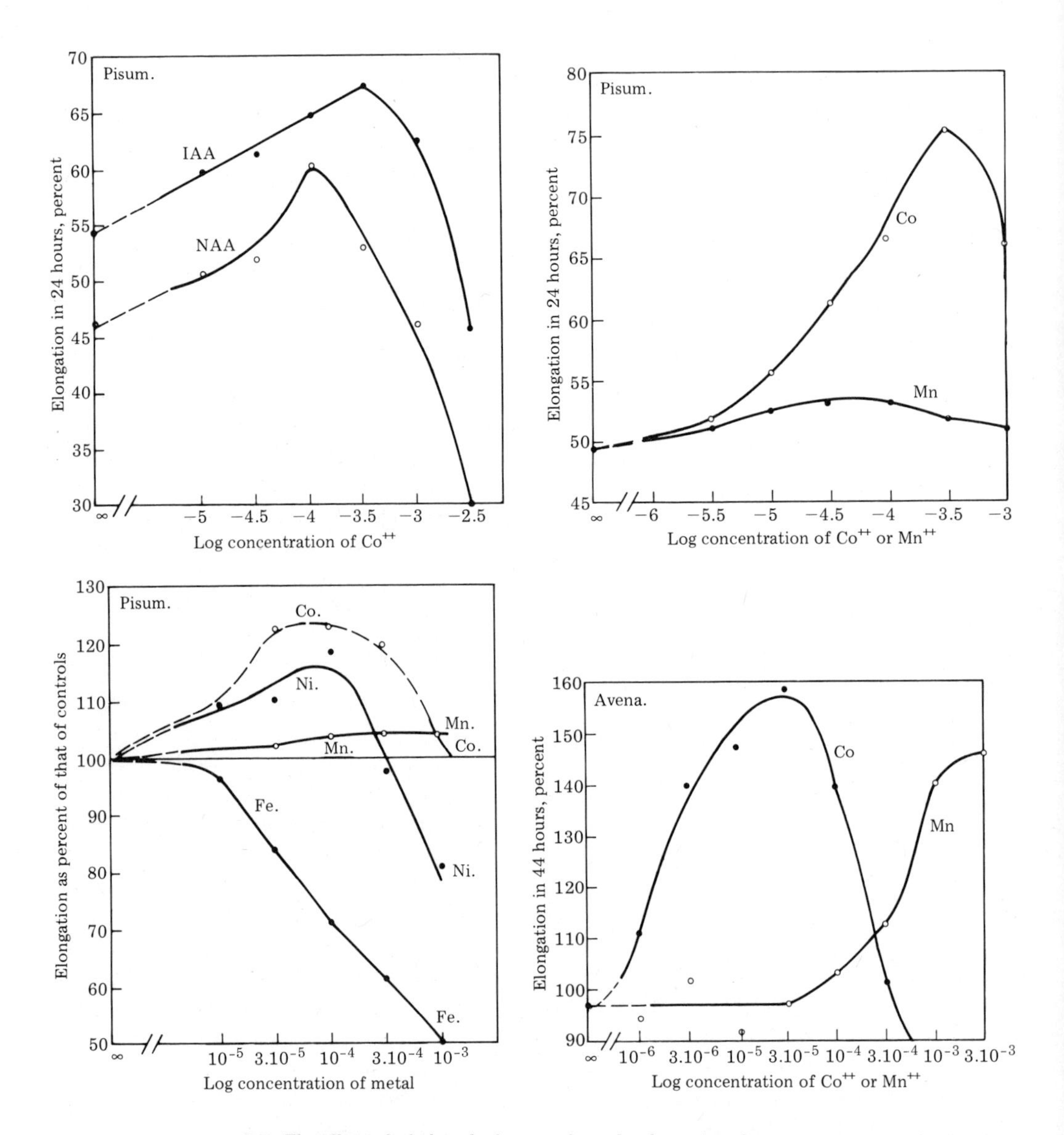

3-5 The effects of cobalt and other metals on the elongation of pea stem segments (above and lower left) and *Avena* coleoptile segments (lower right). Elongation is shown as percentage increase of 10-mm segments, above, and lower right; as percent of the elongation of controls in sucrose alone, lower left. The metal effects vary somewhat from one experiment to another, but the preeminence of cobalt is clear. Auxin IAA in all cases except NAA in top left. From Thimann, 1956.

a single time, 40 hours after the start. These plants were provided with sucrose, potassium, and the optimum auxin concentration. Controls have grown 96%, and those in the optimum concentration of cobalt (3×10^{-5} M) 150%, again a 50% increase. Manganese is not quite as effective and requires about 100 times higher concentrations.

C. THE CURVATURE OF SLIT PEA STEMS

Segments from dicotyledonous seedlings are a little larger and they differ from coleoptiles in that they are solid, while coleoptiles are hollow and only six cells thick, so that the uptake of the auxin, sugar, or ions is rapid. In the case of solid stems, such as those of etiolated pea seedlings, we could not be so sure about the uptake, so my colleague, then Frits Went, slit the stems lengthwise for a centimeter or so to increase the amount of surface beyond the limited area of the short transverse one, and thus insure good uptake.

As a result, when the slit stems were placed in water, the halves curved outwards. This phenomenon has proved to be due to the fact that the outer layers, primarily the epidermal cells and the layer just below, are normally under tension in a rapidly elongating system, so that when the stem is slit the tension is relaxed and they curve outwards. If one does this with extremely rapidly elongating organs—the most remarkable is the daffodil stem, taken just before the flower opens—then instead of merely curving outwards the two halves actually roll up. With a hollow stem, too, there is no central mass of pith to hold the two halves together. On the other hand, if one slits a stem in which elongation has ceased, very little outward curvature occurs. There are a lot of curious engineering considerations in the extension lengthwise of a series of cells each layer of which has different responses; those that are extending most rapidly will naturally draw the others on, while those extending least rapidly will pull the others back. The balance gives the observed growth rate. This phenomenon of differences in tension has a long history, having been observed by Hofmeister back in 1859. Hofmeister was a botanist who had extraordinarily wide interests, and today much of his work is insufficiently appreciated. We are grateful now, knowing what we know about the role of hormones, to look back at some of the things Hofmeister wrote about and to see how close he came to understanding what was going on.

This tension which normally occurs in the growing tissue is released when auxin is applied, because auxin promotes growth, and it promotes the growth of the outer layers (the epidermis and subepidermal cells) more than of the inner layers; as a result the two halves, which at first curve outward, soon straighten up. After two hours or so in auxin the two halves begin to curve inward (fig. 3-6). It was soon found that the final inward curvature, which is very readily measured, can be used as a direct measurement of auxin activity, i.e., as a measure of the elongation caused by the auxin. The angle measured is that between the unslit portion and the tip of the curved portion. It ranges between $-110°$

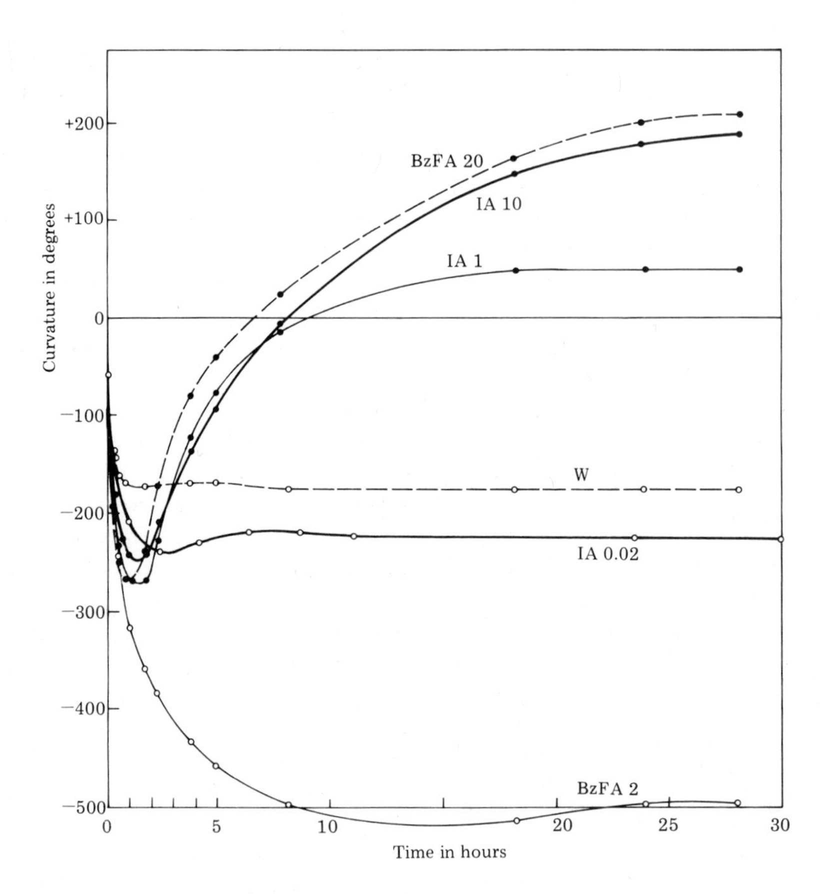

3-6 Development of curvature in slit pea stems (as in fig. 7), in water (W), and in IAA 0.02, 1.0, and 10.0 mg per liter, showing increased outward curvature in first 2 hours, followed, in higher auxin concentrations, by inward curvature. The curves marked BzFA show the atypical behavior of the slit stems in benzofuraneacetic acid. From Thimann and Schneider, 1938 b.

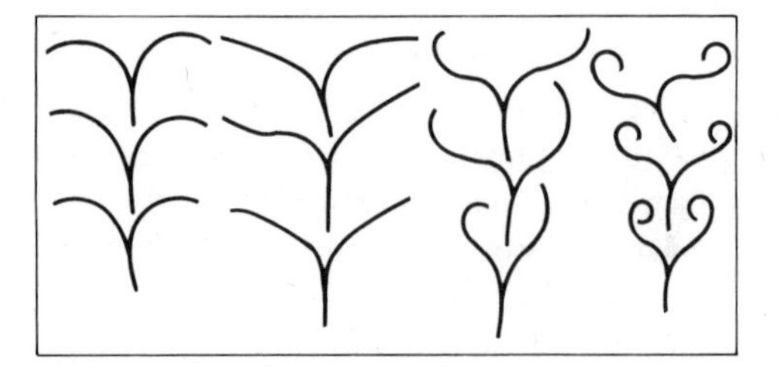

3-7 Typical slit pea stem curvatures after 30 hours. Left to right, water, IAA 0.2, IAA 1.0, and IAA 5.0 mg per liter.

(i.e., 110° outward), the value usually seen in water, and +700° or so (i.e., about two complete 360° turns inward), which is reached in some auxin solutions (fig. 3-7). This "slit pea test" was for a while very widely used for the study of synthetic auxins.

The figure shows a typical slit pea stem test. The single internode used is the most apical internode which is still elongating. These seedlings have the very useful property of growing internode by internode so that at any moment there is only one which is rapidly elongating. The slit sections are further selected by floating on water for two hours, and only those which give good symmetrical outward curvatures are used. Auxin is applied in a logarithmic series of concentrations. Notice the variation between maximum and minimum response; if one selects the plants more strictly one can get somewhat better results than in this figure. The curvature is due to the difference in response between the inner and outer tissue layers (fig. 3-8).

D. THE OXIDATION OF AUXIN

Before the natural auxin was identified as indoleacetic acid, Peter Stark in Germany ground up plant tissues, made them into gels and applied them to coleoptiles, but no curvature resulted. For a long while after the auxin had been demonstrated by the simpler method of placing live tips on agar and just letting them produce auxin, it was not understood why Stark's method had not worked. The answer is very simple. If such ground-up plant tissue is mixed with auxin and allowed to stand, the amount of auxin activity disappears very rapidly. This rapid disappearance is an oxidation and is due to an enzyme which rapidly destroys indoleacetic acid.* The enzyme turns out to be (in most cases) none other than the well-known peroxidase; at first it was thought to be specific and was called indoleacetic oxidase, but ordinary peroxidase will do very well; one can imitate it nicely with horseradish peroxidase mixed with a little H_2O_2. IAA is very rapidly destroyed in plant tissues when they are ground, because all higher plants contain peroxidase, in amounts varying with the age and the species. Etiolated pea seedlings contain plenty. The indole nucleus is rather readily oxidized, and many color tests for indole depend on this fact. But the two commonest synthetic auxins, NAA and 2,4-D, resist the peroxidase. Hence in plant tissues treated with these auxins the amounts present can be easily assayed in the ground-up tissue by the slit pea stem method.

For a long while the oxidation, or peroxidation, of IAA was the despair of workers who hoped to identify auxins in extracts. Of course IAA is stable in ether so that, if tissues are extracted with peroxide-free ether very quickly, one can avoid much of the loss. Subsequently it was found, however, that the action of peroxidase is very sensitive to other materials present and especially to phenols. Phenols have actions of two types, depending on their structure (cf. chap. 8). Monophenols, whether simple or substituted in various positions, generally promote the oxidation. Diphenols inhibit the oxidation, and this very conven-

* We shall abbreviate indoleacetic acid as IAA hereafter.

ient fact can be put to use in a number of different types of experiments; one can protect the auxin with diphenols, or promote its destruction by adding monophenols. Amines are also effective; diamines like benzidine give excellent protection (figure 3-9 shows protection in the standard *Avena* test). This is an interesting group of phenomena that has led to a good deal of study of the mechanism of oxidation of auxin. We need not go into it here since it does not greatly illuminate the normal action of growth hormones in plants, though it does have to be borne in mind when making and testing extracts.

In detail, the oxidation of IAA by peroxidase is not a simple matter, although there are systems in almost all plants which generate a small amount of H_2O_2. For one normally thinks of peroxidase as an enzyme that uses H_2O_2 to oxidize. Not so the oxidation of IAA; here the enzyme uses oxygen, O_2, and only a very small amount of H_2O_2, less than 0.1 mole, even if there is enough present to oxidize a mole of IAA. The reaction is quite well worked out, largely by Peter Ray. The indole structure, as students of organic chemistry will know, is subject to shifts of electrons so that its 5-membered ring is particularly reactive. Under the influence of the peroxidase-oxidase (which is the only convenient way one can name it) there is a shift of electrons towards the carbon atom no. 8 and the nitrogen atom, which has a fractional positive charge and therefore tends to attract electrons. This causes withdrawal of electrons from the side chain and hence releases the CO_2. The first thing one detects in the oxidation is that the acidic group has gone off as CO_2; then the molecule undergoes a series of changes, finishing up with a methylene group instead of the methyl, and an oxygen atom on carbon-2. Not only is this the major mechanism by which indoleacetic acid is destroyed in plants, but as we shall see this plays a large part in horticultural applications. For if, instead of indoleacetic acid, which the plants use, we use a synthetic auxin, like naphthaleneacetic acid or 2,4-D, this oxidation does not occur. The 10-membered ring of naphthalene is quite stable to enzymatic oxidation, and the same is true for 2,4-D, which is metabolized by hydrolysis of the phenoxy group to produce 2,4-dichlorophenol, a relatively slow reaction. As a result the synthetic auxins are very much more stable, both in the plant and in solution, than IAA is.

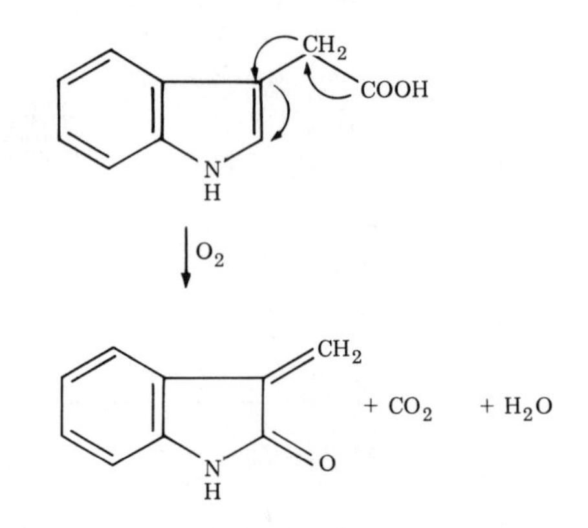

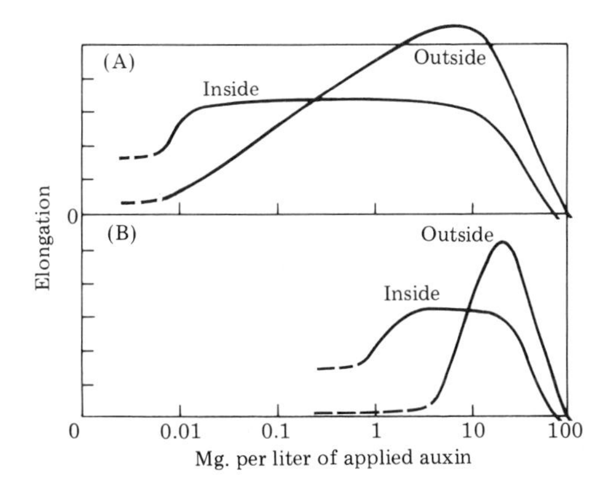

3-8 Schematic representation of the elongation of the outer and inner tissues of a pea stem segment, showing how it is that low auxin concentrations can cause small outward curvatures, while higher concentrations cause strong inward curvatures. (A) IAA; (B) BzFA. From Thimann and Schneider, 1938 b.

3-9 Standard *Avena* curvatures with IAA 0.05 mg per liter, showing no effect of H_2O_2 or peroxidase alone, complete auxin destruction with H_2O_2 and peroxidase together, and protection by benzidine at the same concentration as the IAA. From Platt, 1954.

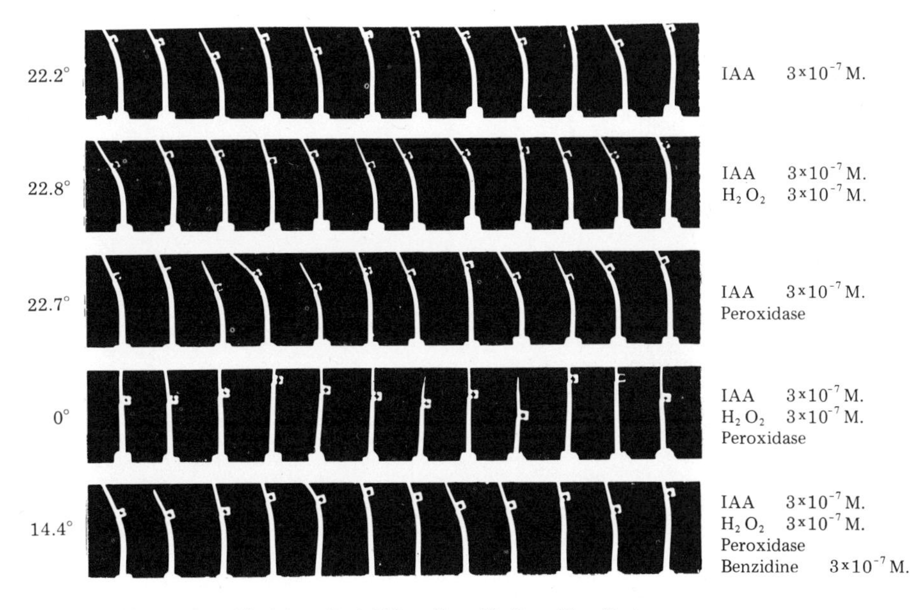

Destruction of Indoleacetic Acid by a Peroxide-Peroxidase System

E. ISODIAMETRIC ENLARGEMENT

An interesting development took place in the 1940s when the Dutch plant physiologist Dirkje Reinders was studying the uptake of water and of salts by slices of storage tissue—generally using potatoes, with a few experiments on other plants. It was known that thin slices of potato will take up water for several days and then reach saturation; in the process there is very little cell division, though there is some tendency for cells next to the cut surface to divide once. What Reinders found was that the uptake of water is greatly promoted by auxin, especially by auxins other than IAA, such as NAA. So what she was studying was not the uptake of water in the ordinary (reversible) sense, but an aspect of *growth*. I have often said that the definition of growth is nothing other than *irreversible increase in volume*; growth does not necessarily require any increase in weight; in fact increases in weight take place during the day's photosynthesis in a leaf that has long since ceased to grow. The increase must be irreversible because otherwise changes in turgor, which are readily reversible, would be considered as growth, which they are not. Figure 3-10, which is from our own experiments, shows that the potato slices after treatment with auxin have in fact increased in volume. The slices were initially 1 mm thick, and these are cross sections. The dark bodies are the starch. In the photograph made at the beginning of the experiment, after the sections had been cut and washed in water for 24 hours, the starch is prominent. After three days in auxin solution the volume has greatly increased, 200% or 300%, and each

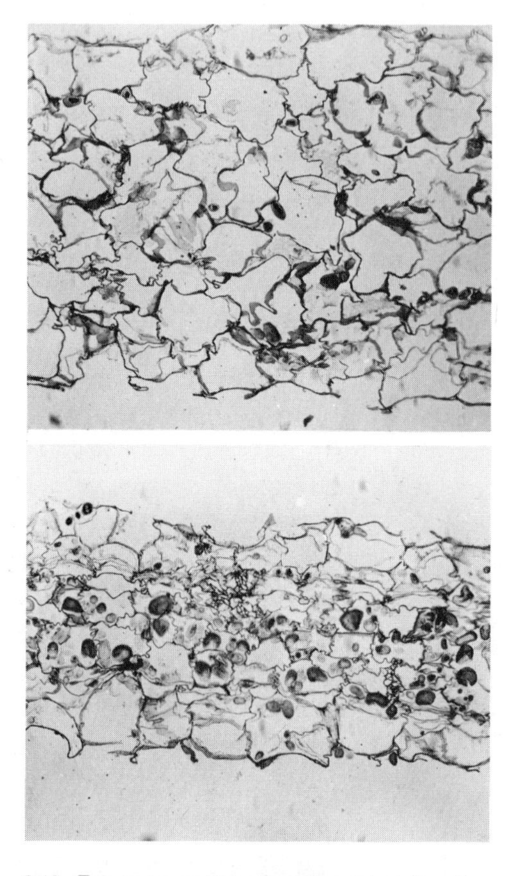

3-10 Transverse sections through potato tuber discs after 8 days in auxin (above) or water (below). The dark granules are stained starch grains, which have largely disappeared with the cell enlargement in auxin. From Hackett and Thimann, 1952 b.

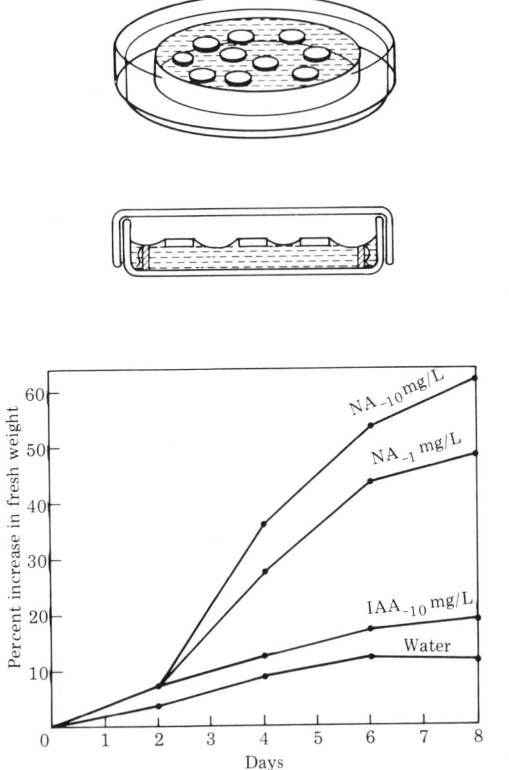

3-11 Above, sketch of method of supporting potato discs at the surface of test solution; below, time course of cell enlargement, measured by fresh weight. For this tissue napthaleneacetic acid (NA) is much more active than IAA (cf. fig. 23, done with artichoke). From Hackett and Thimann, 1952 b.

individual cell has enlarged. There are few visible cell divisions, nothing like a wound cambium such as one can occasionally see on cuttings. Most of the starch has disappeared. Thus as the cells take in water they use up some carbohydrate, partly for energy, but I think largely for cell wall formation. It is easy to show that the amount of walls has increased as the amount of starch has decreased. So this so-called "water uptake" is a system very like elongation, but instead of being elongation in one dimension, it is enlargement in all directions, i.e., *isodiametric enlargement*. It responds to added sugar only to a small extent, for unlike the cells of the etiolated seedling the cells of the tuber have a very good carbohydrate supply. It is very dependent on the supply of oxygen. It also responds to added potassium just as cell elongation does, and, interestingly enough, it even responds to cobalt. The process is much slower than unidimensional elongation, for it continues up to 8 days. For the first 2 days the sections in water and auxin grow at about the same rate; then the sections in water slow down, while growth of the sections in auxin actually accelerates (fig. 3-11). The auxin here was naphthaleneacetic acid, one part per million, which is about optimum for this tissue. When cobalt is added there results about a one-third further increase. Hence there is no doubt that this process is cell enlargement.

F. THE CHELATION THEORY

Other metals besides potassium and cobalt can influence growth. In fact, study of the effects of metal ions on elongation, or on cell enlargement, has led to some rather curious results. Some metals inhibit growth—calcium, of course, very clearly, manganese in high concentrations, and nickel too is extremely inhibitive. The metals that promote growth are limited in number, while a great many inhibit it. Such effects led to a very natural suggestion, that in normal tissue growth may be limited by certain metals, and the action of auxin might be to chelate metals away from the tissue, thereby releasing an increased ability to grow. This was first proposed by Heath at Reading, England; he based the idea on the location of the nitrogen atom in indoleacetic acid, which suggested that a metal might chelate at this position, somewhat as 8-hydroxy-quinoline chelates copper and other metals at that locus. As it turns out, rather unexpectedly, ethylenediaminetetracetic acid (EDTA), a standard and popular chelating agent, does in fact cause a small increment of growth. For this reason, Heath's theory was taken up with great enthusiasm, and several workers began experiments on the supposed chelating powers of different auxins. However, we found that EDTA mainly promotes growth when IAA is already being used as the auxin, that is, it promotes growth more when IAA is present than when it is absent. Now if the auxin were acting as a chelating agent one would not expect a second chelator to be *more* effective in its presence; one would expect it to be less effective. But in fact the action of EDTA is tripled or quadrupled by the presence of auxin. Furthermore, in presence of the synthetic auxins, which cause just as good growth, EDTA does not have this additional effect, that is, it acts *only* when IAA is present. I think, therefore, that what EDTA does is to prevent enzymatic destruction of IAA rather than to act as another auxin. The idea that auxin might act as a chelating agent could not be supported either by a study of the structure of other auxins; 2,3,6-trichlorobenzoic acid, for example, has no particular basis for acting as a chelating agent. Nor indeed does naphthaleneacetic acid. So the role of chelation constituted an interesting side issue which sailed into the field, was exciting for a few years, and then was thrown out.

G. OSMOTIC INHIBITION

Since growth is essentially an irreversible increase in volume, the uptake of water must be a critical part of it, (the other part being the metabolic changes that go on to support and induce growth). What happens if the uptake of water is prevented? This can be done by decreasing the water concentration of the outside solution by adding a high concentration of solute molecules. Experiments were therefore carried out to see what would happen if one decreased the entry of water by immersing the growing system in mannitol solutions. Mannitol was chosen because it is neutral, un-ionized, is not a nutrient, and has no harmful effects, yet is soluble enough to produce fairly concentrated solutions. Figures 3-12 and 3-13 show that as the concentration of mannitol is increased, the elongation of the system in auxin decreases; the two curves of figure 12 show the readings at different times. After a very long time, such as 72 hours, there is a small recovery, suggesting that mannitol can very slowly enter plant tissue. Mannitol is not as potent an inhibitor as calcium (fig. 3-13). With a different plant material, further information has been obtained on what mannitol does in the plant. When potato discs have been floated on mannitol solution containing ^{14}C-mannitol for 3 days and are then transferred to water, the C^{14} comes out within a few minutes, showing that it was only in the intercellular spaces or in the cell walls themselves. Killing the cells with acetone, to destroy their semipermeability, releases only a very small amount more (fig. 3-14). On the whole, therefore, mannitol is recognized as the almost perfect osmoticum,

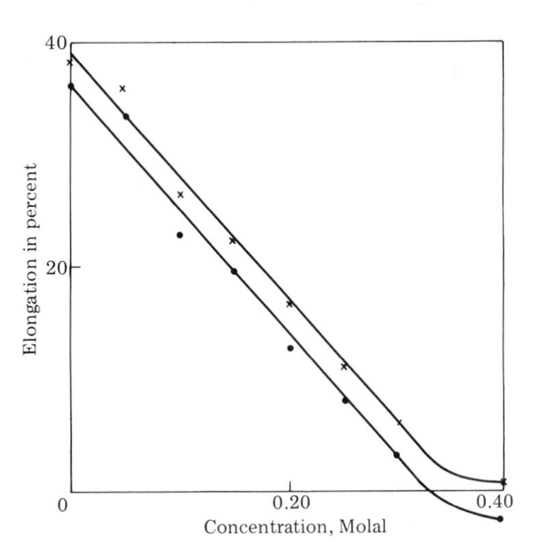

(above) **3-12** Elongation of 3-mm *Avena* coleoptile segments in IAA plus 10 mM KCl (no sugar) plus serial concentrations of mannitol. Circles, at 37 hours; crosses, at 73 hours. From Thimann and Schneider, 1938 a.

(below) **3-13** Experiment similar to that of fig. 12 but with 20-mm pea stem segments. The inhibition by $MgCl_2$ is similar to that by mannitol, but that by calcium ion is about 10 times greater. From Thimann, Slater, and Christiansen, 1950.

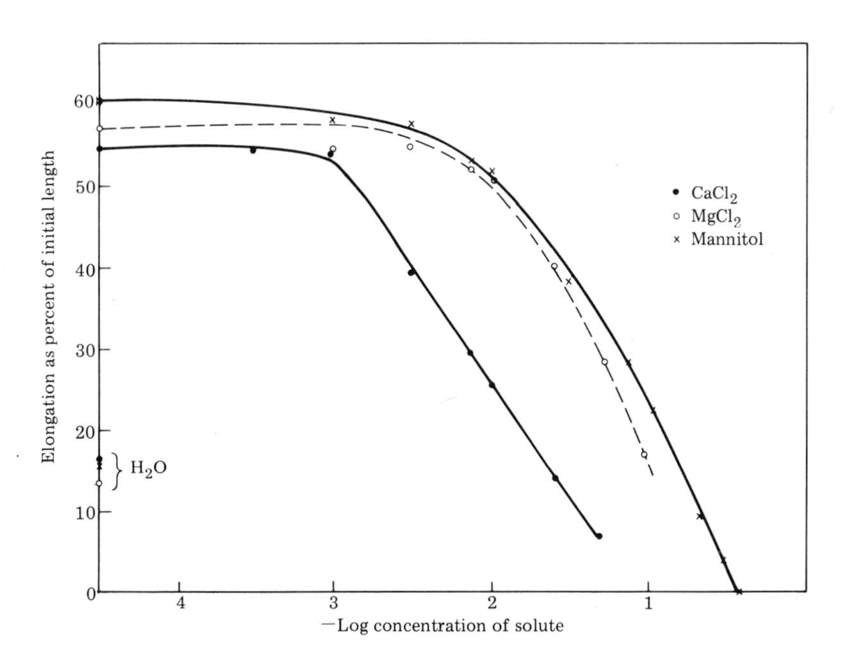

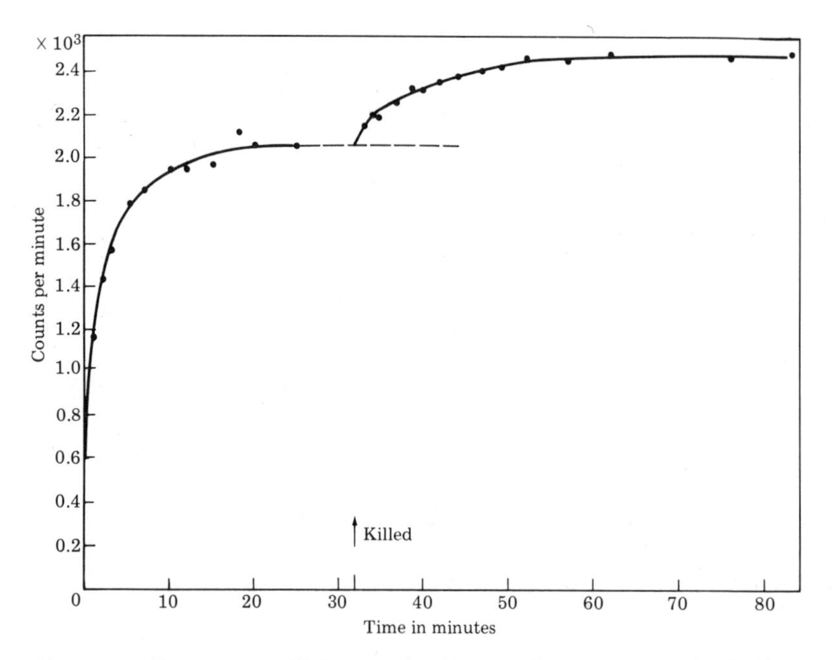

(above) **3-14** The appearance of ^{14}C-mannitol in the outer solution after potato discs had been 72 hours on naphthaleneacetic acid in presence of ^{14}C-mannitol. After the small initial outflow, probably from intercellular spaces, the tissue was killed with acetone; the limited further outflow shows how little of the mannitol had entered the cells. From Thimann, Loos, and Samuel, 1960.
(below) **3-15** Diagram of the loading method for measuring plastic and elastic extension. From Tagawa and Bonner, 1957.

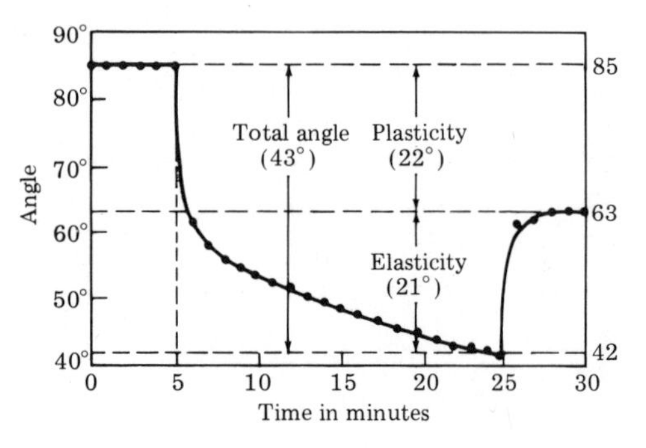

since it does not enter the cells to any appreciable extent. Inhibiting water uptake thus inhibits growth, whether by elongation or by isodiametric enlargement. Furthermore, when the tissues are taken out of the mannitol and returned to water or dilute sucrose, growth resumes. Thus the period of reduced water intake (if not too prolonged) has done no irretrievable damage. In some cases, as with artichoke slices, the tissues on return to water actually grow faster than the controls; after some hours they catch up with the controls, and thus prove to have made up the growth that they missed during the period in mannitol.

H. AUXINS AND THE CELL WALL

So much for the general process of elongation and water uptake. The analysis of growth is a longer story, most of which will be deferred to the last of these lectures, but a summary will be worth giving here. In the Utrecht laboratory where auxins first came to life, Anton Heyn (now at New Orleans) carried out some pioneer experiments on changes in the cell wall associated with growth in the *Avena* coleoptile. Since an extension of the cell wall would probably be, to some extent at least, accompanied by a laying down of new cell walls, one would expect that the mechanical properties of the cell wall might change during elongation. What Heyn did was to stretch a tissue and study the effect of auxin on the stretching. More recently Cleland in Seattle and Masuda in Osaka have greatly extended these experiments. If we support a coleoptile in the horizontal position and lay a small 10-milligram weight on it, this will cause the coleoptile to bend, i.e., to stretch the upper surface, compressing the lower. The resultant bending is partly reversible, for it is mostly just elastic bending which will reverse if we remove the weight. Heyn distinguished the part which is reversible (fig. 3-15), which he called the *elastic* stretching, from the part which remains, and has thus undergone a genuine change, which he called *plastic*. Different terms have been used by other workers, but the simple idea is that part of the stretching is reversible and part irreversible. If now we treat the tissue with auxin (in Heyn's experiments he applied blocklets of agar containing auxin) the maximum stretching increases, and particularly the fraction of it which is irreversible increases. The elastic stretching shows a small increase, but the plastic fraction is very greatly increased. Experiments in which this kind of elongation is measured by a curvature are not wholly satisfactory, because one side is stretching, and the other side is being compressed. Therefore in later experiments actual longitudinal stretching was studied—a delicate business, in which the seedling tissue must be held very firmly and extended mechanically. The result is the same as with the bending, namely that the extensibility greatly increases under auxin, especially the plastic components; the elastic component increases too, but usually somewhat less.

Fortunately for plant physiologists, this kind of experiment is also done by textile researchers who are interested in the extensibility of textile fibers; thus a delicate apparatus which gets a firm hold on fine threads is commercially available. In fact,

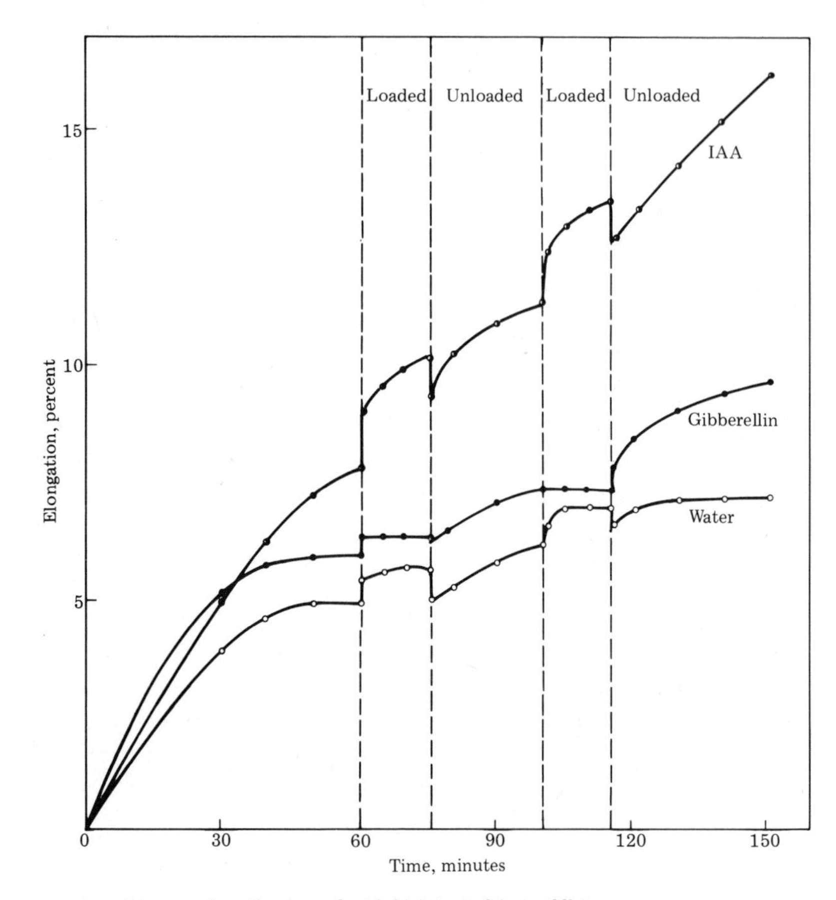

3-16 Plastic and elastic extension of *Avena* coleoptiles treated with IAA (top), GA_3 (middle), or water alone (bottom). From Yoda and Ashida, 1960.

since Heyn's day, with better and more delicate equipment considerably finer measurements have been made, both in the United States and in Japan. Figure 3-16, from Yoda and Ashida in Kyoto, shows typical results. The bottom curve shows the extension in water. A constant tension is applied, extension begins, and after a while it comes to an end. Now we put on a further load, and a rapid further extension takes place, which soon stops as the first extension did. We take the tension weight off and the tissue springs back; there has been virtually no plastic extension, all the observed extension being reversible, i.e., elastic. An additional load has the same result: the extension reverses, though this time not to quite the same extent.

In auxin, however, we see an entirely different picture. When the load is applied, there is a much larger, sudden increase in elongation, and when we take the load off the length by no means returns to its previous value, nor does it even return to a continuation of the same curve. This has been a largely plastic extension. On repeating the load we get similar results. So under the influence of auxin, the mechanical properties of the cell walls have changed; they have become much more plastic. It would be attractive to conclude that this change is the cause, or the controlling influence, of elongation when the elongation is caused by auxin. It probably is (and we shall see more supporting data in chapter 14), but since gibberellin does not have the same effect, the increased plasticity can hardly be the controlling influence in the elongation process *in general*. The middle curve of figure 3-16 shows the effect of gibberellin on the same tissue: we see only a very small increase of length on loading; it is plastic, but under the influence of the load, the extension is considerably less than it is even in water. If auxin makes a wall more plastic, then gibberellin actually makes it less plastic. Elongation under gibberellin will be taken up later, but a very suggestive fact is that, like elongation under auxin, potassium promotes, and calcium inhibits; evidently, therefore, some part of the growth process is common to the action of both hormones.

Figure 3-17 shows experiments of Tagawa and J. Bonner done in the earlier way by measuring the angle of bending under the influence of a rider. Here E is the elastic extension, P the plastic extension (which is measured by taking the rider off again and determining how much it re-

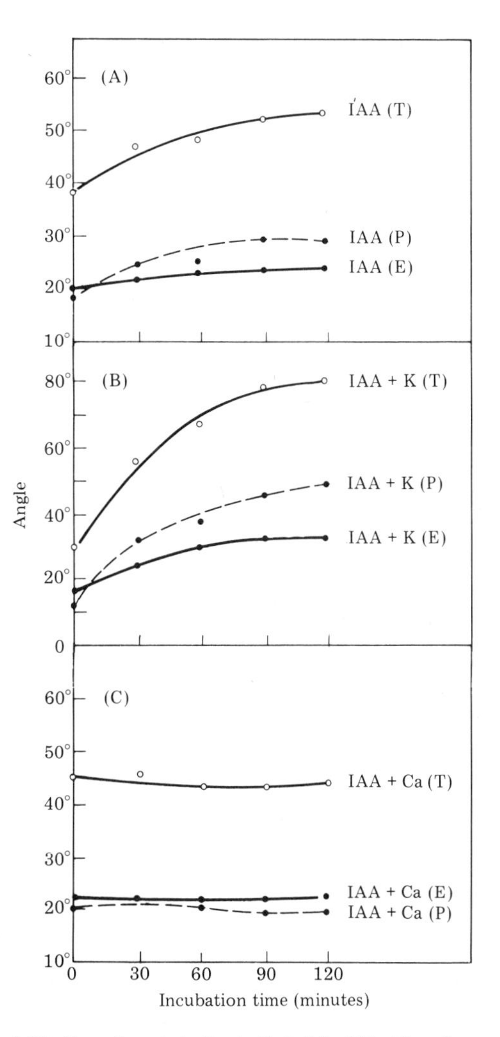

3-17 Experiment similar to that of fig. 16 but based on bending rather than extension. All coleoptiles treated with IAA, the middle group plus KCl, the bottom group plus $CaCl_2$ (both 0.02 N). (P) plastic bending; (E) elastic bending; (T) total. From Tagawa and Bonner, 1957.

turns), and T the total extension. If we compare the results given with IAA along with those given when potassium chloride is added, it is evident that all of the ratios are greater. While the effect on elasticity is not very large, the effect on plasticity is quite clear. On the other hand, in presence of calcium, the opposite is true: the plastic extension is almost zero, and the elastic extension equally is practically zero. Since the load used was the same in all cases, calcium obviously prevents extension, while potassium promotes it. Thus bending under a weight behaves exactly like isodiametric enlargement, as in the storage tissue, for Brauner in Constantinople obtained closely comparable results with cylinders cut out of potato tubers. We conclude that the influence of auxin is exerted in good part on the cell wall, whose properties are a controlling or limiting factor in enlargement. We also conclude that modification of the cell wall is not necessarily the cause of elongation when the elongation results from other treatments.

I. THE ROLE OF OXIDATIVE ENZYMES

The study and analysis of growth produced a second interesting result. It was mentioned that the sections are normally floated on the solution, and that they must break the surface of the solution in order to respond best. Coleoptile or stem segments or potato slices, when immersed only as little as a millimeter or so under water, grow very much less than when breaking surface (fig. 3-18). This means that growth must be very dependent upon the access of oxygen. With potato slices, whose response is slow enough to be measured over a long period of time, growth is a rather straightforward function of the oxygen content. The sections are seen to have grown about 100% in air and their enlargement decreases linearly as the oxygen content is decreased.

Growth can equally well be inhibited by interfering with certain enzymes needed for oxidative processes. Cyanide is the most obvious, and it does indeed inhibit growth successfully, but unfortunately it is not very specific. Of the more specific types, sulfhydryl inhibitors are probably the most important. An effective substance like iodoacetate combines with sulfhydryl groups on many enzymes to eliminate HI and make an inactive complex with the enzyme. This mode of action has been demonstrated with cysteine and also in experiments with pure SH-containing proteins. Arsenite (fig. 3-19) offers a striking example. All of the compounds which react with SH groups are found to inhibit growth. A curious complication with growth experiments is that these inhibitors promote growth at low concentrations, though they inhibit at higher. Such promotion and inhibition is a common phenomenon with whole tissues, though it does not occur with isolated enzymes. Figure 3-20 shows the effect of iodoacetate on coleoptile sections. They were breaking surface and thus were provided with optimum oxygen. Submerged sections are shown for comparison. At low concentrations there is a medium zone of clear promotion. It is indeed remarkable that one can obtain a promotion of 50% in growth by poisoning an enzyme system. Mercapto-ethanol has recently been found by Nicolas and Zarra in

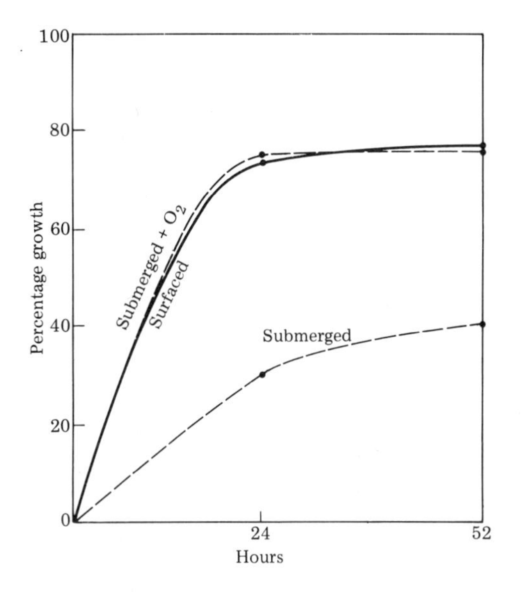

3-18 The difference in elongation between coleoptile segments supported at the surface of the solution and submerged 1 mm below it; also submerged the same distance but with oxygen bubbled through. From Thimann and Bonner, 1948.

3-19 Inhibition of the elongation of pea stem segments in IAA by serial concentrations of sodium arsenite. At above 10^{-4}M the elongation is less than in water, i.e., the auxin effect is nullified. From Thimann and Bonner, 1949.

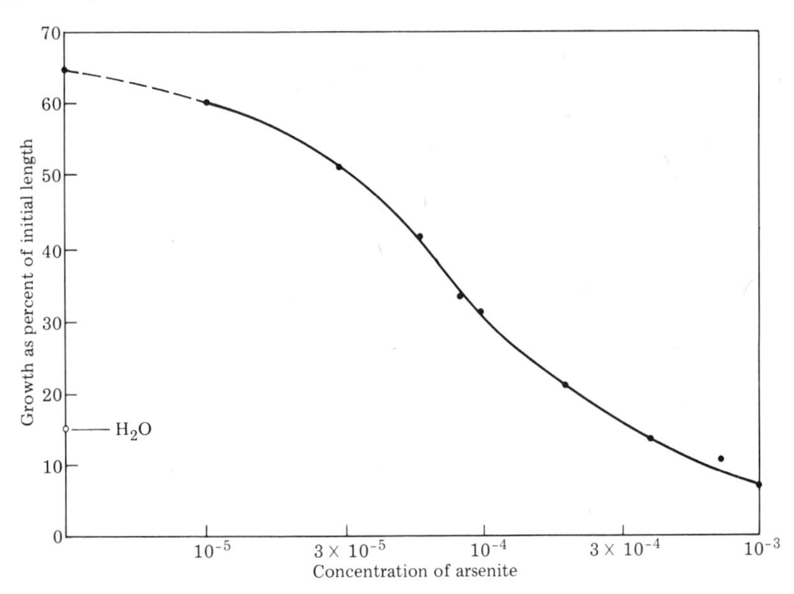

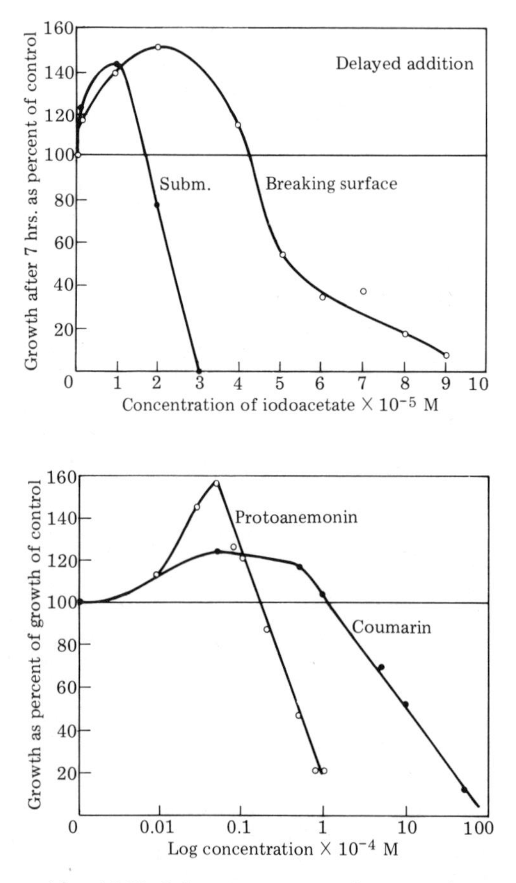

(above) **3-20** Iodoacetate concentration curves for segments from coleoptiles aged 65 hrs (iodoacetate added 7 hrs after the start). From Thimann and Bonner, 1948. (below) **3-21** Experiment similar to that of fig. 20 (breaking surface) but with two unsaturated lactones. From Thimann and Bonner, 1949 b.

Spain to have the same effect on rice coleoptiles. We saw earlier that in the study of the red light effect on germination of seeds, it was found that the naturally occurring lactone coumarin reinstates sensitivity to red light. Coumarin is also remarkable for showing this growth promotion at low levels. Figure 3-21 shows the effect of coumarin as a function of its concentration; over quite a wide range one sees marked growth promotion, an increase of about 25%, while at higher concentration, around ten times that causing maximal promotion, growth becomes strongly inhibited. Another naturally occurring compound related to coumarin is protoanemonin, which is also an unsaturated lactone whose effects on growth resemble those of coumarin but are exhibited over a narrower range. As was mentioned above, many other growth responses show promotion at low levels and inhibition at high. As yet this has not been satisfactorily explained. Although there is some evidence that unsaturated lactones act as reagents for sulfhydryl groups, still their specificity is far from certain. A compound which is well established as reacting specifically with SH groups is parachloromercuribenzoate, $ClHgC_6H_4COOK$. This substance is highly specific, the mercury combining as readily with organic SH as with inorganic sulfide. Mercuric sulfide and all its organic derivatives are very stable and insoluble, so that this reagent combines powerfully with sulfhydryl enzymes, releasing HCl. On the growth of coleoptile segments it has exactly the same effects as iodoacetate. As far as growth inhibition is concerned the differences are smaller, but the extent of growth promotion varies more widely. We can hardly doubt that in the course of normal growth oxygen is consumed, and that the sulfhydryl enzyme plays a critical role. The suggestion was made some time ago by Little that while sulfhydryl enzymes play a role in growth, they may also play a role in some other process, and that by poisoning them this other process than growth is inhibited first, thus liberating some substrate or activity which stimulates the growth process. This remains unconfirmed, however.

We saw above the parallel between growth and isodiametric water uptake, based on the facts that both are promoted by auxin, at about equal concentrations, and both require oxygen to respond. It is significant too that both are inhibited by SH-inhibiting substances, again at surprisingly nearly the same concentrations. Water uptake by potato discs differs from that by artichoke discs in that the potato discs contain a very active peroxidase, and some monophenols; as a result they destroy indoleacetic acid very rapidly. Since artichoke discs do not do this they are the material of choice for use with IAA. In table 3-1. I have tabulated the concentrations needed to cause 50% inhibition of growth, and it is notable that with any given inhibitor the concentrations needed for different tissues are very comparable. Again we must conclude that in isodiametric enlargement (so-called "water uptake") the same processes are going on as in pure elongation. Notice too that strong growth inhibition is produced by dinitrophenol. Dinitrophenol is well known to release the coupling between oxidation and phosphorylation, thus depriving the tissue of its supply of ATP. Hence we deduce that the oxygen needed for cell enlargement is consumed in the production of ATP, which is a direct source of energy for the synthetic processes of growth. Marré, in Milan, has indeed found that in pea stem segments treatment with auxin causes a small increase in the ATP content of the tissue. And very recently Tamás and Bidwell, in Canada, have brought evidence to suggest that auxin acts in the opposite direction to

Table 3-1 Effects of various inhibitors on cell enlargement and respiration in the same tissue

Tissue	Inhibitor	Concn. needed for 50% inhibition of cell enlargement	Inhibition of respiration caused by this concn. of inhibitor
Pea stems	Iodoacetate	$6 \cdot 10^{-4}$ M	26%
	Arsenite	$1 \cdot 10^{-4}$ M	13%
	Fluoride	$5 \cdot 10^{-3}$ M	ca. 0% or slight promotion
Oat coleoptiles	Iodoacetate	$4 \cdot 10^{-5}$ M	10%
	Fluoride	$2.5 \cdot 10^{-3}$ M	9%
	Arsenate	ca. 2 mg/l (ca. 10^{-5} M)	0%
	Dinitrophenol at pH 4.5	2 10^{-5} M (4 mg/l)	ca. 0% or slight promotion
Potato disks	Carbon monoxide	5CO:1O_2	ca. 25%

dinitrophenol, namely that it somewhat tightens the coupling between respiration and the binding of phosphate as ATP.

Nevertheless, the interrelation between growth and respiration is not a very direct one. Those inhibitors which inhibit growth by 50% at the concentrations listed here usually inhibit respiration to a much smaller extent. In the case of dinitrophenol we know that its action is exerted only to inhibit ATP formation, so we need not expect it to have a large effect on oxygen consumption. Indeed, in many instances the release of the coupling allows the respiration to increase. But those compounds which act as sulfhydryl inhibitors, as well as fluoroacetate, which interferes with the Krebs cycle, might be expected to cut down respiration to the same extent as growth. However, they do not. A 50% decrease in growth may be accompanied by perhaps a 10% decrease in oxygen consumption. Fluoride, in fact, may even cause a small increase in oxygen consumption, along with the decrease in growth.

Not only is oxygen, and its consumption in respiratory processes, needed for growth, but there is generally some increase in respiration when growth is promoted. In the growth of coleoptiles or seedling tissue, such increases in respiration following treatment with auxin are generally small. Figure 3-22 compares the increases in growth due to auxin acting on *Avena* coleoptile segments with the simultaneous increases in oxygen consumption. The correspondence is excellent, though the respiratory effects are not large. However there is one case, and the only one so far as I know, in which the changes in oxygen consump-

3-22 The effects of serial auxin concentrations on growth and respiration, both being measured on each group of 30 segments. Solutions contained sucrose 1% and malate 1 mM. From Commoner and Thimann, 1941.

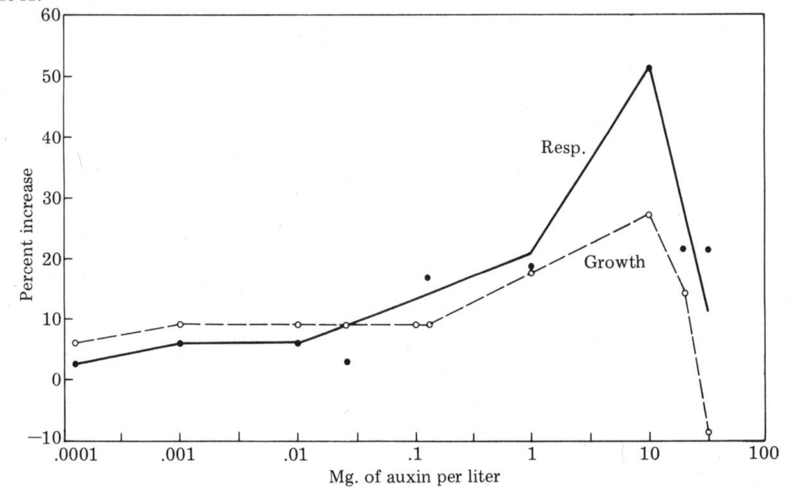

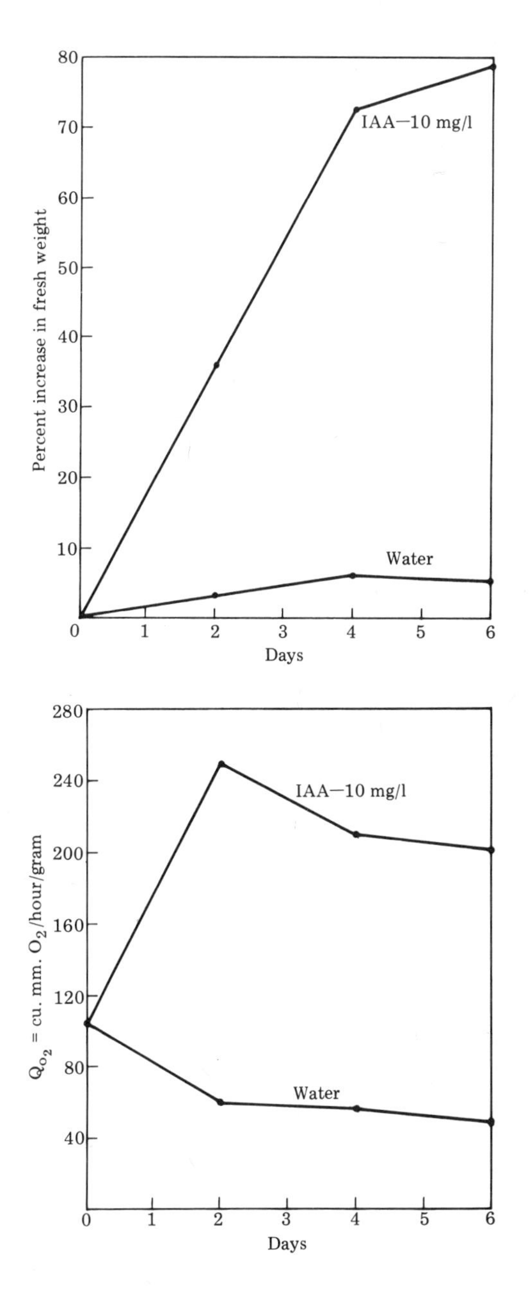

tion are very large: this happens in the discs cut from tubers of Jerusalem artichoke *(Helianthus annuus)*. Figures 3-23 and 3-24 show that when these tuber segments are growing slowly in water, their respiration is low, and after a few days it almost flattens out. In presence of indoleacetic acid there is a large increase in growth (larger than with the similar discs from potato tuber) and there is an increase in the rate of respiration of almost two and one half times. This is the largest known auxin-produced increase in respiration in the plant kingdom, being of the order of five times as great as any other increased respiration due to auxin. It leaves us in no doubt that auxin causes a promotion of an oxygen-consuming process necessary for growth.

Some years ago it was usual to dismiss these observations with the statement that respiration is no more than a *prerequisite* for growth, i.e., that only an actively respiring tissue would be capable of growth. Thus the effects of respiratory inhibitors were due merely to the removal of this basic requirement. But the data presented here point strongly to the conclusion that growth is *of itself* an oxygen-consuming process, separate from the total respiration, to which it makes only a small contribution.

(above) **3-23** The remarkable growth response (measured as fresh weight) of artichoke tuber slices to IAA. From Hackett and Thimann, 1952 a.
(below) **3-24** The respiration of artichoke discs like those of fig. 23. The maximum respiration increase is almost 2.5 times the control. From Hackett and Thimann, 1952 a.

Table 3-2. Content of lipides and of principal nitrogen compounds in pea stem segments before and after growth in darkness

Treatment	Elongation percent	Lipides	Amino acids	Protein	Asparagine
Initial	—	8.95	13.7	16.6	6.7
After growth:					
Water control	20.0	6.64	1.9	23.3	11.0
Auxin (IAA, 1mg/l.)	50.9	6.18	0.6	24.4	11.5
Auxin plus:					
Iodoacetate 0.6mM	25.6	7.44	6.8	17.2	8.1
Arsenite 0.1mM	26.5	8.60	8.7	19.5	7.6
Fluoride 5mM	25.3	9.82	7.5	20.7	7.7

From Christiansen and Thimann, 1950.

There are additional changes in metabolism during growth, many of them not yet cleared up. Perhaps the most extensively studied are those in the internodes of etiolated pea seedlings. These contain, at the start, a quantity of free amino acids, which come directly from the hydrolysis of the proteins in the cotyledons. During growth of these segments (in the dark) the amino acids are converted partly to asparagine, which is commonly found in legume seedlings, and partly to protein. This process of converting amino acids to asparagine and protein goes on in the isolated internodes during the 24 hours of elongation, and is promoted by auxin. Inhibition of growth by sulfhydryl reagents or other types of inhibitors prevents this metabolism to varying extents, even when their concentrations are such as to cause the same percentage of inhibition of growth (table 3-2). The metabolism of neutral lipides is also interfered with to some extent, though it is not greatly promoted by auxin. Each inhibitor appears to operate on metabolism in a different way. We shall return to the problem of the modes of action of growth hormones at the end of this work.

THE PROCESS IN WHOLE PLANTS

A. HORMONES AS CONTROLLING FACTORS

The problem of elongation in whole plants is difficult because the rate is controlled by so many nonhormonal factors like temperature, light, and water supply as well as by hormonal factors. We will consider the latter as best we can. First, there is the natural inhibitor Abscisic acid, AbA. This substance is quite unrelated to indoleacetic acid, being a carotenoid derivative, of the formula shown in figure 7-9 (p. 183). It occurs widely in the plant kingdom, and inhibits a variety of processes, though some only weakly. It is not a simple antagonist of one or another growth hormone but seems generally to have an activity-reducing effect. It was originally thought to be responsible for dormancy and was called Dormin, which could have been appropriate, because Dormin does not necessarily apply to strict dormancy, but rather to "putting to sleep," which is more suggestive of some actions of AbA.

AbA turns out to have an absolutely critical function in the physiology of leaves (chap. 7), but in other cases it merely acts as an inhibitor. In standard growth tests with stem or coleoptile segments, AbA inhibits fairly strongly, and the inhibition depends on the concentrations both of AbA and of auxin. Thus with IAA one micromolar, which is a little below the optimum concentration, 100 μM of AbA will reduce the growth by some 25%. With IAA 10 μM, it takes about 1000 μM AbA to bring about the same inhibition. Thus the inhibition is roughly competitive, though with a ratio of about 1:100. In very short time experiments, the onset of inhibition can be detected within six minutes, so that AbA evidently enters at least as fast as IAA. However, though it is very easy to show growth effects in these isolated segments, it is not so easy to show them in whole plants; even when IAA is applied to whole plants, no appreciable increase in growth results, or at least the effect is relatively minor. Yet it has been known from the early days that a droplet of IAA in a fatty solvent placed on the side of a growing tissue will immediately cause it to curve away from the IAA, just as we have seen in the case of tropisms. This shows that IAA is still in control in rapidly growing whole green plants. The paradox that one can produce a curvature in this way, without being able to get large increases in straight growth, has not really been cleared up.

There is an interesting action of auxin on seeds. When cereal seeds are treated with indoleacetic acid there may be little or no visible effect at the time, yet later, when the seeds are planted, the seedlings may show a considerable acceleration of growth, which may subsequently be retained even by the mature plants. This is a field that has been neglected in recent years, but about 20 years ago it was considered exciting, and workers in several different countries were soaking seeds in auxin solutions in the hope of increasing the yield of crops. The effects are often clear cut, but unfortunately they are rather susceptible to changes in the environment. Figures 3-25 and 3-26 show some of our own experiments done 30 years ago, with oat seedlings which were soaked in IAA solutions for the first 24 hours of their life, then rinsed and planted. After about the first two weeks (for the treatment inhibits root elongation and thus delays the seedlings at the start) the treated plants appear about a week's growth ahead of the controls, since they have about one leaf more. This advance was maintained throughout the life of the plants, and led to a considerable increase in the number of seeds set. Oats rather regularly show increases of about this magnitude. The response claimed the attention of the sugar beet growers in Germany, who obtained a comparable acceleration. Samples of their final beets are shown in figure 3-27. Dr. Amlong, who did the work, assured me that these were fair representative samples from the crop. It so happens that the sucrose content of the larger beets is somewhat lower than proportionate, so

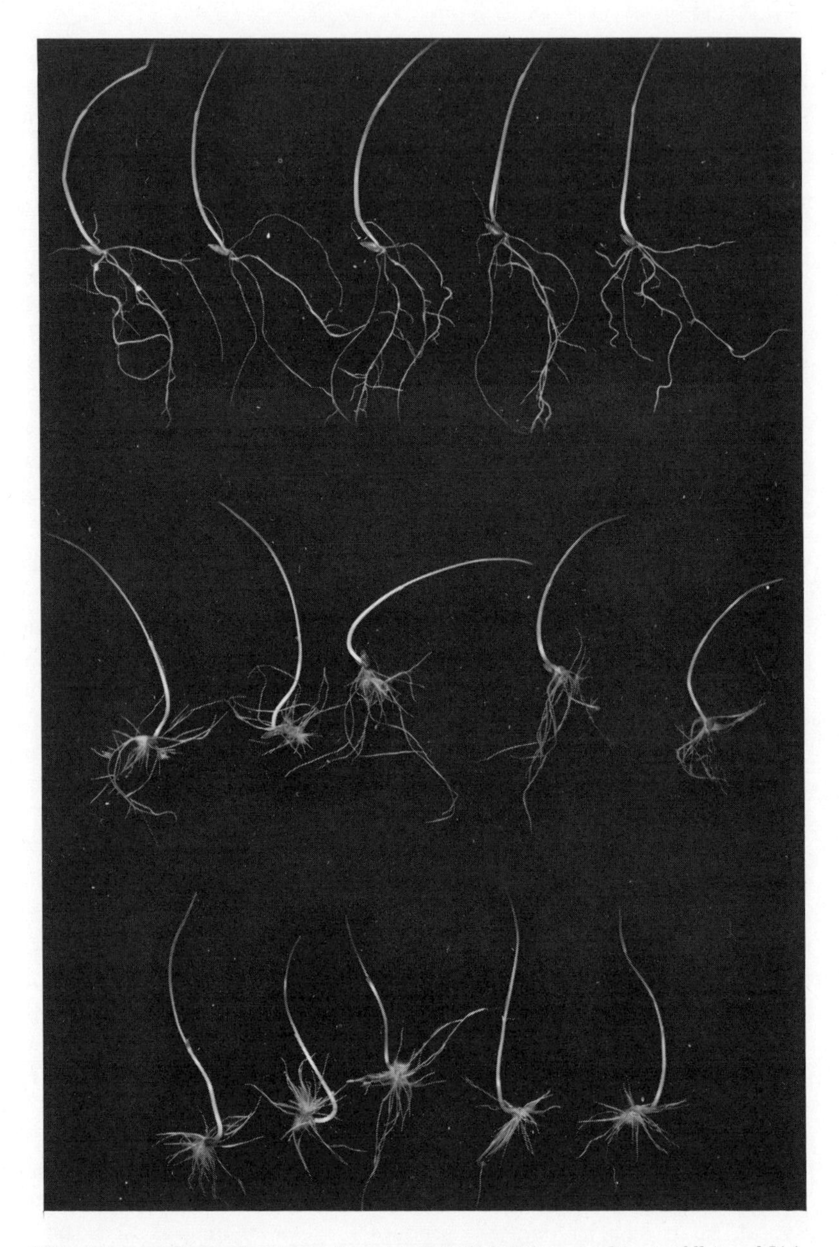

3-25 *Avena* seedlings after 5 days, in water (top); IAA 200 mg per liter (middle); and IAA 1000 mg per liter. Note root numbers. From Thimann and Lane, 1938.

(above) **3-26** Same plants as fig. 25 but at 2 weeks (above) and 5 weeks (below). Left, controls; right, treated with IAA 100 mg per liter. From Thimann and Lane, 1938.
(below) **3-27** Sugar beets grown from seed soaked in water (left), or IAA, 200 mg per liter (right); photographed at end of season. From Amlong and Naundorf, 1937.

there is a smaller increase in sucrose yield per acre. Dr. Amlong found that radish seeds similarly treated grow faster than controls, and can become larger in size. In general the treated plants at optimum auxin concentration show a difference of about 10 days' growth. The effects are fairly reproducible with oats and radishes, but with some crops, including rice, they do not show at all. Dr. Leopold shed some light on this variability by finding that the phenomenon, for peas at least, is very temperature sensitive. Auxin-treated seeds grown at one temperature showed a clear acceleration, which was demonstrated by the number of nodes formed before flowering, or the number of flowers formed by a certain age. If treated at another temperature, however, the effect was even a delay of development. Figure 3-28 shows that at an auxin concentration of 1 part per million, which gives about the maximum effect, they form at 3°C some 18 more flowers than the control. However, at 18°C the same treatment caused flowering to be actually delayed. Such variation of results with temperature, and perhaps with other factors, has discouraged agronomists from adopting auxin treatment of seed as a practical application. The accelerated development is probably related to the fact that auxin promotes the formation of additional roots (fig. 3-25), leading to more rapid subsequent growth.

As we have seen, the discovery of gibberellin shed a fresh light on the growth of whole plants, because gibberellin was discovered through the observation that some rice plants were unusually tall. The name *Bakanae-byo* (which means "foolish seedling") was given because the infected rice

3-28 Development of pea seedlings, as measured by the number of flowers, from seeds soaked in serial concentrations of naphthaleneacetic acid. Note that the aftereffect is wholly dependent on temperature. From Leopold and Guernsey, 1953.

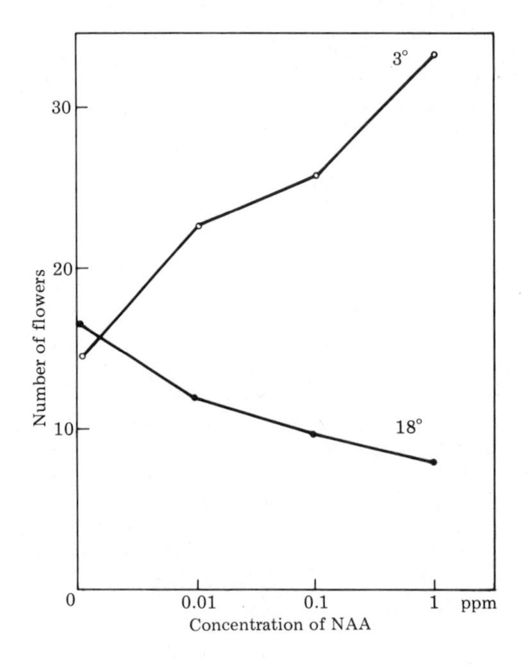

seedlings were not only very tall, but also pale in color, and their yield of grain was greatly reduced. When the fungus which caused the whole plant to grow tall was followed up it became clear that its product, gibberellin, had a dramatic effect on the elongation of the internodes of dicotyledons, and correspondingly also on the leaf sheaths of monocotyledons. In rice seedlings treated with gibberellin the leaves are much longer, and, as it later developed, it is the leaf sheaths which are most sensitive; hence the elongation of leaf sheaths was rapidly adopted as a bioassay. As a result it was soon found that the most responsive plants are those whose controls are normally short. The logical extreme of that is to go to dwarf plants and this Brian and his colleagues did with great success with Meteor peas, which are very rapid growing but very short. At the maximum gibberellin level the plants are about as tall as Alaska, Telephone, Alderman, or other tall varieties.

Many rosette-forming species produce dramatic responses. In a rosette the internodes are excessively short, and a minute amount of gibberellin, given only once in the plant's lifetime, causes every internode to elongate to its fullest extent. In the case of biennials, which are in the rosette form for the first year and elongate during the second year, gibberellin when it brings about this elongation sometimes also causes flowering, thus acting like the second year of a biennial. At one time it began to look as though gibberellin might be the long-sought and elusive flower-forming hormone. However, its flower-forming action is only seen on long-day plants in short

3-29 Growth of normal and dwarf-1 maize at 9 weeks. Left to right: normal, no treatment; normal, GA; dwarf-1, no treatment; dwarf-1,GA. The GA was applied every 2–3 days, totalling 0.06 mg per plant in all. Photo by B. Phinney, cf. Phinney, 1955.

days; it has no such effect on making short-day plants flower, so it is certainly not a general flowering hormone. And even in long-day plants, gibberellin sometimes induces only long nonflowering scapes. One even gets the impression that the flowering is secondary and is brought on in some way by elongation—but this can hardly be true.

A slightly different response is seen with monocotyledons, and especially clearly with corn, of which there are a number of dwarf cultivars. Figure 3-29, from the work of Dr. Phinney at Los Angeles, shows that the tall variety gives little or no response, but the variety dwarf 1, on treatment with gibberellin, grows about as tall as the normal siblings. There are eight or nine dwarf varieties of corn, and of these about five respond like this one to gibberellin, but the others do not. One would like to generalize that a dwarf plant is a plant that does not make gibberellin, but that generalization is unwarranted. It appears to be true for dwarf 1, dwarf 3, and dwarf 5, but not for dwarf 4 or for *nana,* a very similar low-growing form. Using this bioassay, Dr. Phinney measured, by extraction, the amount of gibberellin being synthesized by the different genetic varieties at the same age, and indeed the tall variety produced considerably more gibberellin than the dwarfs. These results looked very satisfactory; but when Dr. Radley carried out the same experiment with her peas, she found little or no difference between the gibberellin produced by several tall varieties and that produced by several dwarfs. Suge and Murakami in Japan came to a similar conclusion in regard to the extractable gibberellin in ten dwarf rice varieties. At one time it was believed the action of gibberellin was to make each internode keep growing for a time longer. Normally, a herbaceous dicot will grow essentially an internode at a time, and as one slows down the next one speeds up; thus, if instead of slowing down each internode kept growing a little longer, a tall plant would result instead of a dwarf. However, Dr. Brian's data show that this is not the case (table 3-3). We see that the gibberellin effect on internode length is very clear, but the number of days that each internode kept growing before it stopped is not increased. If anything the high gibberellin level, which makes the internode much longer, makes it stop growing sooner. This curious paradox at least negates any idea that gibberellin is a kind of juvenility hormone, such as the juvenile hormone of insects that makes them produce larger and larger larvae by preventing pupation.

There have been several attempts to equate the gibberellin effects with some of the varied effects of environment on growth of plants, especially effects centering around the action of light. It is true that in some instances gibberellin does function somewhat like light. For instance, the effects on elongation could be likened to the effects of long days on long-day-sensitive plants. For long-day plants in short days are apt to remain very short, often even to remain in rosette form. Interruption of the night with a small exposure to red light near the middle of the night causes the plant to consider this a long day, and to grow tall in consequence. Short-day plants given the same treatment consider it a long day and are correspondingly inhibited from flowering. Dr. Lockhart (now at the University of Massachusetts) found that the inhibition of growth of bean seedlings by a short red light interruption in the middle of the night could be completely reversed by gibberellin. We might conclude that treatment with gibberellin is just like illumination with far-red light, which reverses the red effect. We could have jumped to the same conclusion with the lettuce seeds in chapter 2. But while the effect of gibberellin is sometimes like that of long day or a

Table 3-3 Mean number of days during which each internode was visibly extending

	Internode no.						
Dose of GA (μg.)	5	6	7	8	9	10	11
0	6.8	7.0	6.4	6.6	6.6	6.6	6.4
0.1	5.8	5.9	5.0	5.3	5.9	5.4	5.4
1.0	5.8	5.4	4.5	5.1	5.2	4.6	4.6
10.0	6.0	5.3	4.8	5.1	5.7	5.0	4.9
Significant difference ($P = 0.05$)	0.7	0.8	0.8	0.7	0.6	0.7	0.7
Significant difference ($P = 0.01$)	0.9	1.0	1.0	1.0	0.9	1.0	1.0

From Brian et al., 1958.

light interruption of the night, sometimes it is not. Hillman, at Brookhaven, who grows many clones of the duckweed *Lemna,* has one clone of *Lemna minor* which can be grown on a synthetic medium providing it receives a very small amount of light. It needs only five minutes of red light every other day in order to grow normally. These five minutes of light, however, cannot be substituted for by any amount of gibberellin. But what is interesting is that it can be substituted for by cobalt, that is, cobalt here acts like red light. Burg's work in other connections has shown that red light commonly acts to inhibit ethylene formation, and Dr. Yang, at Davis, has found a similar action of cobalt. This similarity may explain the similarity of their effects on certain plants and under certain conditions.

On the whole, therefore, the comparison of gibberellin action with light action is not helpful. Indeed, we saw in chapter 2 that with *Phacelia* and *Nemophila* seeds gibberellin acts like darkness. Besides, the action of gibberellin in making dwarf plants grow tall can be likened to etiolation, that is, gibberellin acts on dwarf plants much like darkness. Etiolated peas actually make very little further increase in growth in response to gibberellin; light-grown plants are much more responsive. The effect of etiolation is at least as spectacular as any effects exerted by gibberellin, as the dramatic figure 3-30 shows. In addition, darkness makes the internodes colorless and the leaf blades small and very pale. Gibberellin also tends to make the leaves both smaller and paler and the internodes distinctly paler—not as small and pale as in

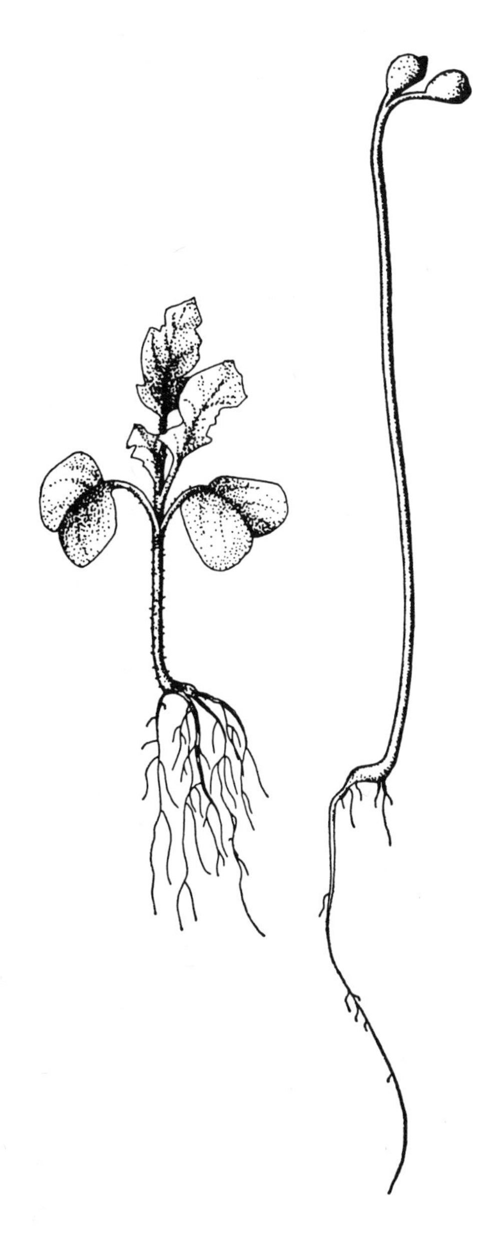

3-30 Effect of light on growth of mustard seedlings (*Sinapis alba*). Left, in white light; right, in total darkness. From Mohr, 1969.

darkness, but certainly in the same direction. There is a curious exception here, shown by beans (*Phaseolus vulgaris*). When beans are grown in darkness, their leaves are not so small as those of peas, and they *enlarge* on treatment with gibberellin; the effect here, then, is in the opposite direction. They can also be enlarged, as Carlos Miller has shown, by cobalt. In this case, then, an unusual effect of gibberellin is paralleled by the influence of cobalt. We noted above an effect of red light (not duplicable by gibberellin) that was imitated by cobalt. Indeed, there are now a number of growth effects known to be exerted by cobalt; and some workers are beginning to feel that cobalt is merely acting to antagonize an inhibiting ion like calcium, or an inhibiting hormone like ethylene, although it is still possible that there is a specific direct cobalt effect.

Another odd parallel to gibberellin action is seen in plants undergoing water stress. If plants that can survive being denied water for a while are kept in a maximally dry condition for many weeks, the internodes become extremely short, and the number of leaves is decreased. The controls, regularly watered, elongate normally, and the contrast, though less dramatic, reminds us very much of the effect of gibberellin on dwarf plants. That, in turn, brings to mind the rapid metabolic water uptake engineered by supplying auxin to tuber slices; it is as though gibberellin has also caused (or forced?) a special rapid uptake of water.

On the whole it appears that there are two kinds of response to gibberellin: (a) that of whole plants, which are short in stature for one reason or another (genetically or due to light treatment), and which elongate spectacularly in response to gibberellins but not in response to auxins; and (b) that of isolated segments or internodes, in which elongation may be considerable but is comparable in many respects with that caused by auxin. In the first type, if one looks at the apical meristem, one immediately sees that under the influence of gibberellin the meristem cells are made to divide. Workers who have counted the number of mitoses can sometimes see 30 or 40 times as many in the gibberellin-treated plant as in the control. Thus the number of cells produced just below the meristem is greatly increased, and when each one of these elongates to the maximum extent we have the characteristic gibberellin effect on elongation. In other words, it is the combination of those two factors, increased division and increased elongation, which causes the spectacular growth of whole plants. In the isolated internodes or coleoptile segments there are few, if any, mitoses. It is basically because this commonly used material for short-term elongation studies does not undergo appreciable numbers of cell divisions that the magnified gibberellin effect is limited to whole plants.

One last point about the long-term type of growth control (that in whole plants) concerns the way in which experiments with added gibberellin apply to the system in untreated plants. Phillips has shown that simple defoliation of the young leaves on a sunflower stem strongly retards elongation, and application of gibberellin restores the growth rate. Auxin does not. Thus the young leaves evidently supply the gibberellin which is needed for normal elongation.

Labeled GA_3 applied to growing leaves of tomato is both transported downwards towards the roots and carried upwards, probably in the transpiration stream, to younger leaves. This type of experiment has, curiously, not been carried out till recently, when it was reported by Couillerot and Bonnemain in France. Thus except for the lack of polarity the role of GA in stem elongation parallels the role of auxin.

In the shorter-term, more analyzable type of elongation, as stated above, the plant material used shows little or no cell division during the test response. The classical 10 mm segment of the *Avena* coleoptile undergoes no mitoses at all during 48 hours of elongation. The internode of etiolated Alaska peas, by my own extensive measurements, undergoes less than one mitosis per cell, perhaps 0.8 mitoses per cell averaged over the different layers, in the 24 hours during which it is elongating. In stem segments of the Azuki bean (*Phaseolus azukia*), which is much like our dwarf *vulgaris* beans, there are also virtually no mitoses during an ordinary growth experiment.

Next in importance is the fact, first reported by P. W. Brian in 1955, that there is a strong interaction between gibberellin and indoleacetic acid. Isolated light-grown pea internodes scarcely respond to gibberellin at all, but they show a strong growth response to IAA. If then, in addition to the IAA, gibberellin is supplied, they now show a growth response to the gibberellin. In other words, the presence of IAA is required in order for them to respond to gibberellin. As an example, such internodes in 30 hours grew an average of 1.75 mm in sucrose, 1.78 mm in GA_3 (i.e., virtually no effect), 3.91 mm in IAA, and in IAA plus GA_3 4.48 mm. Thus gibberellin causes growth in presence of IAA, but not by itself. When whole plants were used, internodes 5 and 6 only being measured, the figures were these: 7.9 mm in controls, 8.1 mm in GA_3, 13.0 mm in IAA, and 17.2 mm in IAA plus GA_3. Internodes of Azuki beans show the same requirement for IAA. With them, gibberellin produces no effect, auxin produces a 55% increase, auxin plus gibberellin an 89% increase. *Ipomea* petioles give similar results; so it is not an isolated phenomenon. These results led to the general conclusion that *auxin must be present* in order for gibberellin to act. But somewhat more sophisticated experiments have been done in the last two or three years which make the explanation a little more complex. In these experiments, we apply the different growth substances in different orders. Having determined first their optimum concentrations, we can apply gibberellin first, then rinse off the gibberellin and subsequently apply auxin, or it can be done in the reverse order. If auxin is applied first and gibberellin subsequently, we get a good growth response, from which again we might conclude that IAA must be present in order for the gibberellin to function. However, both Shibaoka in Japan and Chang and Ruddat in Chicago have observed that if the gibberellin is applied first, it produces little or no growth, but if then it is removed and indoleacetic acid applied, excellent growth results. Often the amount of growth resulting is a function of the duration of the gibberellin treatment. Both workers have come to the conclusion that gibberellin does have an effect by itself when it appears not to; this effect is to *prepare* the tissue for a subsequent response to IAA. Indeed, since there are very few organs on which gibberellin does not have some effect, it would be remarkable if it had no effect leading to elongation. What this preparation amounts to, of course, no one knows. Chang speaks of a "growth potential" which is "accumulated" in the presence of gibberellin, and "released" by indoleacetic acid. This reminds us of Bergson's saying in the last century, that things are alive because they have an *élan vital.*

A possible lead to the nature of this "growth potential" is that certain inhibitors will inhibit the gibberellin-induced growth but will not affect the auxin-induced growth. An example is given by the tannins. Tannins are a large group of glucosides of tannic acids; since they are phenolic it is not surprising that they have an effect on growth, but it is remarkable that though they inhibit the gibberellin-induced growth 100%, they scarcely affect the IAA-induced growth at all. Such effects are hardly likely to be exerted under natural conditions because the ratio of tannin to GA needed is around 1000 to 1; perhaps such a ratio might occur in cells heavily loaded with tannin but such cells are rarely, if ever, still growing. Of more interest, not because it is more likely to occur but because it is more susceptible of a definite explanation, is the recent report of Shibaoka, in Tokyo, that colchicine has a similar effect, inhibiting the GA-induced growth at concentrations and under conditions where it has no effect on the growth induced by IAA. Colchicine inhibits mitosis by disrupting the microtubules, probably through binding to their protein component. In the stem segments here used (as in those discussed above) there are very few mitoses, the stem elongation being almost wholly due to cell elongation. Figure 3-31 and table 3-4 show that colchicine specifically prevents the action of GA but not that of IAA. Other experiments show that GA alone has no effect. Evidently, then, not

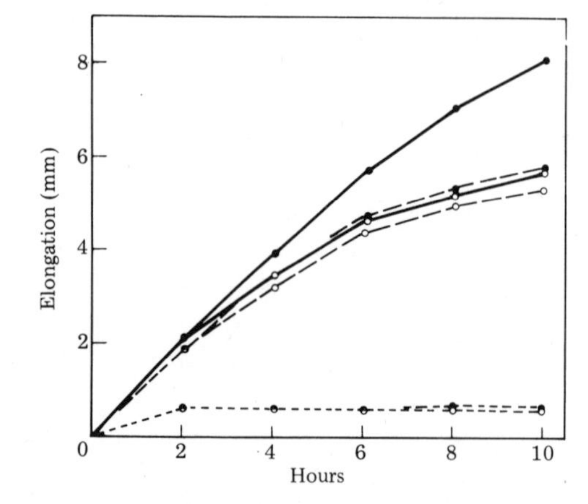

3-31 Effect of colchicine (1 mM) on elongation of stem segments of Azuki bean. Open circles, IAA; filled circles, IAA + GA. Solid lines, alone; dashed lines, plus colchicine. The dotted curves at bottom are without IAA. From Shibaoka, 1972.

Table 3-4 The interaction between gibberellic acid and colchicine in the elongation of stem segments of Azuki bean (*Vigna angularis*)

Treatment 0–1 hr.	1–15 hr.	Percentage elongation
Water	IAA	48
GA_3	IAA	78
Water	IAA + Colchicine	44
GA_3	IAA + Colchicine	49

Concentrations: IAA, 1×10^{-4} M, GA_3, 1×10^{-3} M, Colchicine, 1×10^{-3} M.
From Shibaoka, 1972.

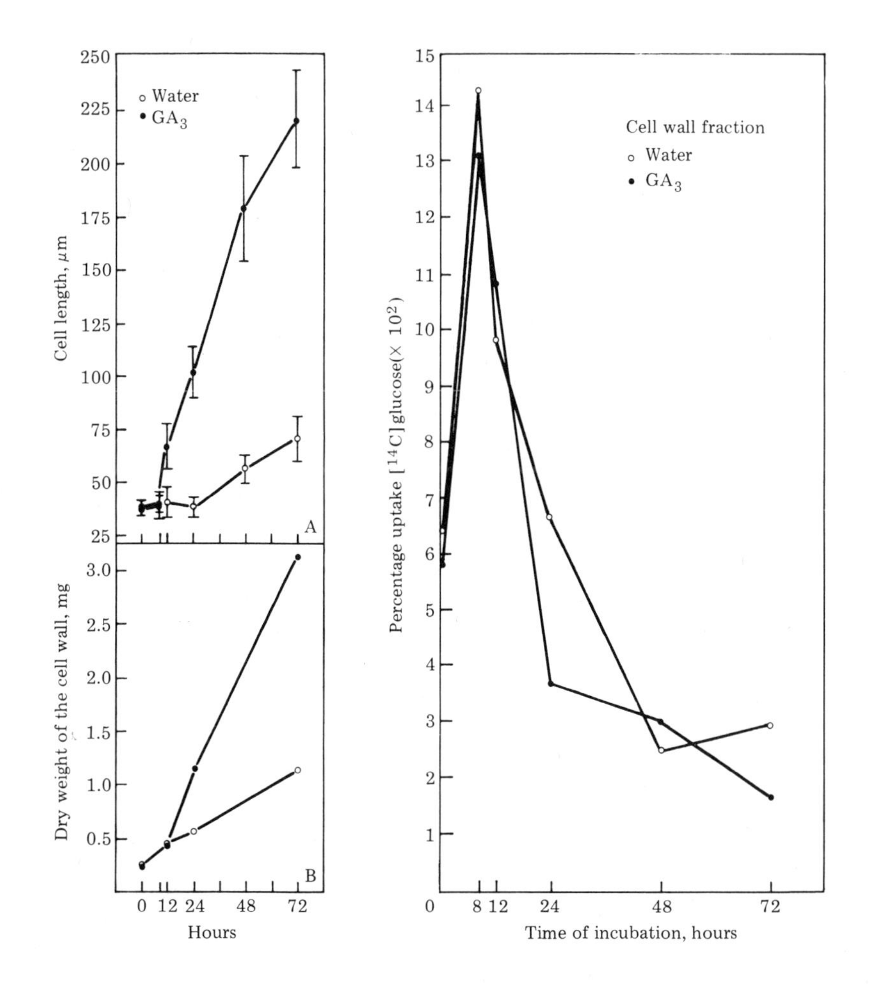

3-32 Effect of gibberellin on elongation and cell wall formation in lettuce seedlings. Left, above, elongation; below, weight of cell walls. Right, incorporation of ^{14}C-glucose into wall fractions. From Srivastava et al. 1975.

only do the two hormones act differently, but the GA action must be followed or accompanied by the other to produce visible growth. Shibaoka interprets this as follows: in elongation as in cell division, the microtubules exert an orienting influence by modifying the arrangement of the polysaccharide microfibrils of the cell wall; GA acts first on these microtubules, thus increasing the effectiveness (probably not the rate) of their orienting action, while auxin acts in some other way not involving the microtubules. This interpretation, while not yet proven, does fit in well with the different effects of these two hormones on wall plasticity, the differences in timing, and their differing effects on different types of tissue.

In regard to the timing, Srivastava of Vancouver has made a curious observation. The hypocotyls of lettuce seedlings give an excellent growth response to gibberellin alone (perhaps because they contain adequate supplies of auxin), but there is a lag of 8 hours between the time that GA is applied and the time the ensuing rapid growth commences (fig. 3-32). During this lag, sugar molecules are incorporated into the polysaccharides of the cell walls, but the GA does not increase the incorporation, nor does it shorten the lag. It is as if some material other than the wall itself has to be acted upon by the gibberellin before visible elongation can begin. Auxin effects on growth seldom require a lag of more than 20 minutes or so (at 25°C) and often much less. Auxin thus seems to act on something which is either already there, or is formed very fast, while gibberellin has to induce the formation of a preceding stage; and auxin can participate in, and profit by, this induction. A logical way to express this would be that gibberellin leads to the laying down of "stretchable" polymers, while auxin changes the existing polymers to "stretchable" form *after* they have been laid down. Clearly this would coincide well with the above microtubule locus of action. However, we must reserve judgment, as we shall see.

There is another curious interaction between gibberellin and auxin which has been observed on a number of different plants and by several workers. It is this: if we measure the rate of auxin production by the growing apex (rather than measuring the instantaneous auxin content by extraction), we find that if it has been pretreated with gibberellin the auxin production is larger—it can be very much larger. In the experiments of Nitsch in Paris with dwarf beans, whose normal output of auxin is very small, the rate was actually 200 times as great as in the controls! In Miller's work on *Hyoscyamus* the ratio was 40 times. At first it was thought that the large increase in auxin might be a secondary result of the great increase in growth. However, the recent experiments of Varga in Hungary show this cannot be so because the increase in auxin is detectable before any increase in growth occurs. In the tissues used by Varga, in the first day gibberellin produces

(above) **3-33** Effect of GA treatment on the IAA content extractable from bean hypocotyls in 24 hours. Solid lines, intact plants; dashed lines, decapitated plants. From Varga and Bitó, 1968.
(below) **3-34** Effect of GA treatment on the diffusible auxin (above) and the elongation (below) of dwarf pea seedlings. Note difference in timing of the effects. From Kuraishi and Muir, 1964.

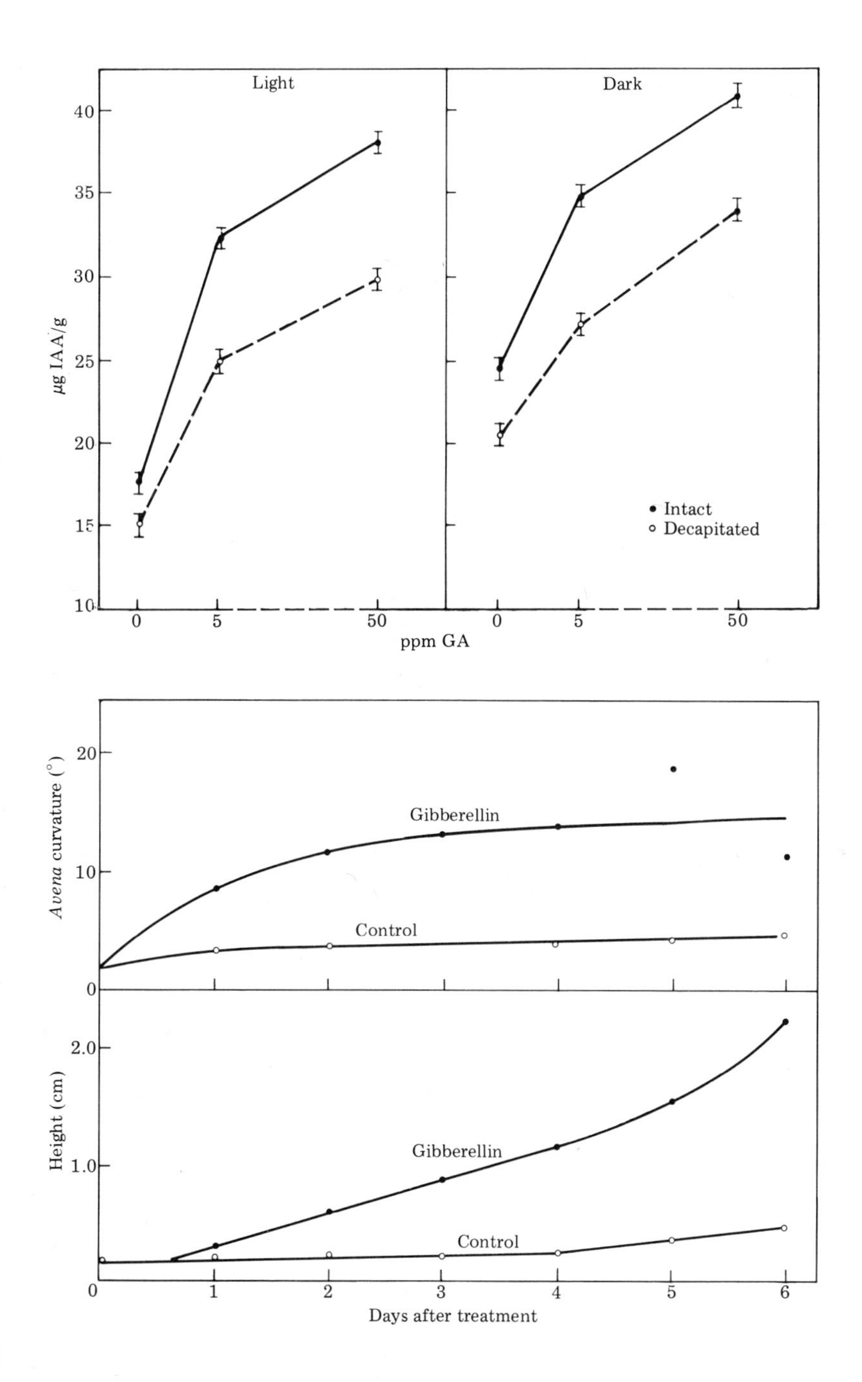
Light
Dark
40
35
30
25
20
15
10
μg IAA/g
• Intact
○ Decapitated
0
5
50
0
5
50
ppm GA
20
10
0
Avena curvature (°)
Gibberellin
Control
2.0
1.0
Height (cm)
Gibberellin
Control
0
1
2
3
4
5
6
Days after treatment

virtually no difference in the length, while most of the gibberellin effect on auxin production, which was measured hourly, had already occurred. The effect took place both in light and in dark (see fig. 3-33). The crucial difference in timing is also seen in figure 3-34. It is not likely that this effect of gibberellin is exerted on the *transport* of auxin. For in many of these cases our measurement depends on the movement of auxin away from the zone in which it is produced. The action of gibberellin on auxin transport is described in chapter 4. It is a curious phenomenon, and it no doubt plays some part in the action of gibberellin on growth, since if the internodes have at their disposal an increased supply of auxin, they will obviously grow more for that reason alone. And at one time Dr. Kuraishi, in Kyoto, felt that the whole effect of gibberellin could be just due to the increase in auxin supply. But that can hardly be, because gibberellin has so many effects that auxin does not have, such as the increased elongation of rice leaves and leaf-sheaths. Indeed, GA does not increase the extractable auxin of rice plants. So although the effect of gibberellin is not *due to* its action on the auxin supply, it may be aided and increased by the greater auxin supply. The obvious, if not very original, deduction from these facts is that the growth of whole plants is an exceedingly complex phenomenon.

Finally it should be mentioned that gibberellin does not enhance all auxin effects. For instance, auxin stimulates root formation, which is antagonized by GA. Nor could the effects of GA in enlarging dwarf fruits be considered as an enhancement of an auxin effect, because auxin has almost no effect on dwarf seedless fruits. They may contain enough auxin so that added auxin has no further effect, but in any event, it does not cause any growth, while gibberellin has a large effect. As far as we know now, therefore, such auxin-gibberellin interaction is seen in only one other function besides elongation, and that is in the growth of parthenocarpic fruits (chapter 11).

REFERENCES

Amlong, H. U., and Naundorf, G. Ueber einige praktische Anwendungen der pflanzlichen Streckungswuchsstoffe. *Forschungsdienst* 4, 417–428, 1937.

Bonner, J. Limiting factors and growth inhibitors in the growth of *Avena* coleoptile. *Amer. J. Bot.* 36, 323–332, 1949.

Brian, P. W., and Hemming, H. G. The effect of GA on shoot growth of pea seedlings. *Physiol. Plantarum* 8, 669–681, 1955.

———, Hemming, H. G., and Lowe, D. Effect of GA on rate of extension and maturation of pea internodes. *Ann. Bot.*, n.s., 22, 539–542, 1958.

Butler, W. L., Hendricks, S. B., and Siegelman, H. W. Action spectrum of Phytochrome *in vivo*. *Photochem. Photobiol.* 3, 521–528, 1964.

Chang, I., and Ruddat, M. Cell elongation in corn coleoptiles. 1 and 2. *Plant Physiol. Suppl.* (50th Anniv.) 304 and 305, 1974.

Commoner, B., and Thimann, K. V. On the relation between growth and respiration in the *Avena* coleoptile. *J. Gen. Physiol.* 24, 279–296, 1941.

Hackett, D. P., and Thimann, K. V. The effect of auxin on growth and respiration of artichoke tissue. *Proc. Nat. Acad. Sci.* (U.S.) 38, 770–775, 1952a.

———. The nature of the auxin-induced water uptake by potato tissue. *Amer. J. Bot.* 39, 553–560, 1952b.

Heath, O. V. S., and Clark, J. E. Chelating agents as plant growth substances. A possible clue to the mode of action of auxin. *Nature* 177, 1118–1121, 1956; *Nature* 178, 600, and 601, 1957.

Kuraishi, S., and Muir, R. M. The relationship of gibberellin and auxin in plant growth. *Plant and Cell Physiol.* 5, 61–69, 1964.

Leopold, A. C., and Guernsey, F. S. Flower initiation in Alaska pea. 1. Evidence as to the role of auxin. *Amer. J. Bot.* 40, 46–50, 1963.

Leopold, A. C., and Kriedemann, P. E. *Plant growth and development.* 2nd. edn. New York: McGraw Hill, 1975, 545 pp.

Mohr, H. Photomorphogenesis. In *Physiology of plant growth and development*, ed. M. Wilkins. London: McGraw Hill, 1969.

Phinney, B. Growth response of single-gene dwarf mutants in maize to GA. *Proc. Nat. Acad. Sci.* (U.S.) 42, 185–189, 1955.

Platt, R. S., Jr. The inactivation of auxin in normal and tumorous tissues. *Année Biol.* 20, 349–359, 1954.

Schneider, C. L. The interdependence of auxin and sugar for growth. *Amer. J. Bot.* 25, 258–270, 1938.

Shibaoka, H. Gibberellin-Colchicine interaction in elongation of Azuki bean epicotyl sections. *Plant and Cell Physiol.* 13, 461–469, 1972.

Srivastava, L., Sawhney, V. K., and Taylor, I. E. P. GA-induced cell elongation in lettuce hypocotyls. *Proc. Nat. Acad. Sci.* (U.S.) 72, 1107–1111, 1975.

Suge, H., and Murakami, Y. Occurrence of a rice mutant deficient in gibberellin-like substances. *Plant Cell Physiol.* 9, 411–414, 1968.

Tagawa, T., and Bonner, J. Mechanical properties of the *Avena* coleoptile as related to auxin and to ionic interactions. *Plant Physiol.* 32, 207–212, 1957.

Thimann, K. V. Studies on the growth and inhibition of isolated plant parts. 5. The effects of cobalt and other metals. *Amer. J. Bot.* 43, 241–250, 1956.

———, and Bonner, W. D., Jr. Experiments on the growth and inhibition of isolated plant parts. 1. The action of iodoacetate and organic acids on the *Avena* coleoptile. *Amer. J. Bot.* 35, 271–281, 1948. 2. The action of several inhibitors on growth of the *Avena* coleoptile and *Pisum* internode. Ibid. 36, 214–223, 1949 a.

———, and Bonner, W. D., Jr. Inhibition of plant growth by protoanemonin and coumarin and its prevention by BAL. *Proc. Nat. Acad. Sci.* (U.S.) 35, 272–276, 1949b.

———, Slater, R. R., and Christiansen, G. S. The metabolism of stem tissue during growth and its inhibition. 4. Growth inhibition without enzyme poisoning. *Arch. Biochem. Biophys.* 28, 130–137, 1950.

———, and Lane, R. H. After-effects of the treatment of seed with auxin. *Amer. J. Bot.* 25, 535–542, 1938.

———, Loos, G. M., and Samuel, E. W. Penetration of mannitol into potato discs. *Plant Physiol.* 35, 848–853, 1960.

———, and Schneider, C. L. The role of

salts, hydrogen ion concentration and agar in the response of *Avena* coleoptiles to auxins. *Amer. J. Bot.* 25, 270–280, 1938.

———, and Schneider, C. L. Differential growth in plant tissues. 1. *Amer. J. Bot.* 25, 627–641, 1938.

Varga, M., and Bitó, M. On the mechanism of gibberellin-auxin interaction. 1. Effect of gibberellin on the quantity of free IAA and IAA conjugates in bean hypocotyl tissues. *Acta Biol. Acad. Sci. Hung.* 19, 445–453, 1968.

Yoda, S., and Ashida, J. Effects of gibberellin and auxin on the extensibility of the pea stem. *Plant and Cell Physiol.* 1, 99–105, 1960.

Polarity and the Transport of Auxin 4

A. DISCOVERY AND EARLY WORK

The curious phenomenon of polarity is a classical topic which is basic to much of the study of morphogenesis in plants. The fact that auxin moves polarly is about the strangest observation in the whole physiology of auxin action. Its discovery dates back to the experiment of Boysen Jensen in 1910, in which seedling tips were cut off and stuck on again; the plants exposed to light then gave a curvature in the part below. After Boysen Jensen's day a number of workers used that as a starting point, including Arpad Páal, who did the critical work mentioned in chapter 1; but Beyer, in Germany, improved on the business of making a cut by making *two* cuts. Instead of just cutting off the tip of a coleoptile and sticking it on again, he cut off in addition another segment below. He repeated this with a second coleoptile, substituted that segment for the lower segment of the original one, replaced the tip, and then illuminated from one side. He got a curvature towards the light, in the expected manner. Hence the inserted segment participated in the response just like the original segment. Then he performed the following critical variation: everything was the same, except that the segment below was taken from an *inverted* coleoptile, so that this middle section now had its apical end at the base, and its basal end at the top (fig. 4-1, *top*). The result was that, when he irradiated from one side, no curvature resulted. Beyer deduced that it was essential that all the tissue have the same apex-to-base polarity; if one segment had the reverse polarity it would not conduct the influence which made plants curve towards light.

When Went, in 1928, was putting coleoptile tips on agar for the auxin to "diffuse" into the agar block, he modified that experiment in a natural way (fig. 4-1, *bottom*): he placed tips on the agar, and after 2 hours placed the agar, now containing the growth substance, on the upper cut surface of a short segment of coleoptile, with a block of plain agar in contact with the lower cut surface; after 2 hours more, the lower block ("receiver") was able to cause curvature on test plants, and therefore it now contained growth substance—auxin. Thus the auxin could be transported right through a segment of coleoptile. Went then did the reverse experiment, namely he placed the agar which was known to contain growth substance on the basal end of a segment of coleoptile; in this way the auxin would be transported along the reverse polarity. Nothing happened; the receiver block did not acquire the ability to produce curvature. So Went deduced that the movement of auxin through tissue is polar, from the apex towards the base, as Beyer's experiment had indicated.

Immediately following Went's work, his colleague van der Weij made an extensive study of the auxin transport. Van der Weij's thesis is a classic in this field. It was published in two parts. First, he confirmed that the transport is strictly polar: little or no transport takes place in the reverse direction. Next, by putting blocks on and taking receivers off at various times, he was able to follow the rate of movement into the receiver block and determine the time it

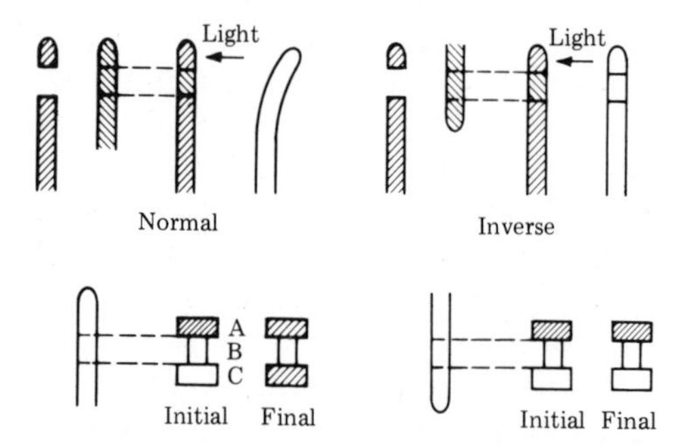

4-1 Above, Beyer's experiment; insertion of a coleoptile segment between illuminated tip and darkened base does not prevent phototropic curvature; but if the inserted segment is inverted no curvature results. Below, the experiment of Went and of van der Weij, much used later; shading indicates the auxin content; auxin travels from the apical donor to the basal receiver, but not vice versa. From Went and Thimann, 1937.

takes to go through a section. When he used a section 2 mm long, he obtained a curve like that shown in figure 4-2. By projecting it back toward the origin, we arrive at the time at which the first molecule of auxin appears at the base of the transporting section. With 2-mm sections, it took 10 to 12 minutes for this point to be reached. The rate of migration was therefore around 10 to 15 mm per hour (most of the time is taken in travelling through the tissue.) Later, more careful measurements showed that at 25°, the optimum temperature, it is more like 15 to 18 mm per hour in oat coleoptiles, and in other plants somewhat less. Subsequent workers have found in *Phaseolus* a rate of 10 mm per hour, and in *Coleus* petioles and stems a much smaller value, only about three mm per hour. But in these tissues also the auxin goes primarily from apex towards base, the only difference being that the polarity is not quite so strict. In other words, in the reverse transport experiment, a small amount may come out into the receiver. In older tissues, though not in coleoptiles, the polarity is frequently less strict.

The polarity observed in a segment, which consists of many cells, may not be so strict in a single cell, for de la Fuente and Leopold have calculated that if the apex-to-base transport in each cell were only 5% greater than that in the reverse direction, then in a 5-mm segment containing perhaps 50 cells end-to-end, the polarity would be almost 100%. One of the most striking observations of van der Weij was that with longer segments, it takes longer, naturally, for the auxin to emerge at the far end, but once it has started to appear, it

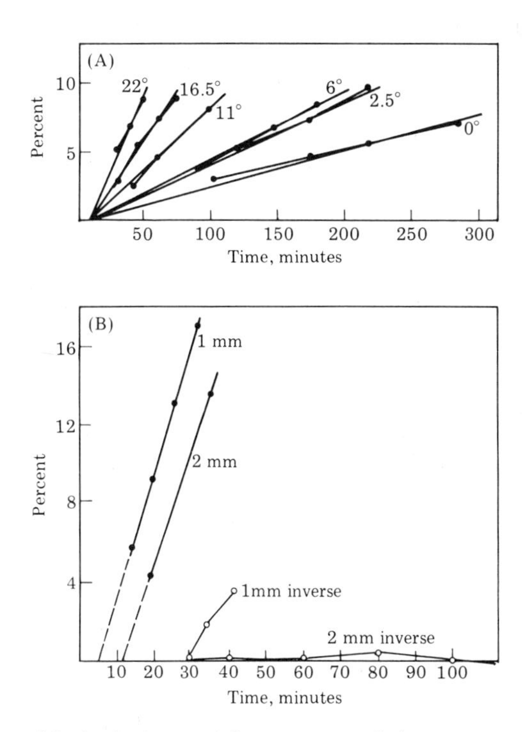

4-2 Auxin transported, as percent of the amount applied, through 2-mm coleoptile segments. Above, time courses at different temperatures suggest that the time required for the first appearance of IAA (the velocity) is nearly independent of temperature. Below, influence of length of segment: closed circles, basipetal movement; open circles, acropetal movement. From van der Weij, 1934.

comes out at the *same* rate as with the 2-mm segments. So if we had a section twice as long, the initial lag time would be doubled but the slope would be the same. With very much longer sections there is a slight slowing, but for short segments the auxin molecules evidently come through at a rate independent of the segment length. Now if they were simply diffusing, the longer the segment the less rapid the diffusion would be, and Fick's Law tells us that the rate of outflow would therefore be inversely proportional to the length. Since the reverse is true here, we are evidently not dealing with diffusion in the ordinary sense. What we have to deal with is something like what happens in a mass production factory, where the products are put on a moving band. At the other end of the band, workers take them off and pack them in packages. Once the flow has started, the rate of arrival of the products at the far end would be independent of the band length. The waiting time before the first objects arrive would (given constant band speed) be proportional to the length of the band, but as soon as they start to arrive they would all be arriving at the same speed. Similarly, in the transport of auxin through tissues, there must be some device whereby the auxin moves at a constant speed and so arrives at the far end irrespective of the length of the section.

Van der Weij's third and perhaps most remarkable finding, further supporting the idea that auxin transport through tissues is not mainly by diffusion, resulted when he had auxin already present in the receiving block. If this contains as much auxin as the donor block, then on any inorganic type of movement there would be no gradient; therefore there could be no movement from the donor to the receiver. But in fact, even if there is as much auxin in the receiver block as in the donor, auxin moves out of the donor block into the receiver. Van der Weij's most striking results in this respect are given in his second paper. He had in the donor, 100 units and in the receiver, nothing; after 2 hours, he found in the donor, 14 and in the receiver, 87. The agreement is excellent, and obviously there has been no loss. In the second group of experiments he started with 100 units in the donor and 100 in the receiver. Although there is thus no gradient, yet at the end of 2 hours he found 14 units left in the donor and 193 in the receiver. Hence the same amount had moved through the tissue, irrespective of the gradient. In a third set of experiments he even had twice as much in the receiver—100 in the donor, 200 in the receiver—yet at the end of the 2 hours he recorded 12 left in the donor and 284 in the receiver. The data are remarkably consistent though they are, of course, the average of a great many experiments. They show strikingly that the amount of transport is independent of the gradient. It is as though molecules were picked up and actively transferred from cell to cell through the tissue until they reach the far end, where some force pushes them into the agar block. Energy obviously has to be expended for this process, and correspondingly, as we shall see later, it is an oxidative process.

Now we know cases similar to this in the uptake of ions by aquatic plants. The cells of Nitella, growing in very dilute solutions, nevertheless reach an internal concentration of potassium or phosphate very similar to that of the cells of higher plants. In the case of phosphate, there is perhaps 10^{-6} M in the external solution, 10^{-3} M inside the cytoplasm; so in spite of a one-thousand times gradient against it, phosphate continues to be brought into the cells. Such solute accumulation is well known to be oxygen-dependent, inhibited by compounds like dinitrophenol that prevent phosphorylation, and therefore ATP-dependent; it presents a classical example of the use of oxidative energy to carry out osmotic work. We deduce, then, that this is the type of process to which auxin transport belongs. We shall encounter influences of other components of the environment on transport in the sections below.

Less is known about the transport of gibberellins. But just recently Jacobs, at Princeton, has carried out experiments with GA_3 similar to those with auxin, using the powerful influence of gibberellins on amylase formation in barley half-seeds as a delicate assay. Figure 4-3 shows that 1 microgram of GA_3 applied to the apical end of a segment of *Coleus* petiole causes gibberellin activity to appear in the receiving block, beginning at 3.5 hours after application. From this time onwards the transport is linear with time for another 2 hours. Since the segments were 5.1 mm long the transport velocity is $5.1 \div 3.5$ or about 1.5 mm per hour; this is about one-eighth the velocity of auxin transport in *Avena*. There was some, but less marked, polarity than with IAA. However, when gibberellin is applied externally to whole plants it is generally observed that transport takes place almost equally in both directions, although this

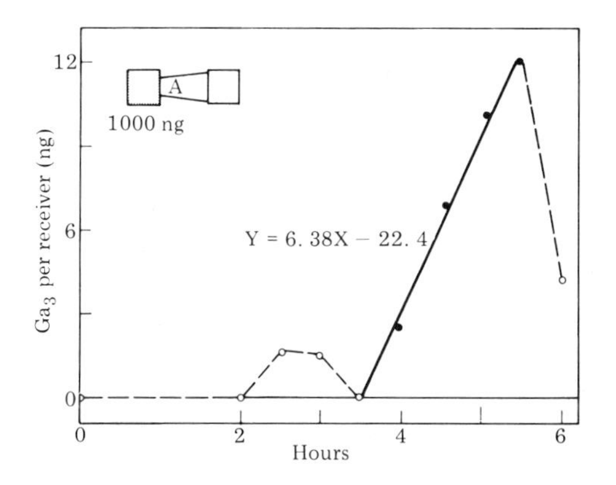

4-3. Transport of gibberellin from 1 μg GA_3, applied to the apical cut surface (A) of a segment of *Coleus* petiole, through to the receiver block. The appearance of GA is linear with time after 3.5 hours. From Jacobs and Pruett, 1972.

could be at least in part due to its entry into the vascular system.

As to the cytokinins, their transport is slower still. Pilet and his co-workers found that not more than 1%–10% of the amount applied to a lentil (*Lens esculenta*) stem reached the next internode in 24 hours. Osborne found a similarly slow movement in petioles. Probably there is diffusion from cell to cell. However, there is upward transport in the transpiration stream, since some cytokinin is produced in the roots. Thus the latter pathway may well be predominant in the intact plant.

B. THE USE OF ^{14}C-INDOLEACETIC ACID

The work we have been describing had to be done by bioassay. This was rather slow and cumbersome because to measure the amount of auxin quantitatively, the best method was to measure the curvature of decapitated coleoptiles to which the auxin in agar had been applied. In the early 1950s it became possible to use radioactivity as a much simpler assay, by synthesizing indoleacetic acid which contained carbon-14. While this seems like a tremendous advantage, for a short while it was a disadvantage, because unless the carbon-14 is present only in the auxin and not in any impurities, the radioactivity will measure the sum of the auxin plus the radioactive impurities. Furthermore, as we saw in the preceding chapter, the oxidative destruction of indoleacetic acid yields CO_2, and the indole residue remains as an oxindole derivative. Thus if radioactivity is present in either of the rings, or even in the CH_2 group of the side chain, and the auxin decomposes, the carboxyl group will be taken off and the remaining nucleus will still contain carbon-14. As a result, the radioactivity will no longer have any biological meaning. It is also essential to maintain the auxin concentrations within the range that the plant normally has to deal with; at unbiologically higher concentrations several secondary processes occur. After a number of these possible mistakes had been made by earlier workers, it was realized that it would be necessary to (a) use concentrations no higher than one would use if one were doing a bioassay, (b) purify the indoleacetic acid rigorously, and (c) have the C^{14} label in the carboxyl group so that if it were released as CO_2 no radioactivity would be left. Accordingly the synthesis of carboxyl-labeled indoleacetic acid was worked out by my colleague Bruce Stowe, and a new series of experiments was initiated on transport, a good many of them at Harvard by Mary Helen Goldsmith (now at Yale).

Figure 4-4 gives the necessary proof that the C^{14} transported from the donor block into the receiver is still biologically active auxin and not any impurity or breakdown product. In this experiment the auxin was provided at a series of donor concentrations, and the concentrations in the receiver block were determined in two different ways: the open circles show radioactivity determinations and the half-circles show bioassay (curvature) determinations. They clearly fall on the same curve. Although the points scatter a little, especially at the higher concentrations, there is no question that the radioactivity and the biological activity go together, i.e., the C^{14} in the receiving blocks represents unchanged indoleacetic acid. From this it was possible to show that auxin *in the tissue* is not necessarily subject to quite the same rules as auxin entering the receiver block. For instance the polarity, which was shown by the failure of auxin to enter the receiver block when moving from base towards apex, is not quite so strict within the tissue. Auxin applied to the basal cut surface does get into the tissue, and some C^{14} is found there. However, if the donor block is removed, after a time the amount in the tissue may decrease again. What has happened is that some auxin enters the tissue, probably by diffusion, and having entered the living cells becomes subject to the polar transport, and is therefore expelled towards the base. Mrs. Goldsmith could show that if, after the donor block is removed, a fresh receiver block containing no auxin is applied, a minute amount of auxin can be detected, coming out into that receiver block. Thus the auxin which at first entered is subsequently expelled.

The radioactivity assay also shows directly that the system transporting auxin through the tissue to a receiver is readily saturated if one supplies too much auxin. The first part of the system, that in which the auxin moves from the donor into the tissue, shows little sign of such saturation. Auxin enters proportional to the amount applied over a very wide range, especially in corn coleoptiles and to a lesser extent in oat coleoptiles. Figure 4-5 shows that saturation is reached in the receiver, while the amount in the tissue continues to go on up to as high a level as was applied. This tells us that the molecules go into the tissue

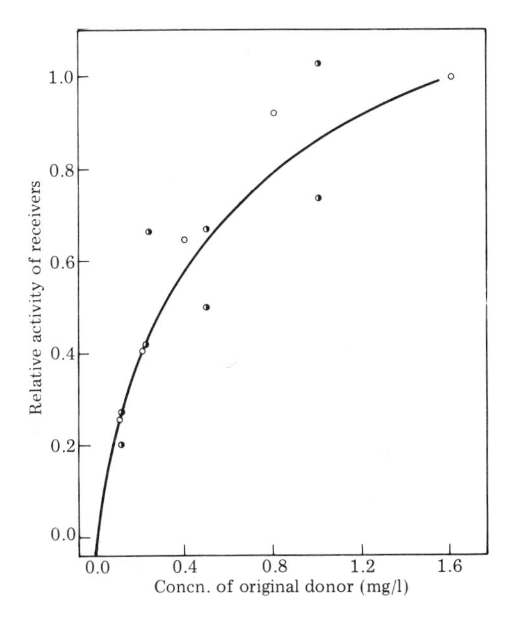

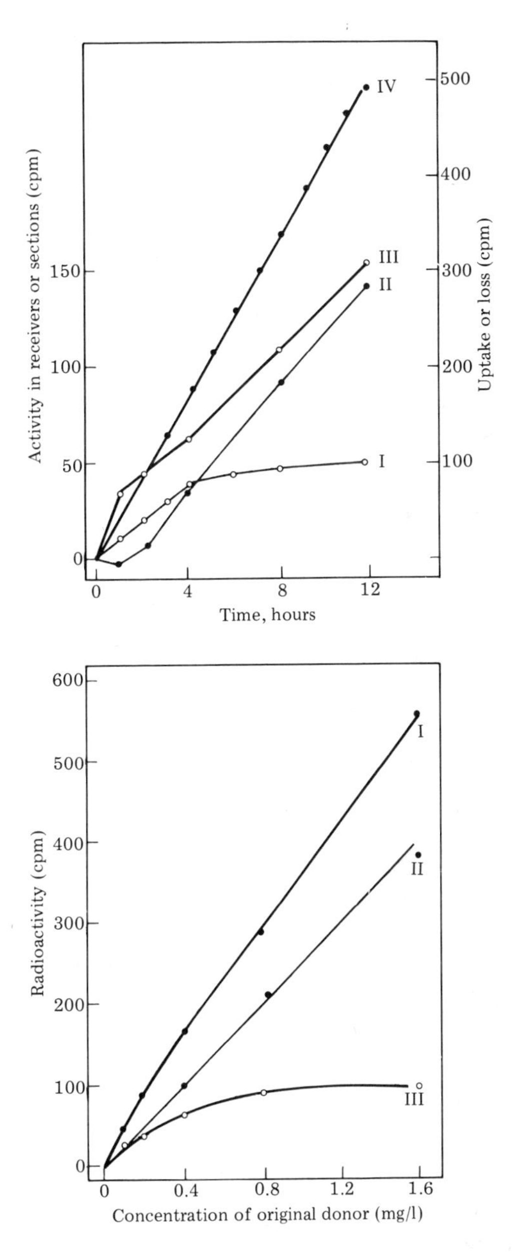

(above) **4-4** Radioactivity (open circles) and auxin activity (half-filled circles) of a series of receiver blocks after 3 hours' transport, using different IAA concentrations. From Goldsmith and Thimann, 1962.

(right) **4-5** Above, radioactivity entering the receivers, I, and tissue, III (both on left ordinate); radioactivity leaving the donors, IV, and that lost from the system, II (both on right ordinate). Below, a similar experiment showing radioactivity leaving the donors, I, found in the tissue, II, and in the receivers, III. Plotted as function of donor IAA concentration. From Goldsmith and Thimann, 1962.

presumably by diffusion (though over a tiny contact distance), and then once in the tissue they are picked up by the transporting mechanism, and that transporting mechanism is easily saturated.

Another important observation is that after some hours, such transport experiments come to a standstill because the cells at the surfaces die out—the cells of both cut surfaces. If we renew the receiver blocks every hour, we usually find that by the fourth hour nothing more comes out. If then we cut off ½ mm at the base and ½ mm at the apex and put the blocks back on, transport continues again normally. So at the cut surface there is a wound influence, which gradually inactivates the cells adjacent to the wound. This phenomenon, though well known, is not understood. It used to be attributed to a wound hormone, especially by Haberlandt, but it appears that we have to deal not with a hormone but with something a little more elusive.

It has been mentioned that the accumulation against a gradient means that auxin is moved by an energy-requiring process; therefore the phenomenon of auxin transport requires oxygen. It is possible that some fermentative process might furnish enough energy for a small amount of transport in nitrogen. The experiment to test this is tricky, because the coleoptiles are hollow and thus contain a good deal of oxygen in the hollow space. The more careful one is, the less transport one observes, but on the other hand it never falls quite to zero (fig. 4-6). In nitrogen the entry of auxin from the base becomes evident, because now there is no reverse expulsion of the auxin molecules; those that go in stay in.

4-6 Percentage of the radioactivity moving beyond 10 mm in the segment, in air (open circles), and in nitrogen (filled circles). This treatment of the data avoids the passive entry into the cut surface cells. From Goldsmith, 1967 a.

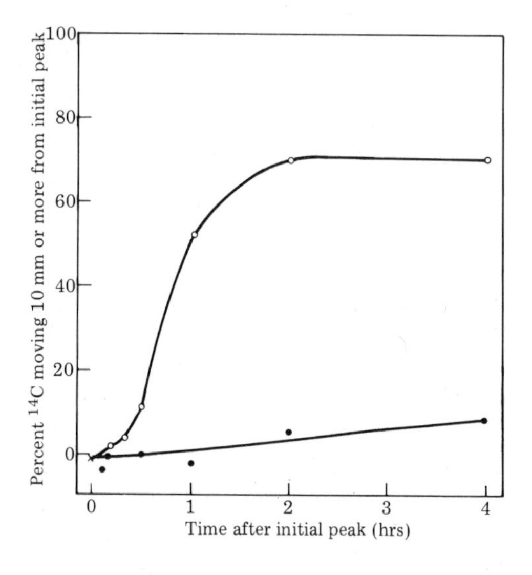

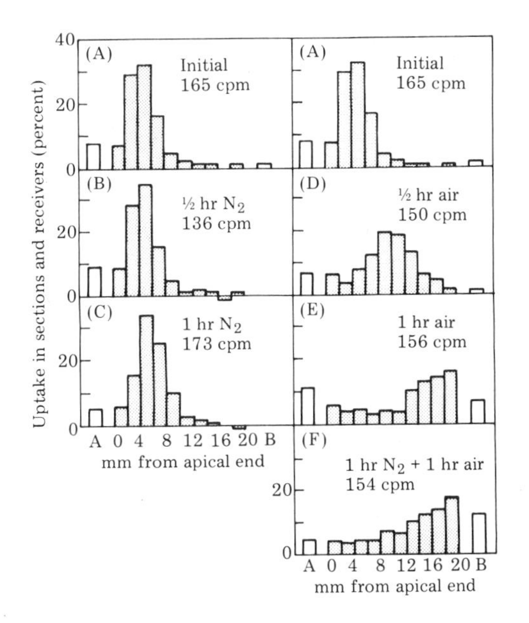

4-7, 4-8 Movement of a 15-minute pulse of ^{14}C-IAA down a whole (decapitated) corn coleoptile. Right (4-7), in air, or (F) in air after 1 hour in nitrogen. Left (4-8), in nitrogen; note the minimal movement of the peak. From Goldsmith, 1967 b.

Hence in nitrogen one observes more entry from the base than in air. But there is no transport through to the apical end.

Perhaps some of the most interesting experiments are those done with very short pulses; ^{14}C-indoleacetic acid is applied, say for 10 minutes, and then removed, and radioactivity counts are made in segments of the tissue every few minutes. In this way one can observe the movement of the pulse of radioactivity down the tissue. In figure 4-7 this method is applied to segments 20 mm long, and after the application they were cut up into 2-mm subsections. The first measurement, 15 minutes after the block was removed, shows a strong peak at about 4 mm down. A half-hour later the peak is still there, but it now centers at 8–10 mm instead of 4–6 mm down; an hour later it has travelled almost to the base. Now we have an appreciable amount coming out at the base of the segment. Thus as the peak travels down it also blurs out, and even at the end of the hour there are still some auxin molecules up near the apical cut surface. These have evidently become bound, and there is evidence below that they bind to protein. Only those molecules which remain free travel on the "moving belt." Figure 4-8 shows the same experiment done in nitrogen. We see the peak after the first 15 minutes, but a half-hour later we still see the same peak; an hour later it is still almost the same. The sharpness of the front is essentially unchanged. In other words, there is virtually no movement. However, if after this hour in nitrogen the segment is transferred to air, transport starts at once and goes as far in the next hour as if it had begun directly in

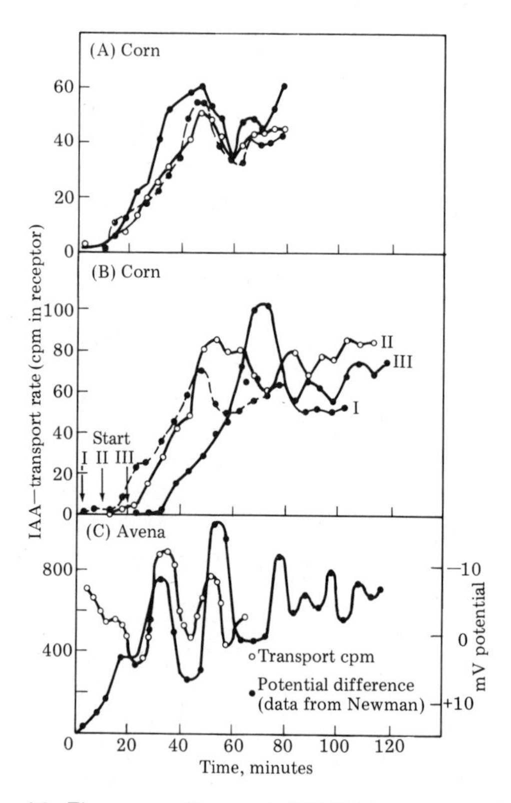

4-9 Time course of transport of ^{14}C-IAA into receivers from individual plants. Above, three parallel experiments: middle, donors applied 10 minutes (II) and 20 minutes (III) later. Below, data compared with Newman's emf. measurements, showing that the emf results from the auxin movement. From Hertel and Flory, 1968.

air. Thus the exposure to nitrogen has done no harm; it has simply held up the whole procedure.

A curious phenomenon is observed if one takes the receivers off every three or four minutes: then there appears to be an oscillation in the amount transported. These experiments were done by Hertel and Flory at Michigan (fig. 4-9). In three experiments with corn coleoptiles, the receiver blocks were taken off every four minutes. There was a period of transport, then a period of none, and then transport reoccurred. All three experiments agreed. Also, three experiments are shown which were started at slightly different times to see if the peaks would be reached at different rates; the distribution of the peaks and troughs supports the real nature of the oscillation. Dr. Newman, who works in Tasmania, has found that in experiments like this the electromotive force between two electrodes across the coleoptile goes into a quite characteristic and remarkable oscillation as the auxin travels down. I think one can say optimistically that the e.m.f. oscillations agree with the auxin oscillations; in other words, every time auxin molecules go past, they modify the potential between the cells, which is recorded on the potentiometer.

During the transport, as was brought out earlier, some molecules stay behind. This phenomenon increases with increasing time; table 4-1 shows an example. We put the block on, measure the total amount of C^{14} taken up, and find how that is distributed between what can be extracted by ether from the segment and what has come out into the receivers. Then we wait one hour and measure it again. Now the amount extractable from the segment has decreased, but the amount in the receivers has not increased much. So some auxin molecules have become unextractable during the process of transport. We shall return to that later.

C. THE INFLUENCES OF ETHYLENE AND LIGHT ON AUXIN TRANSPORT

We come now to the effects of various environmental influences on auxin transport. In the first place, transport is mainly dependent on oxygen, as we have seen. Secondly, it is inhibited by ethylene. This effect is probably of greater importance than we yet know. If auxin-transporting plants or segments are exposed to ethylene for several hours (about three hours at 25°C) there is no change at first; thereafter the transport comes to be more and more inhibited, and after 24 hours the inhibition is nearly complete. However, if we pretreat the whole tissue strongly with unlabeled ("cold") indoleacetic acid, which will not interfere with the transport of the C^{14}, the effect of ethylene is greatly decreased. It seems as though what the ethylene does is somehow to slow down some vital processes, while auxin "vitalizes" the cells and keeps them actively transporting. Thus auxin to some degree influences its own transport.

In the third place, red light inhibits the transport. In coleoptiles the inhibition is only moderate, but in some tissues it is very striking. And far-red light (740 nm) brings about a partial reversal, bringing the transport rate back to about half the original level. Once the tissue has been exposed to red light there is a long-lasting effect which does not reverse in far-red, and a short-lasting effect which can be reversed about 50%. Figure 4-10 shows two different

Table 4-1 Immobilization and export of activity after 1 hour's uptake

Expt.	Time	cpm uptake	% of uptake in section	% of uptake in receiver
		High donor conc. (1.6 mg/1)		
1	1 hr uptake	202	79	13
	1 hr uptake + 2 hr export		52	36
	1 hr uptake + 2 hr export with cold IAA donor present		48	26
2	1 hr uptake	155	85	15
	1 hr uptake + 2 hr export		51	31
Avg.	1 hr uptake		83	14
	1 hr uptake + 2 hr export		50	31
		Low donor conc. (0.08 mg/1)		
3	1 hr uptake	27	71	36
	1 hr uptake + 2 hr export		31	84

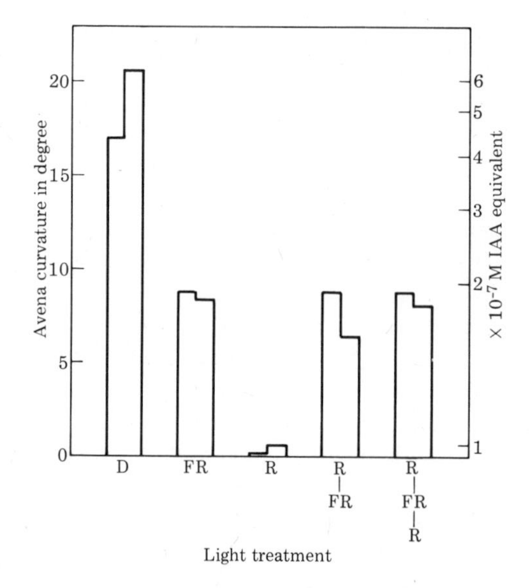

4-10 Sensitivity of polar auxin transport out of the rice coleoptile tip to red light, and its partial reversal by far-red. Plants exposed to darkness, D, red, R, or far-red, FR, for 3 minutes, then darkened and 2 hours later the auxin from the tips assayed. From Furuya et al. 1969.

experiments. In the first, the amount of auxin transport is measured by bioassay, in the usual way, by measuring the curvatures of coleoptiles in the dark. After treatment with red light the transport is drastically reduced, but later it recovers slightly. After treatment with far-red light the rate is also partially reduced, showing that in this case far-red acts in the same direction as red. However, if we follow the red with far-red, *that part* of the inhibition which is due to the red light is reversed and the transport is brought up to the far-red level. It follows that the effect of red is reversible to the extent that the far-red itself allows transport. On the other hand, when far-red is followed by red, the red irradiation does not reinhibit. Evidently the system is not quite as simple as in the classical lettuce seed experiment, although the effects are clear-cut.

The place in which this may be of morphogenetic importance is in the so-called mesocotyl of *Avena,* and indeed of all cereals. At the base of the coleoptile there is a node, below which there is a short stem-like tissue called the mesocotyl. It is actually the first internode to develop, and some prefer to call it simply the first internode. In plants grown in green light or in even a weak red light, that mesocotyl scarcely elongates at all, remaining of the order of 1–2 mm in length, while the coleoptile may be 30–50 mm long. But if the plant is grown in total darkness (and it must be total, because the effect is very sensitive), then the appearance is quite different; the mesocotyl may be 50 mm long while the coleoptile is shortened by only 25%–50% (fig. 4-11). The auxin production is only slightly decreased. Such long mesocotyls

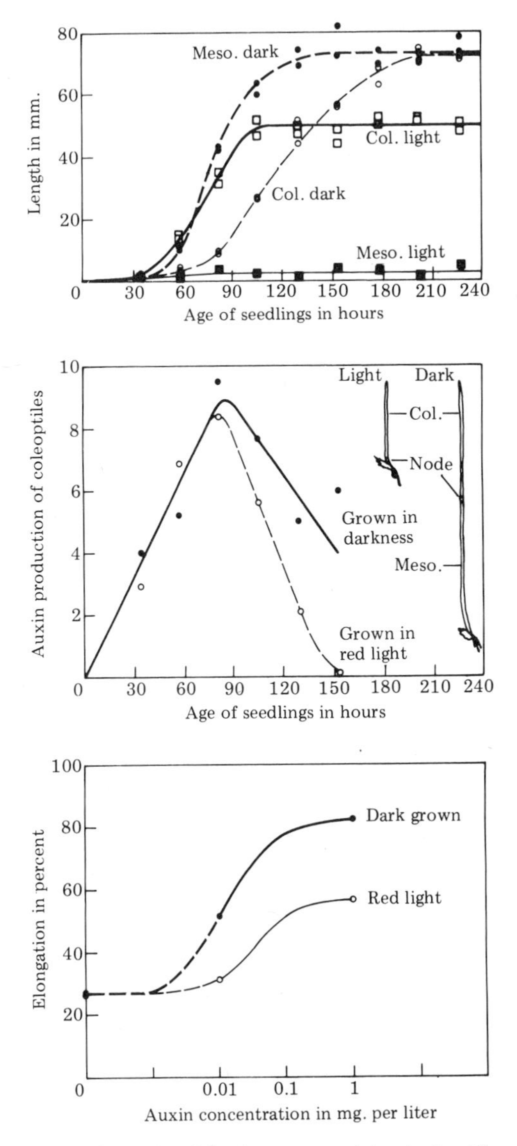

4-11 Elongation of the *Avena* mesocotyl and coleoptile in darkness and in red light. Top, elongation; middle, auxin production from coleoptile tip under the same conditions; bottom, partial inhibition of growth of isolated apical segments of mesocotyls by red light. From Schneider, 1941.

seem to be very unresponsive to gravity, so that they are seldom found straight. Here, therefore, is a tissue that shows extreme sensitivity to light. This sensitivity shows classical phytochrome dependence; that is, the principal inhibiting wavelengths are in the red while far-red (about 740 nm) largely reverses this inhibition. Although this light sensitivity evidently does not depend on auxin production, its relationship to auxin transport has been recently elucidated. It appears that 2 minutes' exposure to red light inhibits that transport by over 80%, and the inhibition lasts for as much as 8 hours after the exposure. Since a very small exposure to light will thus almost wholly prevent auxin from getting through the coleoptilar node, it seems most likely that the growth of the mesocotyl is auxin-dependent, just like that of the coleoptile, but that it is specifically dependent on those molecules of auxin that have made their way all the way down and penetrated through the node. And since this penetration through the node can be stopped for as much as 8 hours by a tiny amount of red light, the mesocotyl growth normally would be entirely inhibited. This is one of the few cases where there is a relatively straightforward explanation for a morphogenetic effect. Incidentally, it has been shown that if the plants are grown (in weak light) in low concentrations of ethylene, the growth of the mesocotyls is promoted (though they do not become as long as in darkness). Since brief exposure to ethylene does not have this effect, the relative promotion in weak light is doubtless connected with the long slow effect of ethylene on the transport system, though that remains to be worked out.

D. THE EFFECTS OF GRAVITY

The last environmental influence to be mentioned is that of gravity. This is an aspect of auxin transport which van der Weij did not get right. The reason was simple: his experiments were too short. The classical experiment, for an hour or two, indicates that the polarity is equally strict from apex to base whether the stem or coleoptile segment stands the right way up or upside down. In other words, gravity does not influence the polarity of the tissue. But this is only true for very short periods. What does happen, and rather quickly, is that the transport against gravity slows down. After a few hours the slowing becomes quite marked, so that the inverted sections transport only about a half, or even a little less than half, of what the normally vertical sections would transport (fig. 4-12). This is of considerable interest in connection with the growth of woody plants. Skilled gardeners often make use of *espalier,* in which they turn branches, which would normally have been growing more or less upwards, into the horizontal position. It is well known to them that the growth of these horizontal branches slows down. Such slowing down of growth may be desired, as they may not want to have the branches spread too far, or they may not want too big a tree. The auxin transport here is not *against* gravity, but it is not reinforced by gravity either. In those rare cases in which plants have been grown inverted (by developing roots at the tip of a cutting), growth slows down even more. Polarity does not reverse, however, for inverted tomato cuttings, rooted at the apex and planted upside down, still transport auxin in the apex-to-base di-

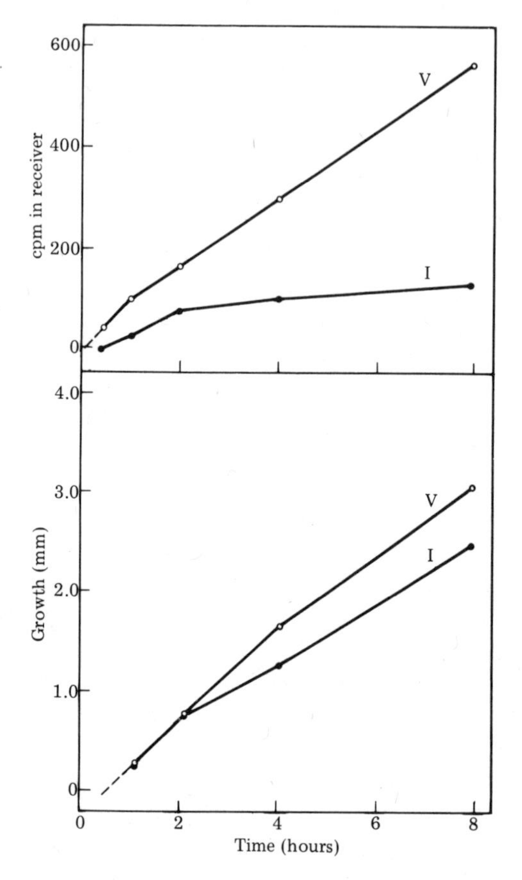

4-12 Above, effect of prolonged inversion on polar auxin transport through *Avena* coleoptile segments. Below, effect of the same periods of inversion on elongation. IAA supplied to apical surface in both. Open circles, segments oriented normally; filled circles, oriented inversely (both vertical). From Little and Goldsmith, 1967.

rection (fig. 4-13). The apparent simplicity, that auxin transport is independent of gravity and independent of gradient, holds only for short segments, for short times, and for certain tissues. Coleoptiles of oat and corn, and hypocotyls of lentil and sunflower, give relatively simple results; *Phaseolus* (bean) stems are not quite strictly polar, as noted above, but the polarity ratio in them is still around nine to one. With increasing age of tissue the picture becomes more complex, as seen in section E below.

In general, the strong polarity means that in the whole plant any source of auxin, such as developing buds or young leaves, will yield a hormone stream which will travel down the plant and finish up at the base of the stem, where it may bring about morphogenetic effects. It may also exert morphogenetic effects on the way down, and these will be to a large extent, like the transport, qualitatively independent of the distance from the source, though quantitatively weakening with increasing distance. Apical dominance, for instance, commonly weakens as the auxin goes down the stem, due to immobilization, destruction and the age effect.

4-13 Polar and non-polar transport of ^{14}C-IAA through 1-cm stem segments of tomato plants which had been rooted at the tip and grown inverted for 85 days. Note very small effect of inversion and lack of acropetal transport. Circles, inverted; squares, control. Open symbols, basipetal (polar) transport; filled symbols, acropetal transport. From Sheldrake, 1974.

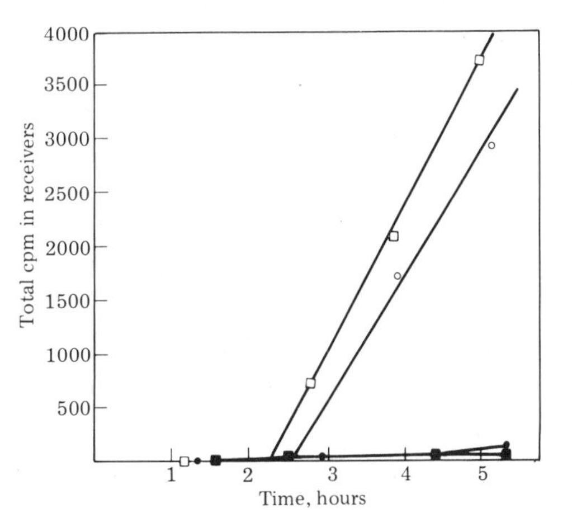

E. INFLUENCE OF THE AGE OF THE TISSUE

All of the above experiments were done with young seedlings, or subapical segments of seedlings, which are often only 2 or 3 days old and still elongating. In older seedlings, especially dicotyledons, growth ceases rather soon in the lower parts, and at this time the ability to transport auxin also decreases. Figure 4-14 shows how the growth rate declines from just below the apex of a *Helianthus* seedling down to the base; the ability of segments to transport auxin declines with a general parallelism.

A second change with increasing age is the tendency to transport a small fraction of the auxin antipolarly, from base towards apex. Figure 4-15 illustrates such a change for segments of *Phaseolus* petioles. This may not mean that the inherent polarity of the living cells is decreasing, but rather that a nonpolar system is being gradually added. When ^{14}C-IAA is applied to a leaf, and the stem and petiole lower down are subsequently sectioned, radioactivity is observed not only in the parenchymatous cells but particularly strongly in or near the phloem. It therefore seems that applied auxin has a tendency to accumulate in the phloem—though whether this is true for the natural stream of auxin from a young leaf is doubtful. Now the movement of sugars in the phloem is known to take place in either direction, depending on the rate of withdrawal into the adjacent tissue (movement "from source to sink" in the jargon of students of transport), which implies that the sieve tubes do not themselves impose any polarity on the movement of their contents. As auxin enters them, therefore, it will be carried along with other solutes. The older the tissue, the more phloem units will have been differentiated, and thus the more antipolar auxin transport may occur.

4-14 The declining gradient of auxin transport down a sunflower seedling. From Leopold and Lam, 1962.

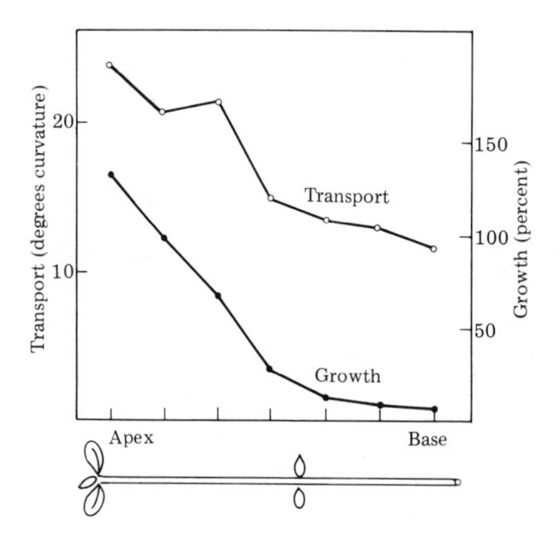

But quite apart from this anatomical consideration, there is a more subtle effect, namely that auxin itself seems to maintain the auxin transport system. When auxin has not been supplied for some hours the ability to transport it decreases, but can be revived by supplying auxin. In Veen's experiments (1969) for instance, with stem sections of *Coleus,* 17 hours sufficed to cut the auxin transport during the next 5 hours by one-half. However, if in that 17 hours auxin had been supplied, the transport in the next 5 hours was just as effective as in the fresh controls. Further, the uptake of applied auxin into tissue segments tends to increase with time, as though "IAA promotes the uptake of IAA." This indicates that the transport system is not a fixed establishment, but is constantly breaking down and being re-formed, the latter process, like other synthetic processes described in the closing chapter, being very much under the influence of auxin.

F. OTHER EFFECTS

In some plants (though not in coleoptiles) the movement in the stem is promoted by

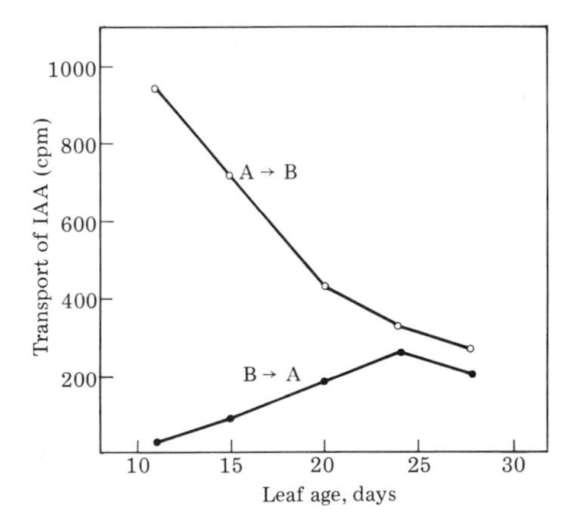

4-15 Changes in the polarity of auxin movement through bean petiole segments with age of the leaf. As apex-to-base transport decreases, base-to-apex transport somewhat increases. From Leopold and de la Fuente, 1968.

4-16 The effect of gibberellin on transport of ^{14}C-IAA through pea stem segments as shown on chromatograms of ether extracts. The ^{14}C content of each region of the chromatogram when IAA alone was applied is shown cross-hatched; that with GA applied as well is shown with vertical lines. Left half, top segment of plant; right half, lower segment. Transport time 2 days. From Jacobs and Case, 1965.

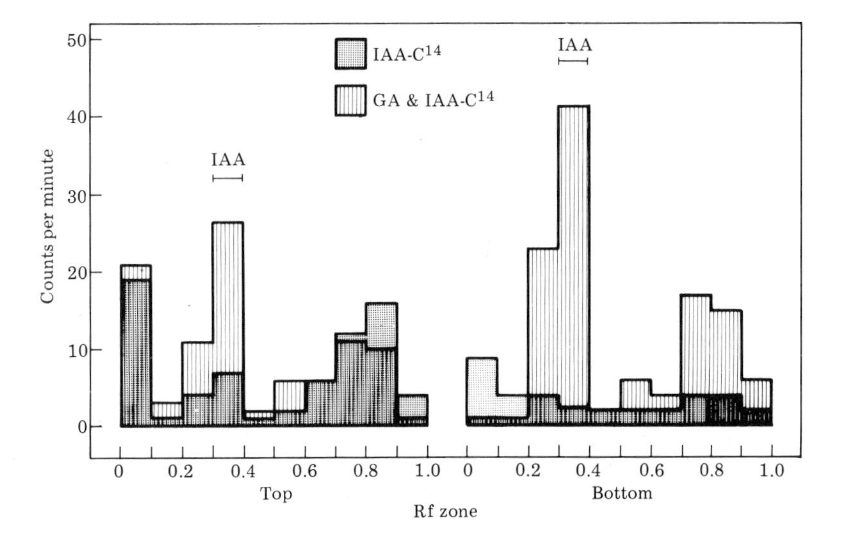

gibberellic acid—another case of interaction between auxin and gibberellin. In a relatively long stem the amount of auxin transported decreases with time, and gibberellin prevents this decrease. Furthermore, on extraction we find that the auxin has penetrated further down in gibberellin-treated stems than in controls. Apparently the gibberellin maintains the transport system and perhaps facilitates it, but does not accelerate the rate. Figure 4-16 illustrates an experiment in which short segments of the chromatographic paper which has been used to demonstrate and measure the distribution of ^{14}C-IAA are tested individually for their radioactivity, i.e., their auxin content. In the gibberellin-treated tissue there is a much larger amount of IAA at the base. There is also more of an unidentified derivative of IAA, probably indoleacetyl glucose, than in the plants without gibberellin. Thus the gibberellin has enabled more auxin to penetrate to the lower parts of the stem, although gibberellin itself has only slight polarity, and is transported rather freely upwards in the transpiration stream. The effect of gibberellin on auxin transport is so strong that for a short time it was proposed that all observed gibberellin effects were due to the increased supply of auxin made available—but this idea soon became untenable.

We saw above that some of the auxin molecules become immobilized. The longer one waits after auxin application, the less auxin diffuses out and the less is extractable with ether. Thus the auxin is not only immobilized from transport, it has actually become bound in the chemical sense. Dr. Zenk, in Munich, has shown that after 2 hours of transport part of the auxin has been converted to a peptide with aspartic acid (which is also prominently formed in roots). Naphthalene-acetic acid forms the same type of peptide, and there is a tendency, though only a tendency, for this ability to increase as one goes up the plant kingdom from mosses to the flowering plants. In the *Avena* coleoptile much of the immobilized auxin is not even readily extractable with water, but it becomes ether-soluble after treatment with proteolytic enzymes, indicating that it has probably become bound to protein.

G. EVIDENCE AS TO THE MECHANISM OF AUXIN TRANSPORT

The general polarity of auxin transport, and the occurrence of transport against the gradient, make it intrinsically unlikely that auxin travels in conducting vessels (with the possible exception noted for older tissues) and would seem to require an active transport from cell to cell. In *Coleus,* which being a Labiate has a relatively square stem, the vascular bundles are located in the corners of the square; and Jacobs, at Princeton, managed to place small blocks of agar on the parts which include no conducting tissue, and other blocks right over a vascular bundle. Comparison of the amounts and rates of transport in these two locations shows little difference, but transport is, if anything, more active in those parts *not* over the bundle. This confirms our a priori assumption that the polar movement depends upon live cells.

If we try to visualize the predominant polar process at the cellular level, we must first note that the cells in general are relatively long, and next that they are vacuolated. The auxin is thus moving in the cytoplasm or along the plasmalemma, until it reaches the cross-wall, and here we have to imagine a little conveyor system which picks up auxin molecules and somehow pushes them through into the next cell. Only an active process like this can lead to accumulation of auxin molecules against the auxin gradient. Thus we have to invoke an energy-requiring process at each cross-wall.

What happens between these cross-walls? Could one assume that the auxin moves in the cytoplasm? It is suggestive that the rate of streaming in these long cells is around 10–12 mm per hour (at 25°C), which agrees fairly well with the observed rate at which auxin is transported. However, recently (1973) Ray, Cande, and Goldsmith have experimented with the drug Cytochalasin B, which brings the streaming of cytoplasm in amebae completely to a stop. It has the same effect in plants, though it is not easy to be quite sure, especially in multicellular tissue, whether or not a very slow residual rate persists. In any event the streaming rate slows down drastically, and it may indeed stop altogether. Yet that scarcely affects the rate of transport of auxin. It causes a small delay at the start but thereafter the amount carried per unit time is absolutely unchanged (fig. 4-17). It does inhibit the growth rate, and drastically (chap. 14), but it may be that auxin is not carried in the cytoplasm, and *within the cell* merely dif-

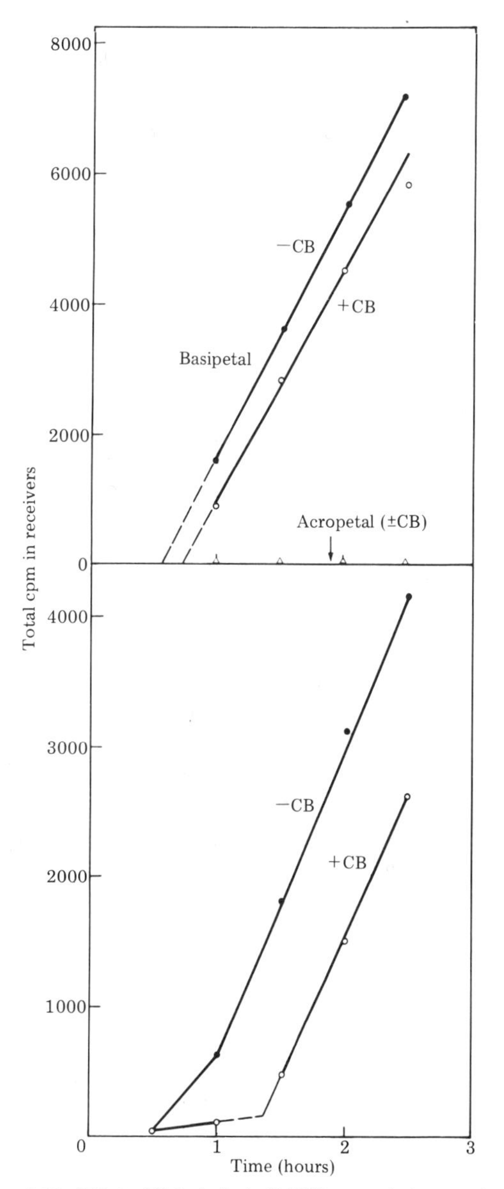

4-17 Effect of Cytochalasin B (CB) on auxin transport through segments of corn (above) and oat (below) coleoptiles. Microscopy showed streaming inhibited by CB before donor blocks applied and remaining inhibited throughout. From Cande et al. 1973.

fuses. What does carry it we do not know.

The transport has three important characteristics. First, it is chemically highly specific. Indoleacetic acid, the natural auxin, is transported faster and in larger amounts than synthetic auxins like naphthaleneacetic acid or 2,4-D. In all cases, whether with coleoptiles, stems or petioles, careful measurement of the rate shows that these synthetic auxins move at one-quarter to one-tenth the rate of IAA. The specificity is shown in another way by the fact that some compounds closely related structurally, but with little or no auxin activity, can strongly inhibit auxin transport. Two such compounds, 2,3,5-triiodobenzoic acid (TIBA) and 1-naphthylphthalamic acid, show this property beautifully. TIBA has some properties of an auxin, and its close relative 2,3,6-trichlorobenzoic acid is in fact an auxin; also the halogen atoms make it very liposoluble, more so indeed than the natural auxin. It has considerable synergistic effect with IAA in some tests. It was first shown by Kuse, in Japan, that if we ring a stem or petiole with TIBA of the right concentration, then the part below the ring acts as though it had received no auxin whatever. Since the inhibition of lateral buds, as shown in chapter 10, is due to auxin, the most striking phenomenon is that the lateral buds below the ring grow out vigorously (fig. 4-18). Elongation slows down below the ring but is unaffected above it. Roots are not formed on cuttings treated in this way. TIBA placed in a ring *above* the source of auxin has no effect on the buds below. Naphthylphthalamic acid acts similarly to TIBA in several ways. Even 2,4-D, which is a true auxin, somewhat inhibits the uptake and movement of IAA.

A second important characteristic of the transport is a requirement for oxygen. Analysis of this need in terms of the steps involved seems to show that the intake of auxin from the agar or solution into the stem is not greatly inhibited by putting the

4-18 Inhibition of auxin transport from the leaf of sweet potato by TIBA, as shown by elongation of the axillary bud. A, ring of TIBA in lanolin; B, point of severance of leaf; C, plain lanolin at cut surface; D, nutrient solution. From Kuse, 1953.

	Control		2% TIBA	
	Leaf blade cut off	Leaf blade intact	Leaf blade cut off	Leaf blade intact
Elongation of bud (mm)	12.0	3.8	14.2	12.2
Ratio	100	32	118	102
Dry weight of bud (mg)	10.1	1.2	11.1	5.6
Ratio	100	12	110	56
No. of new leaves on lateral bud	1.0	0	1.0	0.6

tissue into an atmosphere of nitrogen. Thus it is the movement through the tissue, or the secretion into the adjacent cell, which requires oxygen. Centrifuging the cytoplasm to the upper ends of all the cells interferes with the transport, presumably by taking away the active secreting mechanism.

Third, the uptake is very dependent on pH. From many types of experiments we know that rendering an IAA solution alkaline, so that it is all in the ionic form, lowers both its activity and its transport. Most of this effect is due to a decrease in the uptake. Figure 8-5 shows the uptake of ^{14}C-IAA from solution into floating coleoptile segments, as a function of time. The pH values were chosen thus: pH 4.5 is acid enough so that more than half of the indoleacetic acid is in its free acid form, while at pH 7.5 it is fully dissociated. The ratio between the rates of uptake at these two pH values is very large, around 10 to 1. By contrast, the two curves labeled IAN refer to indoleacetonitrile. This derivative does not contain the carboxyl group, but its nitrile group is readily converted to indoleacetic acid in some plants by an enzyme called nitrilase, which hydrolyzes the -CN to -COOH plus NH_3. But in this experiment, where it has not yet been converted, its uptake is essentially independent of pH. In other words, for a substance without a dissociable acid group pH makes little difference to the uptake. Indeed, in plants that contain the nitrilase, larger auxin effects can be obtained than with the free acid, because the uptake is actually better.

It is probable that the free acid, being soluble in lipide, enters by dissolving in the lipide membrane and diffusing out into the cytoplasm, where it dissociates because of the neutral pH. The dissociated auxin, as we saw above, probably enters and leaves the cells by a specific carrier, and the oxidative energy is needed to make it combine with this carrier. An obvious suggestion is that the ATP converts it to indoleacetyl phosphate, from which it is then converted to indoleacetyl coenzyme A, as amino acids are. However, when an auxin labeled with O^{18} in the carboxyl group was transported through tissue segments, the auxin in the receivers contained the same proportion of O^{18} as in the donors. If it had reacted with ATP and then afterwards hydrolyzed:

$$R\,CO^{18}O^{18}H + ATP \rightarrow RCO^{18}\text{-}\boxed{P} + ADP + H_2O^{18}$$

$$R\,CO^{18}\text{-}\boxed{P} + H_2O^{16} \rightarrow R\,CO^{18}O^{16}H + \boxed{P} + H^+$$

($\boxed{P}$ = phosphate)

the O^{18} content would have been reduced by half. Thus the nature of the auxin-carrier combination, and with it the whole mechanism of polar transport, remains one of Nature's best-kept secrets.

REFERENCES

Cande, W. Z., Goldsmith, M. H. M., and Ray, P. M. Polar auxin transport and auxin-induced elongation in the absence of cytoplasmic streaming. *Planta* 111, 279–296, 1973.

Furuya, M., Pjon, C., Fujii, T., and Ito, M. Phytochrome action in *Oryza sativa* L.3. The separation of photoperceptive site and growing zone in coleoptiles, and auxin transport as effector system. *Development, Growth and Differentiation* 11, 62-76, 1969.

Goldsmith, M. H. M. Movement of pulses of labeled auxin in corn coleoptiles. *Plant Physiology* 42, 258-263, 1967.

———. Separation of transit of auxin from uptake: average velocity and reversible inhibition by anaerobic condition. *Science* 156, 661-663, 1967.

———. The transport of auxin. *Annu. Rev. Plant Physiol.* 19, 347-360, 1968.

———, and Thimann, K. V. Some characteristics of movement of IAA in coleoptiles of *Avena.* 1. Uptake, destruction, immobilization and distribution of IAA during basipetal translocation. *Plant Physiology* 37, 492-505, 1962.

Hertel, R., and Flory, R. Auxin movement in corn coleoptiles. *Planta* 82, 123-144, 1968.

Jacobs, W. P., and Case, D. B. Auxin transport, geotropism and apical dominance. *Science* 148, 1729-1731, 1965.

———, and Pruett, P. M. The polar movement of gibberellin through *Coleus* petioles. In *Hormonal regulation in plant growth and development,* ed. H. Kaldewey and Y. Vardar. Weinheim: Verlag Chemie, 1972.

Kuse, G. Effect of 2, 3, 5-triiodobenzoic acid on the growth of lateral bud and on tropism of petiole. *Mem. Coll. Sci. Univ. Kyoto Ser. B.* 20, no. 3, 207-215, 1953.

Leopold, A. C., and De la Fuente. A view of polar auxin transport. In *Transport of plant hormones,* pp. 24-47. Amsterdam: North-Holland, 1968.

Leopold, A. C., and Lam, S. L. The auxin transport gradient. *Physiologia Plantarum* 15, 631-638, 1962.

Little, C. H. A., and Goldsmith, M. H. M. Effect of inversion on growth and movement of IAA in coleoptiles. *Plant Physiology* 42, 1239-1245, 1967.

Poole, R. J., and Thimann, K. V. Uptake of indole-3-acetic acid and indole-3-acetonitrile by *Avena* coleoptile sections. *Plant Physiology* 39, 98-103, 1964.

Schneider, C. L. The effect of red light on growth of the *Avena* seedling with special reference to the first internode. *Amer. J. Bot.* 28, 878-886, 1941.

Sheldrake, A. R. The polarity of auxin transport in inverted cuttings. *New Phytol.* 73, 637-642, 1974.

Veen, H. Auxin transport, auxin metabolism and ageing. *Acta Botan. Néerl.* 18, 447-454, 1969.

Weij, H. G. van der. Der mechanismus der wuchsstofftransportes. I and II. *Rec. Trav. Bot. Néerl.* 29, 379-496, 1932; 31, 810-817, 1934.

Wilkins, M. B., and Martin, M. Dependence of basipetal polar transport of auxin upon aerobic metabolism. *Plant Physiol.* 42, 831-839, 1967.

Geotropism 5

THE EFFECTOR SYSTEM

A. ORIENTATION IN GENERAL

After the seed has germinated and elongation has begun, the next step is *orientation*. The nutrient supply of most seeds is enough for only a few days' growth, so that it is essential that the shoot reach the surface and the root find the moisture deep in the soil as rapidly as possible. Therefore, this orientation or *tropism* is really a matter of life and death for the seedlings. In fact, seedlings that are disoriented for more than a few days usually do die. Because plants are nonmotile, this orientation is of a permanent type and is, therefore, due to growth—that is, growth in a given direction.

In thinking about the movements of plants, we contrast growth movements with turgor movements. Turgor movements are, of course, reversible, for they depend on the fact that water intake or loss may be enough to change the turgor of a group of cells in a particular position, and this may in turn modify the orientation of an organ of which they are a part. Most people know a classical example of that: if one touches the tip of a leaf of *Mimosa pudica* the leaflets fold, and a minute or so later the petiole bends down at the pulvinus. The cells in the tissue are normally turgid, and the rigidity of the tissue is maintained by the pressure of these cells on one another; at the influence of a touch or, better, of a scorching (strong *Mimosa* movements are obtained by touching a lighted match to them), the permeability of the cells is greatly increased, to the point that some of the cell contents flow out into the intercellular spaces. As a result the cells lose their turgor and the pressure that was maintaining a petiole or a leaflet at a fixed angle is lost. Thus the leaflet or petiole drops down. Usually they recover after half an hour or so, the leaflet or petiole slowly reorienting itself. Among other interesting turgor movements are the famous circadian (i.e., almost daily) movements of plants, in which leaves, as they say, "go to sleep" at night. The circadian movements of the petioles of the common bean are very characteristic; the normal daytime petiole angle is at about 70° to the axis; but at about 7 o'clock at night the petiole angle begins to increase (the petiole sinks down) until the angle is around 160°, and then at about 7 o'clock in the morning it comes up again. The same movements are continued daily. They are not a response to light directly, and if you leave the plants in constant conditions, the period is not quite 24 hours, but the movements still continue, at least for a time. However, since the plant normally experiences daylight every day at about the same time, it adjusts the movement to fit in with the daylength it experiences. These sleep movements, like the movement in the *Mimosa,* are due purely to turgor changes. They are not rare; numerous other plants do the same thing. But the cause of the turgor changes is not really understood except that we know it is not a direct result of the light and dark, since it can continue for a while in constant darkness.

Unlike turgor movements, curvatures due to growth are irreversible; technically they are *tropisms,* and must be distinguished from orientations of freely moving objects like unicellular algae, which will swim toward a given source of light (or, if it is too bright, will often swim away from it). This is a free movement, and such movement is referred to as a *taxis*. There are many papers by respected biological investigators who don't know the difference between these two things; there are papers about the "geotropism" of ants, for instance, referring, of course, to their movement up or down a slope, as influenced by gravity—which is a geotaxis.

Tropisms are usually toward or away from the direction of a stimulus; tropisms toward a stimulus are termed positive and those away are called negative. In seedlings these are usually an all-or-nothing affair—they are strongly positive or strongly negative. And, as I said, it is in seedlings that the tropisms are obviously of crucial importance. Mature plants will show tropisms too; everybody knows that when a potted plant is placed on the window sill, in a day or two it will be curving toward the light, and that it needs to be turned around occasionally to keep it vertical. On the other hand, if the pot is knocked over and one forgets to stand it up, next day there will be a distinct upward curvature in the shoot axis. Thus, mature plants do respond, but they respond more slowly; their tropism is absolutely dependent on their elongation, and they are usually elongating slowly. Also in older plants there are many modifications, not positive or negative but some halfway. In geotropism, for instance (the *geo-* means the earth, toward or away from which there is a turning, or *tropism*), there are a number of varieties. The most basal branch that grows out in many dicots is not negatively geotropic, like the axis; it does not curve directly away from the earth, but grows plagiotropically—that is, more or less at right angles to the direction of the earth. In the potato, the most basal branch even becomes slightly geotropically positive and burrows into the earth, then swells out in the form of a tuber. In the peanut, another famous case, there are flowers in the axils of the most basal leaves, and after the flowers are fertilized, the peduncle elongates and becomes quite strongly geopositive. It burrows into the ground bearing the ovary on it and then, of course, the ovary grows into the one- or two-seeded peanut. That is a rather unusual case, but there are many other modifications of the normal negative geotropism of shoots. One sees very striking examples occasionally in horticultural trees that have been selected for the fact that their branches are strongly plagiotropic or weeping. In the case of weeping willow trees, the branches are not wholly plagiotropic but they often are slightly positively geotropic. The photo (fig. 5-1) shows a famous weeping birch at the Harriman home (Arden House) a little way outside New York City; it is a remarkable tree. If you look carefully, you will see that the branches are positively geotropic. They are not sagging, but are woody and very stiff. The outgrowth from the trunk is not like that in the peanut, which starts out to be negatively geotropic and then reverses; here the angle to the trunk is already more than 90°—in other words, shoots come out already at a slight downward angle. The common weeping willows often show rather similar orientation.

5-1 Weeping birch (probably *Betula pendula* var. *gracilis*).

B. GEOTROPISM AND THE ROLE OF AUXIN

Historically the whole problem of orientation and tropisms has been closely linked to auxin from the beginning, and, as mentioned in chapter 1, the study of phototropism led rather directly to the discovery of auxin. The connection was not immediate; but, after a number of early experiments, in 1927 Cholodny proposed that all tropisms, whether due to light or gravity or any other force, were due to the asymmetric distribution of a growth-promoting substance. This growth-promoting substance, he proposed, became somehow asymmetrically distributed by the action of the stimulus, and asymmetric growth resulted, causing a curvature in the elongating organ. As we have seen, Went very quickly confirmed that concept by his work with the growth substance of coleoptiles—later called auxin—and was able to show that under the influence of light, this material became asymmetrically distributed, with more going to the shaded side. As for auxin and gravity, it was Went's colleague, Herman Dolk, who made the beautiful series of experiments confirming the idea of auxin's role in geotropism. Dolk used the coleoptile of *Avena,* which had already been studied and was known, from Went's work, to secrete auxin into agar: he placed it in a horizontal position, divided its base with a razor blade for a short distance, placed a block of agar in contact with the upper side and another block in contact with the lower side, and left it for a while so that the auxin would be secreted into the agar (fig. 5-2). Sure enough, about twice as much was found in the lower as in the upper block. Dolk's actual figures, the average of all his experiments, were 30.5% in the upper blocks and 69.5% in the lower blocks. In the individual experiments there was considerable variation, for it is a tricky experiment, with the small sizes working against success. The whole distance within which one has to work is 1.1 mm (the diameter of the coleoptile), and unless the division is absolutely symmetrical, the blocks will be in contact with more, or less, than half of the tissue. Furthermore the coleoptile is not solid, but hollow. In cross section there are only 6–8 cells in the coleoptile, so the blocks must be rather precisely oriented. Dolk had a great deal of trouble with this, though the final results came out very well. (Mrs. Dolk, who acted as his assistant, told me that she had no idea there were so many rare swear words in the Dutch language until she worked with him in the darkroom, in the necessarily very weak red light.) Dolk also did certain further experiments, the significance of which was only realized much later. In these, instead of measuring the auxin secreted by the tips, he collected auxin from other tips and applied this to subapical sections 3 to 5 mm long; their bases were split by a razor blade as before, so that the respective agar blocks received the auxin transported through the upper and lower sides (fig. 5-2). Again, the result was a clear asymmetric distribution, not quite so strong as with the tips, but a ratio around 42:58. Thus he could demonstrate not only that the auxin of the tips became asymmetrically distributed, but also that a lower, subapical, section could do the same thing. We have to deduce from both of these experiments that under the influence of gravity, instead of the auxin's moving lengthwise in the tissue as it normally

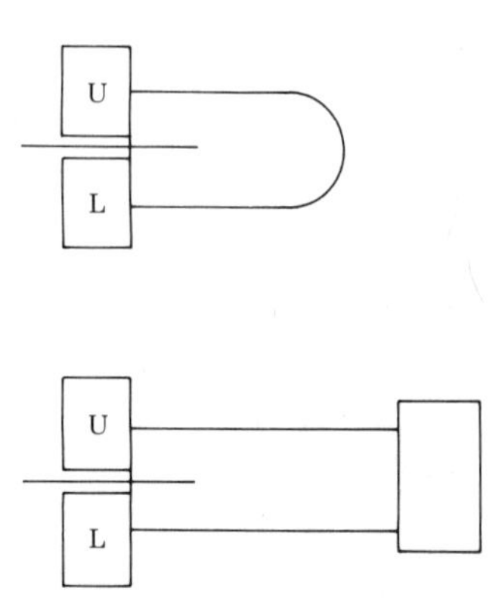

5-2 Dolk's two types of experiments, done in 1930.

would, it is diverted and moves at an angle to the axis so that more accumulates in the lower half.

It was not long following these studies with monocotyledons that people began to study dicots, and one of the Dutch workers, Dijkman, was the first to show auxin asymmetry with the lupin seedling. Lupin seedlings have a small apical bud, rather large cotyledons, and a hypocotyl that elongates considerably. Most of the auxin comes from the terminal bud, but some also comes from the cotyledons. She left the cotyledons on and did essentially a Dolk-type experiment: she cut through at the base of the hypocotyl, placed blocks on either side, and then turned the whole thing through 90°. The controls were, of course, vertical, and they gave an auxin distribution between the two sides of 50 : 50, while the horizontal gave around 40 : 60 or 42 : 58 for the upper and lower sides, respectively. Actually, the differences are not as great in a dicot as in a monocot, but this does not mean that they are less sensitive to gravity. The reason lies in geometry, for the hypocotyl is a solid organ, so that of the total auxin that comes out into the block, while some is really from the lower and some from the upper side, a good deal is from the median part and so does not show asymmetrical distribution. A few years ago Burg showed that if, instead of doing the experiment Dijkman's way, one did it by placing blocks in contact with just the upper and lower quarters, leaving the middle quarters alone, then a much larger asymmetric distribution, something like 25 : 75 or 3 to 1, is obtained. Thus, there is no basic difference between the two types of plants. Subsequently, in an effort to show that what we were dealing with was not only the transport out into a receiver, but the actual distribution in the tissue, Boysen Jensen, who is responsible for many of these crucial early experiments, placed bean plants horizontal and then, instead of receiving the auxin in receiver blocks, cut the whole plant lengthwise. This must have been difficult to do symmetrically. He extracted the upper and lower halves with chloroform and measured the amount of auxin actually present in the tissue. Although the experiment was difficult, it did show a clear asymmetry: always more auxin on the lower side than on the upper side. Just a year or two ago, in connection with a biological satellite that included experiments on the growth of plants in the absence of gravity, Dr. C. J. Lyon from Dartmouth did somewhat similar work with tomato and also with *Coleus*. He could show something like the Dolk experiment using extraneous sources of auxin. He used an apical part of the *Coleus* stem, cut off the bud, applied an extraneous source of auxin (indoleacetic acid) and then placed the shoot horizontal: the usual geotropic curvature resulted, and the auxin received in blocks at the base was asymmetrically distributed. He was able to establish about a 40 : 60 ratio in this work.

Most of the critical experiments (except one or two of the most recent) were done in the 1930s and 1940s; Dolk's work was done in 1930, but it was not published in English until 1936. In the early 1950s carbon-14 indoleacetic acid became available, and several people began to try out the old experiments, now using radioactivity as the criterion instead of biological activity—the procedure is indeed quicker and easier. In a bioassay, it is almost necessary to use the classical coleoptile curvature method because extraction methods mostly give results that vary with the logarithm of the concentration. Looking for small differences in such systems is unsatisfactory. With radioactivity the procedure is simple, and one can measure the radioactivity in the tissue as well as in receiving blocks (as discussed in chapter 4). The strange thing was that when carbon-14 indoleacetic acid was first used the asymmetries of auxin distribution were not found. The first experiments were done with phototropism, where one would look for differences between the amounts of auxin on the lighted and shaded sides. The plants were decapitated, the ^{14}C-IAA was applied, the plants were illuminated from one side, and the radioactivities on the two sides were determined. The experimenters could find no difference, though the results were very variable. In consequence it was suggested that one might not expect a difference, because, since it is the tip that detects the light, the asymmetry could originate only in the natural auxin formed in the tip. Fortunately, the same type of experiment was then carried out with geotropism, and again no difference was found between the radioactivities on the upper and lower sides. But in this case the same argument could not be made, because Dolk had shown that the tip is not needed for geotropic response; a clear geotropic asymmetry can be developed by subapical sections and therefore they should have produced an asymmetry in the radioactivity. Several of us realized that something was seriously wrong and I decided, with two very able graduate students, to reconsider the whole situation carefully.

In experiments of this kind there are a variety of pitfalls that do not appear when measuring biological activity. First, it is essential to supply the auxin at as nearly as possible the same concentration range as that which is present in the plant. When working with radioactivity, on the other hand, one is more inclined to think in terms of counts per minute than in chemical terms of the actual amounts of active compound. As a result, these first workers were giving the seedlings about one hundred times as much auxin as the poor little plants had ever seen under natural conditions, and of course it was illogical to expect the natural system to redistribute these vast quantities. Some other problems were discussed in chapter 4. It is essential to use a substance that, when biologically inactive, will also be radiologically inactive. For this reason, Bruce Stowe (now at Yale) developed a reliable synthesis of IAA that starts with gramine methosulfate: when it is treated with KCN that contains C^{14}, the KCN transfers the nitrile group in place of the trimethylamine, which comes off, and the potassium forms potassium hydrogen sulfate. The rest of the molecule remains cold and the nitrile group contains all the radioactivity. Next, the indoleacetonitrile is hydrolyzed in alkali to produce the acid. Since the indole nucleus is very sensitive to acid but stable in alkali, it does not decompose and there is high efficiency of conversion from the nitrile to the acid, IAA. In this way, one obtains an acid in which all the C^{14} is in the carboxyl group; if this is oxidized, all the radioactivity will go off as CO_2, and the residue will be inactive. So (a) we have to avoid auxin overload, (b) we must have all the C^{14} in a carboxyl group, and (c) we also have to be sure there are no radioactive impurities. Extensive purification was necessary—at least 98% of the radioactivity was finally shown to be in IAA, and with this knowledge the experiments were started again. They left no doubt of the outcome. Table 5-1 shows the first group of experiments done with *Avena* coleoptiles in essentially the same way as Dolk had done them. It gives the counts left in the donor block, the counts in the tissue, and the counts in the upper and lower receiver blocks; it is clear that, although there is a little variation from experiment to experiment, in every case there are less counts in the upper than the lower receiver. The average count comes out to 40.5 in the upper block and 59.5 in the lower block, which is in close agreement with the old data from the biological assay.

As shown in the table, we also measured the amounts of auxin normally transported through these tissues, and we varied the amounts of ^{14}C-IAA applied around this figure. Within a narrow range, no significant differences as a function of concentration appeared. A smaller number of experiments showed that *Helianthus* hypocotyls (i.e., dicotyledonous seedlings) behaved in the same way.

After these data were published, some workers commented that although there is a real difference between the *blocks,* they did not feel that any difference in the auxin *content* of the tissue had been shown. In other words, the suggestion was that when the plant is laid horizontal, the auxin is transported faster through the tissues of the lower side than through those of the upper side, so that more comes out into the lower receiver, but this would not necessarily mean that there is any more auxin *in the tissue* of the lower half. Thus, other experiments were put in hand to test this possibility.

The *Avena* coleoptile is, as was noted

Table 5-1 Summary of the asymmetric distribution of ^{14}C-indoleacetic acid transported through, and present inside, segments of three species of seedlings placed horizontally

		Asymmetry of radioactivity			
		in agar receivers		in tissue halves	
Species	Conditions	upper	lower	upper	lower
Avena	Series I	40.6	59.4	—	—
	Series II	39.1	60.9	47.8	52.2
Zea mays	Amount normally transported	30.4	69.6	40.3	59.7
	Above amount × 0.7	31.5	68.5	38.7	61.3
	Above amount × 0.4	28.9	71.1	—	—
Helianthus	Twice the amount normally transported	45.7	54.3	—	—
	Half the amount normally transported	41.3	58.7	—	—

From Thimann, 1964.

above, small. The *Zea* (corn) coleoptile is much larger. With *Zea* one can allow much larger separation between the blocks. In that case we could show (table 5-1) that not only were the receiver blocks asymmetrically labeled, but the sections of tissue themselves were asymmetrically labeled. Auxin is in fact diverted by gravity to accumulate on the lower side.

The two experiments which will be described here are variations on the old theme. In the first case, in order to avoid some complications in regard to the extraction of tissue, we took a subapical coleoptile section, cut it in half, and for one half substituted a thin sheet of agar. We placed a block on the top containing radioactive IAA, waited for a while, and then analysed the tissue and the agar sheet at the side (fig. 5-3).

Since the radioactive IAA is applied to the cut surface, in normal transport it would simply move directly down. However, statistically a certain number of molecules will tend to move laterally into this side sheet. So we find here 530 counts in the tissue and 100 in the lateral sheet.

When we do the same experiment with about the same amount of activity applied but the tissue placed horizontal, then instead of 100 counts in the lateral sheet, we find over 300 there. Thus under the influence of gravity three times as much auxin has moved across the tissue laterally as would have moved if it had remained vertical.

5-3 Movement of ^{14}C-IAA from longitudinally halved corn (*Zea mays*) coleoptile segments into laterally applied agar sheets. From Gillespie and Thimann, 1963.

	Counts per minutes			%
	Receiver	Tissue	Donor	recovery
			Horizontal Series	
1	166 199	492	2190	90
2	206 228	565	2100	92
3	138 177	407	2410	93
4	108 133	424	2720	100
5	122 118	375	2410	90
avg	319	453	2366	93
			Vertical Series	
1	84 82	743	2220	93
2	78 39	628	2300	91
3	53 47	530	2420	90
4	39 38	455	2720	96
5	27 36	490	2623	94
6	48 34	332	2538	87
avg	101	530	2471	92

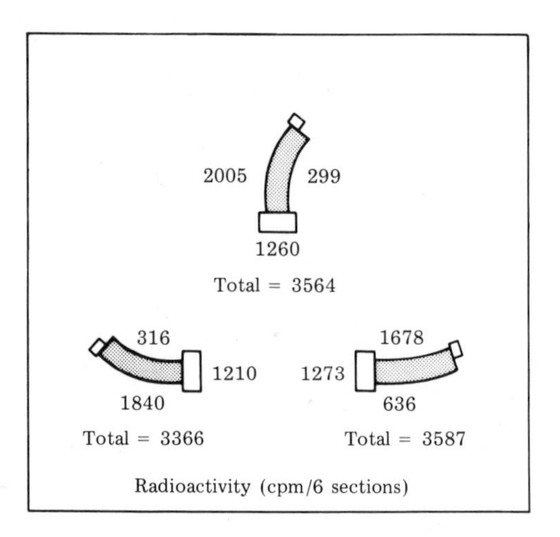

5-4 Movement of ^{14}C-IAA to the opposite side of vertical or horizontal coleoptile segments when applied only to half the cut surface. The data are representative of a group of experiments. From Goldsmith and Wilkins, 1964.

The second variant was done by my former student Mary Helen Goldsmith with Malcolm Wilkins. Its results were equally clear-cut but the procedure was different. The procedure was now to *apply* the auxin asymmetrically (fig. 5-4). It was applied to only half of the coleoptile segment and the distribution between the two halves measured again: when the segment is vertical 300 counts go across into the other side; when it is horizontal, over 600 counts go across to the lower side—more auxin moves laterally than would have done had it been vertical.

In the reverse case the auxin is applied to the lower half, and no excess of auxin goes upwards into the upper half. So there is clearly a gravitational diversion of the auxin from the normal path of transport into a lateral transport.

It should be mentioned that in these experiments not only must one keep the auxin applied within the range that the plant is used to, but it should also be within the range that *growth* is proportional to, because, after all, this is the explanation for geotropic curvature, which means a difference in growth. So it was necessary in all these experiments to show that the actual amount of IAA was in the range proportional to growth. If we plot growth against auxin content, we get curves from which one can readily determine that we are working in this range.

With all these precautions, there is no doubt that the geotropic curvature of shoots is due to the diversion of auxin to the lower side, and since we are in the proportional range where growth is a function of auxin content, this must necessarily cause increased growth on the lower side. Such asymmetric growth will continue until the apical part, or auxin source, has become vertical, after which the auxin distribution will naturally be symmetrical and the curvature will cease.

In 1973 Abrol and Audus showed elegantly that the synthetic auxin 2,4-D also undergoes lateral transport due to gravity, in segments of *Helianthus* hypocotyls. Audus had for many years felt that gravity might cause, instead of lateral transport, accelerated longitudinal transport along the basal side of a horizontal organ. For some years, indeed, he interpreted our results in that way, even after Pickard and I had proved that the asymmetry is directly *due to lateral transport.* Happily, in the work with 2,4-D they had to conclude that the data "constitute no evidence for an augmented basipetal flux induced by gravity on the lower side of horizontal hypocotyls." So another ghost has been laid to rest.

C. SECONDARY INFLUENCES

On the whole, therefore, the fact that gravity modifies auxin flow is fairly well agreed. How the cells detect gravity and respond to it is another matter, to be dealt with in part II of this chapter. Here we must mention that there are a number of subsidiary effects. In discussing auxin transport we saw that the transport is sensitive to changes in the environment: it is affected by light, by ethylene, and to some extent by other factors. The same is true of the auxin transport which participates in geotropism. There is a striking effect of light on geo-

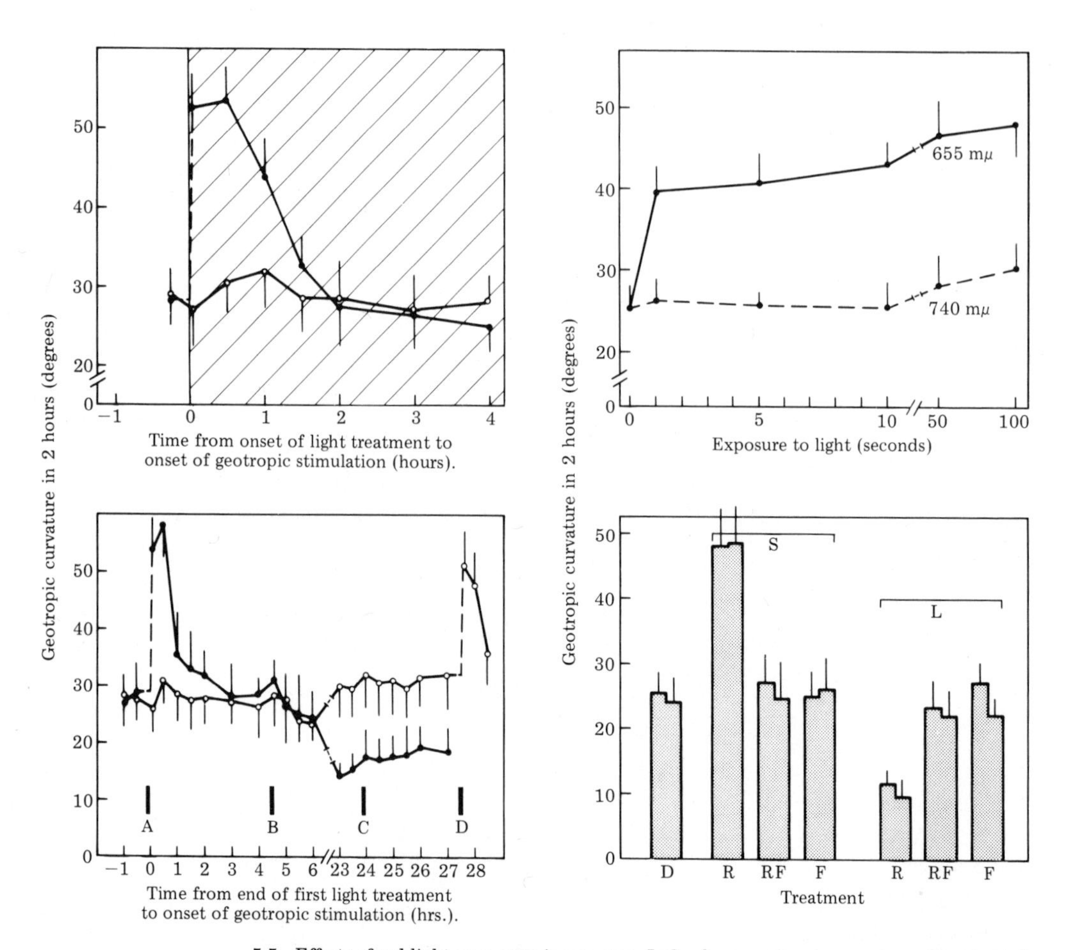

5-5 Effects of red light on geotropic response. Left, above, geotropic curvature of horizontal *Avena* coleoptiles in continuous red light (filled circles) and in darkness (open circles); the sharp increase falls back to the control level in 2 hours. Below, a longer time plot; brief red light exposures at A,B,C, and D; the second exposure causes decreasing curvature, while the first exposure, D, of the previously dark plants gives the expected increase. Right, above, far-red does not cause the sharp increase given by red. Below, far-red, though inactive alone, reverses both the initial increase and the long-term decrease caused by red. From Wilkins, 1965.

tropic sensitivity and there is some evidence that it may involve the red–far-red system. It is not as clear-cut as the effect on germination of light-sensitive seeds, but Dr. Wilkins (now at Glasgow) found that if one irradiates oat coleoptiles with red light and then tests them for their geotropic sensitivity within an hour or two, they give a greater curvature for a given geotropic stimulus than they would have done in the dark. In his experiments, furthermore, if the red light was followed by far-red, the sensitivity was set back to its original value (fig. 5-5). Unfortunately this part of his results has been contested, though the red light effect itself is generally agreed upon. (There is a similar effect of red light (but in the direction of *decreased* sensitivity) in the phototropic response.) What is remarkable in this case is that if one exposes the plants to red light and then waits more than 12 hours, they now become *less* sensitive to gravity. It is as though the red light has released something which was present in limited amounts, which could work only for a short while, because when it was used up it was not replaced; so that the plant becomes less responsive. Far-red light by itself seems to have no effect on the geotropic sensitivity.

There is also an effect of ethylene, and it is a strong one; for ethylene powerfully decreases the geotropic response, both of shoots and of roots. In the case of roots, indeed, it nearly nullifies the whole geotropic response. In the experiment shown in figure 5-6, the geotropic curvature angle of roots is brought down from 42° in air to 2° in ethylene. It should be noted, too, that a level of ethylene between 0.1 and 1.0 parts per million is a common natural level of ethylene in the intercellular spaces of plant tissues. In fruits, for instance, as we shall see in chapter 11, this level operates to cause natural ripening. Thus this effect on the geotropic response is not pharmacological, but physiological.

Judging from the now well-established fact that in several plants red light inhibits the production of ethylene, it would be a reasonable guess that the initial increase in geotropic sensitivity caused by red light would be due to a decrease in the natural ethylene production. This may be the case, but there are some obvious reservations: (a) the *Avena* coleoptile seems to be a good deal less sensitive to ethylene than many other seedlings, and (b) the decreased response to phototropic light seen in figure 6-7 would be hard to explain on this basis. So also would be the decreased sensitivity to gravity after 12 hours in Wilkins' experiments above—although this might well be a secondary effect.

5-6 Effect of ethylene on the geotropic response of pea roots, compared with the effects of (much higher concentrations of) CO_2. From Chadwick and Burg, 1970.

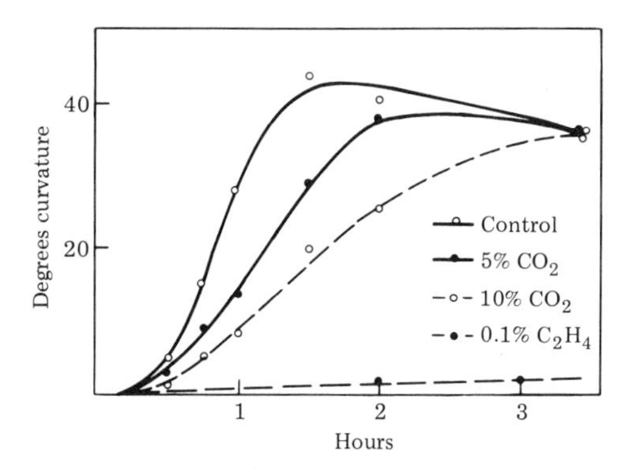

We have seen that ethylene decreases auxin transport; but that is a slow effect, requiring two or three hours before it is even detectable, whereas the exposures in these experiments are much shorter. The conclusion must be that the effect of ethylene on the lateral transport is greater, and more rapid, than its effect on the longitudinal transport. The red light-ethylene-lateral transport interaction is evidently a very close one. In pea stems Kang and Burg found that ethylene production was halved, the lateral auxin transport correspondingly almost doubled, and the geotropic sensitivity also doubled, after five minutes of exposure to red light.

D. THE GEOTROPISM OF ROOTS

The understanding of the basis for the reaction of roots to gravity is in a much less satisfactory state than that for shoots. In the first place the location of the curvature, unlike the curvature of shoots, is very limited. The zone of curvature is usually only 1–3 mm in length, depending on the seedling. It is located a short distance behind the tip, and further behind it is a zone of rapid deceleration, in which the differentiation of xylem and phloem is active. Thus when a root has been placed horizontal, the curved zone gets rapidly left behind as a sharp right angle.

In the study of root geotropic reactions clinostats have been more prominent than with shoots. If correctly used, a clinostat can produce a perfectly straight horizontal plant, which provides the ideal control for geotropic studies. The clinostat is adjusted so that no curvature results, for if it rotates too fast there are secondary effects, due to centrifuging the organelles, while if it rotates too slowly the plant is constantly receiving little gravitational stimuli, and one can detect a series of beginnings of curvatures. At a rate of about one rotation per half minute the gravitational effects just balance out and the plant continues to grow horizontally.

At first it was thought important to determine whether a plant growing on a horizontal clinostat grew at the same rate as a vertical control. It seems that, at least for coleoptiles, this is basically true, and for many shoots it is true for short times. For roots, however, the growth rate generally decreases somewhat on turning horizontal, as was first reported by Sachs nearly a century ago.

Three of the roots whose elongation and tropisms have been most studied are the pea, lentil, and corn. Figure 5-7 shows the structure of the root tips. About 50μ behind the root apex is the zone where the minimum of mitosis, called the quiescent zone, occurs. Beyond that, to a length of 0.5 mm or so, is the root cap, the outermost cells of which continually slough off as new cells are formed from the mitotic zone in contact with the apex proper. The relative dimensions are somewhat different for different plants. All three of these plants give good geotropic curvatures; though some varieties of corn, as well as some other species, respond poorly in the dark, and their sensitivity is greatly increased by red or white light.

In many varieties of corn, the root cap is relatively large, and in at least one cultivar it can actually be dislodged by pulling

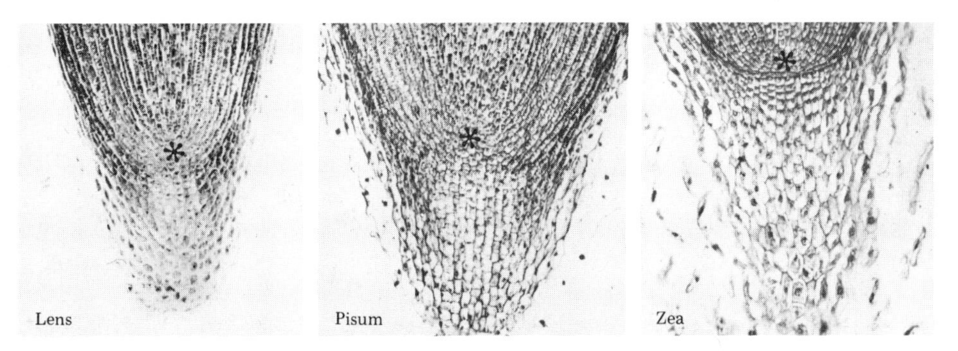

5-7 Sections of root tips of pea, lentil, and corn. From Pilet and Lance, 1967.

gently along the longitudinal axis. From no other plants that I know of will the root cap come off as easily; however, a careful cut at the right place can remove the root cap and little else. Failure to pay attention to the locations of the cap and to the zones of elongation and differentiation has been responsible for a vast amount of inconclusive research in the past. For if roots are simply decapitated one is apt to remove part of the elongating zone, which is only just behind the apex. Nevertheless, if this is done carefully and only the cap zone is removed, then growth can continue normally, and in some well-studied cases it is even accelerated (fig. 5-8). Such decapped roots, still elongating, have lost all of their geotropic sensitivity, though they recover it in 12 to 20 hours.

In classical days Cholodny's theory that all tropisms are due to an asymmetrical distribution of a growth-promoting substance was intended to apply equally to roots as to shoots. Cholodny himself made some critical experiments, one of which was the parallel of Páal's experiment with coleoptiles. Not only he, but later Hawker, Boysen Jensen, and others cut off root tips and replaced them asymmetrically. If the root is vertical, it then curves *towards* the applied tip, that is, the curvature is in the opposite direction to that with the coleoptile. Wilkins, Pilet, and others have done similar work again recently. One needs rather thick roots for this—corn roots are ideal. Roots of some beans are satisfactory, but in cereals like *Avena* the roots are so fine as to be hopeless for this particular experiment, though useful for another aspect of the work.

The same result was obtained by placing a coleoptile tip on a decapitated root: the curvature was towards the tip. We deduce that the tip, whether of root or of coleoptile, produces, and excretes into the part directly below, something which slows down the growth of roots. The slight acceleration on decapitation suggests the same thing. The "something" may or may not be auxin, but it has to be something which is produced by coleoptile and root tip both. Auxin is certainly present, for it has been identified by specific chemical methods as indoleacetic acid; this has been done with the root cap by Rivier and Pilet, and in the root tip more

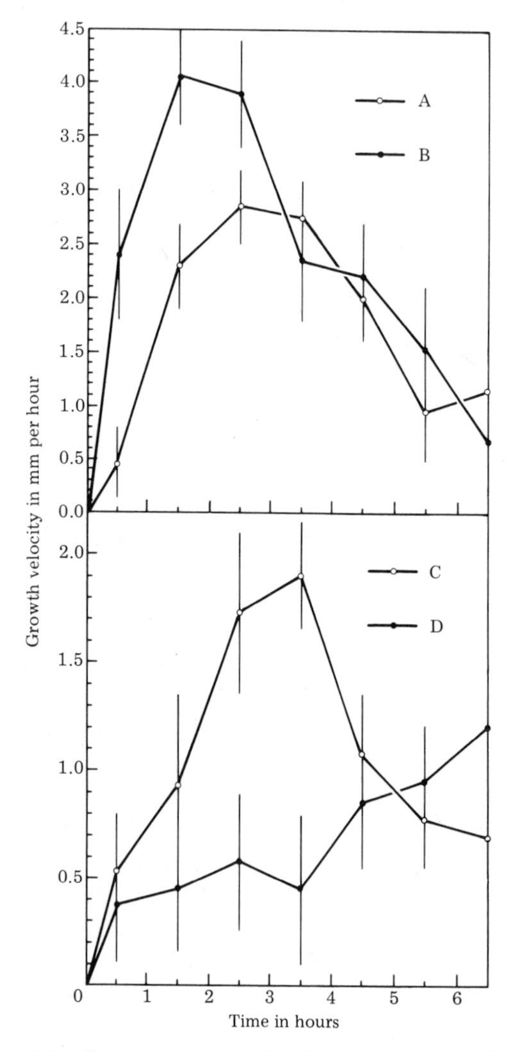

5-8 Above, growth rate of maize root with (A) and without (B) its cap. Below, growth rate of lentil root untreated (C) and with maize cap attached (D). From Pilet, 1972.

generally by Wilkins and his co-workers at Glasgow. However, the amount in the cap was too small to be detected by Kundu and Audus in their extracts of corn.

A complication in cutting and re-tipping experiments is the effect of the wounding, which some workers have felt might be the cause of the slowing down of growth, either of the whole root or of one side. To attack this problem Pilet, in Switzerland, applied an additional root tip to the *uncut* tip of a normal root and it did in fact decrease the growth rate of the normal root. Figure 5-8 shows the growth of his normal roots and of the roots with the extra tip applied. For four hours the growth of the latter is slowed down; still later the curves cross, for reasons which are not clear. Decapped roots showed a similar transient acceleration. So the root tip does secrete a growth-inhibiting influence.

Figure 5-9 shows that the slowing down is due strictly to changes in the rate of cell elongation. The cells on the convex side continue to enlarge, while those on the concave side remain almost entirely isodiametric.

In another attempt to get around the effects of cutting off and replacing, Wilkins simply removes half of a corn root tip as shown. The effect is just like applying a tip asymmetrically and, as expected, the root curves towards the side on which the half tip remains. If the experiment is done in the horizontal position then when the half tip is on the lower side there is an exaggerated geotropic curvature; if it is on the upper side a decreased geotropic curvature results. All of this work shows that we have to deal with a situation in general parallel

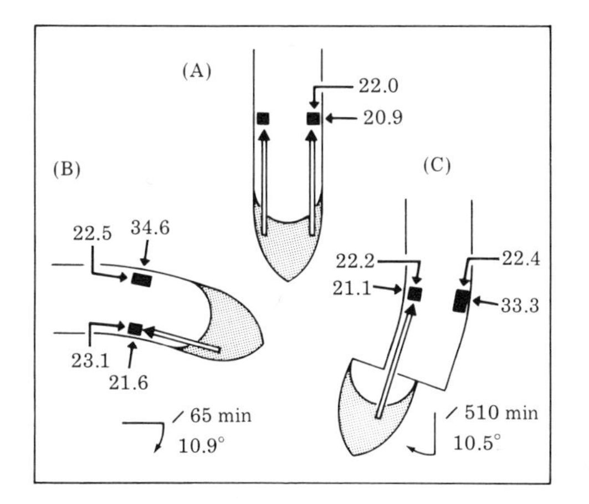

5-9 Diagram of the dimensions, in micra, of the cortical parenchyma cells of corn roots placed vertically (A), horizontally (B), or vertically but decapitated with the tip replaced asymmetrically (C). From Pilet and Nougarède, 1974.

to that in shoots but inverse in direction and modified due to the special dimensions of roots and some other complications.

Now we know that the coleoptile tip secretes auxin, and we know from other types of experiments that auxin powerfully inhibits the elongation of roots, because if one allows the roots of almost any species of seedling to dip into an auxin solution their elongation is drastically inhibited. We know too that root tips secrete auxin, for they can be placed on agar and the agar will subsequently produce normal curvatures (due to growth promotion) of the kind typical of indoleacetic acid, on standard *Avena* coleoptiles.

There are, however, some peculiarities of auxin production by roots which are not seen when auxin is produced by leaves or buds. Some roots will secrete highly active material which gives quite large *Avena* curvatures. Curiously enough, *Avena* roots, which are very thin and small, do this very well, perhaps because they are small enough for many to be crowded on to a small block of agar. But others do not secrete auxin for long. Cholodny showed long ago that corn root tips produce auxin only briefly, but if they are fed with a drop of gelatin containing a little peptone, the auxin yield is continued. The deduction is that these root tips are low in the precursor from which auxin is formed and need a little nourishment. Boysen Jensen obtained a similar result by feeding sugar solution. Although these experiments were mostly done long ago they were carefully done.

So putting all these results together, it seemed reasonable to deduce that the auxin secreted by the root tip becomes asymmetrically distributed in the growing part of the root and inhibits growth more on the lower side than on the upper. Direct evidence for this interpretation was obtained first by Boysen Jensen, who halved the tips of horizontally placed *Vicia faba* roots, placed the half-tips on agar, and tested the agar on decapitated *Avena* coleoptiles in the usual way. Out of 27 experiments, 25 showed more auxin on the lower side; the average difference was about 25% or a ratio Upper: Lower of about 38:62. Hawker applied similar half-tips to blocks which were then tested on decapitated *roots*; the roots curved *towards* the blocks, as shown in figure 5-10, the lower halves producing 30° and the upper halves 11°. However, this result (but not Boysen Jensen's) might have been produced by secretion of a root growth inhibitor other than auxin.

Unfortunately this simple picture has lately been subjected to steadily growing controversy. Much of this is well described by Audus (1975). For one thing the geotropic response of some roots is very weak in total darkness, and is greatly increased by light. This has been held to make a role of auxin improbable. But back in 1938 Segelitz showed that light increases the auxin content of root tips and correspondingly decreases elongation.

A stronger basis for uneasiness rests on the observation noted above that if one decapitates roots so that the elongating zone is not appreciably damaged, or with a suitable corn variety removes the cap directly, the geotropic response is drastically reduced. Figure 5-11 shows the curvature as a function of time; by six hours the controls have reached about 40° downward curvature. With the caps off nothing happens for at least 12 hours, then slowly curvature begins to take place again. If the caps are taken off and then replaced (using Ringer's solution to make contact) the response is more like the normal, and after a lag the curvatures become quite large. If, however, the cap is replaced over a film of oleate there is little response.

The auxin problem raises several questions about roots which do not need to be asked about shoots, particularly in regard to the movement of auxin along the axis. If the root cap is to secrete auxin and gravity is to redistribute this auxin so as to cause geotropic curvature, there must be a more or less polar movement of auxin from the cap into the root tissue. But this is controversial; for one thing, if one applies auxin to a subapical root segment in a typical trans-

5-10 Hawker's experiment of 1932. Upper and lower halves of the horizontal root tip were placed on gelatin blocks for 1 hour and the blocks then applied to freshly decapitated vertical roots. From Hawker, 1932.

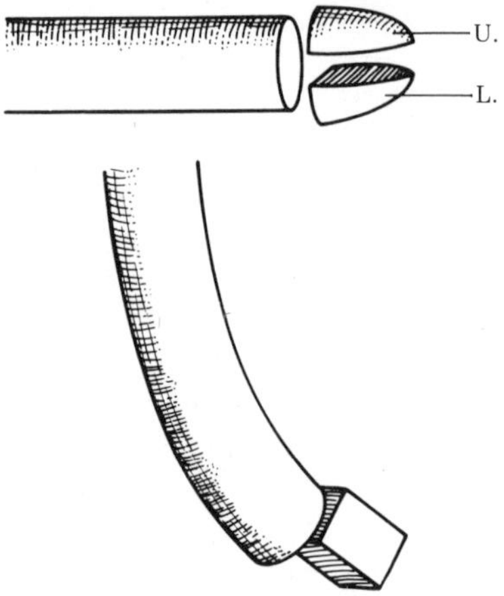

U.
L.

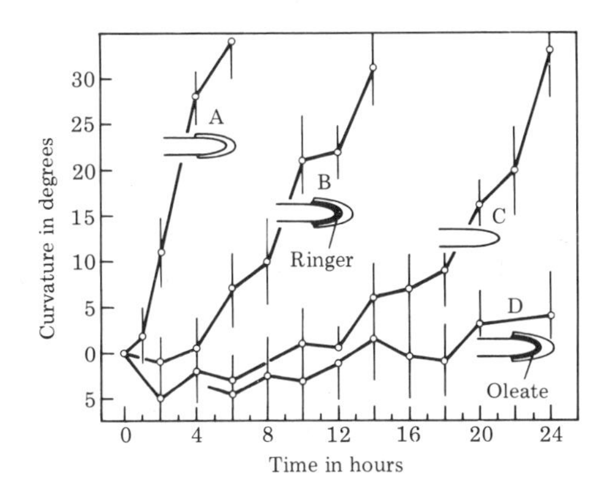

5-11 Time course of geotropic curvatures of corn roots intact (A), de-capped (C), de-capped with cap replaced (B), and with cap replaced over a film of oleate (D). From Pilet, 1972.

port experiment, the results are apt to be completely negative, because little or no auxin appears in the receiving block in either direction. Evidently there is a drastic disappearance of auxin (mainly due to enzymic inactivation), but in any case what little auxin is detected shows no strong polarity. Recently Wilkins in Britain and Scott in the United States have found, using ^{14}C-IAA, that the predominant movement is from the shoot base towards the root tip in pea roots, and this finding has been widely accepted as showing no movement from the cap into the root apex. But the early experiments, in which shoot or root tips applied to decapitated roots caused curvature towards the applied tip, show that at least some auxin can move from the tip or the cap into the root. Indeed, if we remove only the minimal amount of tip material, just 0.5 mm from the extreme tip (mainly the root cap), as both Konings and Pilet have done, we can apply ^{14}C-IAA directly to the cut surface of the apex proper. Konings found very good entry of radioactivity into the first millimeter, less into the second, and very little into the third. It is as if the transport is coming up against a polarity gradient in the other direction, 1 to 2 mm beyond the apex. This would obviously fit with the conclusions of Wilkins and of Scott. Figure 5-12 shows similar experiments by Yeomans and Audus. Soon after auxin has been applied to the root tip we see it entering into the first millimeter. If it is applied in the reverse direction, namely on the fourth millimeter from the tip, again it enters well but less rapidly, and the level reaches a plateau, while coming from the tip it continues to enter at a good rate. Ohwaki, in Japan, finds it to move equally in both directions in *Vicia* roots even when the ^{14}C-IAA is applied as much as 4 to 5 mm from the tip.

After similar transport experiments, Dr. Konings did very careful experiments with ^{14}C-indoleacetic acid in which he removed only ½ mm of the root tip, which is really no more than the root cap, applied a tiny agar block to the cut surface and subsequently measured the radioactivity in the upper and lower halves (table 5-2). Exposures to gravity were of different duration. In the first experiment listed, the most apical millimeter, beginning at the point where the auxin was applied, shows in the upper halves a count of 95 and in the lower halves, 148. One could not ask for a clearer result. After 1½ hours the counts are 112 and 190, respectively. After 2 hours the difference is less, and after 3 hours less still, though still significant. In the first 2 hours the fraction of the total counts found in the upper halves is around 40%, so that the 40 to 60 ratio, which was found in the shoot system, is maintained. I think, therefore, that despite the reservations of some workers, the findings in roots can be reasonably explained on the basis that auxin, if made in the root cap (or in the extreme root tip), is transported to the first 2 or 3 mm of the root, and *if it has been asymmetrically distributed in the root cap,* then where the elongating zone begins there is enough difference between the amounts on the two sides to cause growth inhibition on the lower side.

However, this raises a difficult question. Why should auxin, a cell elongation hormone, inhibit the elongation of roots? It does so at nearly all concentrations, including physiological levels. Only at the very lowest concentrations, even below those recorded *in vivo,* have some workers found it to promote root growth somewhat. Audus and Brownbridge found that both IAA and 2,4-D at 1 part in 10^{11} increased both the growth rate and the rate of geotropic curving by some 20% to 30%. At all more normal concentrations both growth and curvature were inhibited. Åberg considered that growth promotion occurred only with isolated segments, not with intact roots. Burström (in Lund) has shown that there is indeed a brief acceleration of growth when the auxin is first added, but this soon stops. He envisages the action of auxin as bringing root growth to an early stop, whereas in

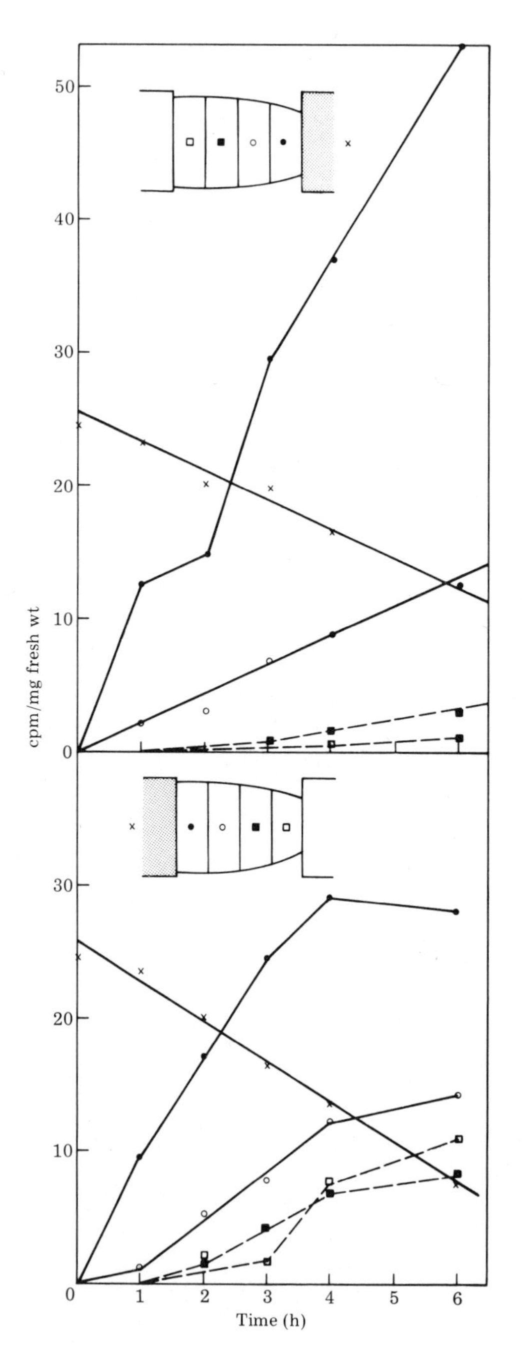

5-12 Time course of entry of ^{14}C-IAA into the first four subapical millimeters of *Vicia faba* roots. Crosses, C^{14} remaining in the donor block; root segments as shown at top. From Yeomans and Audus, 1964.

shoots the accelerated growth rate continues indefinitely. In any case the overall effect, even if this is so, is a net inhibition of growth, so that the question just posed remains to be answered.

So far the most attractive explanation is that proposed by Burg, namely that roots are highly sensitive to growth inhibition by ethylene, and that when auxin is applied to them it stimulates the production of ethylene, as it has been shown to do in several instances in shoots. Thus when gravity causes a modest excess of auxin on the lower side of the root, it stimulates the production of ethylene in this area and the ethylene is the "second messenger" which inhibits the elongation.

In favor of this view is an observation of Konings that most people have overlooked, that although the elongation rate of the lower side slows down, after half an hour or so it slows down temporarily on the *upper* side. The effect is transient but is quite clear. A substance which diffused rapidly across the root, as a gas would do, would give just such a result.

The ethylene explanation involves the assumption that the concentration of ethylene that is normally produced in the root lies only just below the inhibiting level, so that the auxin content needs to increase only by a factor of two or so to make the ethylene reach the inhibiting level. In *Ribes nigrum* (black currant) roots, indeed, the ratio of auxin on the two sides is reported to be only 1.6 to 1. This requires an exceedingly sharp threshold in the response to ethylene, which I for one feel is somewhat unlikely.

There is another complication, however. When less auxin is found in the upper half than in the lower half, one also finds that there is a faster rate of auxin oxidation in the brei made from the upper half. The difference is clear enough to tempt one to believe that the auxin asymmetry is not due so much to asymmetric transport as to asymmetric destruction. Figure 5-13 shows

Table 5-2 The transverse distribution of carboxyl-labeled IAA – ^{14}C in horizontal pea roots at various distances behind the apex

Time of exposure, hours	Sections (length in mm) cut at:	Cpm/mg dry weight		% of total cpm in upper halves
		in upper halves	in lower halves	
1.0	0–1	95.2	148.0	39.1
	1–2	26.6	33.2	44.4
1.5	0–1	112.3	190.0	37.1
	1–3	30.0	42.0	41.6
	3–5	12.7	15.4	45.1
2.0	0–2	94.1	118.8	44.1
	2–4	10.4	13.9	42.7
	4–6	3.2	7.0	31.3
	6–8	0.0	0.0	—
3.0	1–4	63.0	104.9	37.5
3.0	2–4	9.7	25.1	27.8

From Konings, 1967.

how rapidly the auxin disappears from the brei; in a few minutes its disappearance is complete. Evidently the material from the upper side causes a more rapid disappearance than that from the lower side—a ratio of about 2 to 1. This raises the question of whether auxin destruction may also play a role in root geotropism, but one would like first to be sure that it is general, and that other seedling roots, which are equally geotropically sensitive, exert the same type of action.

So much for the role of auxin. We see that there are at least three uncertainties in regard to its role as the mediator of root geotropism: its apparently limited production in the root cap, its transport predominantly (though not wholly) in the wrong direction, and its ability to stimulate adequate ethylene production. For these reasons attention has been turned to the possible role of other hormones.

In shoots, some workers have claimed that gibberellic acid becomes asymmetrically distributed under the influence of gravity, but others have been unable to find any such change. El-Antably has recently shown that, in roots, after 4 hours in the horizontal position there is 5 times as much gibberellin in the *upper* half as in the lower half. However, the elongation of roots seems to be quite unaffected by applied gibberellic acid, although it somewhat decreases their geotropic sensitivity.

The case is different for abscisic acid (AbA), since it does inhibit root elongation quite powerfully. El-Antably, using gas-liquid chromatography, found AbA to show an asymmetric distribution just opposite to

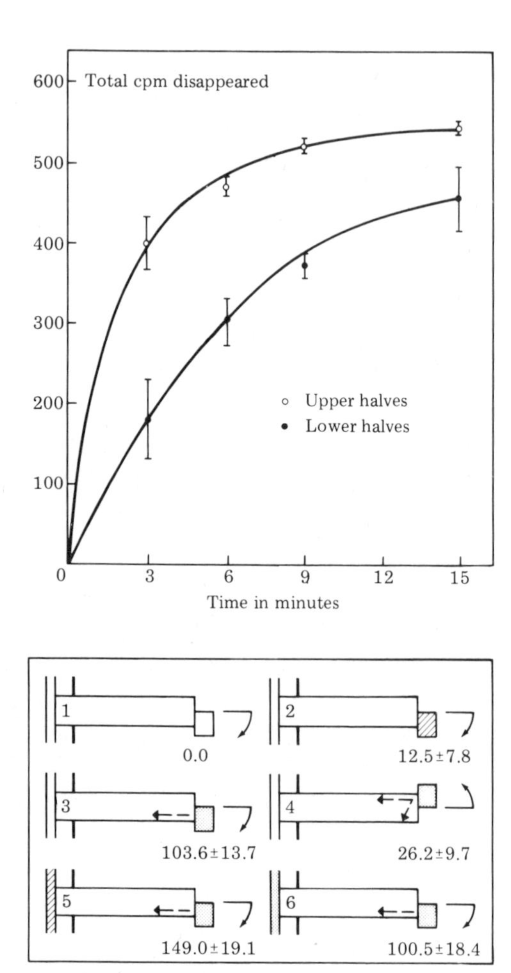

(above) **5-13** The rapid oxidative destruction of ^{14}C-IAA in homogenates from the upper (open circles) and lower (filled circles) halves of the apical 3 mm of horizontal pea roots. From Konings, 1967.
(below) **5-14** Curvatures of horizontal subapical segments of corn roots with blocks placed on half of the cut surface. 1, control; 2, IAA on lower half; 3, AbA on lower half; 4, AbA on upper half; 5, AbA on lower half and IAA at base; 6, AbA on lower half and also at base. Measured after 7 hours. From Pilet, 1975.

that of GA, about 5 times as much being in the lower half as in the upper half. Recently Kundu and Audus have tentatively identified AbA as present in the root cap of maize, but Wain and co-workers find only very small amounts there—not enough to cause appreciable inhibition. However, Pilet has shown that agar blocks containing AbA (10^{-6} M), when applied to the lower half of the cut surface of a decapped maize root, causes extremely large positive curvatures—curvatures due to growth inhibition on the side with the block. These segments of decapped roots gave only a very weak downward geotropic curvature in the 7 hours of the experiment; IAA increased the curvature by 13°, but AbA increased it by 104° (fig. 5-14). What was even more remarkable was that if IAA were applied to the base of the 10-mm segment while the AbA was in the block, the curvature was still further increased, to 149°. The evidence that AbA, symmetrically applied, can migrate to the lower side is still not wholly convincing, but there is a real likelihood that geotropic curvature could be due to the *simultaneous action* of both auxin and AbA, somehow acting in synergism. Either one alone can cause some curvature, and IAA at least does become accumulated on the lower side, but action of the two together may explain the very large and relatively rapid geotropic curvature, the limited zone in which it occurs, the apparent participation of ethylene, and the inability of auxin alone to fit in with all the observations. Taken together with the observations (see part 2 of this chapter) that materials other than auxin can become asymmetrically distributed in horizontal *Vicia faba* roots, we can deduce that gravity affects the distribution of a *number* of cell constituents, not merely hormones. While the matter is still the subject of active research, it is interesting to reflect that here, as in so many cases, the plant's behavior may be under the simultaneous control of more than one substance.

E. PLAGIOTROPISM AND ANISOPHYLLY

Before leaving geotropism let us further consider the special case of plagiotropism. As far as shoots are concerned this is most often seen in trees. In the plagiotropic branch of a tree, the angle of the lateral branches is determined by the presence of the apex, so that the curvature which determines the branch angle is not wholly a geotropic curvature; it is partly a curvature which is related to the main axis. If we remove the apex of a spruce tree (the behavior is different in pines), then one or more of these plagiotropic lateral branches becomes negatively geotropic; that is, it begins to grow vertically, and when it gets long enough, it becomes a new apex. Thus it was potentially geotropic all the time, but its angle to the vertical was depressed by the growing apex. You can see this beautifully on the Olympic peninsula, where large numbers of spruces grow, and where from time to time there are very high winds; one has only to look up to see that although the trees have perfectly good tips each has a little curvature lower down, showing that the original tip was broken off and one of the laterals has taken its place, quickly becoming vertical. In some spruce species, *Picea pungens* (Colorado blue spruce) for example, several laterals in the same whorl will become geotropic, forming a little group of upwardly curved laterals; finally one elongates more than the others and becomes the new apex. The influence of gravity can be balanced out by putting plants on a clinostat, as you know; when this was done with *Coleus* plants the lateral branches, instead of pointing upwards at an angle to the vertical of about 70°, pointed downward at an angle of about 130°. Thus although the effect of gravity plays a part in the curvature, it interacts with the influence of the apex so that the branch maintains a fixed angle to the axis.

It has been suggested, though it has certainly not been proven, that the plagiotropic angle is due to two opposing influences: auxin coming from the young leaves of the branch is diverted in the usual way to its lower side, while auxin coming from the apex and young leaves of the apex travels down to the branch point and then up into the branch, where some other force, which depends on the integrity of the whole plant, works in the opposite direction, diverting it to the *upper* side. As a result, the balance between these two forces holds the branch at a fixed angle, usually about 70°. In some trees this angle changes, being more acute near the top, and gradually increasing lower down, till finally, on the lowermost branches, it is almost 90°. This means that the angle must steadily increase with increasing age, but I know of no experiments that have been done on that.

A related phenomenon is that of anisophylly, which has attracted singularly little attention from experimentalists. In a plant with opposite leaves, each pair of leaves is normally equal in size and rate of development. In an anisophyllous plant the leaves borne on lateral branches are unequal, those on the upper side being the smaller. This again shows that there is an asymmetrical distribution of some growth-controlling factor. It is most noticeable in the pair of leaves which is at the base of the lateral branch and hence nearest to the main axis, a highly suggestive fact which certainly fits in with the explanation just given for plagiotropism.

A small-scale model of the relation between apex and branch angle in trees is given by the potato plant. Here the basal lateral branch becomes a horizontal stolon bearing the tuber. If the main apex is cut off the stolon soon begins to grow upwards, but if IAA is applied to the cut stem the horizontal growth is more or less maintained. Thus in this case, at least, the influence coming from the apex is certainly identified as auxin.

This is as far as we can go now with what zoologists call the *effector* system. It is the system which produces the curvature, or the differential growth. For in a tropism, or any similar problem, there are two types of questions to ask: what is the effector system and what is the detector and how does it function. In other words, for geotropism, how is gravity detected by the plant? For phototropism, how is light detected? We will take up the problem of the gravity receptors first.

THE GEO-RECEPTORS OF PLANTS

A. THE EVIDENCE IN FAVOR OF AMYLOPLASTS

In the old days, that is, in the nineteenth century, the botanists (unlike most botanists of today) read the zoological literature. Some of them, who had thought about the reaction of plants to gravity, were much impressed by the orientation of invertebrates to gravity, and for certain invertebrates it was shown that a small, heavy mass of calcium carbonate is deposited in specific tissue of a multicellular organ. If the orientation of the animal is sideways this heavy mass presses on the side of its cell or organ, while if the animal is vertical it presses vertically downward; this pressure is transmitted to the brain and acts as a signal to enable the animal to right itself. Thus this so-called statocyst is a very simple gravity-detecting device; and Noll in 1892 made the suggestion that plants may have something like a statocyst, though perhaps at the cellular level. He did not try to prove this, but the idea was confirmed simultaneously and independently in 1900 by two famous plant physiologists of the day, Haberlandt and Nemeč. Haberlandt worked on seedling shoots, and Nemeč worked on root caps, and they both showed that in seedlings, but also in some mature plants, certain cells contained relatively heavy plastids full of starch, which are called amyloplasts, or starch grains. When the cells were vertical the amyloplasts were seen on the lower walls, while when the cells were horizontal at least some of the amyloplasts settled on the lateral walls. Haberlandt even showed that when the stems were at an angle the amyloplasts congregated at a corner, where they had migrated both laterally and downward. Of course all these observations were made with stained preparations, but later living cells were observed and one could see that when the cell is placed in a horizontal position, the little round amyloplasts actually roll down on the side wall. One could time the rolling down; Dr. Yocum once made a movie of it. They roll down in curiously close relation to one another, not as though they were single objects, but rather giving the impression of being invisibly connected to each other. In any case their movement is quick, and in many active seedlings when the shoot is turned through 90° it takes only about 1½ minutes for them to roll down from the cross wall on to the lateral wall. Very interestingly, Dr. Nemeč put this idea forward in 1900, and in 1964, no less, he came to the Botanical Congress in Edinburgh and, in the course of giving a paper, mentioned that he had made these experiments a little while ago, and he had now had 64 years to think about it. We shall see below what new idea he had come up with.

If seedlings are placed in a horizontal position and then returned to the vertical, after a while a visible curvature will develop, due to the preceding horizontal exposure. If the horizontal position is maintained for a series of different times, one can determine how much exposure to gravity is needed in order to get the minimum response. The value turns out to be very short: in the *Avena* coleoptile it is between 2 and 4 minutes. In other plants it is frequently somewhat longer; a common time is around 10 minutes. This is the minimum presentation time. It does not tell how long one must wait for the visible curvature to appear, but just tells how long the plants must remain horizontal in order for a curvature to appear subsequently. Following the presentation time, there is a second period called the latent time. This is the time it takes for the curvature to appear. In auxin terms one could imagine that the presentation time is the time needed to establish some sort of transverse asymmetry so that the auxin goes to the lower side, while the latent time is needed either for that auxin movement to develop, or for the existing auxin supply to result in asymmetric growth. The latent time is usually longer than the presentation time, a typical time being 15 to 20 minutes. The value depends on the plant and, among other things, on how near it is to its optimum growth rate and its optimum sensitivity. So there is at least a possibility that the falling of the amyloplasts takes place in a time comparable to the presentation time. Thus, in order to prove that the falling of the amyloplasts represents the detection of gravity, one would have to prove that they move in a time not greater than the presentation time. That fact is correct: in the *Avena* coleoptile they fall to the lower side in about 1.5 minutes. The falling of an object within the cell is, of course, not a simple process because all the time the cytoplasm is streaming around. It is easy to get a static impression of the cell contents from looking at an electron micrograph. One sees all the beautiful little organelles neatly arranged in a sort of mosaic, and one forgets that the whole mass is actively circulating. However, despite this continuous cytoplasmic streaming, the amyloplasts are heavy enough to move, and indeed to move across the direction of the stream. There are many other organelles in the cell and some of these can be seen flowing around with the cytoplasm and making little or no attempt to settle. An intermediate group consisting mainly of mitochondria, and perhaps some Golgi bodies, still goes around with the stream but also shows a slight, very slow tendency to settle. As far as sheer size and weight are concerned, the idea that the amyloplasts are the gravity detectors is satisfactory.

The next question one must ask is, does the location of the cells containing amyloplasts coincide with those parts of the plant that are sensitive to gravity? Haberlandt himself made several studies of this and concluded that with several seedlings the agreement was very good. Miss Prankerd, in England, did the same thing with growing ferns, and concluded that where geotropic curvature occurs, cells containing amyloplasts were always present. Thus a rough but quite good correlation between the zones of geotropic sensitivity and the zones where cells contain amyloplasts was established. More recently, Audus, in London, made calculations for *Vicia faba* roots based on the measured size and density of organelles and the density of cytoplasm, and concluded that to produce a presentation time of less than 20 minutes, the only object that falls sufficiently fast is the amyloplast. He calculated also that mitochondria, which are not much more than 1 μm across, would take 2 hours to fall through half the diameter of an average root-tip cell. This is partly because of their smaller size, and also partly because the amyloplast is filled with starch whereas the mitochondrion contains material of a density more like that of cytoplasm. Even allowing for variation in size of cells, it is clear that mitochondria could not, by rolling down, reach the lateral wall in anything like 20 minutes. The presentation time for *Vicia faba* roots is around 6 minutes. Thus the only object dense enough to give a geotropic response within the known presentation time is the amyloplast. These calculations, and new experiments confirming the old observations, are given by Griffiths and Audus (1964). What they did was to place seedlings of *Vicia faba* horizontally and then fix them after 20 minutes; they then looked at the cells under the microscope, divided each cell in four, and scored them for the number of various organelles in the four quadrants. Then they statistically analyzed the distribution of the amyloplasts, Golgi bodies, and mitochondria in the four quadrants, and confirmed Audus's calculations, namely that it is only the amyloplasts that are statistically enriched in the lower quadrants. They also found something else possibly interesting, namely that there is a slight but statistically significant enrichment of mitochondria in the upper half of the cells, although they are clearly not lighter than cytoplasm. The writers calculate that although there is a slight enrichment of mitochondria in the upper two quadrants, there is no enrichment *relative to* the amount of cytoplasm, because actually there is a little more cytoplasm in the upper quadrants than in the lower. This is because the amyloplasts are large enough to displace a good deal of cytoplasm. In any case the observations show that it is only the amyloplasts that clearly migrate downward; whatever the other organelles may do is not clear, but within 20 minutes they certainly do not settle. Of course the fact that they move into the upper part of the cell might conceivably be physiologically significant. When Dr. Nemeč appeared at the Edinburgh International Botanical Congress, he reported that with careful nuclear staining he had come to the conclusion that in the root caps of horizontally placed roots, the nuclei were clearly displaced toward the tops of the cells; in other words, the nucleus was a little lighter than the cytoplasm. His excellent pictures clearly demonstrated the upward displacement. So one always has to bear in mind that not only are there heavier bodies going down, there are also lighter bodies going up, and the latter might also be acting as geosensors.

We saw above that in Dolk's experiments on the application of auxin in agar blocks, he showed that if such blocks were applied to a subapical segment of the coleoptile, gravity would still produce a displacement of auxin to the lower side. This means that whatever object detects gravity, it must not be limited in location to the extreme apex but must be present subapically as well. Fig. 5-15 shows that such gravity-sensitive amyloplasts are present about 2 mm below the coleoptile tip. Thus their distribution agrees well with the distribution of geosensitivity in the coleoptile.

B. THE EVIDENCE AGAINST AMYLOPLASTS

Some years ago a study was made of the geotropic sensitivity and anatomy of the roots of *Sphagnum,* a plant which populates many square miles of bogs in the North. In this survey Dr. von Bismarck used seven different species, all of which behaved very similarly. They all showed geotropic responses near the tip when the plants had been laid horizontally. The response was slow; the presentation time was of the order of an hour or two and the latent time was of the order of 24 hours or more—sometimes several days. Some of them took as many as 6 days to reach the vertical position of 90°. On the whole the slowness of response is related to the fact that they grow rather slowly, for the geotropic curvature rate naturally reflects the growth rate. Nevertheless, although they are slow, the curvatures are clear-cut and obviously functional. In these plants there is a zone of amyloplasts from 5 to 20 mm back from the tip, and it corresponds roughly with the zone of curvature, which is clearly well behind the tip. Von Bismarck saw that when the roots are placed horizontally these amyloplasts do migrate down on to the lateral walls of the cells; they do this rather slowly, taking about an hour and a half till they come to rest. This again is in line with the general slowness of their geotropic reaction. On the horizontal clinostat, where the influence of gravity is equalized out, von Bismarck noted that after 5 days on the clinostat the amyloplasts had not moved to the side, but were still more or less in the middle of the cell, thus providing a very satisfactory negative control.

Now comes the peculiar observation. He found that when he grew these plants in the laboratory (they did not grow very well) or in the light chamber (which had a very low light intensity), they gradually ran out of starch. He observed that the amyloplasts became smaller and smaller and finally disappeared. After three months in the light chamber no more starch was present. Nevertheless, when tested by being placed horizontally, they produced perfectly normal geotropic curvatures. He also found that in the field during the winter months, that is, when subjected to the very weak light and low temperature of its northern habitat, *Sphagnum* gradually loses its starch, so that by February or March all of the species he examined were completely out of starch, and therefore out of amyloplasts; nevertheless, when brought to the lab they gave a satisfactory geotropic curvature. Table 5-3 shows that some curved 90° in three days,

Table 5-3 Geotropic curvature of starch-free *Sphagnum* plants

Species	Treatment to free from starch	Days to achieve negative (upward) geotropic curvature of 90° at 5°	at 22°
Sphagnum riparium	60 days in lighted icebox	15–18	1–2
Sphagnum squarrosum	60 days in lighted icebox	14	2–3
Both above spp.	Field in February and March	—	2–3

Data of von Bismarck, 1959.

which is well in the range of the normal plants containing starch. This was a very important observation which has been neglected by most workers. In my laboratory we thought that, since *Sphagnum* is a rather difficult plant to grow in the laboratory, the possibility of removing the starch from *Avena* seedlings should be explored.

First, there was a very peculiar property of the amyloplast starch, which was quite different from that of starch in ordinary chloroplasts. For when the plant runs out of carbohydrates by starvation or other treatment it hydrolyzes the starch in the *chloro*plasts, but does not attack the starch in the *amylo*plasts. Thus we found that growing the plants at high temperature in the dark had no effect on the amyloplasts, and we experimented with lowering the temperature for a short time and then raising it, the way one does when studying the sweetening of potatoes (potatoes hydrolyze their starch when held below 4°C for a while), but this also had no effect. Another possible approach was to make them anaerobic, so that the Pasteur reaction would use up their carbohydrates faster than they would by oxidation. The Pasteur reaction might also be induced in carbon monoxide, as it is in some animal tissues. However, none of these procedures caused attack of the amyloplasts. We then considered the fact that, in the germination of barley seeds, gibberellin stimulates the production of amylase. But before we could experiment with gibberellin, Boothby and Wright in England reported that in barley seedlings placed in gibberellin there is a very rapid hydrolysis, not only of starch in the seed but also of starch in the seedling; the plastids containing starch are rapidly hydrolyzed. Dr. Pickard and I found that this procedure does not work well for oats but, on hearing rumors that children in Montana like to pluck the young wheat plants and suck them, we deduced that in wheat the equilibrium between starch and sugar may be more towards sugar than in oats. Changing over to wheat seedlings led to immediate success. Wheat seedlings were treated with gibberellin, especially at high temperatures, and after about 30 hours' incubation, the amyloplasts were largely hydrolyzed. Gibberellin alone was not enough—it clearly acted in the right directions, but the amyloplasts were not removed completely. However, just as in the germination of lettuce seeds, gibberellin and kinetin show a synergism, they show a similar synergism in this system, and when applied together the amyloplasts did disappear totally. It was necessary to work with very young coleoptiles since older ones, even of wheat, do not completely hydrolyze their amyloplasts even under these conditions. Figure 5-15 shows a very young wheat coleoptile: the epidermal cells contain no amyloplasts, then there are two rows of cells which contain amyloplasts and another row which contains none. Thus the situation is very similar to that which we just saw in the *Avena* coleoptile. This plant had been horizontal for an hour before it was stained with iodine and photographed. As pointed out by Dr. Nemeč, the nuclei do tend to be on the upper, outer wall. The plant has been incubated about 36 hours under laboratory conditions, and the amyloplasts are lying mainly on the cross-walls.

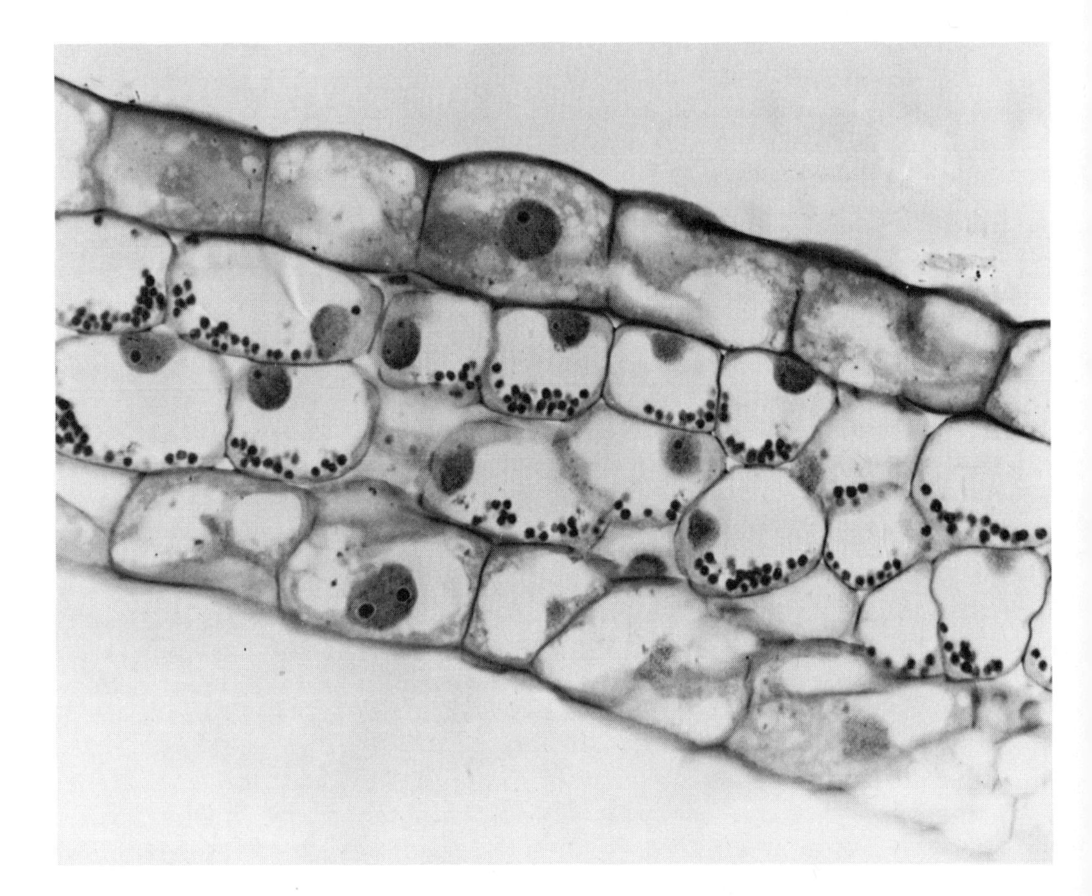

We found that if the plants are incubated in sucrose the amyloplasts become much denser, and there are many more of them (fig. 5-16a). The figure shows their high refractivity and larger size, compared to the normal just below (fig. 5-16c). Figure 5-16b shows a plant of the same age after treatment with the gibberellin and kinetin mixture, and it is evident that the amyloplasts have disappeared. Figure 5-17 is a

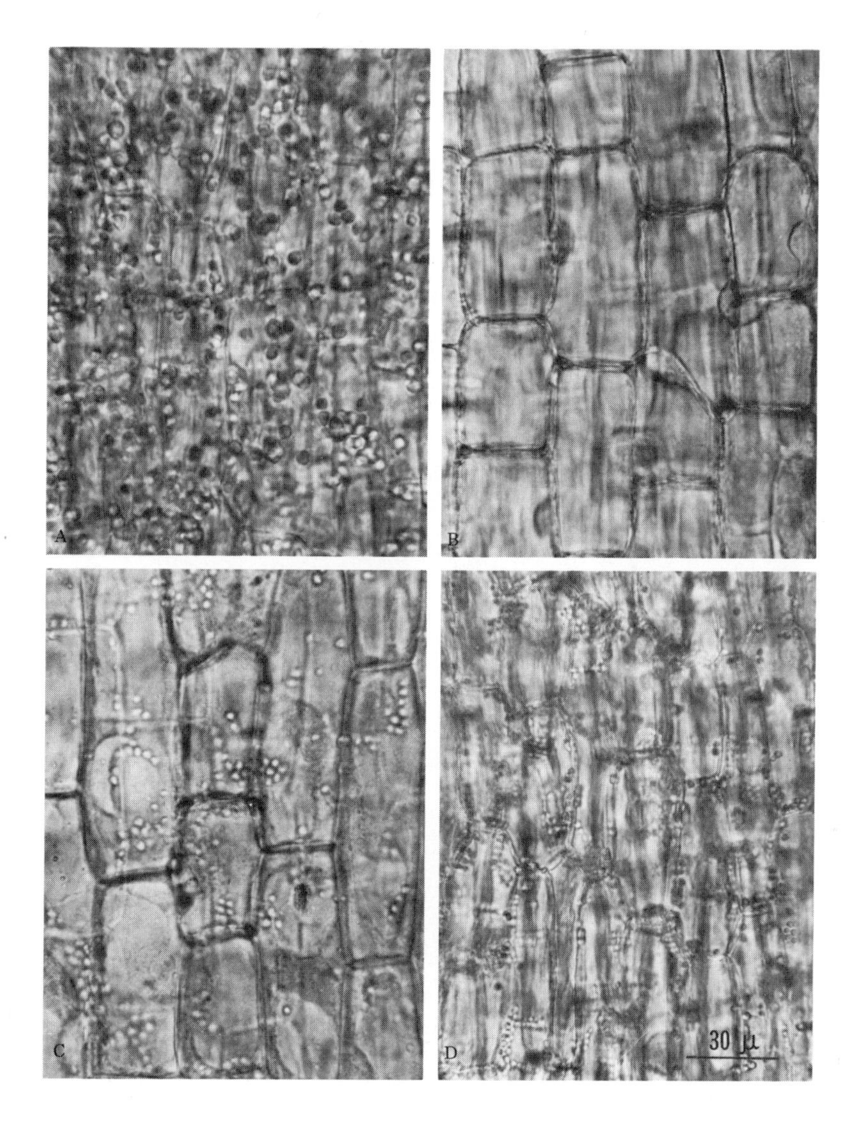

(left) **5-15** Subapical region of a freshly excised wheat coleoptile after being horizontal for 1 hour, showing starch grains on lower wall. From Pickard and Thimann, 1966.
(above) **5-16** Unstained sections of wheat coleoptiles. Left, above, after 34 hours in 1 mM sucrose, amyloplasts dense and numerous. Below, freshly excised. Right, above, after 34 hours in GA plus kinetin. Below, a lower region of a freshly excised coleoptile. From Pickard and Thimann, 1966.

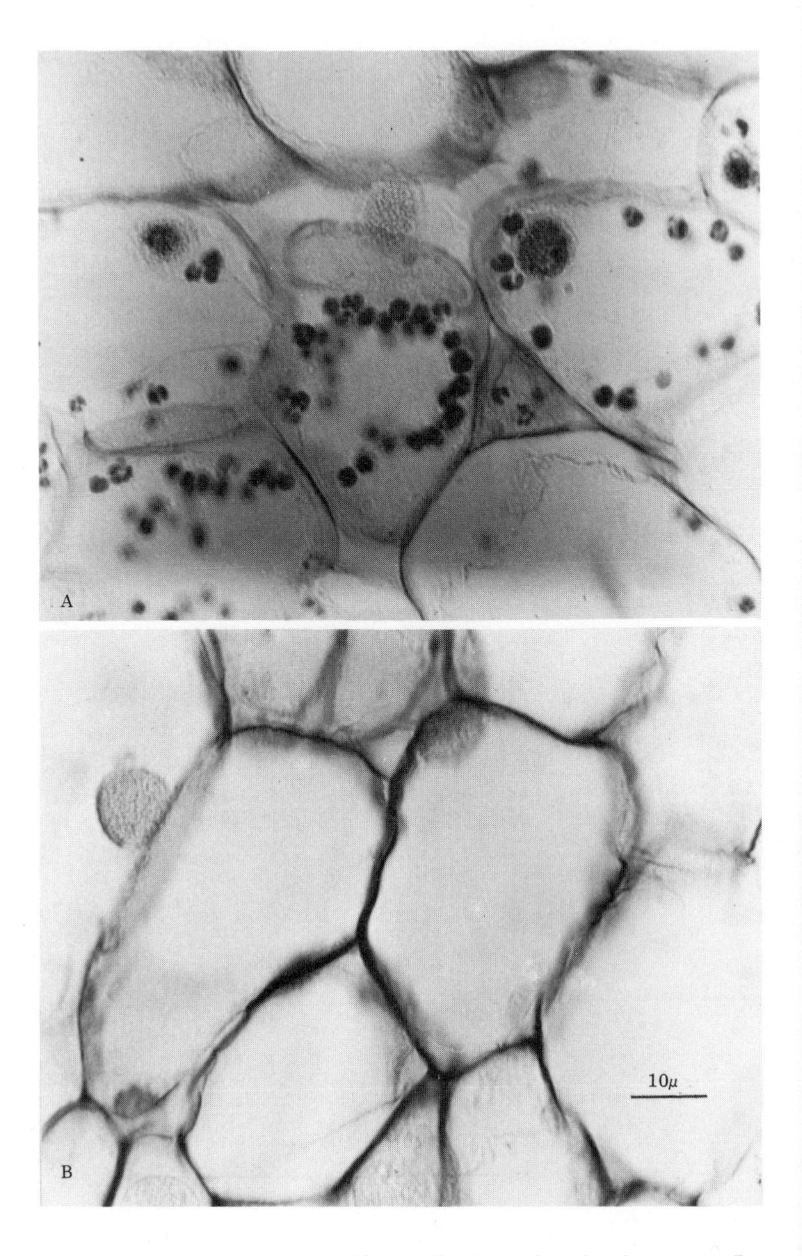

5-17 Above, cells of control incubated in sucrose. Below, cells of similar coleoptile after 34 hours in GA plus kinetin. From Pickard and Thimann, 1966.

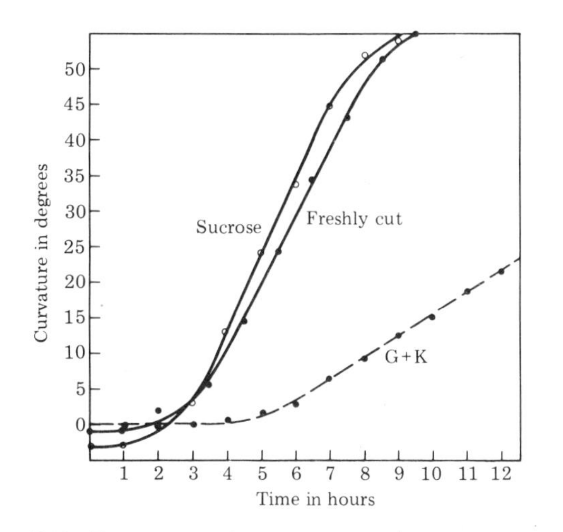

5-18 Time course of geotropic curvature of young wheat coleoptiles like those of fig. 17. G + K, incubated 34 hours in gibberellin plus kinetin. From Pickard and Thimann, 1966.

closeup of some of the cells that are particularly rich in amyloplasts, the second and third layers under the epidermis. There is nothing in their cytoplasm that appears dense under the microscope, and nothing that shows periodic acid Schiff staining,* either. In the next figure (5-18) we see the progress of the geotropic curvature of these plants when placed horizontally. The freshly cut plants curve at a steady rate; after the curvature reaches 45° the rate begins to slow down. The plants in sucrose, with amyloplasts greatly increased both in density and size, show exactly the same rate of curvature, while one would have expected a greater or faster response. The plants without the amyloplasts, after a relatively long lag period (about three times as long as the controls) also develop a quite satisfactory geotropic curvature. The longer latent time is no doubt significant, but the slower curvature may not be significant. These coleoptiles are growing more slowly than the controls; they have been 34 hours at 30°C, which is a very high temperature for them, and they have been incubated in an unusual solution. The table shows the curvature compared to the growth rate. The controls started out averaging −12° and ended up +48°, showing an increase of 60°. There is an increase of 48° in water and an increase of 45°—essentially the same—after the treatment. When we measure curvature per millimeter of growth, there is no real difference.

We must therefore conclude, perhaps reluctantly, that amyloplasts are not necessary for geotropic curvature. This does not

* Periodic acid Schiff (PAS) is a staining method for carbohydrates.

mean that if they are there they will not cause geotropic curvature; they certainly do roll down to the side, establishing an asymmetry. Gordon believes they are constantly hydrolyzing and thus supplying sugar to the cell wall as a source of energy. But in their absence there must be something else, which after a longer presentation time may also fall. This leads to the rather drastic conclusion that the *nature* of the falling objects—their chemical nature—is not important. There is something to add to that too: Dr. Sievers in Halle, East Germany, studied the geotropic response of the rhizoids of *Chara*. *Chara* is a marine alga with very large cells, like *Nitella* in many ways, but marine in habitat. Its long thin rhizoids are very geotropic, only, unlike the coleoptile, they curve downward. In these long single cells there is a single object that looks like a crystal, and as it turns under the microscope one can catch a gleam from its crystalline surface. This object is certainly not starch, yet it behaves very much like a statolith; it falls to the lower side in good time before geotropic curvature occurs, and in those rhizoids which show no curvature, i.e., no geotropic sensitivity, no crystal can be seen. Thus everything agrees with the expectation that this object is in fact a statolith. If a crystal can be a statolith, then anything else might be a statolith, depending on the cell, and the special qualities of starch can be dispensed with.

The long thin sporangiophores of the fungus *Phycomyces* also show sensitivity to gravity; it is rather weak, with about 20 minutes presentation time, but they do curve, and of course they are too small to contain anything resembling the amyloplasts of green plants. They are only a quarter of a millimeter in diameter, and under the electron microscope they do not show any heavy bodies at all, though they do contain mitochondria. So whatever is the georeceptor there it is certainly not starch. Indeed, once we open the door to the idea that there may be many different kinds of georeceptors, there is no end to it, and one begins to think of bracket fungi, for instance, which are highly geotropically sensitive but certainly contain no starch bodies. The conclusion for shoots, then, is that probably where amyloplasts are present they act as georeceptors, but where they are absent some organelle which is smaller or less dense has the same action—slower, yet bringing about a functional geotropic curvature. It is a not uncommon principle that Nature puts in secondary systems to back up possible failure of primary systems.

C. THE DETECTION OF GRAVITY BY ROOTS

There are amyloplasts in the root cap of every root, as we saw in the root cap sections above, and ever since the work of Nemeč in 1900 the idea has been accepted that these starch grains fall under the influence of gravity. Correspondingly if roots are decapitated carefully, without affecting the elongating zone, as described in part 1 of this chapter, it is possible to remove almost completely the geotropic sensitivity and to have horizontal roots continuing to grow at a fair rate without curving downwards. A few years ago Dr. Juniper in England discovered the variety of maize referred to in part 1, from which one can detach the root cap without cutting; a slightly twisting pull, and off it comes, yielding a capless, ageotropic root. There are no amyloplasts in the apex cells nor in those of the elongation zone farther back, for they are all in the root cap. Thus the geotropic sensitivity is wholly located in the place where the amyloplasts are. Since, as we have seen, auxin is formed in the root tip, perhaps in the cap cells, it can be excreted into agar blocks and it can become asymmetrically distributed under the influence of gravity. AbA is also present in the root cap and it also has the effect of inhibiting root growth. So the hormone source and the statoliths are close to one another, and it is natural to deduce that the lateral redistribution of hormones across the root is somehow facilitated by the lateral displacement of the amyloplasts. There is one very strange experiment, however, done by Pilet of Lausanne and, in slightly different form, by Konings of Utrecht: he placed a block of agar containing ^{14}C-IAA on the *side* of a horizontal root, and before any curvature could occur he divided the root into upper and lower halves. He found that, even though the auxin had been placed well behind the tip, nevertheless there was twice as much radioactivity in the lower half as in the upper half. In other words, asymmetrical distribution can take place from a supply of auxin far behind the tip. We have seen in part 1 that there is a tendency for auxin to move in both directions in the root, namely towards the tip for most of the length of the root, and away from the tip for about 2 or 3 mm; so some of the labeled auxin applied further back will certainly migrate towards the tip. It is difficult to explain Pilet's experiment except by assuming that the auxin molecules come down the root and when they reach the tip they go on into the root cap; then they become asymmetrically distributed *in the root cap,* and subsequently find their way back. I see no other reasonable explanation, since there is, as far as we know, nothing in the basal part of the root which has the ability to produce an asymmetric distribution of auxin. However, there is one other possibility: namely, that the root cap might secrete a substance which is neither an auxin nor AbA but which modifies the movement of these hormones—an "asymmetrizing" substance, which moves into the root proper from the cap, and which modifies the hormone transport. This asymmetrizing substance, of course, would be activated by the statoliths in the root cap, since they obviously control the geo-response in it. The asymmetrizing substance need not be a hormone in the true sense.

There is one set of experiments suggesting that also in the root amyloplasts are not essential; these have often been quoted but unfortunately have never been reported in detail. Professor Went, Senior, reported at a meeting of the Dutch Academy in 1909, but without detailed data, that Miss Rutten-Pekelharing had found that if one treats the roots of certain seedlings with potassium alum, they lose their starch. Some of the roots become abnormal, but a great many grow normally enough to curve geotropically and these all give clear geotropic curvatures. This experiment has never been repeated, to my knowledge, and the matter remains open. And here we must leave the problem of the georeceptors, in order to go on to a number of very puzzling phenomena related to the geotropic response. Several of these were done at the Botanical Laboratory in Munich, where the late Professor Brauner worked for more years than I can remember.

D. EVIDENCE FOR MORE COMPLEX CHANGES

The first we must note is the experiment of Brauner and Hager at Munich with sunflower seedlings, *Helianthus annuus.* They put these seedlings, really just hypocotyls, into low temperature, around 4°C, and turned them horizontally. At this temperature they showed no geotropic curvature, largely because they were scarcely growing. However, they now brought these plants, which had been horizontal in the cold, back into the vertical position; after a little while the plants showed geotropic curvature. The curvature was in the direction which would be expected from their having been horizontal. Thus there was a very clear separation in time between the exposure and the response. A remarkable thing is that the plants can be kept at 4°C for up to five hours; during this time, in the absence of any geotropic curvature, they must be detecting the gravity, since when brought back to 20°C they start to curve. The curvature is proportional to the length of time they have been in the horizontal position; Brauner believes it is a logarithmic relationship. Here are some of the data:

Time horizontal at 4°C					
0.5	1	2	3	5	hours
Subsequent curvature at 20°C					
10	18	27	32	37	degrees

Evidently the georeceptor can act at 4° and the "memory" can be stored until the plants are in a condition to respond.

Brauner has also done somewhat similar experiments with plants not chilled but simply decapitated, and then later treated with auxin. The procedure is: (1) decapitate, (2) expose horizontally for a while (during which time no curvature develops), (3) return to the vertical position and apply auxin, then (4) curvature results. Again it is evident that the georeceptor can act without auxin, and the "memory" can be stored until auxin makes the curvature possible.

Professor and Mrs. Brauner found another interesting fact. They carefully measured what they called Saugkraft, which we call suction force, water potential or water entering tendency, and they found that plants placed horizontally develop an increase in this property, as though the mere act of placing them horizontally increases the osmotic content, or decreases the resistance of the cell walls. The effect is detectable after the plant has been horizontal for 2½ minutes, and it increases to a maximum, about a 60% increase, in half an hour. Also there was a slight difference between the values for the upper and lower halves of the horizontal seedling, the lower being always a little larger. In one case they mentioned an increase of 57% in the upper half and of 63% in the lower half. And a very odd fact is that when they poisoned the plant with dinitrophenol, which prevents the cells from having access to active phosphate, or with one of the other metabolic poisons such as azide, the effect was not changed, but the differential between upper and lower halves was greater.

Dinitrophenol and azide both inhibit growth, but growth is not involved here. Thus the metabolic system apparently tends to equalize the slight difference and if that system is poisoned the full difference remains and can be observed. A differential of 3% to 6% between upper and lower halves in the controls became 22% in dinitrophenol, or 32% in sodium azide. That the asymmetry in water potential could be due to asymmetric distribution of auxin is most unlikely, because it is still present 4 days after decapitation when both auxin content and growth rate have fallen almost to zero. Even seedlings that have been decapitated for two days have virtually lost their geotropic sensitivity. The cause of the overall increase in water potential remains unknown, and as to the differential between the upper and lower halves, if it cannot be the result of an auxin differential it might even be thought of as a cause of geotropic curvature. In other words, the georeceptor may bring about some *metabolic* difference between the two halves, and the auxin differential would be a *secondary* result of that primary change in metabolism.

That the auxin differential may result from a metabolic change is perhaps supported by other differences between upper and lower halves of horizontal organs. Back in 1934 Metzner reported that the lower side of horizontal stems and hypocotyls contained more potassium and hydrogen ions and more reducing sugar than the upper side. Bode much later confirmed the potassium asymmetry. Very recently (after these lectures) Goswami and Audus, in a more extensive study (1976) have shown that when *Helianthus* hypocotyls are placed horizontally, potassium and phosphate quickly reach higher concentrations on the lower than on the upper side, while calcium shows the reverse movement. What is more, these asymmetries begin before there is appreciable geotropic curvature. The results after 3 hours in the horizontal position are shown in Table 5-4. Corn coleoptiles gave very similar differentials.

Fabian, in Bucharest, has noted a comparable differential in the sulfhydryl content—both of free -SH compounds and as sulfhydryl proteins, on the two sides of geotropically curving seedlings. The lower side developed a higher content of free -SH groups than the upper (table 5-4, second part). This difference, however, developed only after the plants had begun to curve, and therefore may not be directly connected with the georeceptor system; it may equally well be connected with the difference in growth. Note that the values for the vertical controls lie between the other two, as indeed the growth rates doubtless do also.

My guess is that with the increasing sophistication of microchemical methods, other such differences will come to light. It will be a real challenge somehow to put all these into a picture. Unfortunately, few people, in trying to set up a model or concept of the whole geotropic phenomenon, are prepared to bring all these factors together, most workers preferring to use one set of data that seems interesting and compelling to themselves and quietly preferring to forget all the other facts which may argue in the opposite direction (I have done this myself on occasion).

We might just mention one other curious experiment, also by Brauner. He wanted to compare (a) the amount of curvature which results when a plant is placed horizontally, with (b) the amount of curvature one can make that plant undergo by giving it the same gradient of auxin as the one gravity would have brought about. He placed the seedling horizontally, measured the resulting auxin gradient between the upper and lower sides, and subsequently returned the plants to the vertical position to allow the geotropic curvature to develop. Then, to compare with that, he set up other seedlings vertically and fed one side with one concentration of auxin and the other side with another concentration; the resulting curvature was measured and compared with the curvature resulting from the effect of gravity. He always found that, in a given length of time, the gradient of auxin produced by gravity gave a much greater curvature than the same auxin gradient supplied artificially. For instance, in one experiment he measured an auxin gradient between upper and lower sides in two hours of 2 : 1. The geotropic curvature resulting in two hours was 74°. But an auxin gradient of 2 : 1 induced by differential auxin feeding only produced a curvature of 20° in two hours. Thus in addition to the asymmetric auxin distribution there is a change in the *sensitivity* to auxin brought about by placing the plant horizontally. This result would fit in with the idea proposed above for roots, of a synergism between auxin and AbA; but it would also fit with the increase in suction force in the plant placed horizontally, because then the growth of the cells would be more responsive to the gradient of auxin distribution.

Many of these facts are complications which do not necessarily affect the main problem, and some may even be aftereffects of differential growth and of auxin gradients and not themselves causes. One nevertheless is left with the strong impression that there is a longer or more complex

Table 5-4 Differentials other than auxin between upper and lower halves of horizontally placed seedling shoots

A. Distribution of inorganic ions, as percent of the total, in *Helianthus* hypocotyls, and the corresponding geotropic curvatures, 3 hours after placing horizontally

Ion	Upper half	Lower half	Geotropic curvature
Calcium	64	36	70°
Potassium	35	65	65°
Phosphate	43	57	67°

From Goswami and Audus, 1976.

B. Distribution of free sulfhydryl groups in two seedling shoots, 24 hours after placing horizontally

	Moles of −SH per gram fresh weight			Ratio
	Upper (concave) half	Lower (convex) half	Vertical control	Upper : Lower
Vicia faba	1.05	2.0	1.33	34 : 66
Lupinus albus	1.31	2.05	1.71	39 : 61

From Fabian, 1971.

chain of events in georeception than we once thought. Specifically it seems that the gravitational fall of the georeceptors, whether amyloplasts or other bodies, initiates a metabolic gradient between the upper and lower halves of each cell, which gradually translates into a metabolic gradient between the upper and lower halves of the whole organ; and it is this which so modifies the auxin transport that more of the auxin (or of the AbA) flows through the lower lateral walls of the cells and thus accumulates in the lower half of the organ. Perhaps one of you readers will one day clarify these problems.

REFERENCES

Abrol, B. K., and Audus, L. J. The lateral transport of 2,4-D in horizontal hypocotyl segments of *Helianthus annuus*. *J. Exptl. Bot.* 24, 1209-1223, 1973.

Audus, L. J. The mechanism of the perception of gravity by plants. *Symp. Soc. Exptl. Biol.* 16, 197-210, 1962.

———. Geotropism in roots. In *The development and function of roots,* ed. J. Torrey and L. Clarkson. London, Academic Press, 1975.

Bismarck, R. von. Ueber den Geotropismus der Sphagnen. *Flora* 148, 23-83, 1959.

Brauner, L., and Diemer, R. Ueber den Einfluss der geotropischen Induktion auf den Wuchsstoffgehalt, die Wuchsstoffverteilung und die Wuchsstoffempfindlichkeit von *Helianthus*-Hypokotylen. *Planta* 99, 337-353, 1971.

———, and Hager, A. Ueber die geotropischen "Mneme." *Naturwiss.* 15, 429-430, 1957.

Chadwick, A. V., and Burg, S. P. An explanation of the inhibition of root growth caused by IAA. *Plant Physiol.* 42, 415-420, 1967.

———. Regulation of root growth by auxin and ethylene interaction. *Plant Physiol.* 45, 192-200, 1970.

El-Antably, H. M. M. Redistribution of endogenous IAA, AbA and gibberellins in geotropically stimulated *Ribes nigrum* roots. *Zeit. Pflanzenphysiol.* 75, 17-24, 1975.

Fabian, A. (Changes in sulfhydryl content during geotropic curvature of seedlings of *Lupinus* and *Vicia.*) *Studia Univ. Babes-Bolyai,* Ser. Biol. 71-78, 1971.

Gillespie, B., and Thimann, K. V. Transport and distribution of auxin during tropistic response. 1. The lateral migration of auxin in geotropism. *Plant Physiol.* 38, 214-225, 1963.

Goldsmith, M. H. M., and Wilkins, M. B. Movement of auxin in coleoptiles of *Zea mays* L. during geotropic stimulation. *Plant Physiol.* 39, 151-162, 1964.

Goswami, K. K. A., and Audus, L. J. Distribution of Ca, K and P in *Helianthus annuus* hypocotyls and *Zea mays* coleoptiles in relation to tropic stimuli and curvatures. *Ann. Bot. N. S.* 40, 49-64, 1976.

Griffiths, H. J., and Audus, L. J. Organelle distribution in the statocyte cells of the root tip of *Vicia faba* in relation to geotropic stimulation. *New Phytol.* 63, 319-333, 1964.

Hawker, L. E. Experiments on the perception of gravity by roots. *New Phytol.* 31, 321-328, 1932.

Konings, H. On the mechanism of the transverse distribution of auxin in geotropically exposed pea roots. *Acta Botan. Néerl.* 16, 161–176, 1967.

Lyon, C. J. Auxin transport in leaf epinasty. *Plant Physiol.* 38, 567–574, 1963.

Ohwaki, Y., Tsurumi, S., and Nagao, M. Auxin transport in Vicia roots. In *Plant Hormones* 1973. Tokyo, Hirokawa Pub. Co., 1974; pp. 1071–1078.

Pilet, P. E. Abscisic acid as a root growth inhibitor: physiological analyses. *Planta* 122, 299–302, 1975.

———. Géoperception et géoreaction racinaires. *Physiol. Vég.* 10, 347–367, 1972.

———. Root cap and root growth. *Planta* 106, 169–171, 1972.

———, and Lance, A. Structure histologique et catabolisme auxinique des méristèmes radiculaires de *Lens, Pisum* et *Zea*. *Bull. Soc. Bot. Suisse* 77, 156–172, 1967.

———, and Nougarède, A. Root cell georeaction and growth inhibition. *Plant Sci. Letters* 3, 331–334, 1974.

Sorokin, H., and Thimann, K. V. The Plastids of the *Avena* coleoptile. *Nature* 187, 1038–1039, 1960.

Thimann, K. V. Studies on the movement of auxin in tissues and its modifications by gravity and light. In *Régulateurs naturels de la croissance végétal,* pp. 575–585. Paris: C. N. R. S., 1964.

Wilkins, M. B. Red light and the geotropic response of the *Avena* coleoptile. *Plant Physiol.* 40, 24–34, 1965.

Yeomans, R., and Audus, L. J. Auxin transport in roots. *Vicia faba*. *Nature* 204, 559–562, 1964.

Phototropism 6

THE ROLE OF AUXIN

A. GENERAL AND HISTORICAL

Phototropism, like geotropism, is at its best in seedlings, because there it is a matter of life and death. If seedlings fail to reach the light in a very short time, they are finished. For this reason the curvatures towards the light on the part of very small plants are nearly always much more marked, and much easier to study, than those in mature plants. But of course mature plants also curve towards the light, as everyone knows who puts a potted plant in the window. They usually curve more slowly than seedlings, taking several days where seedlings take hours, but the phenomenon is basically the same.

Roots tend to curve away from the light, but this is seldom a very well-marked phenomenon. One sees it more in roots of *Cruciferae* than in other plants; these roots are often negatively phototropic. (The usual terminology is maintained: that is, towards the stimulus, *positive,* away from the stimulus, *negative.*) But the negative phototropism of roots has been little studied because it is weak, slow, and limited to certain plants.

Thus the only conclusive research has been with the phototropism of shoots. Going back in history, we recall that it was Cholodny in 1927 who said that all tropisms must be due to the asymmetric distribution of growth substance, because they depend on asymmetric growth. Then Went, the very next year, was able to show, as we have seen, that if he placed a coleoptile tip on two blocks of agar, separated by a razor blade, and then illuminated from one side, more auxin diffused out on the shaded side than on the lighted side. Went's procedure was really more qualitative than quantitative; he was concerned to show that qualitatively there is a difference, and a fairly marked difference. The curvatures of test plants were about twice as great with the blocks from the shaded side, and this ratio is commonly found. There are a great many variations due to different times of illumination, different times on the agar, and different species of plants.

B. THE TWO TYPES OF PHOTOTROPIC RESPONSE

An unfortunate, though interesting feature of phototropism is that there are two different phototropic reactions which complicate our understanding of the process. Figure

6-1 shows these two kinds of response. *Avena* coleoptiles have been illuminated for a brief time and then photographed every half hour without being moved. Those at the top left have been given a very brief illumination, 500 ergs per square centimeter, of white light. After the first half hour, curvature begins and is restricted to the extreme tip. Then in subsequent half hours, not only does the angle of curvature increase but the region curving increases; in other words, curvature spreads down the plant. But about half of the plant is not involved in the curvature, the basal part remaining straight throughout. The plants at right have been given a long exposure, 10 minutes of light of the same intensity as the other. Even in the very first picture, half an hour after the zero point, curvature has begun over a long zone, at least to halfway down. In subsequent periods the curvature increases, and by the end of the 2 hours, the curvature has extended almost to the base. The early workers referred to these as tip and base type curvatures, but those terms are misleading. The tip curvature starts in the tip and spreads downward. The so-called base curvature does not start only at the base, but it starts all the way down, tip and base included. So we have preferred to call them first and second positive curvatures. The plants in the lower row are included to show how the first- and second-type curvatures compare when they show equal angles. The left-hand pair of plants has been given ultraviolet light. It is of some interest that the ultraviolet light type curvature apparently does not start in the tip but is of the second positive type which extends all the way down.

Long ago, in dealing with photochemical systems, Bunsen and Roscoe formulated the so-called reciprocity law (also called the Bunsen-Roscoe law). What it says is that light is only effective to the extent that it is absorbed. For a given percentage of absorption, the effect exerted by light depends on the *total amount* of light (irradiance). So the Bunsen-Roscoe law essentially states that the photochemical reaction (whatever it is) will be dependent on the product of the intensity of the light and the duration of exposure—the two multiplied together, $I \times t$, giving the total irradiance, or the total amount of energy received. Many early workers set about to see how phototropism behaved in regard to the Bunsen-Roscoe law and found that, at least for low light intensities, the response, as measured by the amount of curvature after a fixed time, is indeed proportional to the product of intensity times time.

Figure 6-2 shows how oat coleoptiles respond to a logarithmic series of exposures, measured in total amounts of energy (ergs per square centimeter). This energy was obtained partly by varying the time, so that the product $I \times t$ is what is plotted. Although two different wavelengths are shown, consider only the black points first, referring to exposures to blue light, 436 nm. The very lowest energy which gives a measurable curvature is around 4 ergs per square centimeter and it gives around 5°. From there up, the curvature is proportional to the log of the irradiance, that is, $\log I \times t$, till it reaches a maximum (in these experiments about 27°), whose actual value depends on the conditions and the length of time allowed for curvature in the

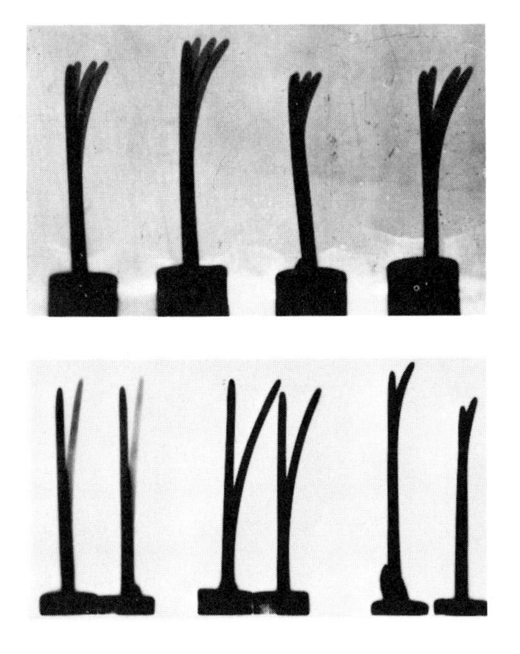

6-1 Above, the two types of phototropic response. In the pairs of plants below, the right-hand and central pairs (second and first positive curvatures, respectively) show the same angle (ca 23°).

dark. Thereafter as we increase the light energy, the curvature decreases. This would not happen in a simple photochemical system. It goes down all the way to zero and even goes below the line; in other words, curvature becomes negative (i.e., away from the light). Generally these negative curvatures are small, around 5°, but under some conditions they can reach up to almost 15°. They never become as great as curvatures *towards* the light. Finally, as we further increase the light dosage, we begin to see that type of curvature (cf. previous figure) which begins all over the whole plant. The response may or may not be log-linear because of the practical limitations of the amount of light that can be administered.

Thus we have three kinds of response to deal with: the regular positive phototropism or the "first positive"; curvatures away from the light or "negative" curvatures; and finally the "second positive," which occurs only at very high light dosages. Let us discuss the first positive curvature.

C. AUXIN REDISTRIBUTION IN THE FIRST POSITIVE CURVATURE

Went's type of experiment on the first positive was done in 1928, was repeated by others, and was tried out with the dicotyledonous seedlings *Helianthus* and

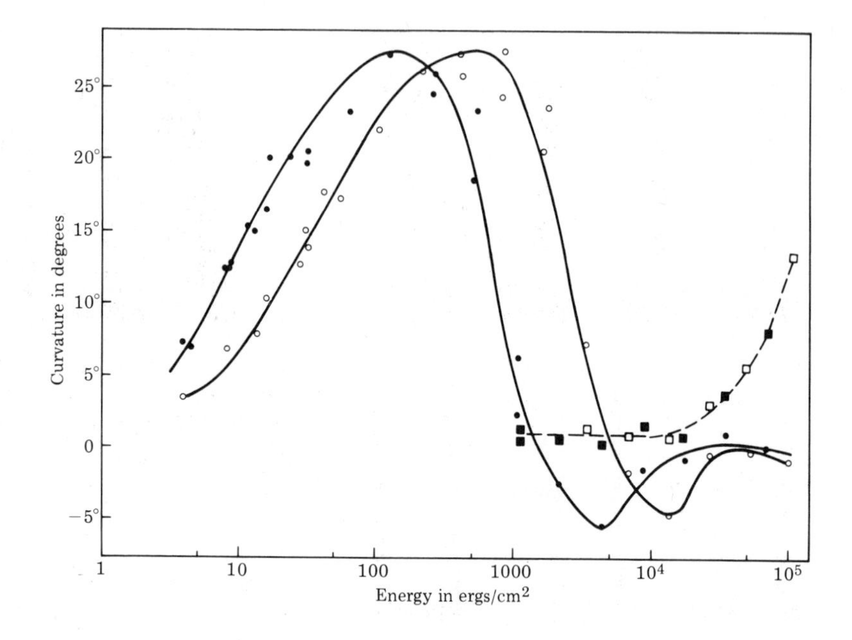

6-2 Responses of the *Avena* coleoptile to light dosage (log I x t) in ergs/cm². Ordinates, curvature after 90 minutes in the dark. Filled circles, 436 nm, tip reaction; filled squares, 436 nm, base reaction. Open circles, 365 nm, tip reaction; open squares, 365 nm, base reaction. From Thimann and Curry, 1960.

Raphanus. Because more auxin was found on the shaded side, and because we know growth in these organs is proportional to the auxin applied, therefore there must be greater growth on the shaded side and hence curvature towards the light. For a long time, therefore, the Cholodny-Went theory of tropisms was accepted. Only in the 1950s did it begin to be doubted. It was doubted partly because people did not care to do these rather cumbersome experiments, but doubted more for two other reasons. Firstly, it was thought that the light may not simply be causing a diversion of auxin to the shaded side, but it may instead be destroying the auxin on the front side, photochemically. This mechanism is much easier to visualize than the idea that light somehow causes a *lateral* movement. Secondly, radioactive indoleacetic acid was brought into the picture in the 1950s. This was done either by completing the indole ring in the synthesis with a C^{14}, or by making the acid group through the labeled nitrile, to give carboxyl labelling (cf. chap. 5). It is also possible to make it through another route to give labelling in the methylene group. The initial experiments in the 1950s, especially in Germany by Reisener and others, seemed to show that when one took off the extreme tip of the plant and replaced it with radioactive auxin, no asymmetrical distribution resulted. However, the problems that are involved in this kind of experiment have been discussed earlier. They are: (1) if the breakdown products of IAA are also radioactive, then the measurement of C^{14} does not necessarily measure the auxin itself; (2) the amount of auxin normally in the plant is very small, and if it is overloaded with artificial sources the plant may not be able to deal with it; and (3) especially with phototropic experiments there was an additional complication, namely, that in order to introduce the ^{14}C-IAA into the plant the tip was cut off; but we know that if the extreme tip is removed there is at low irradiances no phototropic reaction. If one tries to get around this by applying the block externally, little auxin enters because of the cuticle; so rather high specific activities would be required, and these were not at first available. There is another complication too, in that during the experiment, since it takes some time to get redistribution, the plants grow somewhat and thus force their way into the agar. As a result the agar is then being applied to both flanks instead of just to the tip; and so a sort of stream of radioactivity goes down both flanks. If there were any asymmetry, that would obscure it still further. (I believe no one thought about this, nor did Barbara Pickard and I till we came to do the experiment.) So all in all there are a number of pitfalls and it is difficult to do the experiment correctly. We decided that the only way to do it was the hard way, using intact tips and rather high specific activity, which fortunately was obtainable in later years. Bruce Stowe succeeded in making the indoleacetic acid with the labelling limited to the carboxyl group, so that when the substance is oxidized the $C^{14}O_2$ comes off, and no radioactivity is left in the tissue to confuse the result.

With all those precautions the work was repeated, and the results were entirely different from those of the German workers. In table 6-1 are shown exposures of 1000 meter-candle-seconds, which is equivalent to 3 or 4 hundred ergs per square centimeter, and gives a curvature well towards the top of the curve. This 1000 meter-candle-seconds is reached either as 100 meter-candles for 10 seconds or 50 meter-candles for 20 seconds; it makes no difference because we are dealing with only the product $I \times t$. We see that the radioactivity in the receiver on the shaded side is always larger than that on the lighted side by a factor of 2 or more. The experiments are arranged in order of the amount of IAA applied, because one must be careful not to overload, and we know roughly how much IAA is normally transported in the plant tissue. For the data where the amounts of auxin are low, we find an average of only 24% or 25% on the lighted side. With larger amounts of auxin one begins to see the overloading effect coming in. Thus, as we increase the auxin loading the asymmetry can be decreased, but it is also evident that the asymmetry is there and is even larger than the biological experiments had indicated.

The first reaction to the publication of these data was the supposition that on the lighted side transport is being inhibited; normal transport continues on the dark side, but in the plant the amounts are *not necessarily* different. However, that is answered by another procedure in which the tips themselves are halved. It is not easy to halve these tips in such a way that the weights of the two halves are equal, since the coleoptile is only 1.1 mm thick, and we have to work in very weak red light. But the results are essentially the same (table 6-1), the radioactivity in the lighted half being about 35%, and that in the dark half about 65%, of the total. It may be significant that the ratio in the tissue is not as great as the ratio of the radioactivity coming out into the blocks, no doubt because a certain amount of auxin becomes fixed on the way down. We know that there is auxin fixation; it apparently becomes attached to a protein and this is not the biologically active fixation, it is a secondary fixation that no doubt obscures the results.

We can conclude that phototropism is very like geotropism in that it is due to a lateral movement of auxin in the tissue, more auxin being diverted to the dark side. The fact that the original proposal of 1928 could be corroborated by an entirely different method adds great strength to the conclusion. So much, then, for the first positive curvature.

D. AUXIN REDISTRIBUTION IN OTHER CURVATURES

Now for some of the other phenomena. The negative curvature is very small; it is fairly elusive and not always obtained with different species. It has peculiar kinetics which are quite different from that of the positive curvature and will be discussed below. The best data on auxin distribution were obtained in 1939, by a Ms. Wilden. It was obviously her thesis but was not published in full; she published only a two-page paper. She first used different light exposures, following a curve like the one we saw above, to determine what light exposure it takes to get negative curvature, then varied the conditions a little to get large negative curvatures, about 10°, and then proceeded as follows: she illuminated the plants and then cut off the tips and applied the tips asymmetrically to fresh test plants,

Table 6-1 Summary of the asymmetric distribution of ^{14}C-IAA in *Zea mays* coleoptiles caused by unilateral light

No.	Light dosage	Plant Material		Amount of IAA transported × normal	No. of expts	Asymmetry found	
		Illuminated	Analyzed			Lighted side	Shaded side
1	First positive	Tips	Agar receivers	*ca* 0.1 ×	4	24	76
2	First positive	Tips	Agar receivers	0.15–0.4 ×	4	25	75
3	First positive	Tips	Agar receivers	0.5–1.2 ×	5	34	66
4	First positive	Tips	Tissue halves	0.1–0.25 ×	8	35	65
5	Second positive	Tips	Agar receivers	*ca* 0.5 ×	20	34	66
6	Second positive	Subapical section	Agar receivers	1.4 ×	5	46.1	53.9
7	Second positive	Subapical section	Agar receivers	0.5 ×	7	46.9	53.1
8	Second positive	Subapical section	Agar receivers	0.1 ×	6	46.0	54.0
9	Second positive	Subapical section	Agar receivers	0.02 ×	12	41.8	58.2

Data of Pickard and Thimann, 1964.

in such a way that in one case the lighted side was in contact with the test plant, while in the other case it was the dark side that made the contact (fig. 6-3). When the irradiance was in the range of the first or second positive curvature the distribution of diffusible auxin, as determined by the different curvatures of the test plants, was close to what was expected, namely, showing ratios dark to light of 4.9 : 1 and 1.8 : 1, respectively. But with an irradiance causing relatively large negative curvatures the mean ratio was 1 : 1.6—which was in the right direction for a negative curvature and also of about the right magnitude (compared to the other two values) for a curvature of 10°. That the polarity of lateral auxin transport should thus be reversed at an intermediate value of the light dosage is hard to understand, and indeed so far no one has succeeded in giving a satisfactory explanation of it. Brauner has tried injecting india ink into the hollow space in the coleoptile, to increase the light gradient between the two sides, but the results were inconclusive.

The second positive curvature has also been relatively little studied till recently. Contrary to expectation, curvature in this range is not proportional to I × t. For a fixed light intensity, I, the curvature is indeed proportional to t, but for fixed time, t, the curvatures show only minor variation with I; they pass through a maximum at intermediate I values, as shown in figure 6-12. Thus they do change with changing I but they are not at all proportional to it. Eike Libbert and his co-workers at Rostock have shown that a curvature of second positive type is given by many dicot seedlings

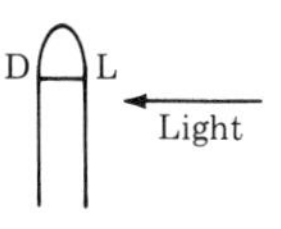

Curvature after 4 hours

First positive,	+ 50°
Negative,	- 15°
Second positive,	+ 20°

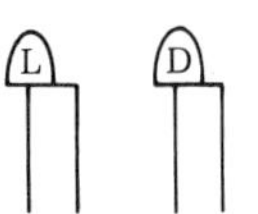

Auxin distribution after 130 minutes

Light side, L		Dark side, D
17	:	83
62	:	38
36	:	64

6-3 Asymmetry of auxin distribution in positive and negative phototropism. Data of Wilden, 1939.

as well as monocots. We decided to repeat the above procedure with ^{14}C-IAA, using second positive curvatures, although the curvature depends not only on duration of exposure and wavelength, but also on plant species and on the conditions supplied, and although the light requirement seems to be different in different plants. Using individual wavelengths like 436 nm, which is very effective, it is possible to get reliable second positive curvatures on *Zea,* and the results are included in table 6-1. We found that 3 hours at 500 meter-candle-seconds, a very long exposure, brings the plants well into the second positive range. As we found later, good second positive curvatures can be obtained at 1/6 the duration or 1/10 the light intensity, but we had a special reason for wanting to use very high irradiances. The average counts per minute come out about 34% and 66%. With lower counts, the ratios still come out the same. So, curiously enough, and contrary to what many workers were thinking at that time, the second positive curvature depends upon the same kind of auxin asymmetry as does the first positive.

In the second positive system curvature occurs all the way down the plant, and that gave rise to the thought that perhaps the light causes auxin asymmetry not only in the tip but all the way down the coleoptile. We saw that in Dolk's experiments on geotropism he found that subapical sections could also redistribute applied auxin, in fact almost as well as the tip. In contrast, however, subapical sections, either of coleoptiles or of dicotyledonous plants, produce only very slight auxin asymmetry under the influence of light. Table 6-1 shows a small asymmetry with *Zea,* when exposed to light of very high intensities or for long times, like those in the second positive range. These data are for subapical sections with a 2-mm tip removed and are the mean values of a large number of experiments. The ratio of 46% on the lighted side to 54% on the shaded side is real and significant. The old experiments of Charles and Francis Darwin showed that when the seedlings had their tips covered with a plastic cap no curvature resulted, although they were exposed to light; however, they did not use irradiances as high as this. This degree of asymmetry would have given rise to a curvature of around 10°, which they might not have detected either. In general a distribution of around 46 : 54 is about what is found in dicotyledons. *Helianthus* seedlings in a few experiments with radioactivity showed an auxin distribution ratio of 45 : 55. With both light and gravity it can be shown that if we take the diffusate only from the extreme edges, the gradients are somewhat larger, i.e., the transverse gradient probably varies continuously across the whole solid dicotyledonous tissue. Since there have been few studies of auxin distribution in dicotyledonous plants, conclusions are none too certain.

E. AUXIN DESTRUCTION IN PHOTOTROPISM

Now we come to the role of photo-oxidation. In 1949 Galston, at Yale, showed that indoleacetic acid is destroyed in light in the presence of riboflavin; oxygen is absorbed, and CO_2 is given off. The products were not identified. The light intensities needed were rather high, but the auxin was clearly destroyed. This was the main reason why phototropism was ascribed to photo-oxidation, especially in the 1960s. Indeed, a review by Reinert of phototropism and other reactions in the *Annual Review of Plant Physiology* simply assumes that the Cholodny-Went theory is disproved in the case of phototropism, and that the curvature is only due to photo-oxidation. This, of course, was supported by the facts that (a) there is riboflavin in practically all plant tissues and (b) that Reisener had found no C^{14} asymmetry. Nevertheless, Wilden's data, published earlier, had made this particularly unlikely because she had to use long exposures in order to get negative curvature and yet found more auxin on the *lighted* side. Several other studies pointed strongly away from a role of photo-oxidation in phototropism.

Firstly, Dr. Briggs, at Stanford, using the biological method, compared the total

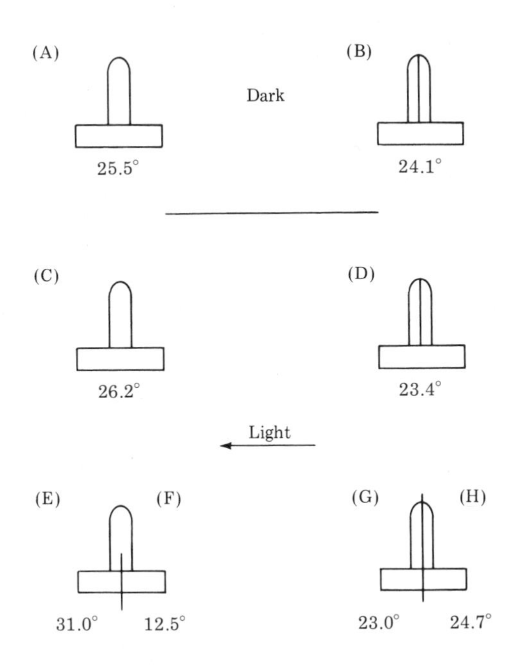

6-4 Transport ("diffusion") of auxin out of corn coleoptile tips into agar in 3 hours, measured as curvature on standard test coleoptiles. (A) 3 intact tips, dark; (B) 3 totally split tips, dark; (C) 3 intact tips, light; (D) 3 totally split tips, light; (E) and (F) 6 tips split at base, light; (G) and (H) 6 tips totally split, light. Light from the right side. From Briggs et al. 1957.

auxin on the two sides of a lighted tip with the sum of the auxin on the two sides of an undivided tip in the dark. His experiment is shown in figure 6-4. Coleoptiles were placed on agar and the agar blocks then assayed on test plants. From tips in the dark the average curvature was 25.5°. When the tips were slit lengthwise, as a control for the light experiment, the curvature was only slightly lowered, so the operation of slitting has little effect on the auxin yield. From coleoptiles in the light, irradiated all around, he found the same curvature as in the dark. When the tips were halved in the light, there was again a small decrease, suggesting that the wound effect is real, though small. However, when he carried out the Went type of experiment, receiving the auxin from the lighted and shaded sides, the asymmetry was strong (E and F), yet the average of the two sides was 22°, comparable with the above. Then, lastly, he pushed the dividing razor blade all the way up so that the tip was divided completely into two. This was an old experiment of Boysen Jensen and Nielsen, back in 1910, but repeated now quantitatively. This procedure prevents the asymmetry, and the value again on each side was the same as in the controls. Indeed the shaded side was in fact much darker than the control because there was a metal razor blade in place, and the lighted side was correspondingly brighter by reflection. Yet these exposures made no appreciable difference to the amounts of auxin determined biologically. It seemed to me that that experiment was clear-cut.

Secondly, we made experiments with C^{14} auxin, using exposures to light strong enough for the second positive curvature, with the results shown in table 6-2. Here we added up the activity left in the donor block at the end of the three-hour exposure, the activity that remained in the tissue, and the activity found in the receiver block. The sum of these three shows essentially complete recovery of the radioactivity we started with. The results were the same in the dark and in the light. A repeat of the experiment with 2½ times as much auxin gave the same result, 100% recovery. So even with 500 meter-candles for three hours, which is a very large exposure, there is no decrease in the amount of C^{14} and therefore in the amount of indoleacetic acid, because the C^{14} was in the carboxyl group.

As a matter of fact there may be a very small amount of photodestruction in Briggs' experiments, but it is of the order of 1° on 24°, or 4%, and these were very high exposures. Thus the small irradiances that produce large positive curvatures could not possibly produce enough photodestruction to cause any appreciable curvature.

Another approach to this problem can be made independently of any measurement. If there is appreciable photodestruction or photo-oxidation, one should be able to show a relationship between the amount of light energy absorbed and the amount of auxin destroyed. To match a given phototropic curvature, we know how much auxin would have had to have been applied asymmetrically to get the same curvature, because that can be deduced from calibration experiments. So we can make a little calculation of the amount of auxin destruction needed on the lighted side, using the equation of Max Planck, $E = h\nu$. We saw that light of 436 nm wavelength gave a very good curvature. At this wavelength 1 erg is equal to 2×10^{11} quanta. From the phototropism plot the minimum curvature of around 5° requires 4 to 5 ergs per square centimeter. The coleoptile tip does not provide a square centimeter by any means, being only 1 mm across, and the tip is mostly hollow; the essential area is really about 0.2 sq mm. So if 5 ergs per square centimeter fall on 0.2 sq mm, each tip receives 0.01 ergs, or 2×10^9

Table 6.2 Effect of white light (500 meter-candles for 3 hours) on the total recovery of C^{14} from *Zea mays* sections. Averages of several series of experiments in each case.

Treatment	Counts per minute					Recovery (percent)
	Applied	Found				
		left in donor	in tissue	in receiver	total	
Illuminated, 3.7μM IAA	407	189	87	148	424	104%
Dark, same	414	174	85	147	406	98%
Illuminated, 9.3μM IAA	1 009	586	214	219	1 019	101%
Dark, same	1 004	574	205	225	1 004	100%

Data of Pickard and Thimann, 1964.

quanta. This is the amount of light energy needed to cause 5° phototropic curvature. From calibration experiments, in which we apply indoleacetic acid to decapitated coleoptiles, we know that to get 5° curvature needs 0.03 mg IAA per liter. In practice we use an agar block whose volume is 10 cubic mm, which is 10^{-5} l, so that one block contains 3×10^{-7} mg IAA actually in the agar block. Now from experiments in which agar blocks are applied, produce a curvature, and are then taken off and applied to new plants to cause a second curvature, we know that only about 15% of the applied auxin actually goes in during the 110 minutes of a curvature experiment. The curvature is proportional to the *concentration* in the block, not to the *amount* of auxin there. Hence 15% of 3×10^{-7} mg of IAA actually goes in. Now we evoke Avogadro who tells us that there are 6×10^{20} molecules in a millimole, and the molecular weight of IAA is 175. The resulting arithmetic gives us 1.5×10^{11} molecules. Thus, in order to get that curvature phototropically, the light would have to destroy 1.5×10^{11} molecules. But in order to do this, the tip received only 2×10^{9} quanta. So each quantum would have to destroy nearly a hundred molecules of IAA. We know from most photolytic or photochemical experiments that it is very rare for the quantum yield to be more than 1; it is usually less than 1. On rare occasions where there is a chain reaction values higher than 1 are known, but values like 100 are unknown. Thus we can confirm theoretically by these rather simple calculations what can be shown biologically and radiologically with the ^{14}C-IAA.

One point remains to be cleared up, namely, the real but very small photodecomposition of indoleacetic acid. It is more marked in the ultraviolet, as we shall see below, but it does occur to a minute extent in the visible. After exposures to high irradiances, collected experiments tend to show small decreases on the order of 1°–2° in a curvature of 20°–30°, suggesting that there is statistically a small auxin breakdown. To study the timing and extent of this, Thornton and I carried out a rather troublesome experiment. What we did was to bring together a selected group of plants with ^{14}C-IAA applied to the tips, and every five minutes take the tips off some of them and put them on agar blocks to measure the ^{14}C-auxin which they export. We used the system in which the tip was intact and the agar placed on top so as to make sure we were not interfering with the natural auxin production. The absolute amount of auxin applied was small. In figure 6-5 we see that the light is turned on at 35 minutes and off at 50, so that it is on for 15 minutes. Immediately the auxin yield goes down, and shortly after the light is turned off it comes up again, and tends to go slightly above the initial level. Notice the scale; the decrease is actually only 20% or 30% and is not maintained for long.

In the normal phototropic experiment we measure the auxin output for some 2 hours, so that a small drop like this, amounting to 20% or 30% for perhaps 20 minutes, would only show up as a decrease, over the whole two hours, of a few percent. This agrees with what has been detected.

Long ago Blaauw, one of the early students of phototropism, showed that when plants are illuminated there is usually a

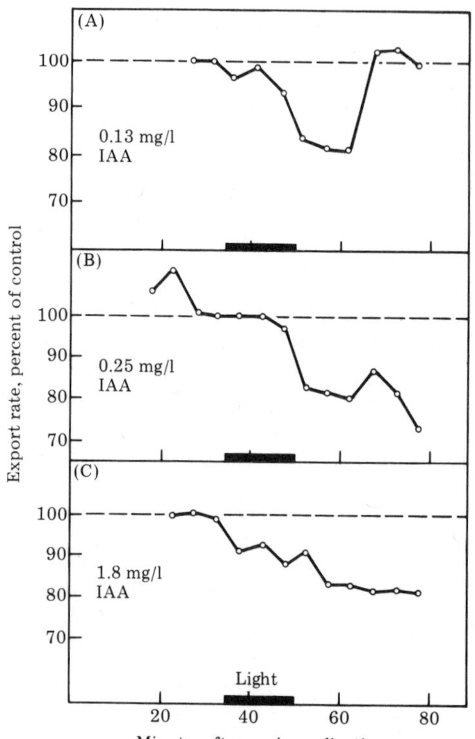

(left) **6-5** The effect of blue light (bar on abscissa) on the auxin export rate at three donor concentrations. Data as percent of the rate in dark controls. From Thornton and Thimann, 1967.

(below) **6-6** Above, the "light-growth reaction"; growth rate measured with a travelling microscope in minimal red light. Below, a similar reaction, and one with decapitated coleoptiles with IAA applied in apical agar blocks. From Thornton and Thimann, 1967.

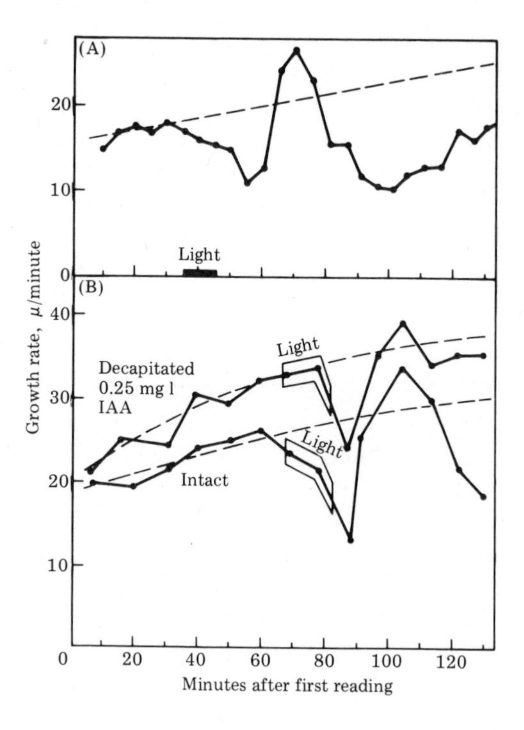

decrease in growth followed by an increase. He called this the "light growth reaction." The growth slows down, then accelerates, then slowly comes back to normal. Several other workers in Holland observed this. An example from our own data is shown in figure 6-6, which presents the growth rate in μ per minute. It goes down, then comes up above the level for a short time, then down again, then if continued long enough would slowly come back to the normal rate. That is the characteristic light growth reaction. Blaauw in his day (1914) believed that this reaction was somehow the cause of phototropism. That belief has not been supported, but the changes in auxin export rate do roughly match the changes in growth rate. In other words these plants duplicate very well the light growth reaction of the classic workers. Decapitated plants supplied with auxin show the same thing, so that it is not limited to the tip but is a direct effect of light on the rate of growth. The decrease, which occurs almost immediately after, or even during, the illumination, agrees very closely with the small decrease in auxin export shown in figure 6-5. The decrease can hardly be due to photodecomposition of the auxin, for if the plants were being deprived of the auxin we would not expect the growth rate to go above the normal. What it is, more probably, is that on irradiation, auxin, instead of being transported downwards, is transported laterally. If the molecules are reaching either the receiver block of agar or the lower part of the plant at a constant rate in the dark, then when we irradiate, some move laterally instead, and hence the rate at which they arrive is temporarily decreased. In other words what we are seeing here is the sequestering of some auxin from its normal path of transport into lateral transport, due to the action of light. That remains to be proved but it is a satisfying explanation.

In summary, the effector system in phototropism is now fairly well understood. One cannot be so optimistic about the photoreceptor system which, as we shall see, remains unsure despite a great deal of careful work.

PHOTORECEPTORS FOR PHOTOTROPISM

A. PHOTOTROPISM IN THE VISIBLE AND NEAR ULTRAVIOLET

We have discussed the effector system (the auxin system) and will now summarize what we know about the photoreceptor. First let me remind you of figure 6-2, which showed the plot of phototropic curvature against $I \times t$, the product of intensity times time. There are two wavelengths of light represented here, one at 436 nm in the blue and one at 365 nm in the very near ultraviolet. If we follow the variation in phototropic curvature with light intensity, using one wavelength, 436 nm, this curve results. For a large part of this curve, as we increase the energy we increase the curvature; this is a log-linear relationship which can be used to compare the effectiveness of one wavelength with another. With this second wavelength we also have a log-linear relationship. Thus we can determine the *action spectrum* of the process, i.e., the *absorption spectrum* of the photoreceptor, by measuring the relative effectiveness of different wavelengths. In general, phototropically sensitive higher plants do not respond to the red; they only respond to the wavelengths at the short end, the blue and green end. Naturally, therefore, one does most of one's handling using red light. This entails a complication, which at first we were not really aware of, but whose importance has subsequently been much stressed. If we follow the dose-response curve like the one above with plants that have been kept in the dark, and if we compare it to one with plants that have been exposed to red light, we find that the latter have lost a great deal of their sensitivity (fig. 6-7). General properties of the reaction have not changed—the shape reproduces the same curve, but it is shifted over by nearly a whole log unit of intensity. Thus, although the plants do not curve towards the red light, they have "seen" the red light and have responded by reducing their sensitivity by about a factor of 10. A part of this reduction of sensitivity can be reversed with far-red, but not all of it. Briggs has made some study of the reversal, and he points out that the coleoptile does in fact contain quite a lot of phytochrome, especially the coleoptile tip. Indeed when plants are treated with red light and provided with auxin, their sensitivity to the auxin is also decreased. Thus the effect of red exposure on phototropism may be explained in terms of less sensitivity to the effector system; that is, the plants are simply less sensitive to the auxin redistribution which light brings about. Since Briggs and Siegelman, in other studies, have found that the red light sensitivity of different seedlings is roughly proportional to their content of phytochrome, it is probable that the effect described here is mediated by the phytochrome in the coleoptile. In any case, one needs to know about the effect and to take appropriate precautions. The plants must be exposed to as weak as possible red light, or even better, to weak green light (though that is more dangerous), and the duration and intensity of exposures kept nearly constant so that all plants are comparable. When this is done, one can determine more precisely the amount of light, the I × t product, necessary to produce a given curvature. Since 15° is about in the middle of these linear ranges, we determined how much light it takes to produce 15° curvature for different monochromatic wavelengths. The answer appears in figure 6-8.

There is some variation between plants and between experiments, but by taking large numbers of points (which are fairly

(right, above) **6-7** The desensitizing effect of one hour's pre-exposure to red light before unilateral blue light. The light dosage required for 10° curvature is increased from 3 to 15 ergs/cm^2. From Curry, 1957.

(below) **6-8** Action spectrum for the tip response of the *Avena* coleoptile, deduced from the integrated results of 3 large series of experiments. The large square indicates the 436 nm reference point; the small dots represent single readings (12 plants), the others from 4 to 8 times as many. From Thimann and Curry, 1960.

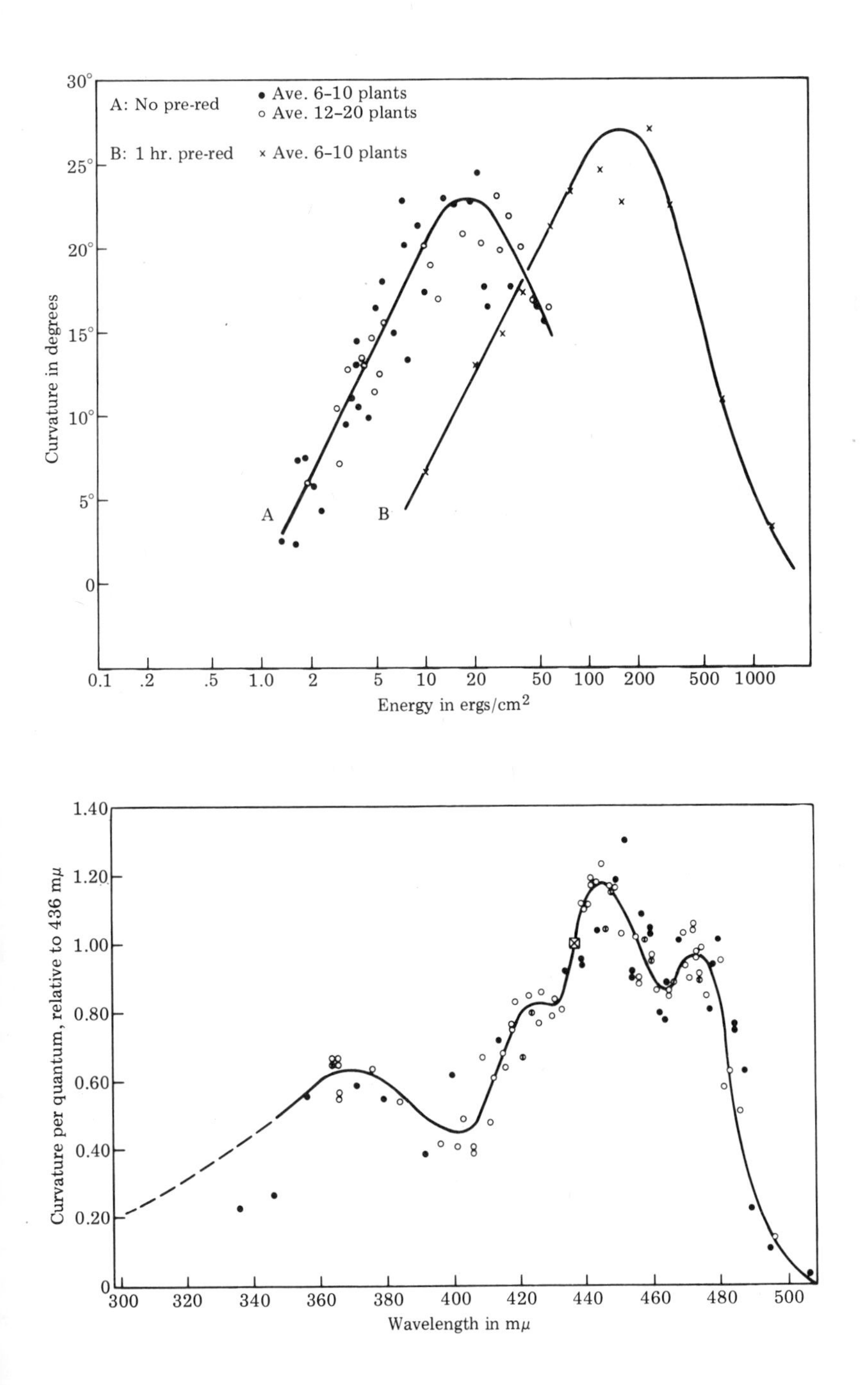
A: No pre-red
• Ave. 6–10 plants
○ Ave. 12–20 plants
B: 1 hr. pre-red
× Ave. 6–10 plants
Curvature in degrees
Energy in ergs/cm2
A
B
Curvature per quantum, relative to 436 mμ
Wavelength in mμ

easy to repeat) one derives a good action spectrum. This represents the number of mole-quanta of light needed to produce a given amount of curvature. This action spectrum, which is thus the absorption spectrum of the photoreceptor, has about five characteristics: first, it shows some sensitivity in the near ultraviolet, falling off towards shorter wavelengths; second, it has a trough at the very edge of the visible spectrum, a small subsidiary peak at about 425 nm (420 in some experiments), and a major peak near 445 nm, in the center of the blue region. There is a trough and a third peak at 475 nm, and then it falls with great steepness, reaching zero at 510 nm. Thus it is a very characteristic absorption spectrum. The other thing to be noted is that the height of the peak in the ultraviolet, at about 370 nm, is a little less than half that of the peak in the visible light at 448 nm. The part in the visible with three peaks is very characteristic of the carotenoids, which are the most important yellow pigments in the plant. Of course when a substance absorbs only in the blue, to the eye it would appear yellow. So the obvious deduction is that this part, at least, represents a carotenoid. Indeed there is a considerable quantity of carotenoid visible and readily extractable in these coleoptiles.

In one set of experiments the wavelengths of the peaks were about 422, 445, and 475 nm. A hexane extract of the carotenoids (mainly lutein) shows peaks at about 420, 442, and 475 nm. One could hardly have more perfect agreement between a physiologically obtained set of data, on the tropism, and a chemically obtained set of data, from the hexane extract of the carotenoids. For this reason we did not hesitate to conclude that this curve shows that the absorption of the photoreceptor, in the visible at least, is due to a carotenoid, probably to the lutein which is the principal carotenoid in *Avena* coleoptiles.

What about the ultraviolet absorption? Neither lutein nor alpha- nor beta-carotene, all of which have groups of 3 peaks close together in the blue, has this absorption in the ultraviolet. Carotenoids, as you probably know, have *cis* and *trans* forms. The *cis* forms are generally highly reactive, but the *trans* forms are the ones that are present in largest amounts in the plant. Beta-carotene has a *cis* form which does have an absorption in the ultraviolet, but it shows a somewhat narrower peak, and unfortunately it is at about 345 nm, not at 365 nm. The most attractive candidate for the 365 nm peak is riboflavin. Riboflavin has an absorption spectrum in water with two peaks, at 450 and 365 nm, but of different form from those in fig. 6-8. Not only are both of these broad and smoothly rounded, but they are equal in height; while, as we saw, the peak of phototropic response at 360 nm is a good deal lower—only about half the height of the peak at 445 nm.

When we had the temerity to say in print that the action spectrum indicated a carotenoid, there were immediately workers lined up on two sides of the fence, those who believed in carotenoids and those who believed in flavins. The riboflavin group very soon pointed out that Dr. Harbury at Yale had found that some flavins, like methyllumiflavin, when dissolved in fatty nonaqueous solvents, show a spectrum

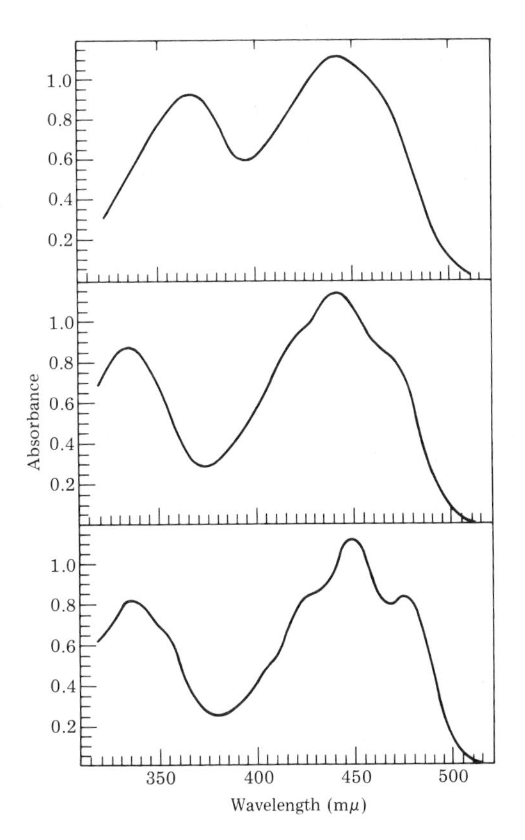

6-9 Absorption of 3-methyllumiflavin. Top, in water; middle, in dioxane; bottom, in benzene. From Harbury et al. 1959.

broken up into three peaks. The central peak remains at about the same position as the smooth rounded peak, but two peaks appear on either side of it so that it does come to resemble the action spectrum, and it has absorption at around 360 nm (fig. 6-9). The two peaks are still unfortunately of about the same height, so that it is difficult to make the argument that the action spectrum could be due to riboflavin. Riboflavin itself, because of its content of the sugar alcohol ribitol, could not be readily brought into a nonaqueous solvent. The argument was thus more or less at a standstill until recently. We did not know of any one pigment which showed this complete action spectrum, though we did know of a pigment (lutein) which shows exactly the action spectrum in the visible. One possibility is that, since there is riboflavin in practically all living tissues (in the forms of flavoprotein enzymes), there is riboflavin in the coleoptile tip, and it absorbs in the UV and *transfers energy* into the region absorbed by the lutein. Such energy transfer is a very common phenomenon in plant tissues. We know that, in the blue-green and the red algae, wavelengths absorbed by phycocyanin and phycoerythrin are transferred with great efficiency to chlorophyll *a* and energize it to carry out photosynthesis. Another possibility would be that this UV absorption produces fluorescence (the flavins are highly fluorescent) and that the fluorescence is absorbed by the carotenoid. But this is not an attractive theory because fluorescence usually takes place with only low efficiency—not more than a few percent. For an action in the UV which is 50% as efficient as in the visible peak, one would require a 50% transfer of light energy. Thus I doubt that fluorescence could account for the effectiveness of the UV; but direct resonance transfer, as in photosynthesis, might account for the facts very acceptably.

Fortunately there is still a third possibility, brought out by the recent work of Hager in Germany. What Hager has done with carotenoids is the analogue of what Harbury did with flavins. He has shown that several carotenoids, including lutein, if dissolved in a semipolar solvent, such as a mixture of alcohol and water, develop a peak of absorption in the UV exactly at 370 nm. He finds that with lutein he can match our action spectrum quite precisely (fig. 6-10). Certainly there is no good reason why the lutein in the coleoptile should not be present in some material—perhaps a lipoprotein membrane—which is of semipolar type. Thus we can conclude that at present the lutein hypothesis is very well supported.

B. PHOTOTROPISM IN THE SHORT WAVELENGTH ULTRAVIOLET

In addition to the responses in the visible and at 360 nm, there is phototropism in the extremely short ultraviolet region. This appears to be a different kind of phototropism. It has more in common with the second positive type than it has with the first positive. It takes place particularly at around 290 nm. What is surprising about it is that this type of phototropism can be obtained without the tip. In other words, it resembles the second positive curvature in that light energy is detected all down the plant. The shorter wavelengths of ultraviolet are known to be able to destroy IAA. So it is theoretically just possible that this curvature, unlike the normal phototropic curvature in which there is no destruction, is actually due to auxin destruction on the lighted side. The action spectrum (fig. 6-11) is a curious one; it seems to have two very close peaks. Where it approaches zero the points are rather uncertain, but this is about where normal phototropic curvature of the positive type begins. Thus there is not much overlap between the two. The dashed line in the figure is the absorption spectrum of indoleacetic acid itself. It consists of two peaks close together, a small rise, which is very characteristic of indole derivatives, on the long wavelength side, and a pretty steep descent to reach zero quite some distance before the edge of the visible. It would be attractive to think, therefore, that the action spectrum for this response represents the absorption spectrum of the auxin, shifted over about 20 nm. Such a shift would be the kind that would take place if the auxin were attached to a protein or some other binding substance. Commonly in such attachments carotenoid-proteins show a shift of the peak of 20 or even 30 nm, so it is of the order of magnitude to be expected. To get proof of this hypothesis should not be difficult, but would require a very long and wearisome series of experiments, which nobody has as yet undertaken.

C. THE SECOND POSITIVE CURVATURE

This type of curvature, of course, takes place in the visible. It has been rather neglected, although it has certainly been

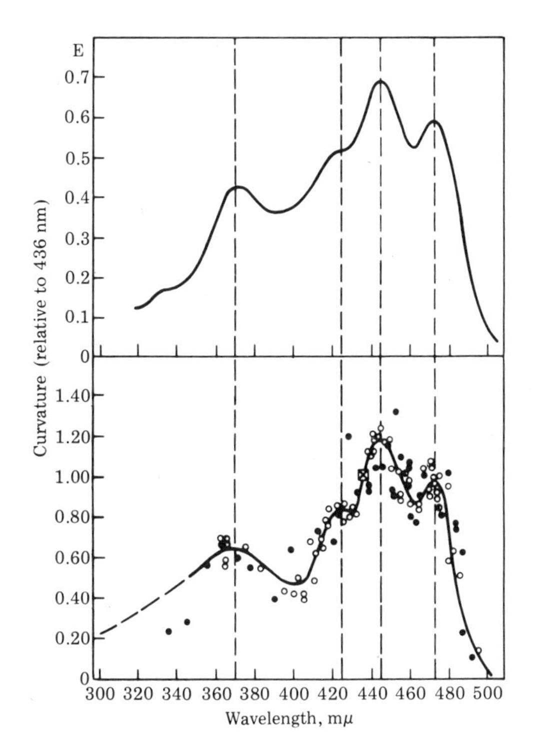

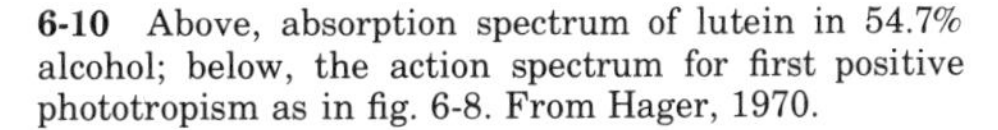

6-10 Above, absorption spectrum of lutein in 54.7% alcohol; below, the action spectrum for first positive phototropism as in fig. 6-8. From Hager, 1970.

6-11 Action spectrum for the base curvature of *Avena* coleoptiles (circles and solid lines) in the ultraviolet. The absorption spectrum of IAA is the dashed line, and the action spectrum for photo-oxidation of IAA in solution, brought to the same relative value of 1.0 at 280 mμ, is given by the black squares. From Curry et al. 1956.

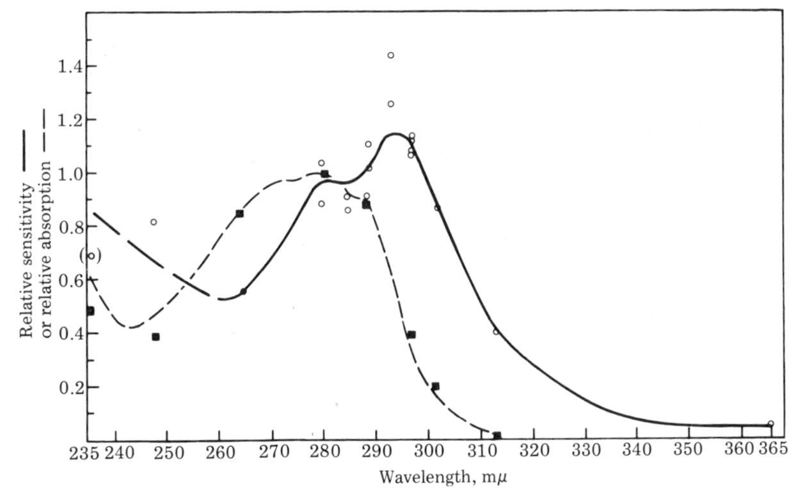

known to exist for forty years or so. We saw above that in the second positive curvature the asymmetric distribution of auxin was of about the same degree as in the first positive; in other words the effector system is the same. Obviously the photoreceptor is not the same, because curvature takes place all the way down the plant and is not limited to the tip. There is a more important reason, which is brought out by studying the kinetics.

If we plot log I × t against curvature, using a *very* wide range of intensities, and of necessity also a very wide range of times, we get a family of curves, only one of which is the one we saw above. As the I decreases, the t of course has to increase. Briggs and Zimmerman found that with very long times (or very low intensities), the onset of the second positive curvature comes nearer and nearer. With extremely long exposures, the second positive curvature begins already close to the peak of the first positive response. Even these four points suggest that with still longer exposures one would get a continuous curve, so that the dual character of the response would have disappeared. This means then that the second positive curvature is not a simple function of I × t, but, for very long times, comes on at much smaller light doses. Hence it is not really dependent on intensity at all, but is dependent on the *time*. We found that we could work with a second positive curvature by giving a normal log I × t to bring the curvature down to about zero. About 5000 ergs per square centimeter suffice to bring the plants back to the unresponsive level in this way. If we give our plants that exposure and then expose them to further light

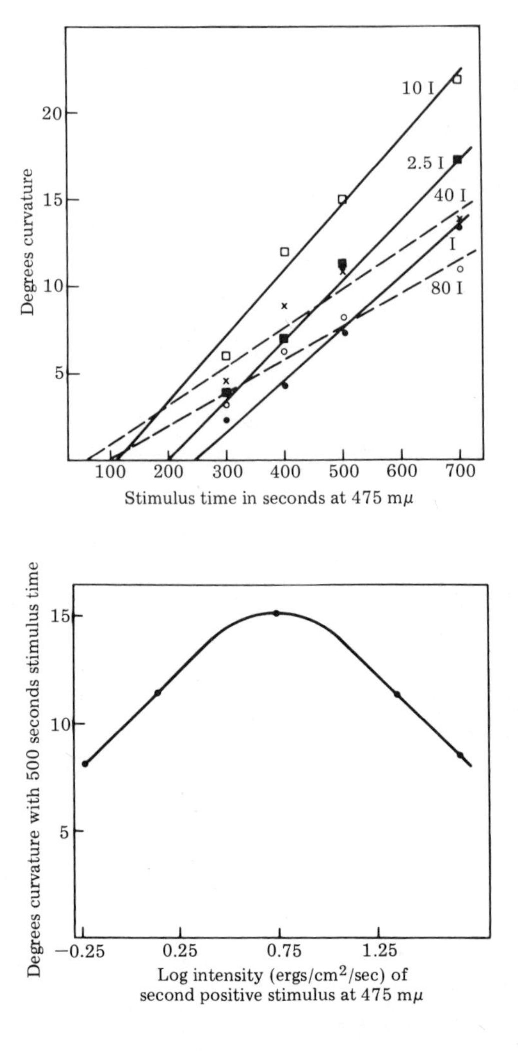

6-12 Above, second positive curvatures of *Avena* as function of duration of exposure. For all intensities the first 100 seconds comprised 4800 ergs/cm² blue light at 475 mμ. The intensity I was 0.6 ergs cm^{-2} sec^{-1}. Below, curvatures with fixed duration (500 seconds) as function of light intensity. From Everett and Thimann, 1968.

we get essentially pure second positive curvatures. With this procedure it becomes clear that the second positive response is not primarily a function of intensity. Figure 6-12 shows a simple plot of second positive curvature against time. Take first the initial intensity, indicated as I to simplify matters; there is a nice linear reaction against time. Using 2.5 I, or 10 I, we find nearly the same thing; there is only a small effect of intensity. Working at still higher intensities, 40 I and 80 I, we return to curves which essentially bracket those at low intensity. Thus there is a family of curves, each of which is linear with time, but whose absolute values depend slightly upon intensity, showing a maximum at around 10 I and somewhat decreased sensitivities at both higher and lower intensities. But even at the maximum, the response to intensity is only by a factor of two. This is another reason to think that, although it uses the same auxin system to express itself, the second positive is quite different from the first. It does not require the tip, although better curvature results when it is present—the response occurs all down the plant, and its kinetics are clearly different.

Keeping in mind this influence, one can choose an intensity and then measure the response to different lengths of exposure at different wavelengths. In this way one can obtain an action spectrum like the action spectrum for the first positive. The experiment is much more troublesome than for the first positive, since the light must be controlled carefully to bring the plants to the zero point. For this reason we have obtained fewer points than for the first positive, but the result (fig. 6-13) is most interesting: (1) in the region about 370 nm we see a smoothly rounded peak; (2) there is a trough at the edge of the visible, at about 400 nm; (3) there is clear evidence of a secondary peak at about 425 nm; (4) there is a main peak at about 450 nm; (5) there is some suggestion of a secondary peak at about 470 nm; and (6) it goes down very steeply on the long wavelength side, apparently to reach zero at 510 nm. In other words, it is quite a remarkable imitation of the action spectrum for the first positive curvature. We were unsure as to whether there is a marked second peak at 470 nm, because the xenon lamp that gives a smooth light output over this region unfortunately shows a break at between 470 and 475 nm, so that it is difficult to be sure of intensity values. Some other points suggest that there is indeed a lower response in between

6-13 Action spectrum for the second positive phototropic response. The reciprocal of the energy (delivered in 400 seconds) for a 15° curvature is plotted against wavelength. From Everett and Thimann, 1968.

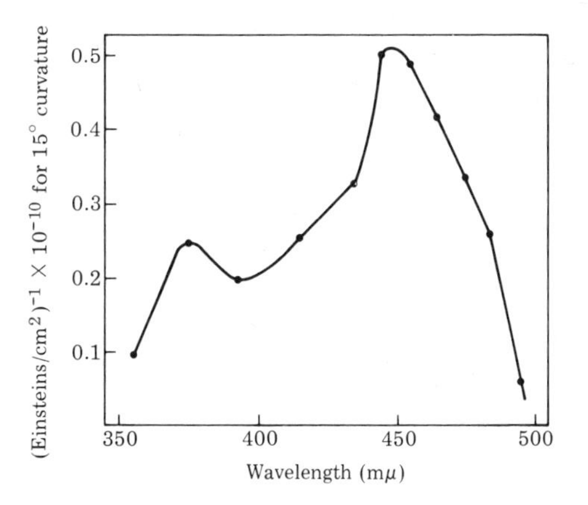

these two values and therefore a small subsidiary peak at 475 nm. My colleague, Dr. Everett, who went to a lot of trouble over those points, is convinced that there is a very small peak there.

We have, therefore, a strange situation. We have seen that the *effector* system for the first and second positive curvatures is the same—an auxin asymmetry of about 3 to 1. This spectrum suggests very strongly that the *receptor* system of the second and first positive is also the same. Yet nevertheless the two types of curvature are different, and they have different kinetics. Is it possible that they both operate with the same photoreceptor, causing some sort of lateral polarization which leads to the auxin movement? This would mean that there are two different paths between the photoreceptor and the effector—a relatively simple one for the first positive, and a time-response path for the second positive. That seems like an unusual arrangement. I am encouraged, though, when I think of the phenomenon of vision, because in the eyes of animals we have two kinds of photoreceptors and two kinds of detector systems, the rods and the cones. The pigments are not quite the same, though nearly so. The cones operate primarily in high intensity light, like the second positive, and the rods operate in low intensity, like the first positive. One might imagine that in the more advanced creatures a system resting basically on this principle has evolved still further to give different sensory responses; that is, the rods evoke only black and white, while the cones, due to color filters operating with them, give the impression of three colors; however, although it is very important *to us,* in point of fact this is not a very profound difference. One can easily induce three-color responses by just interposing three different pigments, so that in itself is not a profoundly different phenomenon. The fact that these two visual systems operate at different ranges of intensities is certainly very suggestive.

D. PHOTOTROPISM IN PHYCOMYCES

It is often a good rule in biology that when one seems to have gone as far as one can and would still like to know more about the phenomenon, it can be enlightening to turn to another object. For this reason we have experimented with a quite different object, the fungus *Phycomyces,* which is well known to be sensitive to light. *Phycomyces* was studied by Blaauw long ago, around 1914. He showed that not only does it curve towards visible light but it shows a light-growth reaction. This light-growth reaction resembles that of *Avena:* when a light shines on it, its growth is accelerated, and slowly the growth rate comes back to normal, perhaps hunting a little below and above before returning. *Phycomyces* shows, besides the light-growth reaction, a sensitive phototropism. The sporangiophores are very long and thin—about 300 μm across—and they emerge at different ages, so that in any one culture they reach different heights. At the tip of the sporangiophore is produced a small black mass, which is the sporangium containing the spores. When the plants have been illuminated from one side for a few minutes, they develop very well-marked phototropic curvatures (fig. 6-14).

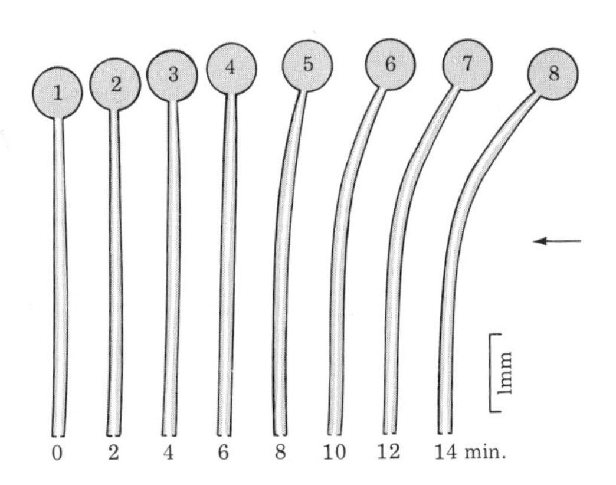

6-14 Phototropic curvature of *Phycomyces* sporangiophores, from photos taken at 2-minute intervals from the beginning of unilateral illumination. Note 6-minute delay before curvature begins. From Castle, 1962.

Along with the response to visible light is a comparable response to ultraviolet light. Figure 6-15 shows a series of photographs taken 10 minutes apart. The beginning of curvature appears at once, and curvatures develop very rapidly. After reaching 90° some have gone beyond that by even as much as 20°. Then they return, and in their enthusiasm return too far, so that they have to curve back again. Thus it is an extremely reactive system. Perhaps the most interesting part of all is that the light in this figure came from the *left.* In other words these are curvatures *away from* the light—negative phototropisms. So here we have a stimulating parallel with *Avena,* for *Phycomyces* is also a plant which shows both positive and negative phototropic curvatures. It is not directly comparable to *Avena,* for the negative curvature is a response to ultraviolet light while the positive is a response to the visible. But another suggestive point is that the negative curvature has different kinetics from the positive—it is much more rapid. The curvatures reach 90° and go beyond, while in the visible light they do not reach 90° and they curve more slowly, so there are two different reactions.

The theory of these curvatures is controversial. Blaauw showed that when visible light shines upon the plants their growth is accelerated. Nevertheless they curve *towards* the light. One would think that if the growth is accelerated, the acceleration would be greater on the side towards the light, and that this would make them curve away from the light. To explain this problem the so-called lens theory has been put forward and supported by several

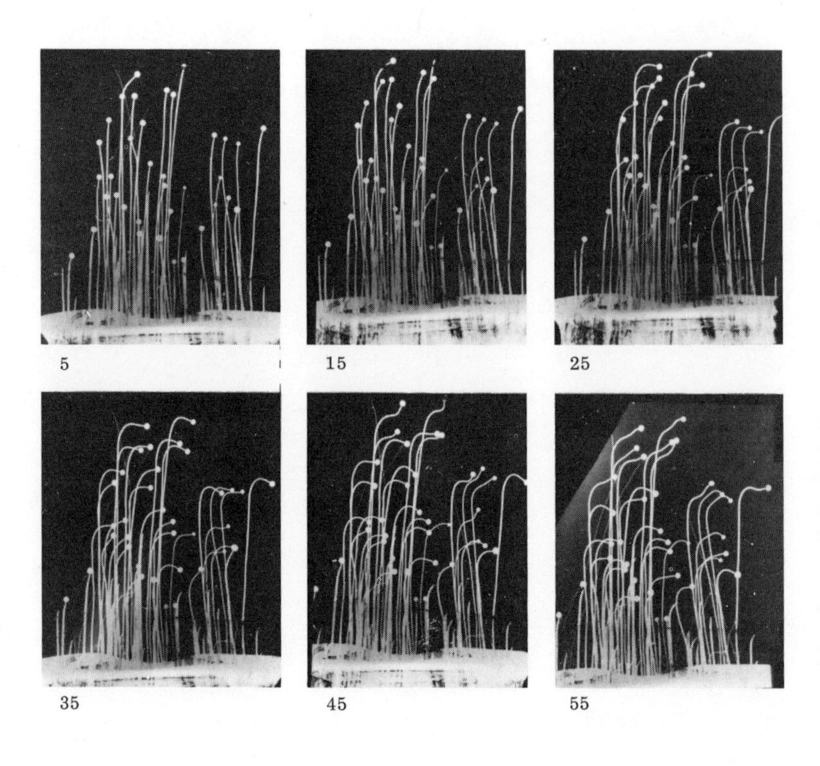

6-15 *Phycomyces* sporangiophores illuminated from the left with continuous ultraviolet light of 280 nm, showing development of negative curvature. Left to right, top row, 5, 15, and 25; lower row, 35, 45, and 55, minutes after beginning of exposure. From Curry and Gruen, 1957.

workers, especially by Max Delbruck and his colleagues at Pasadena.

The lens theory postulates that because the light falls on a translucent structure, that structure acts as a lens; so that at the back of the sporangiophore there is a very brightly illuminated but narrow zone, while in the front the zone is more evenly illuminated by the incident intensity. The center of this front zone is exposed to the highest intensity because it receives the light at normal incidence; the zones further round are exposed to lower intensity, until at a certain angle there will be total reflection, and no light will get through (fig. 6-16). A large part of the back side will be in darkness, which is a rather peculiar fact; at the back there is a brightly illuminated central stripe and then an area of darkness on either side (fig. 6-16). Now the lens theory states that, because this part is more brightly illuminated than the opposite zone on the front, it will show the light-growth reaction more, accelerating more than the front side; and as a result curvature will take place toward the light. As against that we have to remember that there is certainly some absorption in the cell contents, for, although it looks clear to the eye, an electron micrograph shows that each cell is packed full of organelles. Since there is inevitably some absorption and scattering within the sporangiophore, the *total amount* of light on the back half must of necessity be less than the total amount on the front part. To add to the absorption in the body of the material, there is also some total reflection. Another complication results from those parts that are completely dark, so that even if the bright stripe were accelerated by the light, the adjacent dark parts will be growing only slowly, at the dark rate, and would resist the acceleration. This would set up a series of tensions in the plant, which would probably prevent the acceleration of the narrow brightly lit zone. All in all, the lens theory is a good deal less than convincing.

We have an additional complication, namely, the negative curvature in the ultraviolet. Now the optical properties of 340 nm would not be extremely different from those at 445 nm; nevertheless the curvature is in the opposite direction. One might deduce, of course, if the lens theory is true, that ultraviolet must cause a retardation in growth rate. This retardation would again be different on the two sides, so that the sporangiophore would curve away from the light. However, in fact ultraviolet does not retard; it causes an acceleration of growth. If sporangiophores are growing at a steady rate, when we turn on the ultraviolet light from both sides the growth rate goes up from about 23 to about 48 units per minute, and comes down again very quickly. Each time the illumination is repeated, the response is a little less, but always the response is clearly an acceleration of growth, which lasts roughly for the first half of the duration of the negative phototropic curvature. Thus during the negative phototropic curvature there is an acceleration of growth in the first half and then a return to normal in the second half, so that there is no correlation at all between the absolute growth rate and the phototropic curvature. In an effort to explain this and still hold on to the lens theory, Carlyle has made the suggestion that the *Phycomyces* sporangiophores contain so much tannin, and the tannins absorb in the ultraviolet, that in this case the light influence is reversed and in fact they get more light on the front side. As a result the acceleration is on the front side and so the sporangiophores curve away from the light. Still, the timing does not agree. In fact the timing of the visible light growth reaction does not agree with the timing of the positive phototropic curvature either. It seems to me most unlikely that the amount of absorption in the ultraviolet would be enough to produce so steep a gradient, and light absorption data have not been offered in support. My personal conclusion is that phototropic curvature is probably *not* due to the light growth reaction, any more than it is in *Avena*.

Why is there a curvature? Is it, as in higher plants, due to an asymmetric distribution of auxin, or some other growth substance? A natural experiment would be to apply auxin to one side of a sporangiophore to see if it would produce a curvature. But before we look at such experiments, let us look more closely at these sporangiophores. We find there are three types. The normal sporangiophores, which are about 300 μm thick, have a very characteristic structure; the sporangium rests on a slightly tapered zone. The second kind of sporangiophore is formed especially in young cultures; it is similar but much thinner, hence called a fine sporangiophore. These are about half or a third the normal thickness and they bear tiny sporangia, but are otherwise basically comparable. The third kind of sporangiophore has a different shape: it has about the same thickness as the normal type, but instead of tapering it widens out at the top and bears a rather big sporangium; these are called giant sporangiophores. We shall come back to the three types—the fine, the giant, and the normal. The normal and fine types show phototropism; the giants seem to be phototropically insensitive. Of course all sporangiophores develop from hyphae, and there is usually a cross-wall at the base. The hyphae seem quite phototropically insensitive, too, growing neither towards nor away from the light.

My student, Dr. Gruen, tried a number of ways to place auxin on the side of a normal sporangiophore. If one just sets a small agar block on the end of a spatula and sticks it on, then the agar block and the spatula remain in contact; surface forces do not allow the block to slide off. On the other hand the block remains in contact with the sporangiophore, so that one cannot remove the spatula without the whole sporangiophore becoming pulled over and broken. So the only way to do it was to bring the droplets through the air on to a sporangiophore. For this Gruen developed a fine-tipped micropipette, from which protrudes a glass thread. When a droplet comes out of the tip and slides along the glass thread, a little puff will make it jump off the glass thread and fall on the sporangiophore. It is an extremely tricky operation, but the results were of great interest. The droplet, when it does alight on the sporangiophore, does in fact produce a curvature. The curvature is a function of the position of the droplet on the sporangiophore. This is not at all surprising, because we know from previous measurements that most of the growth in the sporangiophore takes place in the up-

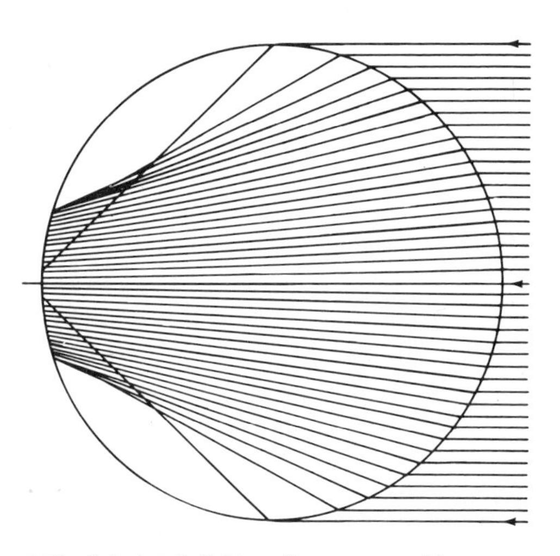

6-16 Calculated light paths across a *Phycomyces* sporangiophore illuminated (with a parallel beam) from one side. From Castle, 1965.

permost half-millimeter. The growth rate falls off very sharply in the second half-millimeter, and in the third half-millimeter it is almost zero. Naturally there would be no use putting a droplet on a nongrowing zone. Roughly, therefore, the curvature resulting from the droplet is a function of the growth zone to which it is attached; an entirely reasonable result. What is unfortunate, however, is that the curvatures are the same whether or not one puts indoleacetic acid in the droplet.

Figure 6-17 shows marked curvatures, but they are produced by H_2O. Figure 6-18 summarizes two experiments with two different auxin concentrations. The dashed line shows the curvatures that can be produced by water droplets, while the crosses show curvatures produced by auxin droplets. It is necessary to group together results with sporangiophores whose growth rate fell within the same range, otherwise they are not comparable. The plots show that the curvature is a function of the distance: the nearer to the sporangium the larger the curvatures. Excellent curvatures can be obtained—60° or more in half an hour—but there is no promoting effect of auxin. Indeed, if there is any effect, it is to produce a slight inhibition with the higher concentrations. Although this inhibition is actually statistically significant, one hesitates to draw many conclusions from it. As a matter of fact, the sporangiophores, like many fungal fruiting bodies, turn out to contain quite a lot of indoleacetic acid, and no one has yet been able to show that it has a function there. They also contain what is probably indole-carboxylic acid, a substance which is inactive as an auxin.

The fact that the curvatures are not due to indoleacetic acid but to water demonstrates that the elongation of the sporangiophore in this growing zone is very sensitive to the water supply. The rate at which water comes in from the hyphae evidently determines the rate of growth and therefore the rate of curvature. Gruen has (very delicately) pulled off sporangiophores from the hyphae and immersed their bases in pure water: they go on growing at about half the normal rate. They are not receiving any nutrients, and so far we have found no peptone, amino acid, or salt mixtures which will reinstate the normal growth rate. With the water supply they continue for many hours, getting somewhat finer and very thin-walled, but they do not stop. Hence we know that the rate of supply of water is a major limiting factor in this elongation.

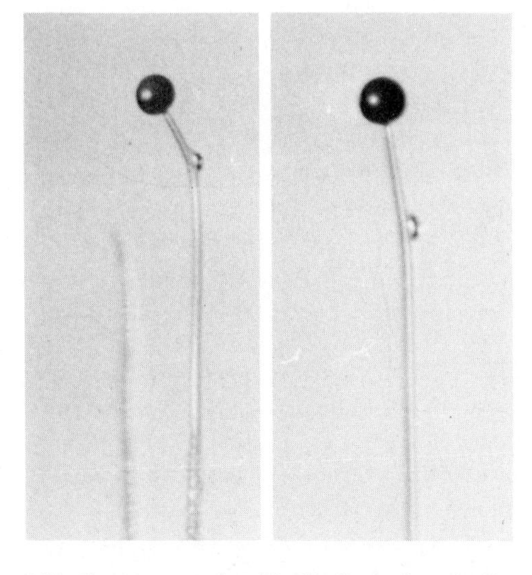

6-17 Curvatures produced in 15 minutes by a droplet of water on one side of a *Phycomyces* sporangiophore. Left, droplet 0.58 mm below sporangium, curvature 32°; right, droplet 1.02 mm below, curvature 5°. From Gruen, 1956.

E. THE ACTION SPECTRUM FOR PHYCOMYCES PHOTOTROPISM

Another approach was to try to determine the action spectrum. This can be done without too much difficulty, because one can determine the range in which the plants are linearly responsive and expose them to various I × t values. It was long ago discovered by Blaauw that they are responsive to the product of intensity and time, like the first positive curvature in *Avena*. We did just that and determined the action spectrum (fig. 6-19). It shows some familiar features. It has a rounded peak around 370 nm, a trough not quite at, but close to, the edges of the visible, evidence of a small peak at about 425, a major peak at 445, a second peak at 475, and a drop to zero at

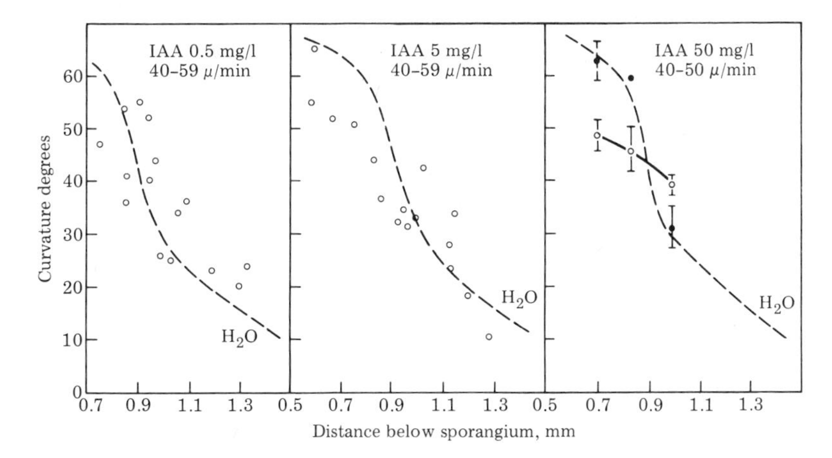

6-18 Plot of curvatures of *Phycomyces* sporangiophores produced in 15 minutes by droplets of liquid placed at different distances from the sporangium. The dashed lines show curvatures produced by water, the points show curvatures produced by IAA at three concentrations. Growth rates (shown at top) all comparable. From Thimann and Gruen, 1960.

510. In other words it is the same, to all intents and purposes, as the action spectrum for *Avena,* a higher plant. There are some slight points of difference: the location of the base of the trough is nearer 410 nm than 400; the height of the shoulder on the higher plant at 420 is a little greater and in *Phycomyces* it appears nearer to 430. It is by no means certain that these differences are significant; if they are, one might suggest that perhaps the same pigment is involved but that it is either in a slightly different optical system or in a slightly different chemical combination, so that the positions of the peaks are slightly shifted. In any event the two systems are either identical or almost so. The large squares in figure 6-19 are means of much larger numbers of observations than the others and were done with particular care to establish that second peak. Delbruck and Shropshire have carried out somewhat similar experiments using the light-growth reaction, following the growth acceleration as a function of wavelength. They come out with a comparable curve, though with less fine structure.

We made a brief study of the action spectrum for the ultraviolet reaction of *Phycomyces* (fig. 6-19). It shows a peak nearly where there is one in *Avena,* at about 280 nm, and the response comes down steeply on the long wavelength side and goes up steeply at very short wavelengths. There are not enough points to make a firm deduction as to the nature of this photoreceptor, but it shows a great deal of similarity to the ultraviolet responding system of *Avena.*

Meanwhile several other action spectra similar in type to figure 6-19 have been reported. Cell division in the protonema of a fern (*Adiantum*) shows, after exposure to red light, an action spectrum with a main peak at 459 nm, secondary peaks at 487 and

375, and a shoulder at about 430 (Wada and Furuya 1975). The light-induced formation of perithecia (the sexual stage) in the ascomycete *Gelasinospora* shows a main peak at 458 nm and secondary peaks at 475, 420, and 365 (Inoue and Furuya 1975, fig. 6-20). Photo-induction of the sexual stage of another ascomycete, *Nectria Hematococca,* shows a main peak at 440 nm, with secondaries near 360 and 480 (Curtis 1972). Earlier the photo-induced movement of the chloroplasts of *Lemna* had been found to show main peaks near 460 and 385 nm. It is evident that there is either a closely related family of pigments, or one single one that is widely distributed in both higher and lower plants and is responsible for many light-induced reactions.

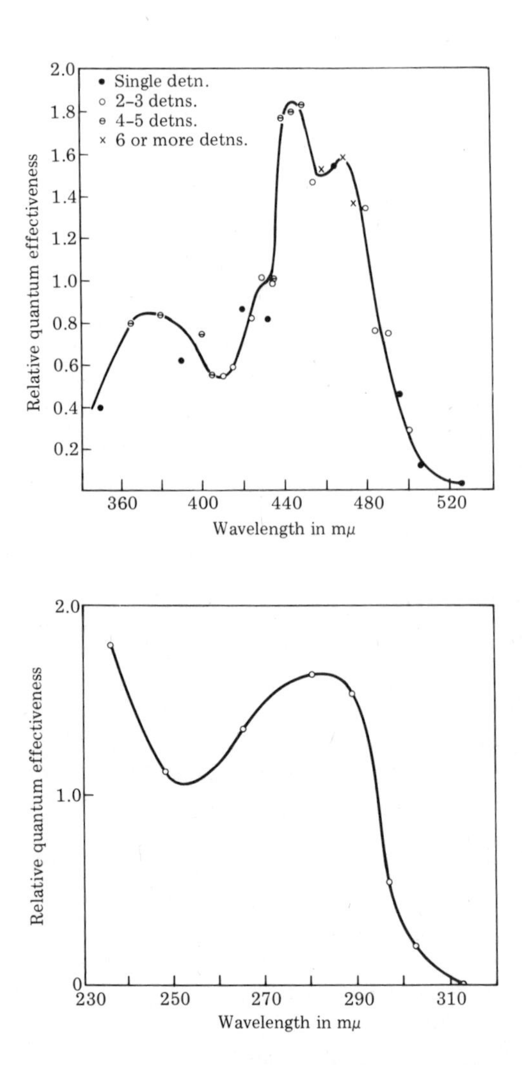

6-19 Action spectra for phototropism of *Phycomyces.* Above, positive curvature in visible light. Below, negative curvature in ultraviolet light. From Curry and Gruen, 1959.

F. POSSIBLE PHOTORECEPTORS

The last approach is the search for a possible photoreceptor by visual means; that is, can one detect anything visible in the cells that might be a photoreceptor? We reported long ago that in *Avena* there are some curious bodies, more or less cubical in shape, having a lattice structure, and enclosed in a membrane which shows no other granularities. We called them crystal bodies, and they resemble objects which other workers have called microbodies. They are found with particular frequency in *Avena* coleoptiles, but they are also present, with less frequency, in the leaves. We have considered them as at least possible candidates for the photoreceptors (fig. 6-21). The crystal is probably a protein.

Figure 6-22 shows an electron micro-

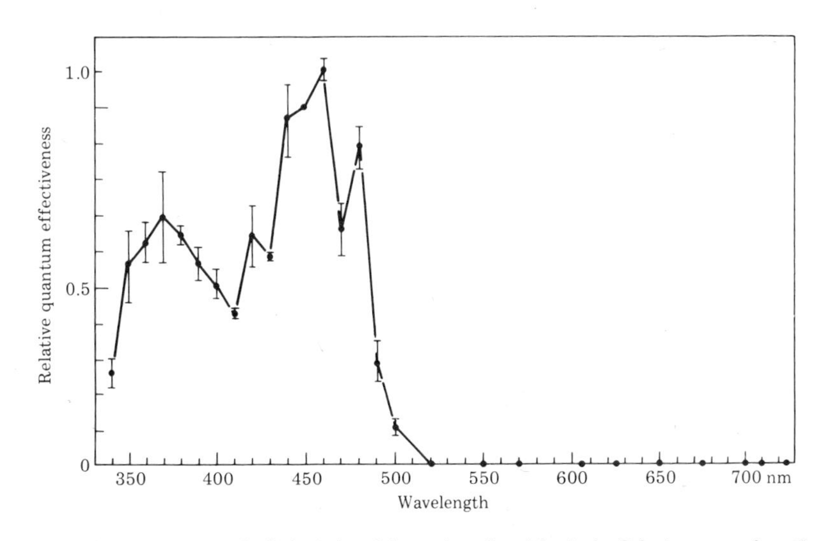

6-20 Action spectrum for light-induced formation of perithecia in *Gelasinospora* after 48 hours in the dark. Data are relative to the value at 450 mμ. From Inoue and Furuya, 1975.

graph of a *Phycomyces* sporangiophore, and again there are crystal bodies present; they normally have a membrane. Lower down in the sporangiophore they find their way into a vacuole. One gets the impression that they are being eroded away, as though they have performed a function and can then disappear. Such an interpretation is purely speculative, but it is a remarkable fact that these crystal bodies in *Phycomyces* bear a considerable resemblance to those in the higher plant. Their appearance does seem to correlate with the appearance of phototropism.

My student and colleague, Dr. Thornton, has compared the occurrences of different kinds of bodies in all the structures of *Phycomyces:* the tip of the hyphae, the mature hyphae, storage vesicles, the dwarf sporangiophores, a salmon-colored mutant bearing sporangiophores (all objects which show no phototropism), normal sporangiophores—very young with no sporangia on top—normals with the fully formed sporangia on top, and the fine sporangiophores which are also phototropic. In table 6-3 we see the various vesicles and organelles that one can make out on an electron micrograph. The one thing that is interesting is that the crystal bodies are absent from every one of the phototropically unresponsive organs. They are present in every one of the photoresponsive organs, and they are the *only* bodies which show such a distribution. They show exact correlation with phototropic sensitivity. Thus it is certainly possible that those bodies do contain the photoreceptor. So far we have not succeeded in preparing them pure in sufficient quantity for analysis.

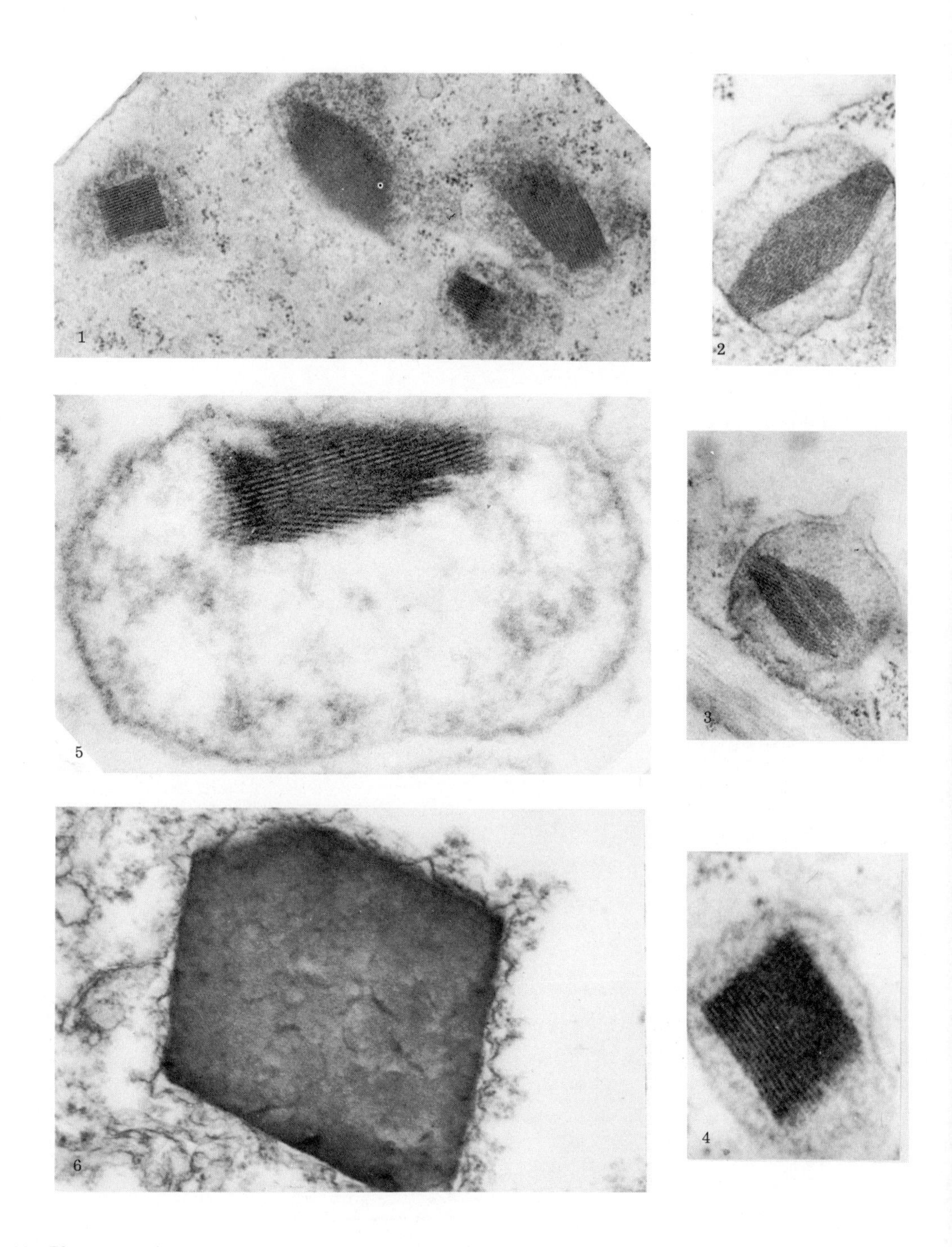
1
2
5
3
6
4

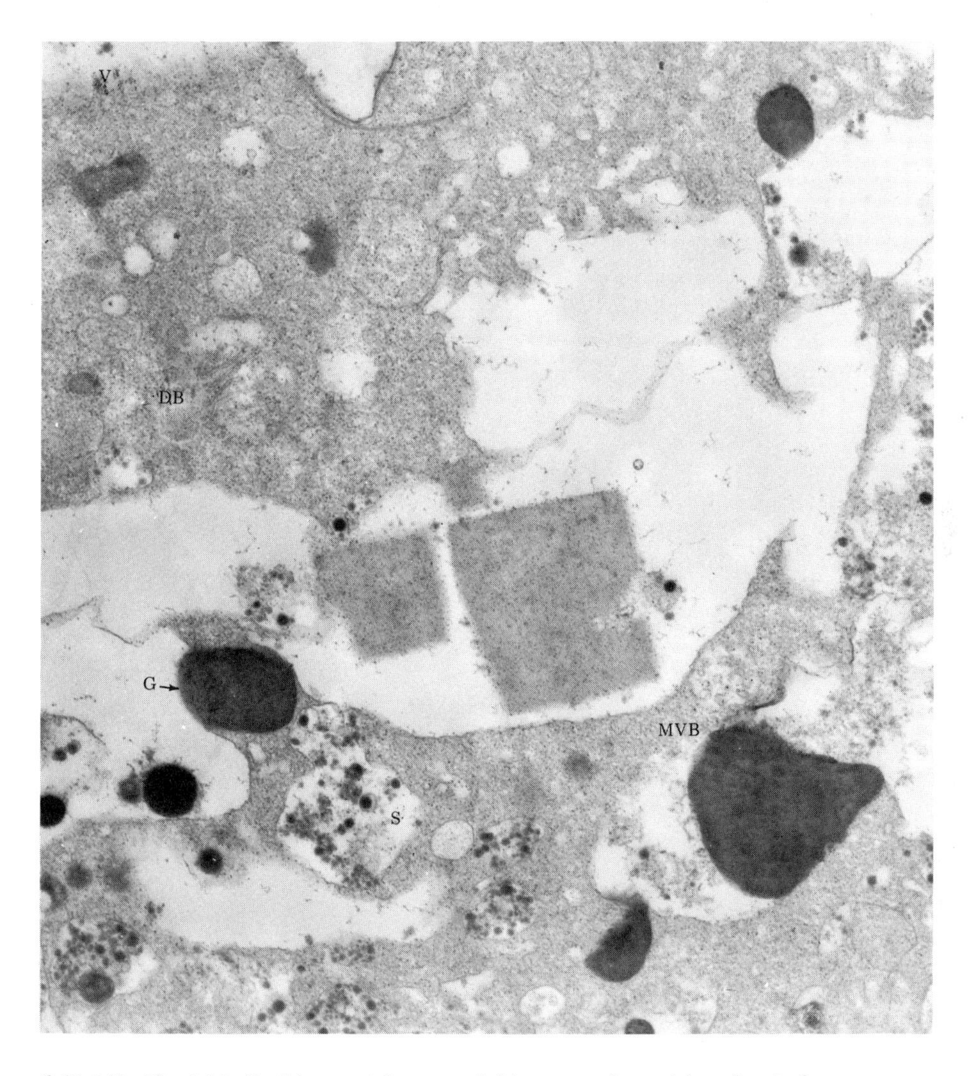

(left) **6-21** "Crystal bodies," i.e., crystals surrounded by a granular matrix and a single membrane, in the subepidermal cells of *Avena* coleoptiles (42000 X) (Nos. 1-5) and, No. 6, in a *Phycomyces* sporangiophore (46000 X). From Thornton and Thimann, 1961.
(above) **6-22** Crystals interpreted as "crystal bodies," from which the membranes have been dissolved, apparently undergoing dissolution in the vacuole of a *Phycomyces* sporangiophore, about 250 micra below the base of the sporangium (27000 X). From Thornton, 1966.

G. CONCLUSIONS

In closing, I would like to draw a few generalities. We have discussed two kinds of tropism—geotropism and phototropism. In geotropism there are probably two types of georeceptor bringing about the same reaction. Both work through the auxin system. One is probably the amyloplasts, which work relatively rapidly; and the other type consists of those organelles that are left when the starch has been hydrolyzed away. There are doubtless smaller bodies which have the same function. In phototropism we also have two different mechanisms for bringing about an auxin response, even though both use apparently the same photoreceptor. Thus in geotropism there are apparently two receptors and one effector (the auxin system); while in phototropism there is one receptor and one effector but two different linkages. Also there is the curious ultraviolet response, which may or may not be important in nature. The other surprising fact is that while *Phycomyces* belongs to the Phycomycetes, which, except for the Myxomycetes, is generally regarded as the *lowest* order of fungi, *Avena, Zea* and *Hordeum* (barley) and other such higher plants are in the Gramineae,

Table 6-3 Light sensitivity and fine structure of *Phycomyces*. Presence or absence of organoids in photoresponsive and non-photoresponsive forms or organoids.

	Granular Mass	Endoplasmic Cisternae	Glycogen Masses	Free Globules	Dense Bodies	S-Vesicles	H-Vesicles	Apical Nuclei	A-Mitochondria	B-Mitochondria	Microtubules	CRYSTALS	Multivesicular Bodies	Lomasomes and "Myelin Figures"
Non-photoresponsive														
Hyphal Apex	−	+	−	−	+	−	+	−	−	+	+	−	+	+
Mature Hypha	−	+	+	+	+	−	+	−	−	+	?	−	+	+
Storage Vesicle	−	+	+	+	−	−	+	−	−	+	?	−	+	+
Dwarf sporangiophore	−	+	+	+	+	+	+	−	+	+	+	−	+	+
Salmon Variant sporangiophore	+	+	−	+	+	+	?	+	+	+	+	−	+	+
Photoresponsive														
Normal Stage I sporangiophore	+	+	−	+	+	+	?	+	+	+	+	+	+	+
Normal Stage IV sporangiophore	−	+	−	+	+	+	?	−	+	+	+	+	+	+
Fine Stage I sporangiophore	+	+	−	+	+	+	?	+	+	+	+	+	+	+

From Thornton, 1966.

which are monocots and, except for the orchids, are very nearly at the *highest* end of the higher plants. Apparently we have the same (or almost the same) photoreceptors all through the entire plant kingdom. What does that suggest? The obvious suggestion is that to be a photoreceptor of phototropism requires some very special properties; any yellow pigment will not do. Probably what is needed is such a tricky combination of photoreceptors in a special location, that when Nature hit on it she carried it through the whole plant kingdom. Indeed, we see the same phenomenon with chlorophyll, since all plants, from the lowest algae to the highest orchids, have chlorophyll *a*; they may have sundry additions but all have chlorophyll *a*. Only in the bacteria is there any difference, and even there the bacteriochlorophyll does not differ greatly from chlorophyll *a*; it has one different side chain and that modifies the effective wavelengths. Thus in these light reactions the system required is so special that there is not a wide range of possibilities, and almost the entire plant kingdom is organized around the same few substances.

REFERENCES

Briggs, W. R. Mediation of phototropic responses of corn coleoptiles by lateral transport of auxin. *Plant Physiol.* 35, 237-247, 1963.

———, Tocher, R. D., and Wilson, J. F. Phototropic auxin redistribution in corn coleoptiles. *Science* 126, 210-212, 1957.

Castle, E. S. Phototropic curvature in *Phycomyces*. *J. Gen. Physiol.* 45, 743-756, 1962.

———. Differential growth and phototropic bending in *Phycomyces*. *J. Gen. Physiol.* 48, 409-423, 1965.

Curry, G. M. Studies on the spectral sensitivity of phototropism. Thesis, Harvard University, 1957.

———, and Gruen, H. E. Negative phototropism of *Phycomyces* in the ultraviolet. *Nature* 79, 1028-1029, 1957.

———, and Gruen, H. E. Action spectra for the positive and negative phototropism of *Phycomyces* sporangiophores. *Proc. Nat. Acad. Sci.* 45, 797-804, 1959.

———, Ray, P. M., and Thimann, K. V. The base curvature response of *Avena* seedlings to the ultraviolet. *Physiol. Plantarum* 9, 429-440, 1956.

Curtis, C. R. Action spectrum of the photoinduced sexual stage in the fungus *Nectria hematococca* Berk. et Br. var. *cucurbitae*. *Plant Physiol.* 49, 235-239, 1972.

Delbruck, M., and Shropshire, W. Jr. Action and transmission spectra of *Phycomyces*. *Plant Physiol.* 35, 194-204, 1960.

Everett, M. S., and Thimann, K. V. Second positive phototropism in the *Avena* coleoptile. *Plant Physiol.* 43, 1786-1792, 1968.

Gruen, H. Growth and curvature in *Phycomyces* sporangiophores. Thesis, Harvard University, 1956.

Hager, A. Ausbildung von Maxima im Absorptionsspektrum von Carotinoiden im Bereich um 370 nm; Folgen für die Interpretation bestimmter Wirkungsspektren. *Planta* 91, 38-52, 1970.

Harbury, H. A., La Noue, K. F., Loach, P. A., and Amick, R. M. Molecular interaction of isoalloxazine derivatives. 2. *Proc. Nat. Acad. Sci.* 45, 1708–1717, 1959.

Inoue, Y., and Furuya, M. Perithecial formation in *Gelasinospora reticulispora*. *Plant Physiol.* 55, 1098–1101, 1975.

Pickard, B. G., and Thimann, K. V. Transport and distribution of auxin during tropistic response. 2. The lateral migration of auxin in phototropism of coleoptiles. *Plant Physiol.* 39, 341–350, 1964.

Thimann, K. V., and Curry, G. M. Phototropism and phototaxis. In *Comparative Biochemistry,* vol. 1, ed. M. Florkin and H. S. Mason, pp. 243–309. New York: Academic Press, 1960.

———, and Gruen, H. E. The growth and curvature of *Phycomyces* sporangiophores. *Festschrift Albert Frey-Wyssling, Beiheft z. Schweiz. Forstvereins,* No. 30, 237–263, 1960.

Thornton, R. M. Comparative study of phototropism in *Phycomyces* and *Avena*. Thesis, Harvard University, 1966.

———, and Thimann, K. V. On a crystal-containing body in cells of the oat coleoptile. *J. Cell Biol.* 20, 345–350, 1966.

———, and Thimann, K. V. Transient effects of light on auxin transport in the *Avena* coleoptile. *Plant Physiol.* 42, 247–257, 1967.

Wada, M., and Furuya, M. Action spectrum for the timing of photo-induced cell division in *Adiantum* gametophytes. *Physiol. Plantarum* 32, 377–381, 1974.

Wilden, M. Analyse der positiven und negativen phototropischen Biegungen. *Planta* 30, 286–288, 1939.

The Development of Leaves and Roots

7

SOME HORMONAL ASPECTS OF THE GROWTH AND DEVELOPMENT OF LEAVES

Leaves are the most important of all objects on the face of the earth, bar none. We all live from leaves, in the form of, for example, spinach, in the form of beef or milk, or in the form of seeds whose growth is made possible by the leaves. We know a certain amount about leaves, but considering their importance we have surprisingly inadequate knowledge—more than normally inadequate—in the field of hormonal control. We know a great deal about photosynthesis; indeed the whole field of photosynthesis has burst wide open in the last decade. We know a good deal about the anatomy of leaves; but we know rather little about how this anatomy arises and is controlled.

A. MORPHOLOGICAL DEVELOPMENT

For one thing, the leaf is a curious structure. We have to think of a shoot apex which is, after all, a *symmetrical* growing organ, a mound. In dicotyledons at least, this apex is producing *asymmetrical,* dorsiventral structures continuously, and it is producing them from the side, not from the top where one might expect the apex to be producing subsidiary organs. Spread over the surface of the mound is a tunica, a layer of nondividing epidermal cells. Underneath this is the meristem, in which many of the cells are dividing steadily (fig. 7-1). Mostly they are dividing asymmetrically, so that the lower of the two daughter cells tends to elongate; as a result, a little way below the zone of rapid division they will form an elongating "rib meristem," which pushes the whole thing up. But at the base of the mound, cells are dividing tangentially and giving rise to a little sheet parallel to the surface of the mound, and because of that shape this develops into a thin flat sheet of tissue which overarches the mound. As the elongation at the base continues, pushing the whole apex up, these *leaves* are continuously formed along the outside of the apex. The next one will form at some position further round the base, the actual location depending on the species. In other words, it will be just inside the last one, though still on the outside of the growing apex.

In the case of grasses the base of the leaf is typically not a thin flat sheet, but a cylinder. It is produced in a somewhat similar

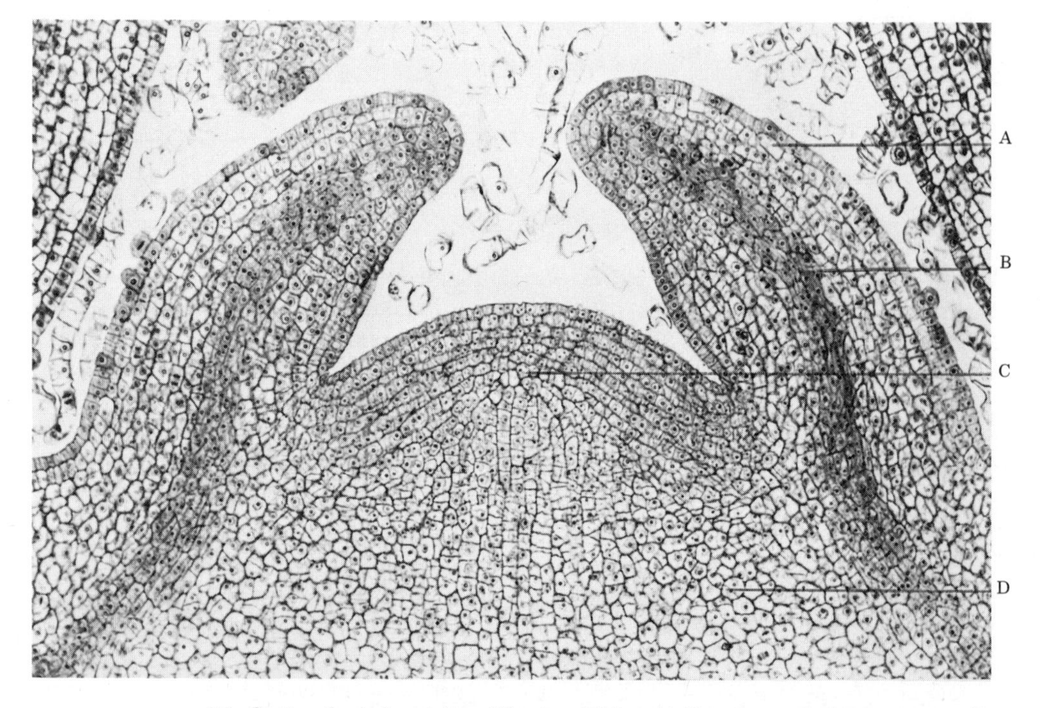

7-1 Section of apical meristem of *Fraxinus* (white ash). Note the continuing transverse cell divisions in the cells towards the bottom center (and left) which, when they expand, will push the meristem upwards. (A) Embryonic leaf; (B) Procambium; (C) Apical meristem; (D) Ground meristem. From the botany textbook of Wilson, Loomis, and Steeves; Holt, Rinehart and Winston, 1971. (Photo supplied by Dr. Wilson.)

way, though the monocot apex is apt to be a little flatter. As the growth activity gives rise to a sheath, and as the sheath grows up, the stem elongates inside it until there are a series of cylinders, each attached to the stem at the node below it. After giving rise to a series of such sheaths with leaves attached (in our common grasses often about seven), the stem elongates further to produce the leafless spike of flowers. In some plants, like the big arborescent bananas, the stem itself elongates very slowly, giving rise to the series of cylinders growing up inside one another. This produces what is called a pseudostem, which looks just like a stem but is really nothing but a series of these leaf sheaths inside one another. At the base there is a little stem apex producing all these organs in disproportionately large amounts.

Note that the coleoptile is a sort of leaf sheath with no blade attached; instead it is closed over at the tip. Like a leaf sheath, it bears stomata on both inner and outer surfaces.

Monocot leaves are dorsiventral too, but in the grasses this is true only in the upper part, the blade section. For although the basal part is typically a cylinder, this cylinder opens out above into a flat sheath comparable to that of the dicots, except that the venation, as you well know, is very differently arranged. Furthermore, there is a point about leaf structure which anatomists often fail to stress, namely, that both monocot and dicot leaves consist of two parts: in the monocot a cylindrical sheath and then a flat blade; in the dicot, typically, a solid petiole and then a flat blade. (Inevitably there are exceptions, for some dicots have sessile leaves, i.e., without petioles, and many monocots bear all their leaves direct from the basal crown.) In general, though, petiole and blade can be compared with sheath and blade. The comparison even extends to some of the nonflowering plants, for the fern leaf similarly comprises a central rachis and the pinnae, which resemble the leaflets of a compound leaf.

Thus one part of the typical leaf is radially symmetrical and the other flat. A flat structure is ideally suited to absorb light, for it can lie at an effective angle to the direction of the sun and thus absorb a good deal of the incident light. However, the flat structure is not as well suited for water retention because it provides a maximum of evaporating surface, while a petiole or stem exposes relatively little surface in proportion to its mass. Although the leaf needs the surface to absorb light well, because of the large surface it loses much water; so plants have a grave problem of water supply from which animals in general do not suffer.

In ferns, though not in higher plants, if one operates early enough to separate the leaf primordium from the apex, leaving it connected at the base, it will develop into a radially symmetrical organ, a little branch apex. It usually does not develop much, but it elongates somewhat so that one can see that it is a solid cylinder instead of a flat, leaflike structure. Both Steeves and Sussex have shown this with ferns. Another rather special property of fern leaves is that a very young but well-developed primordium, if placed on a nutrient medium containing sugar, tiny amounts of auxin, and the usual balanced mineral solution, will grow into a leaf. It will produce a rachis and then pinnae in the usual way. Hence in this primordium there must be all the "orders"—all the RNA arrangements—for the production of the whole leaf. The resulting leaf is unfortunately never of full size, but usually 1/5 or 1/10 the natural size; nevertheless it is recognizable and normal in its relative proportions (fig. 7-2). DeRopp has tried the same procedure with stem tips of rye, and Ernest Ball with several higher plants, especially Lupinus, but it just does not work. An isolated apex with several leaf primordia formed on it can be placed on every kind of medium, but all that happens is that the existing leaf primordia will enlarge a little and the apex may elongate somewhat, but growth soon comes to a standstill. Evidently the flowering plant apex does not have within it the requirements for making leaves, as the lower plant has. The addition of coconut milk, auxins, various vitamins, kinetin, etc., has no effect. Some "influence" which is present in the apex of ferns is absent from the apex of many plants (but cf. Chap. 9).

In 1882 Julius Sachs came to the conclusion that every part of a plant is formed by the action of a specific organ-forming substance. For leaves this substance would be *phyllocaline*, for roots *rhizocaline*, and for stems *caulocaline*; he may have had two or three more in mind. Indeed Sachs was no wild theorist; in each case he had very good reason for believing that there were these special forces. The kind of biochemical differentiation which must have occurred in the *Lupinus* and other flowering plant apices is evidence for something like a phyllocaline, that is, for a substantial influence which causes an apex to develop into leaves. Sachs knew one very suggestive thing: that roots promote the growth of leaves. This can be seen in some callus cultures in which, if they form roots, leaves may appear; that is, the apices will differentiate leaves. Unfortunately this behavior is far from general. It can hardly be due to formation of a cytokinin, because no cytokinins, at least none that Ball had available, had any effect on his apices. So, as far as the differentiation of the apex is concerned, we are still very much in the dark.

Now we come to the development of the shape of the leaf. Here there has been considerable study. In the majority of simple dicotyledonous leaves, of which tobacco is a good example, the young primordium is a relatively long narrow structure, while the mature leaf will in general be much wider relative to the length. It may have elongated 5 times but it has grown in width about 10 times, so the growth in width does not bear a constant relationship to the growth in length. This was investigated many years ago by Huxley, who called it *heterogonic growth*. D'Arcy Thompson described the same thing very interestingly in flat fish. This heterogonic growth, Huxley found, can be described in a relatively simple equation. If the width is equal to y and the length equal to x, then:

$$y = \mathrm{a}x^{\mathrm{b}}$$

where a and b are two constants. By choosing the constants appropriately, he could describe very accurately the growth of many leaves from primordia to maturity. It is easy to test this, because if we take logarithms on both sides and plot log y

7-2 Six stages in the growth of a leaf of the fern *Osmunda cinnamomea* developing from an excised primordium on nutrient medium. Above, the enlarging primordium; below, left, in the "crozier" phase, X5; center, uncoiling, X1.8; right, mature frond with typical fern morphology, X1.0. Scale in mm. From Caponetti and Steeves, 1963.

against log x, it should yield a linear plot. Several workers have in fact obtained beautiful linear plots, from the tiniest leaf in the primordium to the full-sized mature leaf, with all the points nicely falling on a straight line (fig. 7-3). Generally the exponent b is not a great deal more than 1, perhaps 1.1 or 1.2, so the difference in the rate of change in the two dimensions is not great, though it is quite important. Of course such an algebraic description is an oversimplification. There are often minor changes in leaf form; for instance, many leaves will elongate more in one part than in others, producing long slender tips, or rounded cordate forms, etc. Thus there are places where the dimensions do not change according to a simple law. But, in any case, such a description does not help to understand what is going on.

Parallel to the growth from primordium to mature leaf is the growth of the leaf of an etiolated plant, in which it is no more than a scale, to the green normal form in light. The change of proportions in this case is not very different from what happens in the normal, lighted primordium. The light that is needed to convert etiolated plants into green plants is basically red light. Red promotes and far-red inhibits, as they do in several other plant responses. The effect is not primarily due to photosynthesis. It is difficult to produce full-sized, mature, outdoor-type leaves from etiolated plants, since it requires very long exposures. Curiously enough, very bright light is not needed; it requires time rather than intensity. The reaction saturates at about 1/10 the light intensity which saturates photosynthesis.

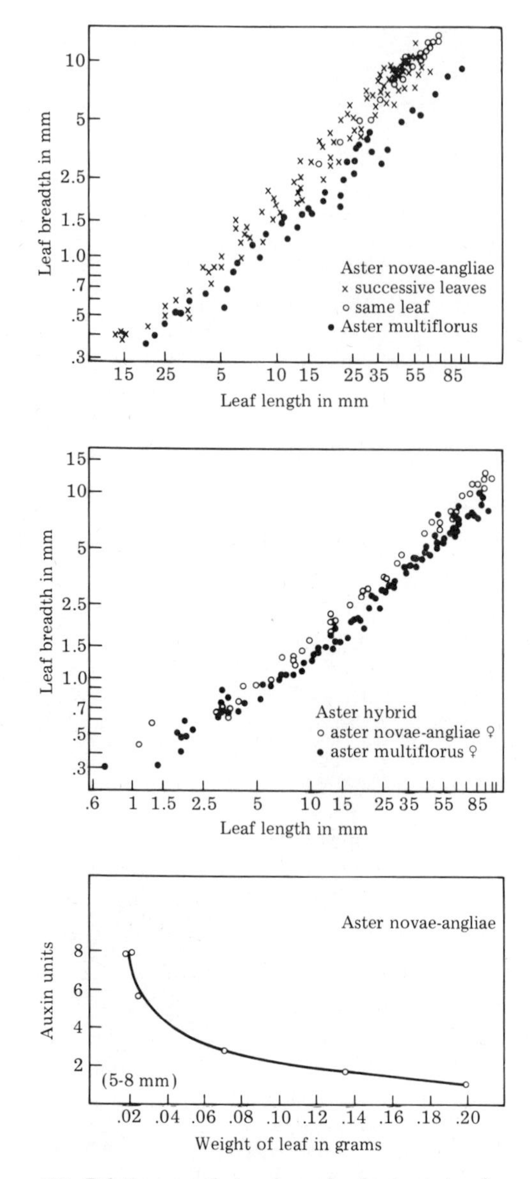

7-3 Relative growth (as shape development, i.e., log breadth vs. log length) of leaves of 2 spp. of *Aster* and the two hybrids between them. The curve below shows auxin output of leaves of *Aster novae-angliae* as function of leaf weight. From Delisle, 1938.

B. AUXIN RELATIONS

As stated at the outset, our understanding of the role of hormones in leaf growth and development is not impressive. Among the earliest of the hormone relations to be studied was the production of auxins. If leaves are placed on agar and then the agar tested on standard *Avena* coleoptiles, reasonable amounts of auxin are usually found. A few good generalizations can be made from a large number of such experiments. Firstly, if a leaf is divided into subsections, the amount of auxin diffusing out is least from the topmost section and increases toward the base. This is partly due to the existence of polar transport within the leaf, just as in the cotyledon or petiole. In fact the whole structure—leaf and petiole—behaves as one in regard to transport. The gradient becomes intensified with age. With a full-grown tobacco leaf George Avery obtained the results shown in figure 7-4. The results are in so-called "plant units." The gradient from tip to base is relatively steep. Because the production of auxin falls rapidly when the leaf ages, the gradient can be ascribed in part to age differences in the tissue, for the tip is the oldest part of the leaf. In a leaf that is still growing there is a meristem along the margin region, and the basal region tends to continue growing later than the rest. The apex, the oldest part, has stopped growing earliest. Thus the supply of auxin roughly parallels the growth rate.

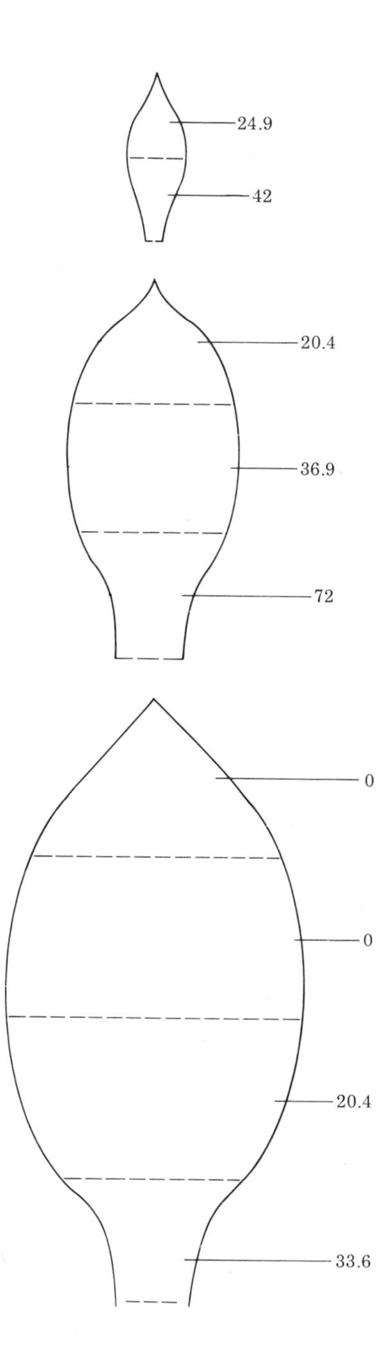

7-4 Auxin yields by diffusion from portions of tobacco leaves of three ages. Note the gradient from tip to base. From Avery, 1935.

Secondly, the total production of auxin decreases very rapidly with age. This has really never been explained. The tiny primordia can produce about as much auxin, absolutely, as a whole leaf. Richard Goodwin followed the production with age in *Solidago* (goldenrod) leaves, and calculated it as auxin units per square centimeter area (table 7-1). Leaf no. 1 was the first leaf that was big enough to handle. We see that the ratio between the amount of auxin per square centimeter in the first leaf and the seventh leaf is extremely high. Other workers have found that the yield of auxin per gram fresh weight varies inversely as the weight. All the data suggest that the production of auxin in these very young leaf primordia is from something already there. It seems unlikely that they are deriving it from any other organ, and it seems still more unlikely that they are making it themselves. Sheldrake has proposed that the formation of auxin results from the formation of xylem, which means from the death of cells. Considering the proportion of growing cells to xylem-forming cells in these small primordia, the data do not favor this view. More probably some precursor enters the apex and the primordia convert it into auxin.

Table 7-1 Auxin production per unit area of leaves as a function of leaf enlargement in *Solidago*

Leaf number	Auxin in units per cm²
1*	35
2	11
3	3
4	2
5	0.7
6	0.3
7	0.2

* No. 1 was the youngest.
From Goodwin, 1937.

In perennials and woody plants the age effect is shown in a different way, namely in the seasonal development. As the buds open, the young leaf primordia begin to produce auxin and then gradually stop as they mature, so that in trees there is a wave of auxin production in the spring. This is nicely correlated with the well-known wave of cambial activity, which has been recorded by earlier workers as moving down the tree, from the young twigs towards the base, at the rate of a few decimeters per day—a rate that corresponds fairly well with the observed rate of auxin transport. Figure 7-5 shows the auxin production of the bud and growing leaves of an apple twig. As the leaves enlarge, their auxin production increases steadily; then it falls rather quickly as they mature.

Figure 7-6 shows a young *Vicia faba* plant. These numbers shown are the plant units of auxin produced per hour in absolute amounts (i.e., not corrected for size). The axillary basal buds, in the axils of leaves that have not grown out, produce only tiny amounts. The rather large, but not yet fully grown, leaves also produce only small amounts. A leaf in the middle of its expansion phase produces 1.5 units. A young leaf, rapidly growing, produces 2.2 units. But the apex with two or three minute leaf primordia produces 12 units. So again the ratio between the productivity of tiny leaf primordia and fully grown leaves is very large.

The third major controlling influence is that of light. All of the above experiments

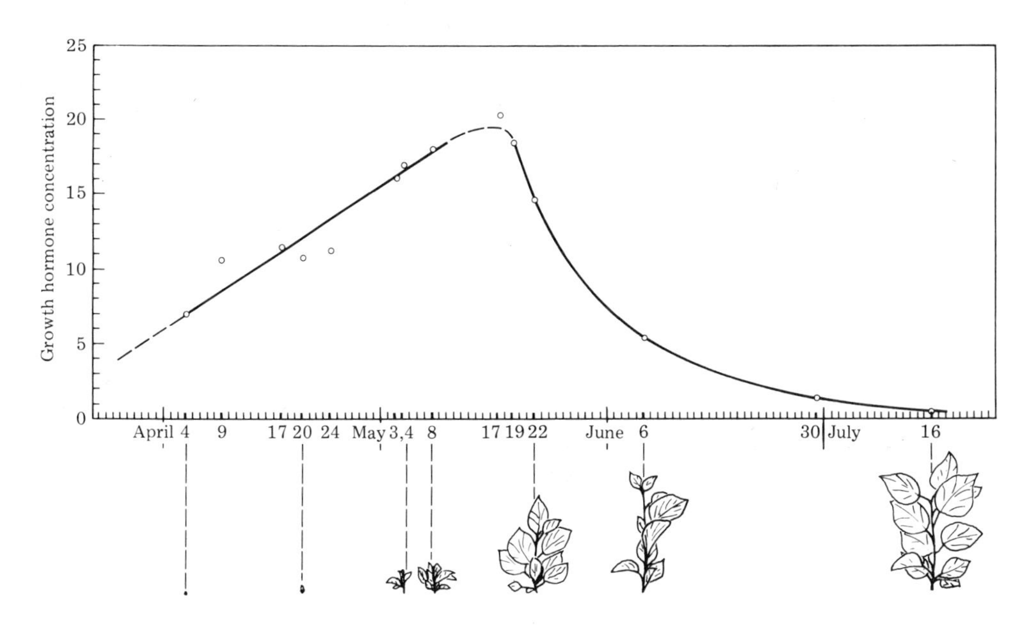

7-5 Auxin yields by diffusion from buds and young shoots of the McIntosh apple throughout the spring. Here and in figs. 3 and 4 the yields are expressed as degrees curvature of standard *Avena* plants. From Avery et al. 1937.

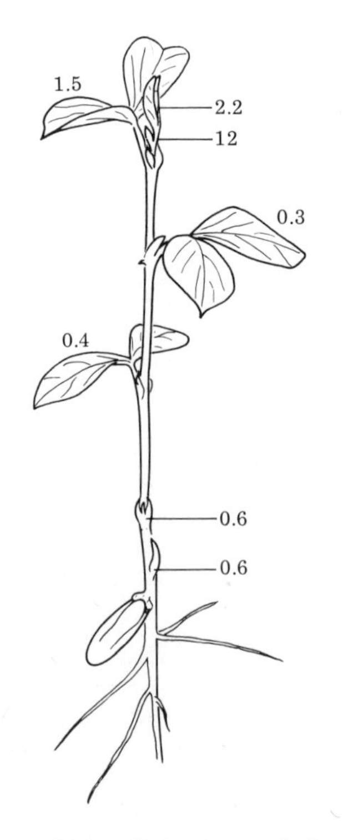

7-6 Auxin yields by diffusion from the buds and leaves of a young *Vicia faba* plant, grown in light. The absolute data are not directly comparable with those preceding. From Went and Thimann, 1937.

were done in white light with green leaves. But if one puts a series of green plants into the dark and measures the diffusible auxin day by day, it falls off very rapidly, reaching zero in about 5 days, more or less. Correspondingly the diffusible auxin yields from etiolated pea plants are much lower than from light-grown plants of the same age, as Scott and Briggs showed some years ago. It is a curious relationship that on the one hand auxin is produced in light, or the production is *maintained* in light, but on the other hand, as we all know, the elongation of the stem below the apex is much greater in dark than in light. Thus as the auxin production falls the sensitivity of the tissue to auxin must increase greatly, since the etiolated plant is far taller than the light-grown plant. It might be, of course—and this is just a speculation—that, although the auxin is needed, the limiting factor in stem growth is not auxin but gibberellin, and that the production of gibberellin (unlike that of auxin) actually increases in the dark. So far as I know, there are no data to support or deny this speculation. The appearance of gibberellin-treated, light-grown plants is, however, suggestive of the appearance of etiolated plants, at least in peas.

As might be expected, the amount of auxin in the stem tends to increase towards the base, usually passing through some maximum, though not always. Thus if we let auxin diffuse from segments of the stem, or if we extract them with ether, the amount of auxin available per millimeter of stem segment commonly increases towards the base. Sometimes the amounts coming from the apex are quite small, and the contribution from the young growing leaves is larger. The steady increase towards the base is shown also in the sensitivity to differentiation of cambium into xylem in the petiole. Samantarai and Kabi at Cuttack, India, recently found that the cross-sectional area of xylem occupied 1.2% of the total area at the apex of the petiole and 3.5% at the base. Applied auxin (indoleacetic or indolebutyric acid, the latter acting at lower concentrations) increased these figures to 2.8% and 7.9% respectively. Comparable results were seen in leaves of two other species. Inhibitors of auxin action on growth inhibited this action on vascular differentiation too.

Figure 7-7 shows how growth and auxin distribution vary in shoots of *Ginkgo*. The marks show the elongation that each initially marked millimeter has undergone. The tip, of course, is growing most rapidly, and the amounts of auxin are quite low there, in spite of its being produced at a good rate. The amounts pile up towards the base but then begin to decrease near the extreme base. In the two particular branches for which detailed data were obtained, a peak in the amount of auxin occurs about three-fourths of the way down. Thus the yields of auxin are not parallel to the growth rate—very far from it, the peak of auxin content coming where growth has about stopped. Evidently, then, the auxin level in the stem is not the controlling factor, as noted above.

The production of auxin in this *Ginkgo* plant shows interesting relationships with the development of the bud. In figure 7-8 the buds have been divided into arbitrary stages of development, with number 5

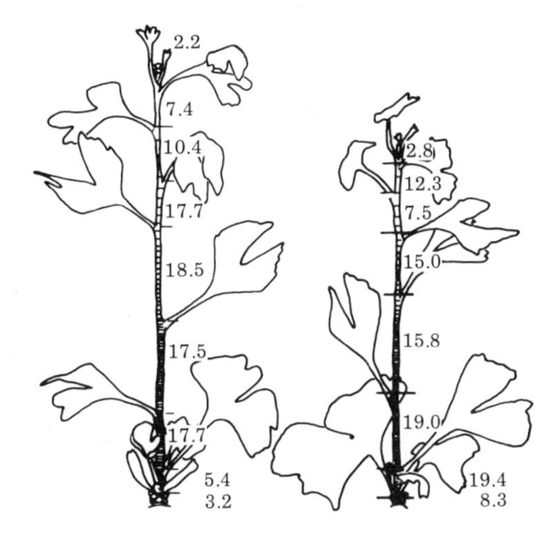

7-7 Elongation of the internodes of two long shoots of *Ginkgo biloba* and the auxin diffusing from the base of each internode. From Gunckel and Thimann, 1949.

showing measurable small leaves. In the first group the auxin yield goes through a clear peak at buds of medium size whose primordia are still very small and then slowly tapers off. In a plant like *Ginkgo*, some shoots elongate, and some do not—they just produce their five leaves and then remain stationary. In those shoots that are going to elongate ("long shoots") one finds the same sort of early peak and decrease, but then there comes another very large increase. This increase is correlated with the elongation. Here it seems that auxin is indeed the limiting factor. The production of the auxin from the leaf, especially from the young leaves, can thus have very dramatic effects on the elongation of the stem. This development of short and long shoots is taken up again in chapter 10, section E.

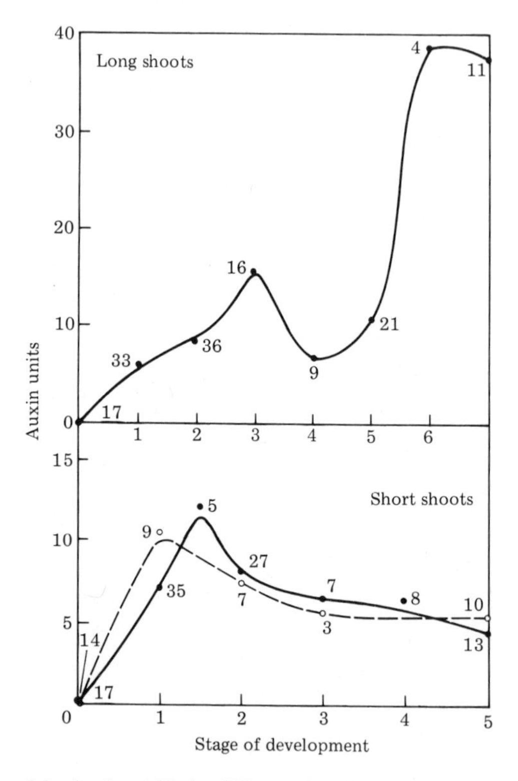

7-8 Auxin yields by diffusion from buds which will produce short and long shoots respectively of *Ginkgo biloba,* showing the second wave of auxin associated with elongation in the latter. The figures show the number of determinations at each point. From Gunckel and Thimann, 1949.

It is worthwhile to review the several effects of the auxin production in the young leaves. Firstly, together with gibberellin, it causes the elongation of the parts below. Secondly, as we shall see later (chap. 9), there is a bud, a remnant of the apical meristem, in the axil of every leaf. The auxin coming from the leaf inhibits the development of that bud to varying extents in different plants, and may modify the subsequent fate of the buds after they unfold. The removal of the leaf, especially of the blade and at least part of the petiole, will stimulate the opening of the bud in the axil and may lead to its elongation. (Removal of the terminal apex has a much stronger effect.) Thirdly, the auxin coming from the leaf prevents abscission of that leaf (see chap. 13), so that leaves deprived of their auxin supply will abscise. Some leaves abscise just the blade, and the petiole stays on; others abscise at the base of the petiole. No doubt, then, a normal function of the auxin is to prevent the abscission of the leaf. Fourthly, the auxin promotes the formation of the xylem in the petiole.

The fifth function, which has only been shown in a few plants, is rather unexpected: the auxin coming from the leaf modifies the type of flower formation on the stem. This was first shown in plants like cucumber or hemp, which form male and female flowers. The balance between male and female flowers commonly shifts during the season, male flowers tending to be formed in the spring and female flowers later. It has been shown now in a number of cases, first by Laibach, that the change from maleness towards femaleness is promoted by the auxin coming from the leaf. Control is not limited to auxin, for it can be promoted to some extent by high nitrogen and to some extent by change of day length; long days favor femaleness, short days, maleness. But the auxin effect is very clear. Probably the best work on this has been done by Galun in Israel, with tissue cultures of cucumber. These cultures will continue to produce flowers, on the right medium, for many weeks. He found that of the first eleven flowers, none were female, all 11 were male. But when he supplied the tissue with auxin, 12 out of 13 of the first flowers were female. Though similar effects have been seen in other plants, this represents the largest effect observed. Both indoleacetic acid and synthetic auxins would act in the same way. Hence the stream of auxin from the leaf has, as a fifth function, a modifying influence on the development of flowers.

A curious correlation is that in general, where it has been tested, the ability to oxidize auxin goes parallel with the yield of auxin obtainable. This may not be physiologically important, because nearly always the ability of a given tissue to oxidize auxin is far greater than that which would destroy all the auxin it produces. In our own tissue cultures we came to the conclusion that if the culture really destroyed auxin with all the oxidase system that one could demonstrate to be present, it could destroy 100 times the amount of auxin that was available to it. Since the culture grows and yields auxin by diffusion, it follows that the auxin oxidase-peroxidase is not acting on the auxin of the tissue. The enzyme must be sequestered in some organelle which separates it from its substrate, so that it only gets to act in exceptional circumstances or as a result of wounding.

While the effects exerted by the auxin that comes out of the leaves are fairly clear, the effects of auxin applied to leaves are usually rather disappointing. Doubtless the leaves have an adequate supply of auxin, so that it is not a limiting factor. The growth of major veins, especially the mid-vein and a few of the largest basal veins, can be promoted by auxin. If the veins stick out on the back side, auxin applied to them can produce appropriate curvatures, so that the growth of these veins can be easily accelerated. Long-continued applications produce many growth abnormalities. Old professor Dostál at Brno, Czechoslovakia, who retired about 30 years ago, has made many experiments of this type. (Only since he retired has he had time for extensive experiments; unfortunately in many universities in Europe a retired professor has no laboratory of his own, so Dr. Dostál installed a small lucite greenhouse on the window of his office and does his experiments there, some of them very interesting.) As I have seen on two visits to him, what he does is to treat small plants continuously with very low concentrations of auxin for weeks or months at a time, renewing it daily or frequently. He also uses auxin transport antagonists like triiodobenzoic acid. In this way he can imitate many peculiarities of diseases, especially modified apices; leaf primordia that would normally have been separate structures are made to grow together into a continuous cuplike form. More basal leaves sometimes develop a lateral tissue which joins them together so that decussate leaves appear as a continuous sheet around the stem. It is hard to say what these mean from the point of view of normal physiology, but they do help to explain the abnormal growth of diseased plants.

C. HORMONES OTHER THAN AUXIN: GIBBERELLIC ACID

The gibberellin relationships of leaves are less clear than those with auxin. Gibberellin does not generally diffuse well out of leaves—the amounts are small, and the extraction method is unfortunately seldom decisive. If one applies gibberellin to young leaf primordia it tends to make the leaves somewhat longer, and they are usually paler green, since it does not promote the formation of chlorophyll. There is a net increase in area but a marked change in the slope of the growth curve. In monocots, as we have seen, the effect is very strong, both on the blade and on the sheath—though stronger on the sheath, as it is indeed stronger on the petioles of dicots than on the blades. Again the sheath and the petiole, which are anatomically quite different, are physiologically somewhat parallel. Many of the early bioassays of gibberellin were based entirely on the elongation of the sheath. The assays were very simple: one placed a droplet of gibberellin at specific locations on a young dwarf monocot; corn is often used. Some days later the length of one or more of the leaf sheaths is measured. With dwarf corn the length of the second leaf sheath is particularly sensitive to the concentration of gibberellic acid. One can measure with this procedure a tenth of a microgram of gibberellin, sometimes less, depending on the conditions. The test is reliable and very sensitive. The response of the coleoptile is not so striking. One could conclude on anatomical grounds that a coleoptile is really just a leaf sheath, one that bears no blade, but that conclusion is not borne out as far as elongation in response to gibberellin is concerned.

Evidently the normal growth and development of leaves requires interaction between the two hormones, auxin and gibberellin. In leaves of the sweet potato (*Ipomea batatas*), whether isolated or attached to the plant, Kabi and Sarma in India have shown striking synergism, exerted in the increase both in leaf-blade area and in petiole length. The critical data are given in table 7-2. Since each hormone was at its optimum concentration, it seems that they must be exerting their actions on two different systems.

A second kind of hormone balance is that between gibberellin and cytokinin. Here the elongating effect of gibberellin is brought under control. The balance was elegantly worked out on tobacco tissue cultures recently by Engelke, Hamzi and Skoog (1973). Tissue cultures are well adapted for such studies because the test substances can be readily supplied in the growth medium. Briefly, the yield measured as weight of tissue increases with increasing concentrations of both cytokinin and gibberellin; with either one alone it is very low. But the form of the leaves depends on the *ratio* of the two hormones; with GA at 125 times the concentration of cytokinin the plantlets were tall and the leaves very long and narrow, while with cytokinin alone the plants were short and the leaves almost circular. In all the mixtures, cytokinins tended to promote expansion of the blade tissue and to inhibit elongation, while gibberellins act in just the opposite sense. Tronchet and co-workers in France showed very similar trends by application of hormone solutions to young leaves of several different dicot genera. Thus, unlike the auxin-gibberellin interaction, there is no synergism here, but a delicate balance. Indeed one wonders how it is that Nature can produce such highly species-specific and reproducible leaf shapes when the controlling hormone relations are so multifarious.

Table 7-2 Percentage increases over the growth of controls in *Ipomea* leaves

	Blade area		Petiole length	
Treatment	Detached	Attached	Detached	Attached
Auxin[a]	42.5	41.2	74.0	88.4
Gibberellin	54.7	53.2	110.4	129.8
Both	88.8	85.4	155.4	162.5

[a] The auxin was indolebutyric acid 2.5 ppm, the gibberellin was GA_3 50 or 100 ppm. Initial blade area was 27 sq cm, initial petiole length 5.0 cm.
From Kabi and Sarma, 1973.

The interaction may involve the synthesis of gibberellin too. For when the roots of tomato plants are flooded, which means they are deprived of oxygen and therefore of their ability to carry out synthetic processes, the gibberellin content of the shoots decreases. It can be restored by spraying with cytokinin (Reid and Railton, 1974). So it looks as if cytokinin formed in the roots controls the synthesis of gibberellin in the leaves.

One clear-cut action that gibberellin exerts in leaves is seen in those plants which have two leaf forms—adult and juvenile. Gibberellin strongly influences the change from adult to juvenile. Strangely enough, in two different plants it works in opposite directions. In ivy, when quite small amounts of gibberellin are applied, the juvenile leaves stop being formed and adult leaves are formed instead. The plant changes its form, becoming more shrubby, less of a vine; there are less rootlets on the stem, and it usually flowers if it remains in the adult stage. On the other hand in *Ipomea* the opposite is true; gibberellin causes adult leaves to stop being formed and juvenile leaves are formed. No one knows what this means physiologically. It implies a fairly important change in the way the apex is cutting off cells, but has not yet been studied in anatomical detail.

Gibberellin also modifies the type of flowering in the opposite direction from auxin. We saw above that auxin tends to promote femaleness. In the experiments of Galun, if gibberellin was added as well, the number of female flowers decreased almost to zero. In one experiment gibberellin plus auxin reversed the auxin effect, and the culture bore no female flowers and 9 male flowers. Thus the balance between the hormones coming out of the leaves may play a very important part where maleness and femaleness is delicately balanced. Such balance occurs in a number of trees—the red maple, for instance, where some trees tend to form only male flowers and some mainly female flowers. I suspect that the balance is hormonally controlled in the same way.

D. ABSCISIC ACID

Now we come to a remarkable discovery about the hormone balance of leaves. During the progress of hormone studies when Bennet-Clark, Nitsch, and others began making extracts and chromatographing them, they commonly found that, while auxin would appear at certain points on the chromatogram, at other points, usually moving faster in the solvent, an inhibitor would appear. As a result of a great deal of work, this substance, called inhibitor-β, began to be recognized as a normal constituent, especially of mature tissues. In dormant tissues, such as potatoes, it would appear in good amounts, but then as the potato got nearer to germinating the amount of inhibitor-β would go down. In trees, inhibitor-β tends to be found in leaves extracted near the end of the season. Dr. Wareing, in Aberystwith, noted, as others have, that after a sharply defined time in the late season leaves stop being formed, and only bud-scales are formed. With birch (*Betula*) and sycamore (*Acer pseudo-platanus*) the change from forming regular leaves to forming bud-scales can be hastened by transferring the young trees into short-day conditions, or just by putting one branch of a tree in short days. On the other hand, when he started removing the late season's leaves, new leaves continued to be formed for a longer time, delaying the onset of bud-scale formation. Evidently the late-formed leaves promote the transition from leaf to bud-scale. Believing that perhaps some hormonal influence was at work, Wareing and his students extracted the late-season leaves and obtained an extract which, when placed on the apical bud, hastened the development of bud scales. In some experiments it stopped the formation of leaves after only one or two more had formed, and the formation of bud-scales began even though the plant was being held in long days. Thus the mature late-season leaves produce something which hastens the onset of adaptation to the winter condition. From this work a large chemical development took place and after some years the group isolated a pure substance, which he called dormin, since it leads to dormancy. Milborrow, at the Shell laboratories in England, was able to establish its structure; it turned out to be a carotenoid derivative. The formula shows an aliphatic ring, two gem-methyl groups as in carotene, one double bond only, and an unsaturated chain bearing an acid group. Because of that acid group it was sufficiently water-soluble to be extracted in at least semipolar solvents.

By a strange coincidence, Addicott, in

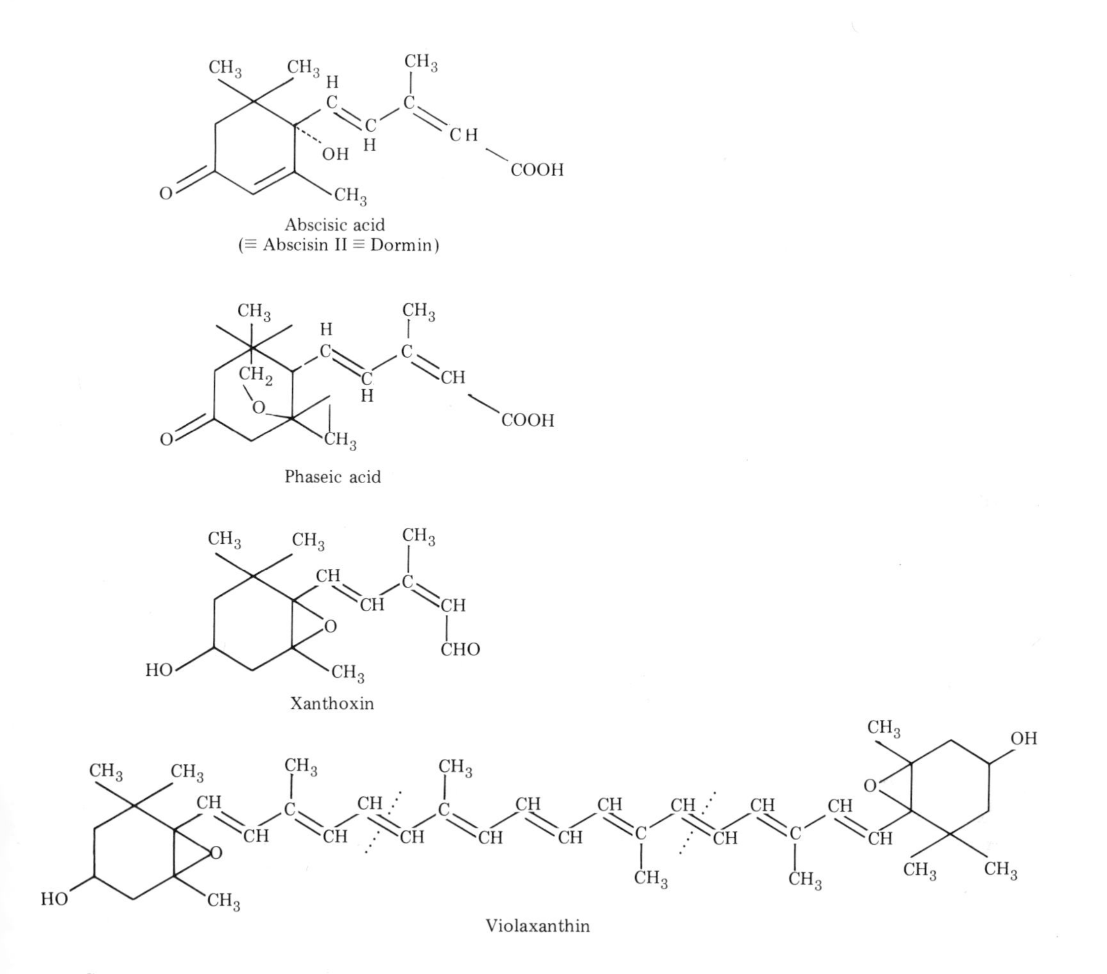

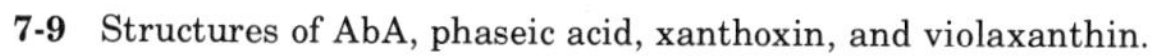
7-9 Structures of AbA, phaseic acid, xanthoxin, and violaxanthin.

California, had been working on the abscission of cotton—a subject of great interest to cotton growers, since many of the young bolls abscise before reaching maturity. Because this represents a large loss of potential crop, the cotton growers had been supporting work on the physiology of abscission for some years. Addicott had as co-worker a clever Japanese, Okuma, who, to cut a long story short, had made extracts which promoted the abscission of cotton bolls. They worked out the chemistry, and to everyone's astonishment it turned out to be the same compound. It was called abscisin II. At the growth hormone conference in Ottawa in 1969, there was much discussion as to what this substance should be called, the Englishmen feeling that it was primarily dormancy-inducing and the Americans that it was primarily abscission-inducing. But since the Addicott group had been the first to recognize this hormone, it was called abscisic acid, AbA. No sooner had abscisic acid been identified than it became clear that it had more than one role. It is produced mainly in mature and in dormant tissues and in some fruits, and its role in abscission may not be its most important one, as we shall see below.

The first development was that other related compounds quickly came to light. In 1969 MacMillan identified phaseic acid in *Phaseolus*. Phaseic acid has the same basic structure as AbA, but instead of the double bond in the ring there is a 3-membered ring at this point (fig. 7-9). It is apparently formed in dormant bean pods and accumulates there. It is also formed by oxidation of AbA. Weight for weight it is not quite as biologically active as AbA. The following year (1970) xanthoxin was identified by Taylor and Burden from seedling leaves of wheat and dwarf bean. Again the structure is related, though, instead of the hydroxyl or the oxygen in a 3-membered ring, it has an epoxide. The side-chain is the same except that it bears an aldehyde group instead of an acid. There is a special interest in this compound which probably applies to the other structures too—xanthoxin has exactly the structure of part of the carotenoid violaxanthin. In figure 7-9 the dotted lines show where the molecule would be divided, doubtless by oxidation. Indeed it could be shown by Taylor and Smith in England that when violaxanthin is exposed to ultraviolet light xanthoxin is formed. Perhaps these three compounds are only the beginning of a family of substances, derived either directly from the oxidation of carotenoids or from a closely related synthetic pathway, having a number of functions in the plant and being particularly characteristic of older or mature tissues or organs.

The special interest of these compounds for the physiology of leaves is this: if we extract mature leaves and measure the amount of abscisic acid present, by determining the growth-inhibiting power of the extract on a rapidly growing test organ, such as coleoptile segments, we find a fairly repeatable amount. But if we first let the leaves wilt for an hour or two in a stream of air, the amount of abscisic acid goes up dramatically and very rapidly. In one of S.T.C. Wright's experiments with wheat leaves, for instance, he calculated from the test that the leaves contained 44.7 micrograms of AbA per kilogram. If he wilted the leaf for 105 minutes he obtained 234 mg—6 times as much. Bean leaves in a similar experiment reached 200 mg of AbA per kg in an hour's wilting. It is hard to believe that so much AbA can be being synthesized so rapidly in the wilted leaves; it seems more likely that it is present as a complex, perhaps with a carbohydrate residue, and is released in the wilted leaves by a hydrolytic process. There is reason to believe that hydrolysis sets in very rapidly on wilting, as in aging (chap. 12).

What is more remarkable is that AbA has a specific function in controlling the wilting process. Suppose we take a young bean plant with just its simple leaves and apex, and blow warm dry air on it: the leaves will wilt very quickly, but after an hour or two they will begin to recover. If we examine the leaves we find that when wilting began the stomata began to close. Indeed, stomatal closing usually follows wilting. When the stomata become tightly closed, enough water accumulates to reestablish turgor so that the leaves recover. This closing of stomata parallels exactly the production of abscisic acid. As the leaf begins to wilt the AbA level shoots up, and the stomata begin to close. Wright's paper in 1973 establishes a perfect parallel between the amount of AbA produced, or liberated, and the closure of the stomata. Some data collected by Wain in a special lecture are shown in table 7-3. The very rapid action of AbA in stopping the transpiration and the uptake of CO_2 is shown equally by applying it to excised leaves with their petioles in water (fig. 7-10). Phaseic acid acts similarly, but also seems somehow to interfere with photosynthesis.

There can be little doubt, therefore, that AbA acts as a stomatal-closing hormone. This is extremely interesting because the physiology of the opening and closing of stomata has been obscure for so long. Classical work by Scarth at McGill in the 1930s showed that the opening and closing of stomata of isolated leaves could be brought about by pH changes. On leaves, or epidermal strips from leaves, placed in alkaline media, the stomata nearly always open; in acid media they close, and close very rapidly. Furthermore they close in high CO_2 concentrations, especially in the dark. This, by the way, is an unfortunate fact which has made it difficult for nurserymen who want to supply extra CO_2 to increase the growth of greenhouse plants. Extra CO_2 added to the air causes more photosynthesis, but it also causes the stomata to close, so that after an hour or two the plants are no better off than before. In spite of this, some nurserymen do add CO_2 to the air in closed greenhouses and do derive some benefit.

This work led to the idea that, since CO_2

Table 7-3 Changes in the abscisic acid content of leaves on becoming wilted or waterlogged

Plants	Abscisic acid content, micromoles per kg fresh weight of leaves		
	Control	Wilted	Water-logged
Maize (two European cvs.)	2.5	18.8	—
Maize (drought-resistant cv.)	13.1	50.2	—
Tomato	68	—	524
Dwarf bean	30	—	190
Wheat	31	—	252
Rice	16	—	5

From Wain, 1975.

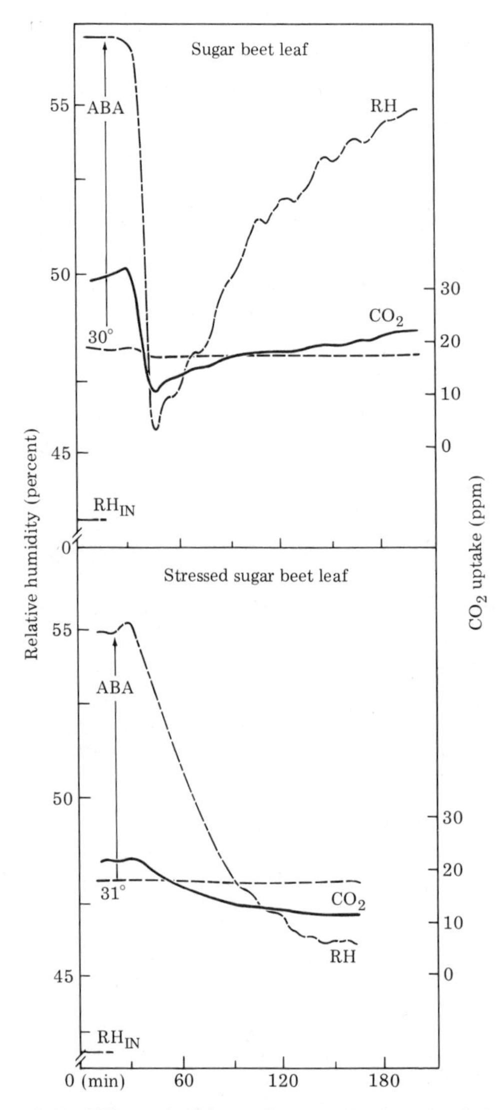

7-10 Effects of AbA on the transpiration (dot-dash line) and the CO_2 uptake (solid line) of an excised sugar-beet leaf. The transpiration was measured as the relative humidity (RH) of the air passing over the plant in a closed system. In the lower figure the leaf had been deprived of water for 4 hours before the AbA was applied. From Kriedemann et al. 1972.

is an acid, normal opening and closing as a function of light and dark may be due to the photosynthetic removal of CO_2. As soon as light falls on the leaves (and the response is extremely sensitive), the stomata begin to open. We know that the guard cells contain chlorophyll, and if they immediately begin to remove CO_2 there would be a slight change in pH towards alkalinity, which could make the stomata open. It now appears that the mechanism is more complicated, since the opening involves the uptake of potassium ions, which are brought in very rapidly. Probably also there is excretion of hydrogen ions to balance the electronegativity, and the dependence on pH is the result.

Thus in AbA we have a substance which very rapidly causes closure and is a stronger acid than CO_2. The reason that its action in causing stomatal closure is more powerful than that of CO_2 or externally applied acid buffers might simply be that this happens to be an acid which is produced extremely rapidly and is produced inside the guard cells. However, its effect is doubtless more specific than that, since Mansfield et al. have shown that it inhibits the potassium uptake also. It seems to require the presence of CO_2 for this action. In any case it looks as though the British workers have finally uncovered the mechanism of the opening and closing of stomata. One can only guess at the enormous agricultural importance of developing a real control of the opening and closing of stomata in crop plants.

While this elucidation is very satisfying, the whole reaction raises a number of fresh problems. Dr. Milborrow of the Shell Laboratories in England points out that the response of a leaf to a period of water stress must involve three separate messages: (1) start biosynthesis of AbA; this comes almost immediately, certainly within a few minutes; (2) stop the biosynthesis; this follows within about an hour, depending on the extremity of the water shortage; (3) begin AbA decomposition; this may come as soon as the water stress is relieved by stomatal closure or water is supplied. However, growth inhibitions due to AbA may persist for 2 days or so. It is too early to speculate on the origin and nature of these messages.

The fact that AbA inhibits growth does not rest wholly on its ability to antagonize auxin or gibberellin in the growth of young elongating segments, for it inhibits the elongation of roots as well as that of shoots. In the case of roots, elongation is promoted only by excessively low auxin concentrations, as we have seen, so that AbA must act against the elongation process, rather than as an antagonist to the growth-promoting hormones. That is an important distinction, which will be critical in working out its mode of action. The effect of abscisic acid is observed quickly, probably in 4 hours. This suggests again, as with the reaction of stomata, that AbA enters rapidly from the external solution and brings about a general slowing down of several kinds of reactions.

In connection with stomata, I must mention that a number of years ago Professor Bünning and his colleague Sagromsky, at Tübingen, showed, with very young leaflets just emerging from the bud, that if one spreads auxin paste on the surface, the formation of the stomata is prevented. This is a peculiar reaction, which does not seem to parallel any of the other effects of auxin that we know. To produce a stoma requires cell division in the epidermis. Two cell divisions occur, the first to produce a pair of guard mother cells, and the second to cut off the guard cells and thus produce a stoma. If auxin is spread on very early, the first divisions occur and then they stop; nothing further goes on and thus no stomata are formed. Bünning calls these little groups of cell divisions *meristemoids*. They occur all over the leaf in a standard pattern, at a certain distance from each other. Evidently each meristemoid exerts an inhibiting influence on the adjacent cells, discouraging them from wanting to become meristemoids. Only cells at a certain distance can form a new meristemoid, an arrangement which results in a fairly evenly spaced pattern. The inhibiting secretion cannot be readily transported, or else it is very unstable. Bünning had the idea that it may be auxin and sought to find out what would happen if he applied auxin. He did not actually change the pattern, but what he did was to stop all of them from going through their cell divisions. One might go so far as to deduce from this that each formation of guard mother cells liberates a little auxin, which diffuses from all these centers. But why did these centers suddenly become auxin-producing areas? Unfortunately, further work has not been done on this problem; it is ripe for some plant morphologist to take it up.

THE FORMATION AND GROWTH OF ROOTS

A. FORMATION OF THE ROOT SYSTEM IN INTACT PLANTS

The most commonly used criterion of the germination of seeds is the emergence of the radicle, or embryonic root. Since the rapid uptake of water plays so prominent a part in seedling growth, the development of functional roots has to be very rapid. Where one main root is produced, as in *Pisum, Zea,* or *Lens,* this root has become a major subject for study, both for hormonal function and for anatomy. Members of the grass family have no main root but a spreading, fibrous root system, and correspondingly the young seedlings quickly develop several slender roots; wheat, for instance, produces 4, oats produced 4, barley often produces 5. Thus every seedling has its group of seedling roots, which are produced very quickly. After they have been produced there is often a slight pause as the shoot develops, and then the shoot begins to generate additional roots. In the mature plant, the largest number of roots by far are these that are secondarily formed.

When I was in Australia, the problem was being discussed of raising wheat varieties which would survive in the central part of the country where the soil is porous and the rainfall is very erratic, so that for cereals everything depends on their getting their roots deep enough. The studies that have been made there of some resistant varieties of wheat showed that most of the secondary roots only penetrate a foot or so into the soil; but 2 of the 4 seedling roots continue growing throughout the life of the plant, and these may get 20 feet down in the soil. As a result, for such varieties, the whole plant is dependent upon these two seedling roots. Only if they get down deep enough before the upper part of the soil is thoroughly dried out will the plant survive. This leads to some interesting calculations on the rate of uptake and movement of water in these two very fine roots 20 feet long. With the several long leaves transpiring at a great rate in the very dry atmosphere, and the total cross-section of the xylem not much over 100 microns diameter, the rates of water transport must be up to 50 cm per second

It is the continuous addition of secondary roots in the mature plant that makes up most of what we observe as the typical "root system." Occasionally the formation of roots by the mature plant is spectacular, and the most striking examples are those in some of the trees, like the *Ficus* species, which go on forming roots, not only underground, but from the base of the branches or from the trunk.

Figure 7-11 shows a very remarkable tree; this is a single individual and there may not be another to equal it in India. This *Ficus* is in the grounds of the Theosophical Society just outside of Madras. What looks like a forest of trees is seen on closer observation to be all roots formed from branches. This particular species forms roots, not from the base of the branches or from the trunk, but from far out along the branches; and the roots, as soon as they have reached the ground and branched out, obviously begin transporting,

7-11 The great number of massively thickened roots of the banyan tree (*Ficus benghalensis*) at Adyar, Madras.

as roots should. Thus they develop very marked secondary thickening—there is an active cambium in these roots—and they provide physical support for the branches, which then can continue to grow. This photograph shows what has resulted. There are so many roots and they cover so wide an area—several acres—that public meetings are held here, with 5,000 people all sitting under the shade of what is really one tree. The main trunk is far off to the left and the picture is taken near the periphery of the tree. Notice how thick and solid these roots are, even high up. No one knows the age of this individual, but it is probably no more than 200–300 years old.

Although this ability to form roots at a great distance from the main trunk is special for *Ficus*, to produce additional roots from near the base of a moderately large plant is not rare. If you walk through a cornfield in the Midwest, you will commonly see what are called prop roots, emerging a few centimeters above the surface of the ground and growing at a slight angle to the main stem.

Such roots are very prominent in *Pandanus* species, a group of common tropical trees in the Caribbean, South America, and the South Pacific. On these plants, roots form at the base of the major branches, not high up, and also on the trunk. They grow at a slight angle to the vertical or the main stem, usually around 15°–20°, and they continue unbranched until they reach the ground. When they enter the soil, and not before, they produce numerous branches, forming a nice secondary root system just below the surface. In their aerial parts there is generally no branching at all. Why this is I have no idea.

B. THE INITIATION OF ROOTS ON STEMS

This phenomenon is, like many other plant processes, somewhat mysterious. One has only to think of the minutiae of root formation—the initiation of root cells in apparently perfectly normal shoot tissue, their outgrowth, their characteristic vascularization, and the control of the angle of their development—to see that there are many unexplained things happening here.

The first real attempt to construct a theory was, as we noted earlier, due to Julius Sachs, who in 1882 proposed the idea of specific substances which act to form leaves, roots, or stems respectively. These substances would move polarly in specific directions in the plant. Both Vöchting (1854) and Beijerinck (1856) discussed the problem, but for 40 years it was nutritive factors that dominated the thinking in this area. Sachs's root-forming substance would be formed primarily in leaves and would migrate to the lower parts of the plant, there stimulating shoot tissue to produce roots. Although this was a concrete proposal, no real attempt was made to disprove or even test the theory for many years, although there were some studies of the formation of roots on cuttings and, of course, gardeners continued, quite ignorant of any proposals by Julius Sachs or any other academics, to root cuttings according to time-honored procedures. It was not until 1924 that Van der Lek of Holland made the first experimental study, a beautiful piece of work which was partly his thesis but was continued on for some years after he received his doctorate. It was a serious attempt to relate physiological experiments to anatomical structure. Such a viewpoint is not uncommon now, but at that time it was original.

He made cuttings of numerous plants, examined their rooting ability, and then, after they had formed roots, studied the anatomy of the roots. Before any roots were formed he also studied the anatomy of the stem, in order to see whether, in those plants that root readily, root initials were already present. He found indeed that in some plants (the raspberry is a classical example) quite large numbers of root initials are already present in the stem and all that happens when they are placed in moist soil is that they grow out; no particular stimulus is needed. In other cases, however, it was clear that the formation of roots was dependent upon a stimulus, which typically was supplied by the leaves. With cuttings of willow and grape, Van der Lek was able to show three important things: firstly, that the formation of roots on cuttings was dependent on the presence of leaves; secondly, that it always took place below the leaves, never above, indicating that whatever constituent came out of the leaves moved downwards; and thirdly, where roots were formed in only small numbers, they were formed on the same side of the cutting as the leaves.

Figure 7-12 shows Van der Lek's grape

7-12 Rooting of cuttings of grape (*Vitis vinifera*) whose buds have been kept covered with plaster beforehand to ensure rapid growth. Note location of roots directly below the young leaf or shoot. From Van der Lek, 1925.

cuttings, made from the previous year's growth. Where there is one enlarging leaf, all of the roots are formed not only below the leaf but on the same side. With an elongating lateral branch the same thing is true of the young leaves on it. The conclusion is clear that a stimulus comes from the leaves, travels directly down with relatively little lateral movement (of course the cuttings are supported in a vertical direction), and stimulates the stem to form roots below.

Following Van der Lek's work, which had been done in Utrecht, Went, also at Utrecht, who had been working on auxin, picked up the idea; but at this time he went to Java, as so many Dutchmen did. In Java he and Ray Bouillenne from Belgium worked out a related test in which they used plant extracts. They used the common tropical shrub *Acalypha Wilkesiana* which roots very readily from green stems, even of the same season's growth. As with Van der Lek's experiments, rooting was very dependent upon the presence of leaves, and if one defoliated the cuttings completely few or no roots formed. Some aqueous extracts of leaves, applied to leafless cuttings, produced a small but real increase in root number.

Since he was in Java, Went's thought was bound up with the history of Java, which, as every plant physiologist knows, means the discovery of thiamine. For it was in Java that Eijkman in 1904–05 discovered that the disease beriberi could be cured by feeding people the rice polishings that they normally discarded. Then Jansen and Donath, organic chemists from Holland, came to Java, extracted the rice polishings and crystallized thiamine (1926). Perhaps not surprisingly, some extracts of rice polishings (called Dedek) were actually at the botanic garden at Bogor, in bottles which had been put away during Jansen and Donath's time; so it was natural for Bouillenne and Went to see whether their defoliated cuttings would respond to extracts of rice polishings as they had done to extracts of leaves. As it happened they did, and formed very good numbers of roots. Thus the "influence" or "stimulus" coming from the leaves could be replaced by a substance which is extractable, concentrated, and can be preserved.

No biochemical work was done in Java, but later Went came to California and we worked together to isolate this root-forming substance. Since rice polishings extract is not very easily come by in California, many other sources were investigated. One of the best sources of this so-called rhizocaline was the medium in which *Rhizopus suinus* had grown, which was being used at that time as a source of auxin. Parallel with the purification of auxin, therefore (that is, with bioassays for *Avena* curvature), the purification of rhizocaline was carried out, using bioassays for root formation. In order to do this, Went invented a bioassay for root formation which used seedlings instead of woody cuttings which are so variable. Etiolated pea seedlings, derooted and decotyledonized, then decapitated and placed with their basal ends in a test solution, there develop after 1 or 2 weeks a very few fine roots. But if we first *invert* them, with the apex in the same test solution, for some hours, then reverse them and put their bases in sugar solution (etiolated plants do not contain much carbohydrate)—then, if the test solution is active, one sees clear root formation. It follows that the extract material enters much better through the apex than it does through the base. On the other hand, the nutrients such as sugar are better taken up from the base.

As a fine point we found that if one slits the apical part that is in the test solution, to obtain a larger cut surface and therefore more uptake of extract, these slit ends curve outwards in water and inwards in auxin. Thus one of the widely used bioassays for auxin (chapter 3, section C) came as a side result of this root-forming test.

With this method it turned out that the *Rhizopus suinus* medium was a very rich source of rhizocaline, and as the auxin was concentrated and purified, so the rhizocaline activity came in the same fractions. The dissociation constants determined by bioassay gave for auxin 4.75 and for rhizocaline also 4.75. Finally when the auxin was crystallized and identified as indole-3-acetic acid, this substance was tested, and it produced large numbers of roots in the pea seedling bioassay. To make sure we were not misled by a highly active impurity, Joe Koepfli and I made some indoleacetic acid synthetically and showed that it was fully active. These results were a great surprise, since it was not expected that the same substance would have two such different effects. Auxin at that time was thought of as primarily an elongation hormone. We have seen earlier how Snow showed that it could also activate cambial divisions, but it was thought to be still *primarily* an elongation hormone; whereas the formation of roots, of course, is *anything* but an elongation phenomenon. It is a stimulus to a change of growth type altogether, a loss of chlorophyll, a loss of negative geotropism, and a change in anatomical structure.

We had in the laboratory at that time Dr. William Cooper, who worked at the nearby USDA experiment station on citrus physiology; since indoleacetic acid could be readily synthesized, he immediately tested cuttings of oranges and lemons to see if they would respond by rooting. The lemons responded beautifully, and it was obvious that here was a procedure that could be used by nurserymen in general. Figure 7-13 shows my own first experiment with woody cuttings. Taking a leaf from the book of Van der Lek, I also used grape cuttings, made a little early in the season, and photographed after two weeks. Two groups of cuttings have been placed in sand side by side. The water controls, dipped in water, have well-developed young leaves starting, but there are only two roots on this group. The lower group have been treated with indoleacetic acid at the beginning for 24 hours; this group have developed very large root systems, and furthermore, the root systems now show no relation to the position of the leaves. Evidently, therefore, the stimulus from the leaves must have been largely the production of auxin, which can be added externally to substitute for it.

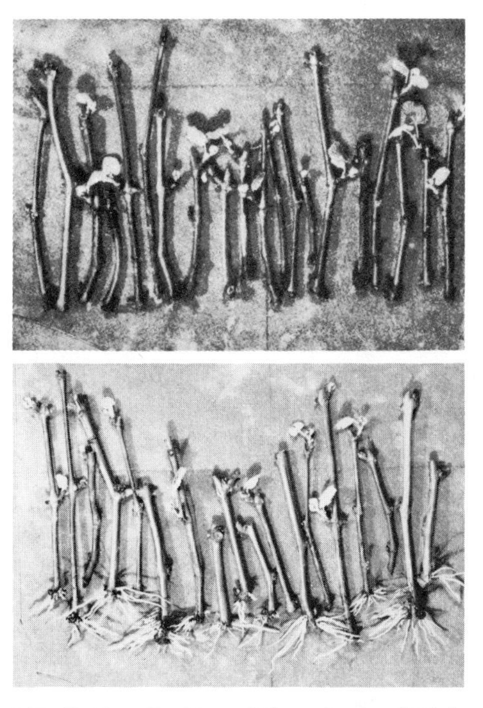

7-13 Rooting of cuttings of Concord grape (*Vitis labrusca*) after 14 days in sand. Above, controls; below, cuttings dipped in IAA solution (200 mg per liter) for 24 hours.

Continuing the history for a moment, Dr. Went visited the Boyce Thompson Institute and gave a seminar there on the identification of the rooting hormone with indoleacetic acid. They later took out a patent for rooting of cuttings, using synthetic auxins, and from this patent arose such products as

Hortomone, Rootone and other commercial preparations. Incidentally it was found that, actually, unnatural auxins like indolebutyric acid and napthalene-acetic acid are often more potent root-forming substances than indoleacetic acid itself. I think this is undoubtedly due to the fact that the oxidase is widely distributed in plants, so that indoleacetic acid will be destroyed fairly rapidly; while the others, especially napthalene-acetic acid, are very resistant to this enzyme, so that their activity will be more stable.

Various methods of application are used. Much higher concentrations than in the original bioassay, it turned out, can be applied to the base of the cuttings. Concentrations of the order of 100 mg per liter can be applied to the base, and (especially if the cutting is a little dry) the solution will be taken up in good enough amounts to form excellent root systems. Also there was devised a powdered preparation of the auxin mixed with an inert material such as talcum; the procedure with this is to moisten the cuttings and dip them into the powder so that they pick up a small dosage of auxin. Unfortunately the procedure is not very quantitative and consequently the results are variable. However, for nurserymen's purposes this is often acceptable. A fine collection of practical examples is given by Hubert et al. (1939).

C. THE ANATOMY OF ROOT DIFFERENTIATION ON STEMS

The peculiar changes whereby a root initial develops inside stem cortical tissue were first studied by Van Tieghem back in the 1880s. Van Tieghem, professor at Ghent in Belgium, made many nice anatomical studies. In a few cases, which include several aquatic plants, roots may be formed from the epidermis; but in most cases they arise internally. The first change to be seen is the beginning of tangential cell divisions in cells at the innermost part of the cortex, close to the phloem, usually in the pericycle. It takes place at rather specific points in the circumference of the cutting, and usually at a little distance above the base. The first picture of figure 7-14 shows these initial divisions (in cuttings of *Coffea arabica*); notice the large cell below, which no doubt is going to divide shortly. A later stage appears in the next picture. A large population of tiny cells develops, and with it a vascular system soon forms, usually branching from a nearby sieve tube and xylem vessel. In this way the mass is supplied with carbohydrates and amino nitrogen from the upper part of the plant. Here we also begin to see the formation of the epidermal layer of the future root. This develops from the endodermis cells of the stem, which are just outside of the initiating cells of the pericycle. A few days later there is clear organization and the formation of a recognizable root cap. Notice too that just outside this root mass there are empty areas, which have given rise to the idea that the root excretes a small amount of hydrolytic enzyme on its surface, so that as it pushes its way through the cortex it may actually digest some of the cells. Whether or not that is true, if we break open the stem at this time, we usually find that the stem tissue comes away suspi-

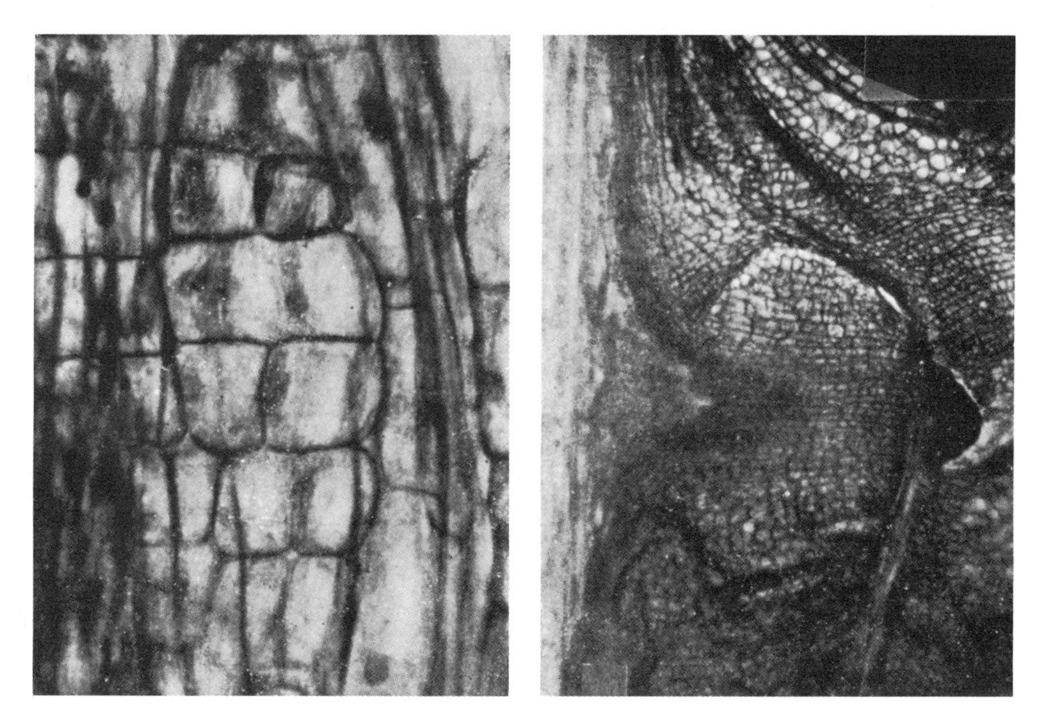

7-14 Two stages in the formation of a root on a cutting of coffee (*Coffea arabica*). Left, the first tangential divisions in the pericycle; right, the root initial ready to elongate through the cortex. From Reano, 1940.

ciously cleanly from the little mass of root initial under the surface.

First, then, there is the stimulus of cell division, leading to the formation of a mass or nodule of small cells; then there is a second stimulus causing that nodule to be supplied with vascular tissue; and finally there is control of *direction*. The nodule, which is a root initial, not only begins to organize itself, but it begins to grow in the direction perpendicular to the axis. Its location is in general close to the vascular bundle, as Van Raalte and others have recently shown. All of these developments are quite unlike anything that has gone on in the shoot up to this point.

After emergence the root usually changes its direction of growth, becoming more or less positively geotropic. It first tends to come out at an angle near 90°, and then gradually positive geotropism asserts itself. Some roots, like those on pine cuttings (fig. 7-18) may even grow vertically upwards for a while.

I have said little about environmental influences, primarily because they vary so much with the species. Many roots are intolerant of anaerobic conditions and die if the soil gets waterlogged; yet marsh plants flourish in watery soils, and rice is a classic example of a crop plant whose whole life is spent in waterlogged soil. Similarly with temperature, plants of warm soils have root growth optima around 20°C, while the roots of arctic plants grow best at 5°–10° (Chapin, 1974). Recent work at the Wisconsin "Biotron" suggests that the new Alaskan oil pipeline, by raising soil temperature a few degrees, will favor the root growth of certain arctic grasses in its vicinity. For the rooting of most cuttings, gardeners use a heated bed at around 25°, providing the so-called bottom heat.

In treating cuttings with auxin, we often see that the maximum number of roots per cutting is nearly constant. When one varies the auxin concentration, the number of roots formed is roughly proportional to the logarithm of the concentration, up to an optimum where the curve flattens off (fig. 7-15). More auxin may have a deleterious effect and sometimes will kill the cortical tissue. This may be due to ethylene formation locally, and in any case high auxin toxicity is the basis for using synthetic auxins as herbicides. The base of the cutting must in almost all cases be well aerated.

With maximum auxin treatment, other materials will sometimes increase the number of roots. Besides the access to oxygen I am assuming that carbohydrate is present; and etiolated material naturally needs sugar to be supplied. Most woody cuttings contain enough starch to supply the necessary carbohydrate, and added sugar seldom increases root number, though partial defoliation may well decrease it. (The closeness of root initials to vascular bundles certainly suggests dependence on food supply.) Root number is also sometimes increased by materials other than nutrients. Vitamins B, especially thiamine and pantothenic acid, have a marked effect in some plants. In etiolated pea cuttings biotin has a dramatic effect (fig. 7-15). Given optimum auxin concentration and optimum sugar, these cuttings form about 10 roots per cutting; and with optimum biotin added, they can form as many as 18 roots per cutting. This effect is larger than most effects of B vitamins. Wounding will also sometimes increase root number. Naturally there is always a wound made in preparing a cutting, but when one wounds the cortex longitudinally near the base, say for 2–3 mm up from the base, there will usually be some increase in the number of roots formed. So the number of roots seems to depend on several internal limiting factors, which can be partially substituted for by external materials. Auxin remains the best known and the only really internal one of these.

Secondly, as is always seen in any rooting experiment, no sooner have the roots emerged and begun to elongate than their elongation becomes negatively reactive to auxin. Auxin inhibits the elongation of roots under all circumstances except at concentrations below 10^{-8} molar, and even then, as we have seen on previous occasions, the zone of auxin concentrations causing promotion is usually very small. There is often a temporary acceleration, as Burström has shown, but this is followed by a sharp curtailment of the time during which elongation continues. So, as soon as

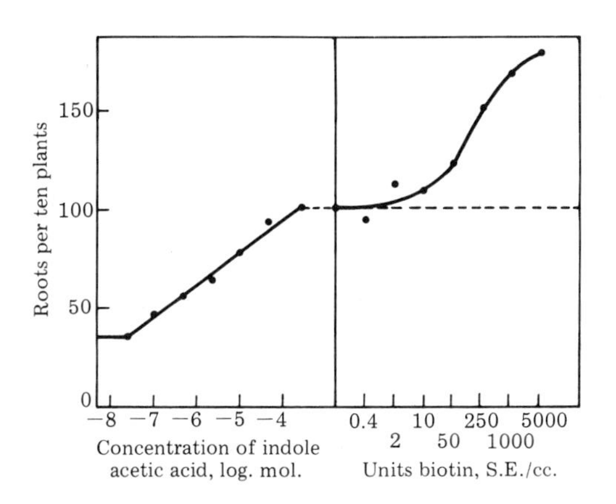

7-15 The effect of auxin and biotin on the number of roots formed by excised etiolated pea stems with their bases in sucrose solution. Left, the response to IAA; right, keeping IAA constant at the maximum value (3×10^{-4}M), biotin is added to the base. From Went and Thimann, 1937.

roots have been stimulated to form, the same auxin which stimulated them to form inhibits their elongation. For this reason it is absolutely necessary, in experiments of this type, to remove the cuttings from the auxin as soon as the root initials have emerged and to place them in a non-auxin medium where they can elongate.

There are some curious effects interrelated with pigmentation. Van Overbeek first found this in Puerto Rico when he worked on the rooting of *Hibiscus* cuttings. Most *Hibiscus* cuttings respond well to auxin plus a little sucrose, and nothing else promotes their rooting. He found that although this was true of the colored varieties of *Hibiscus,* which are mostly red or orange, in the white Hibiscus there was *no* response to auxin alone, and the only way rooting could be obtained was by supplementing the auxin with sources of nitrogen—especially ammonium salts or arginine. Subsequent to that, a number of plants have been found to improve their rooting in the presence of added nitrogen sources. Some beans respond very well to asparagine. There is a particular oriental shrub, *Michelia* (related to *Magnolia*), which responds best to peptone, another nitrogen source. However, the point was that in *Hibiscus* this was true only of the white flowered varieties, while colored flower varieties responded adequately to auxin. It does not mean that the red anthocyanin indicates a supply of nitrogen; for recently, at Yale, Bruce Stowe and Eric Bachelard have tested the rooting of a number of maple and eucalyptus varieties and species, and clearly the anthocyanin-forming varieties or species will root well in response to auxin alone, while the nonanthocyanin-forming varieties root either very poorly or scarcely at all. The explanation of this curious correlation may perhaps be as follows: in order to form anthocyanin, plants must synthesize several enzymes of an enzyme system needed to produce the anthocyanin. The synthesis involves two different light reactions, one using phytochrome and one using high light intensity, and is slow. Plants whose DNA is "flexible" enough to produce these enzymes when so induced may also be "flexible" enough to produce other enzymes of the kind needed for roots when induced by auxin and cutting. Probably the connection is no more direct than that.

Occasionally the presence of a particular organ is a help in rooting. Some petioles will root on treatment with auxin, some will root on treatment with auxin plus sugar, but some will root on treatment with auxin only *if* they have attached a small piece of the stem. In *Phaseolus,* for instance, the simple leaves root very poorly from the base of the petiole, but if attached to a very short section of stem they will root well. The "mallet" cuttings which are used commercially to root rhododendrons and other shrubs similarly have a short segment of stem attached to a single leaf. Perhaps in these plants the necessary pericyclic cells are absent (or scarce) in the petiole.

There is also an influence of light. Rooting in many plants is strongly inhibited by light. A classical study on this done by Dr. Shapiro of the University of Massachusetts showed the great sensitivity of cuttings of *Populus* to light. Light decreases the number of roots and delays their appearance. In some seedling material, it appears that green and blue light are the most effective wavelengths in inhibiting. In *Cissus,* a tropical climber which forms aerial roots, light does not inhibit completely, but it delays rooting and decreases the number of roots that develop. Though actually studied with only a few species, the inhibiting effect of light is probably widespread.

Although all of this information gives the nurseryman a convenient procedure, there are a certain number of plants which are extremely resistant and will not form roots at all under most conditions. The application of auxin and these other principles to so-called difficult plants (i.e., difficult to root) has engaged a number of workers. In the first place it can be said that some so-called difficult-to-root plants are merely very slow rooting. A second and more important general principle is that the property of not responding to auxin is one which increases with the age of the tree. In a difficult-to-root plant like *Pinus strobus,* the white pine, cuttings from 2-year-old seedlings will root very well, indeed about as well as many ordinary herbaceous plants. Cuttings of the same size from 10-year-old trees, however, will scarcely root at all, given the same auxin treatment, the same external conditions, and using exactly the same type of cutting, namely, last year's growth in each case. The response is thus not a function of the age of the cutting, but a function of the age of the plant from which the cutting is taken. In some cases we have found that this age effect is quantitative over a considerable period of years. With sugar maple, for instance, two-year-old cuttings from 2-year-old trees root 100%, with the right auxin treatment; the rooting decreases steadily year by year to reach zero percent at about 8 years.

The slowness of the rooting phenomenon with woody plants is sometimes remarkable. Cuttings treated for 24 hours with an auxin and then left in a rooting bed for 2, 3, or 4 months may form numerous roots. Figure 7-16 shows the rooting of cuttings from Canada hemlock (*Tsuga canadensis*). They were treated for just 24 hours and then left in the rooting bed for 3 months, at which time there were no roots. They were examined again every month thereafter, and the photograph shows them after 4 months. So a single treatment for 24 hours had its effect 4 months later. The white fir, *Abies alba*, showed no rooting at 6 months but rooted beautifully at 7 months.

Some time ago when I was in Italy I came upon the opposite phenomenon: instantaneous rooting. You perhaps know the story of Daphne and Apollo. Apollo was a great woman-hunter in his day, and Daphne was a beautiful young lady whom he decided to capture. Unfortunately Daphne's father was the god of one of the local rivers. He, of course, had the powers that gods have, and when he saw that Apollo was about to catch Daphne, he turned her into a bush, and the figure (7-17) shows the result: her feet are demonstrating instantaneous rooting. She is now the sweet-smelling bush *Daphne odora*.

The rooting of conifers has some interesting peculiarities:

(1) As we saw, the roots are initially plagiotropic; but sometimes this is seen in its most extreme form—the roots are actu-

7-16 Roots formed on cuttings of Canada hemlock (*Tsuga canadensis*) after 4 months. Left, six controls; center, treated with IAA 100 mg per liter; right, with IAA 200 mg per liter. All treatments at base for the first 24 hours.

7-17 Daphne and Apollo. Marble statue by Bernini in the Galleria Borghese, Rome, Italy.

ally negatively geotropic. They emerge from some pine cuttings vertically, like shoots, and after a growth of some centimeters (which means of course quite a few weeks) they gradually regain their geotropic sensitivity (fig. 7-18).

(2) In white pine particularly, the individual needle bundles, which are really short shoots or brachyblasts, root better than the stem cuttings. This seems like the opposite of that phenomenon in *Phaseolus*, where a segment of stem helps; in this case the brachyblasts root very much better when stem tissue is not present. As a practical matter one would like to use such short shoots for regular propagation, but the problem is that the bud in the needle bundle shows prolonged dormancy, and it has not yet proved practical to stimulate it into growth.

(3) The geotropic behavior of the shoot sometimes survives after it has been made into a rooted cutting. Often the branches are essentially plagiotropic and the apex is negatively geotropic. If we root the apex we obtain a vertically growing plant. But if we root the laterals, the result depends upon the genus used. In *Picea* (spruce)—at least in all the species I know of—rooted laterals will grow plagiotropically, that is, they will produce creeping plants, from which upright plants are only produced later from a lateral bud. The same is true in *Taxus* (yew). On the other hand, in *Pinus* (pine) rooted laterals will grow vertically; so whatever plagiotropism is in them is lost when they are removed from the parent tree. How the tendency to plagiotropism can be somehow *fixed* in *Picea* and *Taxus* is unknown.

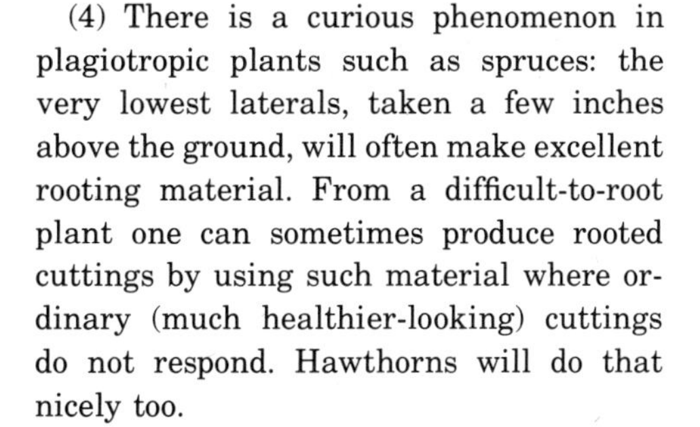

(4) There is a curious phenomenon in plagiotropic plants such as spruces: the very lowest laterals, taken a few inches above the ground, will often make excellent rooting material. From a difficult-to-root plant one can sometimes produce rooted cuttings by using such material where ordinary (much healthier-looking) cuttings do not respond. Hawthorns will do that nicely too.

(5) Rooting ability tends to vary inversely with growth rate. When confronted with difficult-to-root plants, the grower has always one possible recourse, namely, to try to find a dwarf form. It is a remarkable generalization that the dwarf forms of difficult plants will generally root easily, whereas the tall rapidly growing forms of the same species will not. This is unfortunate for the tree breeders, because they naturally look for tall, rapidly growing forms; and having found them (and not wishing to go through the troublesome genetics of breeding trees which do not flower until they are 10 or 12 years old) they would like to reproduce them from cuttings. But those are the very varieties that will not reproduce vegetatively.

It might be suggested that gibberellin could be incriminated as the agent causing excessively tall growth. And correspondingly we do find that GA tends to inhibit rooting in a great many plants. Percy Brian's Glasgow group, working on GA in the early days, studied the rooting of a number of cuttings, all of which under normal greenhouse conditions were inhibited by gibberellins. Curiously enough, in one case studied recently by Gautheret, GA actually promotes rooting in tissue cultures

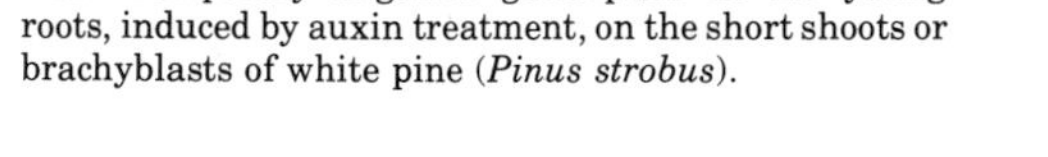

7-18 Temporary negative geotropism in the young roots, induced by auxin treatment, on the short shoots or brachyblasts of white pine (*Pinus strobus*).

of Jerusalem artichoke tuber. This result occurs only in the dark; in the same tissues kept in the light, GA inhibits just as it does on many ordinary angiosperm cuttings; so does kinetin (table 7-4).

Dr. Kefford of Honolulu has recently observed, as we saw in part 1, that this effect of GA can be antagonized with a substance called EL 531 (a product of Eli Lilly) which is believed to be a gibberellin antagonist. As the EL concentration is increased the number of roots increases, but if we then add gibberellin (GA_3) in successive concentrations, the number of roots decreases until finally the top concentration very nearly wipes out the promotive effect of the EL 531 gibberellin antagonist. The fact that EL *can* promote rooting in controls is probably an indication that the cuttings are naturally well supplied with their own gibberellin. Among other conclusions, the idea that EL is indeed a gibberellin antagonist seems well supported by these experiments. The rooting in the presence of optimum EL 531 is really spectacular (fig. 7-19).

Let us go back to the relation between root formation on cut stems and root formation in the intact plant. The action of auxin is certainly the same. Long ago Roy Lane and I germinated oat seeds in auxin solutions instead of in water, and we found that the elongation of these seedling roots was inhibited, but the number of roots formed was greatly increased (fig. 3-24). Also the initiation of lateral roots on the main root is certainly promoted by auxin under very similar conditions to that of shoots, though generally with lower concentrations. In the absence of auxin, the root tip exerts an inhibiting influence which travels a certain number of millimeters back from the root tip. Isolated pea roots will produce laterals as a function of the auxin concentration, beginning at a certain distance back from the tip. Lateral roots will not form close to the tip unless the actual tip is removed, then laterals are formed right down to the cut surface. It seems that there is an inhibitor in the tip, and an opposing auxin influence which is mainly exerted from the base, i.e., from the shoot in the intact plant. Since cytokinins inhibit lateral root formation, it is likely, though by no means proven, that the inhibiting influence of the root tip is due to its production of a compound of this type.

For lateral root formation on roots of etiolated plants, sugar is sometimes promotive, and several B vitamins have been reported also to promote—especially in longer roots, further from the cotyledons.

Table 7-4 Modification by gibberellin and kinetin of the auxin-induced root formation by illuminated rhizome segments of Jerusalem artichoke

First treatment	Second treatment	Mean Number of Roots per piece
NAA 4 days*	None	6.5
NAA 4 days	Kinetin 6 days*	3.3
NAA 4 days	None	9.9
NAA 4 days	Gibberellin 4 days*	5.5
None	NAA 4 days	7.9
Gibberellin 4 days	NAA 4 days	5.1
None	NAA 4 days	6.3
Gibberellin 10 days	NAA 4 days	1.1

* Concentrations: Naphthalene acetic acid (NAA) 1 part per million, Kinetin 1 part per million, Gibberellin (GA_3) 3 parts per million. The segments were well washed between treatments.

Summarized from Verhille, Saussay and Gautheret, 1974.

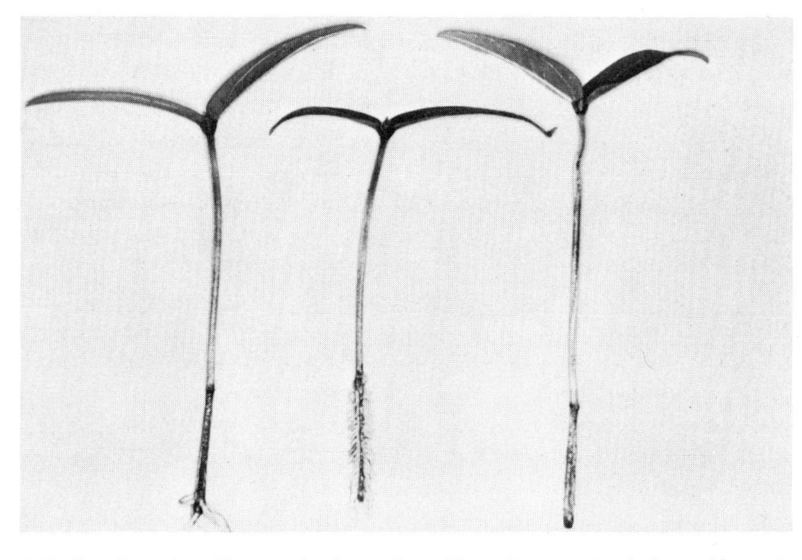

7-19 Root formation of hypocotyls of mung bean (*Phaseolus aureus*) as influenced by auxin and by EL 531. Unpublished photograph (by R. W. Tetley) from N. P. Kefford. (cf. Kefford 1973).

John Torrey has found mixtures of thiamine and B-6 to be active on peas; nicotinic acid also has an effect. These promotions depend on the conditions, age, presence or absence of cotyledons, etc. Cytokinins, on the other hand, powerfully inhibit the formation of lateral roots.

It was not mentioned in chapter 4 that the polarity of auxin transport in shoots can be antagonized by excessively high auxin concentrations. Concentrations about 100 times the physiological will poison the polarity, and thus such auxin concentrations are effective when applied at the base of a stem cutting. If the tissue at the extreme base is killed, the roots will be formed just above the killed zone. Their outgrowth may be delayed, but the initials can readily be detected. This does not occur in intact plants, because such auxin concentrations do not naturally occur.

Elsewhere I have made the point that many—perhaps all—growth phenomena are the result of interaction and balance between several hormones, rather than the apparently direct response to a single such substance. While auxin is certainly the dominant factor in root formation, the limited data discussed above go far to suggest that such interaction and balance actually occur in this process too.

REFERENCES

Avery, G. S. Differential distribution of a phytohormone in the developing leaf of *Nicotiana*, and its relation to polarized growth. *Bull. Torrey Bot. Club* 62, 313-330, 1935.

———, Burkholder, P. R., and Creighton, H. B. Production and distribution of a growth hormone in shoots of *Aesculus* and *Malus*, and its probable role in stimulating cambial activity. *Amer. J. Bot.* 24, 51-58, 1937.

Caponetti, J. D., and Steeves, T. A. Morphogenetic studies on excised leaves of *Osmunda cinnamomea* L. *Can. J. Bot.* 41, 545-556, 1963.

Chapin, E. S., III. Morphological and physiological mechanisms of temperature compensation in phosphate absorption along a latitudinal gradient. *Ecol.* 55, 1180-1198, 1974.

Delisle, A. L. Morphogenetical studies on the development of successive leaves in *Aster. Amer. J. Bot.* 25, 420-430, 1938.

Dostal, R. On Integration in Plants. (O Celistvosti Rostliny). Cambridge, Mass.: Harvard Univ. Press, 1967.

Engelke, A. L., Hamzi, H. Q., and Skoog, F. Cytokinin-gibberellin regulation of shoot development and leaf form in tobacco plantlets. *Amer. J. Bot.* 60, 491-495, 1973.

Goodwin, R. H. The role of auxin in leaf development in *Solidago* species. *Amer. J. Bot.* 24, 43-51, 1937.

Gunckel, J. W., and Thimann, K. V. Studies of development in long shoots and short shoots of *Gingko biloba*. 3. Auxin production in shoot growth. *Amer. J. Bot.* 36, 145-151, 1949.

Hubert, B., Rappaport, J., and Beke, A. Onderzoeking over de beworteling van stekken. *Meded. Landbouwhoogesch. Gent.* 7, no. 1, 103 pp. (English summary), 1939.

Kabi, T., and Sarma, R. An analysis of IBA, GA_3 and CCC interactions in the growth of leaves of *Ipomea batatas* L. *Ind. J. Plant Physiol.* 16, 140-145, 1973.

Kefford, N. P. Effect of a hormone antagonist on the rooting of shoot cuttings. *Plant Physiol.* 51, 214-216, 1973.

Kriedemann, P. E., Loveys, P. R., Fuller, G. L., and Leopold, A. C. Abscisic acid and stomatal regulation. *Plant Physiol.* 49, 842-847, 1972.

Lek, H. A. A. van der. Over de Wortelvorming van houtige stekken. Dissertation, Utrecht, 1925.

Plich, H., Jankiewicz, L. S., Borkowska, B., and Moraszczyk, A. Correlations among lateral shoots in young apple trees. *Acta Agrobotanica* (Poland) 28, 131-149, 1973.

Reano, H. C. (Root formation in cuttings of *Coffea arabica*). *Philippine Agriculturist* 29, no. 2, 87-110, 1940.

Reid, D. M., and Railton, R. D. The influence of benzyladenine on the growth and gibberellin content of shoots of waterlogged tomato plants. *Plant Sci. Letters* 2, 151-156, 1974.

Samantarai, B., and Kabi, T. Role of hormones in the initiation of cambium and formation of secondary xylem in the petioles of some herbaceous dicot leaves. *J. Indian Bot. Soc.* 53, 58-64, 1974.

Scott, T. K., and Briggs, W. R. Recovery of native and applied auxin from the light-grown and dark-grown pea seedling. *Amer. J. Bot.* 49, 1056-1063, 1962; ibid. 50, 652-657, 1963.

Thimann, K. V., and Delisle, A. L. Notes on

the rooting of some conifers from cuttings. *J. Arnold Arboretum* 23, 103–109, 1942.

Verhille, A-M., Saussay, R., and Gautheret, R. J. Nouvelles expériences sur la néoformation de racines par les tissus de Topinambour cultivés in vitro; persistance des actions exercées par quelques facteurs de la rhizogénèse. *Compt. Rend. Acad. Sci.* 278, 1199–1204, 1974.

Wain, R. L. Some developments in research on plant growth inhibitors. *Proc. Roy. Soc. B.* 191, 335–352, 1975.

Went, F. W., and Thimann, K. V. Phytohormones. New York: MacMillan, 1937.

Chemical Aspects of the Hormones 8

THE NATURALLY OCCURRING AUXINS AND THEIR CHEMICAL RELATIVES

Now that our seedling is past its initial stages and responding freely to the external stimuli of light and gravity, it is a good moment to divert our attention from the seedling and take a much-needed look at the general biochemistry of the auxins. The subject is of some interest, not only for its own sake, but because it shows how the general biochemistry of a growth substance becomes intertwined with many other aspects of the biochemistry of a growing plant. As we go along, we shall see how many different approaches to plant biochemistry have become involved with the nature and action of auxins.

A. THE GRADUAL EMERGENCE OF INDOLEACETIC ACID

In the early days of the study of auxins there was a strange episode which confused the field for at least a decade, and which has, in part, never been thoroughly cleared up. Two rich sources of auxin were discovered when chemical work began. Since it was clear from calculations that to extract and purify the auxin from oat coleoptile tips or such material would be almost impossible, a search had been made for much richer sources. These two richer sources were (1) human urine and (2) certain media in which fungi had grown. These were worked on in different laboratories. The human urine gave rise to the strange interlude mentioned; for there was isolated, in Kögl's laboratory in Utrecht, a pair of substances, called auxin A and B, which were considered to be cyclopentene derivatives. They had two isobutyl groups and then a 5-carbon chain ending with a carboxyl. Of the two substances, one was a trihydroxy acid and the other was the corresponding ketoacid with two less hydrogen atoms. This was, for the thirties, a remarkable piece of organic chemical research, because at that time it was very difficult to establish structural formulae from milligram quantities of substances.

At about the same time, however, in other laboratories the fungus preparation was being studied. This was from the fungus *Rhizopus suinus,* which is an animal parasite, though a number of strains of *Aspergillus niger* were actually just as good. But *Rhizopus suinus* has the advantage that it is grown at 37°C and therefore acts more rapidly. The substance from *Rhizopus suinus* which had all the auxin activity, as far as could be told, was found to be indoleacetic acid (IAA). This was an already-known substance, whereas auxin A and B were not; it had been isolated from human urine and from the bacterial decomposition of proteinaceous materials by Salkowski as long ago as 1885. But it was clear that this substance had all the auxin activity which the fungus produced, and shortly afterward Koepfli and I synthesized it and showed that the synthetic product had full activity in several different biological tests. Because it seemed so different from auxin A and B it was at first labeled as "heteroauxin," literally, a *different kind* of auxin. The first experiments with oat coleoptile tips seemed to indicate (by biological means) that the *Rhizopus* "growth substance" had a rather high molecular weight, about 300, certainly not that of IAA, whose molecular weight is only 175. The auxin diffusates from corn coleoptiles and broad-bean root tips gave the same value, about 330. On the other hand, diffusion experiments at different pH showed that we had to deal with a weak acid, readily oxidized, having a pK of 4.75, which, as it subsequently turned out, is exactly the pK of indoleacetic acid. To make the confusion worse one of the workers in Kögl's laboratory isolated the auxin from yeast and obtained indoleacetic acid too. Thus it looked at first as though mammalian preparations had one kind of auxin—auxin A or B—while microbial preparations produced the other kind of auxin. However, there were more and more difficulties with auxin A and B. First, the activity which was at first ascribed to them later on was found not to be present; to explain that, it was proposed that the substance is so unstable that even in the solid and crystalline form it slowly decays. A study was made of the action of light on it in this respect, and light was found to inactivate it rapidly, producing a substance called lumiauxin.

Indoleacetic acid was not so labile as that; in fact in solid form it keeps perfectly, though in solution it is susceptible to the action of mineral acid, as indeed most indole derivatives are. The auxin diffusate from coleoptiles, however, was first reported to be labile in alkali, where indole derivatives are stable. However, this fact was not confirmed; and, more and more, the auxin action of the cyclopentene compounds could not be confirmed. When an attempt was made to synthesize them, at first it was decided that it was not practical because there were too many possible isomers to make it worthwhile; but finally when a compound was synthesized it turned out that it would not occur in the free acid state but would form a lactone. In any case, the substance obtained did not have any auxin activity. So, gradually over about 15 years, the growth substance community made up its mind that auxin A and B were false leads; either they did not exist or, if they did exist, they were not the active compounds. It was a difficult decision to come to, because a lot of apparently good chemical work had gone into the study, but nevertheless they had to be dropped.

Indoleacetic acid, on the other hand, appeared to have all the properties of plant auxins except for the molecular weight.

More and more preparations from plant material began to show the properties of indoleacetic acid. Indole compounds in general are very characteristic; they are labile to acid, and they react strongly with Ehrlich's reagent, giving blue to purple colors. They also show, in the case of indoleacetic acid, indolecarboxylic acid and one or two others, the characteristic Salkowski reaction with ferric salts in the presence of very strong acid. The reaction is best carried out with perchloric acid and ferric chloride and the reagent is kept in the dark, otherwise it loses sensitivity. It gives a crimson color at very high sensitivity with indoleacetic, indolecarboxylic, and indolepyruvic acids and with almost nothing else. It gives other colors with some other indoles but the crimsons are extremely characteristic. In fact, one of the early findings in testing the extracts from fungi was that the most biologically active fractions gave this Salkowski reaction. Furthermore, it was found (1935) that in order for the fungus to produce heteroauxin, as it was called then, a certain kind of peptone had to be in the growth medium, and this turned out to be the kind that was rich in tryptophane. Tryptophane, like all alpha-amino acids, is very readily oxidatively deaminated by microorganisms, its amino group thus becomes a ketone group and the resulting keto acid in turn is oxidizable to indoleacetic acid. Now everything began to make sense.

As more and more extractions were made from different kinds of plants, their behavior on the whole tended to agree with that of "heteroauxin." Unfortunately many of the crude preparations turned out not to have been worth the effort, because the crude preparations were often tested by crude bioassay. As we saw earlier, many growing organs like coleoptile and stem segments are responsive not only to indoleacetic acid but also to potassium ions, to sugar, and to some organic acids such as malic acid. Thus, unless one is careful either to have all these materials present in the control medium or else finds a bioassay organ which is unresponsive to these materials, one can be badly misled. In the 1940s it became clear that semipurified extracts could be purified by making use of the dissociation constant—that is, by extracting with ether and then shaking out with bicarbonate, so that only the weakly acidic compounds remain, then re-acidifying and re-extracting. This gives biologically active crude extracts which often give the Salkowski reaction directly. However, the Salkowski reaction may be prevented by impurities, because it depends on high acidity and ferric ions; if there are any reducing agents present which can act in high acidity they will reduce the ferric ion and the reaction will not occur. There have been one or two cases where people have extracted and purified an auxin and concluded, because it did not give the Salkowski reaction, that it must be a new auxin. Probably they had highly reducing preparations containing ascorbic acid or similar impurities.

Finally even the auxin from oat coleoptile tips was extracted by Haagen Smit and Bonner and shown to give a crimson Salkowski reaction, and to have the right dissociation constant. Many years later, Ohwaki very carefully made similar extractions: he modified his thin layer chromatography to use cellulose instead of silica, which makes it more sensitive, and with the extract from only a few dozen coleoptile tips detected a substance which moved to the same place on the chromatogram as indoleacetic acid. Indeed he was even able to show that if he laid the coleoptiles horizontally and cut them in half more of this substance came out of the lower half than the upper half. Thus there was little doubt that it was indoleacetic acid. From pines, tomato plants, and several others extractions have now been made which clearly contain indoleacetic acid, and, from many other plants rough extractions have been made which generally agree in properties. Thus we cannot doubt that indoleacetic acid is a universal auxin, common probably to all plants. Phenylacetic acid, which has much lower activity, is also present in a number of plants and doubtless may contribute to observed auxin phenomena in them. Indeed, Wightman, in Canada, has recently found much larger amounts of this acid than of IAA in several plants.

As to the whole auxin A and B episode, we have to forget it. Of course one cannot prove a negative, and occasionally someone comes up with a new preparation which appears not to be indoleacetic but still seems to be a powerful auxin. A striking new auxin, which came upon the scene a few years ago from Dr. Marré's lab in Milan, is called fusicoccin; it is a bacterial preparation and is not indoleacetic acid. Although it is an active auxin it is totally unrelated to IAA and has the complicated formula shown. However, since as yet it has not been found in higher plants, we may not have to consider it from the strictly botanical viewpoint.

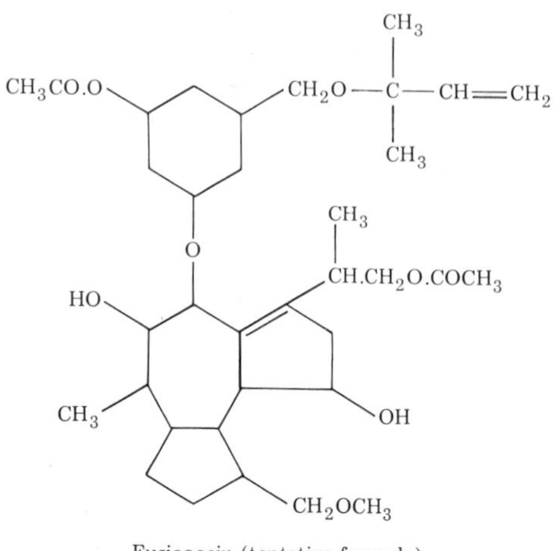

Fusicoccin (tentative formula)

B. NATURALLY OCCURRING COMPOUNDS CONVERTIBLE TO IAA

What is especially interesting about indoleacetic acid is that we find more and more of its relatives present in plants, and many of these are convertible to a bio-active auxin. Firstly, consider indolepyruvic acid (IPyA, in fig. 8-1). This substance would of course be produced directly from tryptophane by oxidative deamination and no doubt is the source of much of the IAA produced by microorganisms. It has been obtained from corn several times, and it gives a crimson Salkowski reaction; so at first it might be mistaken for IAA, except for the fact that IAA is present too on the chromatograms at a slightly different place. There has been some confusion about indolepyruvic acid. At one time it was thought that it could not really be of any

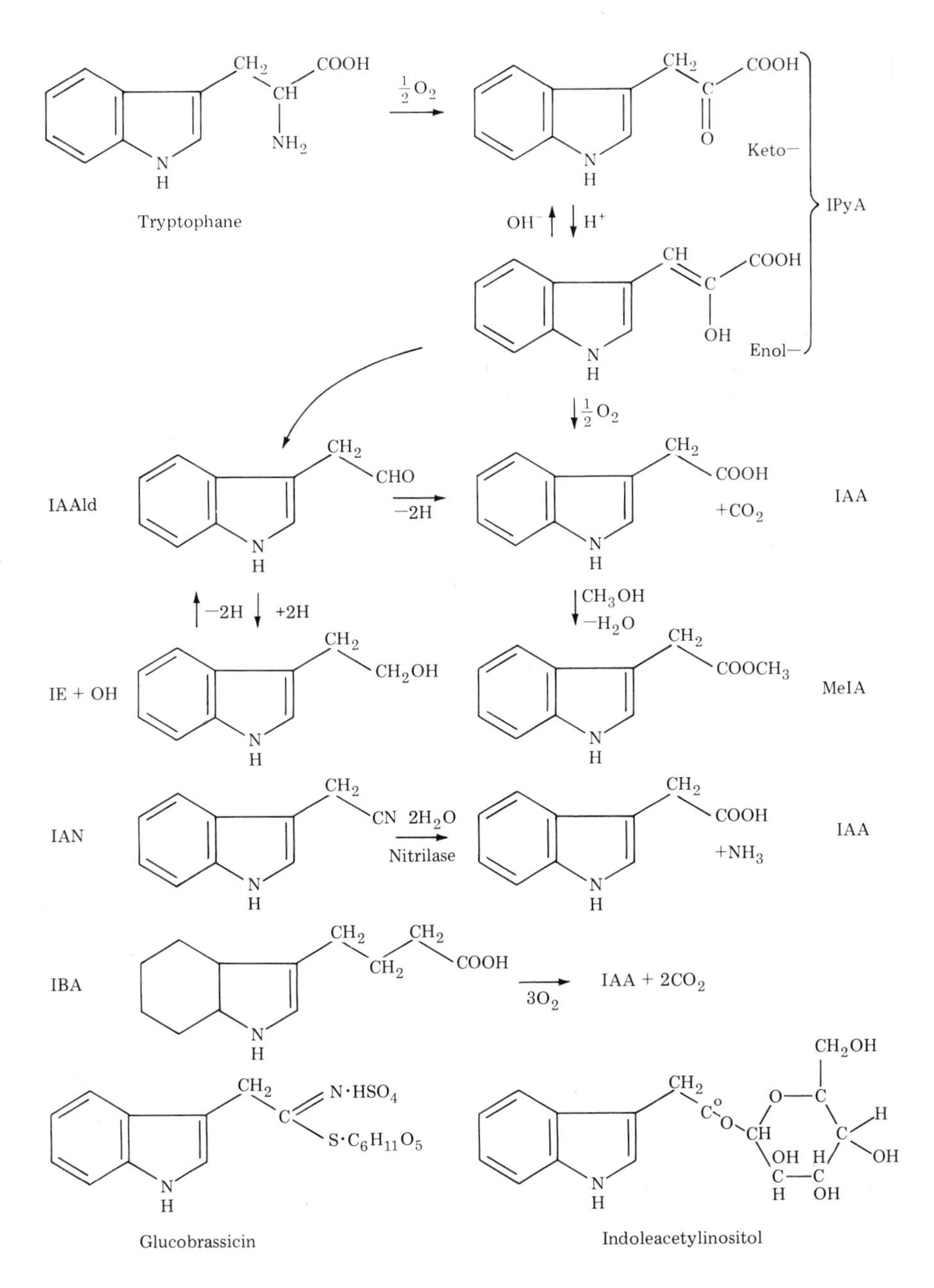

8-1 Some naturally occurring relatives of indoleacetic acid and their conversion to IAA.

importance because it decomposes so rapidly. However, what happens is this: the compound as isolated is not the keto acid but is in the enol form. It is the enol form that is stable and that crystallizes. In the enol form, furthermore, the double bond of the indole is conjugated with the double bond on the side chain, and as a result there is strong absorption in the ultraviolet. With a fresh solution the absorption in the far ultraviolet, due to the conjugation, slowly disappears as it swings over to the keto form. Several workers concluded, therefore, that it was decomposing. Indeed, its behavior is quite at odds with what is known otherwise in keto-enol reactions. The enol form in acetoacetic acid, for instance, is the one that is favored by alkali, the OH having a weakly acidic property; the keto form is favored in acid solution. With indolepyruvic acid the opposite is the case, alkali favoring the keto form, so that in an alkaline solution it appears to be decomposing. Furthermore, the enol substance reacts with ammoniacal silver nitrate and the keto form does so much more weakly, so that the silver nitrate reaction seemed to disappear too. In point of fact IPyA is reasonably stable and the two forms are normally both present. However, if a solution is left to stand for some hours in air it does slowly oxidize, and one of the products is IAA. It is only one of the products, and if a strong solution of this substance is left in air for a day or two and then chromatographed, besides the two spots for IPyA and one for IAA, half a dozen spots appear. Of the possible reactions, one would obviously be the simple oxidation to indoleacetic acid and CO_2, requiring ½ O_2. Another would be a Cannizzaro reaction in which one molecule is oxidized to indoleacetic acid and to compensate for that another molecule is reduced, forming indolelactic acid; and that has been identified on certain chromatographs too, especially by Kaper and Veldstra. Then the substance can decarboxylate, to lose CO_2 without oxidation, producing indoleacetaldehyde, which has been found several times in studying the action of bacteria on tryptophane. Indoleacetaldehyde can then undergo a Cannizzaro-type pair of reactions, giving indoleacetic acid and indolethanol, called tryptophol. In addition, Dr. Schwartz, who works in Brazil, found that one of the chromatograms from indolepyruvic acid contained an acid spot which gave no Salkowski reaction, and turned out to be formed from still another reaction which gives indolealdehyde plus oxalic acid. This was an unexpected reaction but a small amount of oxalic acid appeared on the chromatogram and thus confirmed it. There may be still more such reactions; when Kaper studied the action of *Agrobacterium tumefaciens* on tryptophane he found all these spots and two or three more unidentified ones as well. The molecular weight of IPyA is only 203, so it does not help us in accounting for the biological indications of a molecular weight of over 300. It has been suggested that two molecules of IPyA could combine, with elimination of H_2O, to make a dimer which would have a molecular weight of about 380. However, there is a better explanation (below) for the apparent high molecular weight.

The compound indole-ethanol (IEtOH, in fig. 8-1) or tryptophol (so called because it is an alcohol obtained from tryptophane) has actually been found as a growth substance in cucumber plants. When cucumber seedlings were extracted, they gave such a disappointingly low yield of indoleacetic acid that Rayle and Purves looked to see if another auxin was there, and they isolated indolethanol in modest amounts (2.5 mg from 93 kg of plants). For bioassay they were using the cucumber seedlings that they were extracting. These give about as good a growth with indole-ethanol as they do with IAA. But when tested on *Avena* coleoptiles, pea stems or even on the closely related squash *Cucurbia pepo,* it produced little or no growth. Thus not only did cucumber seedlings contain this substance, but they also contained the dehydrogenases necessary to convert it to indoleacetic acid, to which the growth is actually due. Alcohol dehydrogenase would produce indoleacetaldehyde and then aldehyde dehydrogenase would yield indoleacetic acid. Indeed, aldehyde dehydrogenase turns out to be common in plants; but alcohol dehydrogenase, though often present in plants, is not so widespread nor so powerful. As a result, the effectiveness of indolethanol as a growth substance varies very much with the species, depending on presence of these enzymes in high activity.

Not only indolethanol but indoleacetaldehyde (IAAld) have also been found in several plants, notably by Larsen in Denmark. In a fine piece of work (1945), Larsen was extracting the auxin from a number of etiolated seedlings, especially *Pisum* and *Helianthus.* He was using the regular curvature test, putting agar blocks on decapitated coleoptiles, but, being a student of Boysen Jensen, he grew his Avena seedlings in little vials of earth, instead of in water as at Utrecht or in the United States. Thus when he dropped an agar block it fell upon the earth, and careful Mr. Larsen naturally recovered his blocks from the surface of the soil and put them on the test plants. To his astonishment they were now much more active than they were before. This led him to the discovery that the auxin that he was extracting was very weak biologically—was in fact indoleacetaldehyde, a neutral compound. But in soil, especially in good garden soil rich in bacteria, there is plenty of aldehyde dehydrogenase, the so-called Schardinger enzyme. He then prepared Schardinger enzyme from milk (a convenient source) and showed that it converted his auxin quite quantitatively to a substance of much higher activity which was no longer neutral, and in fact was indoleacetic acid. Subsequently the aldehyde was obtained from pineapple leaves by Gordon, who found that the auxin in pineapple leaves was a neutral compound, and was inactivated by sodium bisulfite, which is known to combine with aldehydes to form insoluble aldehyde-bisulfite compounds. So in pineapples, in etiolated sunflowers, peas, beans, radishes and several others, part at least of the auxin is in the form of indoleacetaldehyde. As was pointed out, aldehyde dehydrogenase is commonly present in higher plants, but the rate of conversion is slow.

Another group of compounds that have been obtained on several occasions comprises the esters of indoleacetic acid. On several occasions a neutral auxin has been obtained that is soluble in ether but which, after being warmed with alkali, gives much higher activity. Here we have to deal with esters, and unfortunately it is not clear that these esters were in fact present in the plant. The trouble is that the esterases are enzymes that catalyze a reaction of very low energy exchange. Thus, although ethyl acetate is normally hydrolyzed by esterase, the reaction is in fact reversible; so that a mixture of acetic acid and ethanol, with minimal water present, will form some ethyl acetate if an esterase is added. So if in the auxin extraction there is a step in which alcohol is used, and the last traces of plant material have not been removed, some indoleacetic ethyl ester will be produced. This is especially true where the plants are extracted with alcohol as the first step; a number of times workers using alcoholic extracts have come up with indoleacetic ethyl esters, or occasionally methyl esters (MeIA, in fig. 8-1). They have some activity because they are readily hydrolyzed in the plant, but it is not absolutely certain that the ethyl ester occurs naturally in higher plants.

Now we come to perhaps the most interesting variation of all. A number of years ago Professor E. R. H. Jones' laboratory in Oxford became interested in auxins and sought for a rich source of auxins in plants. They ran across an earlier account, by Linser in Germany, who obtained surprisingly high yields of an auxin from a particular cabbage variety; Brussels sprouts were also very good. So the Jones group worked up Brussels sprouts: they brought them fresh from the field and froze them with chloroform until they had accumulated a large enough quantity to work up, and then extracted them. They obtained a neutral, ether-soluble compound which was not an aldehyde and was not an ester either, but which turned out to be indoleacetonitrile, ICH_2CN (IAN). It was surprising not only that indoleacetonitrile was active on *Avena* segments but also that it was *more* active, in two of the tests, than indoleacetic acid. Normally these convertible relatives of IAA are less active, since the degree of conversion controls the increase in activity. In the Oxford work they used tests with *Avena* coleoptiles, either growing in solution or giving curvature in the classical way, but did not use the slit curvature method with pea stems.

The slit curvature method was described in chapter 3. In essence, a rapidly elongating organ—in the case of 7-day-old pea seedlings the fourth internode—is slit lengthwise, leaving the two halves joined at the base (fig. 3-6). The tissue tension in the epidermis will, if the section is placed in water, cause the two halves to diverge and produce an *outward* curvature. However, if the slit sections are placed in an auxin solution, the auxin causes further growth, which not only releases the tension but brings the section into the rapidly growing state. As a result the outer tissues now grow more rapidly than the inner tissues, producing an *inward* curvature. The curvatures are easily measured with a protractor and one need only measure the angle to the nearest 10° to obtain a nice logarithmic response.

When this method was applied to the neutral auxin obtained by Jones and his co-workers, with slit *Avena* coleoptiles, the result was as shown in figure 8-2. In-

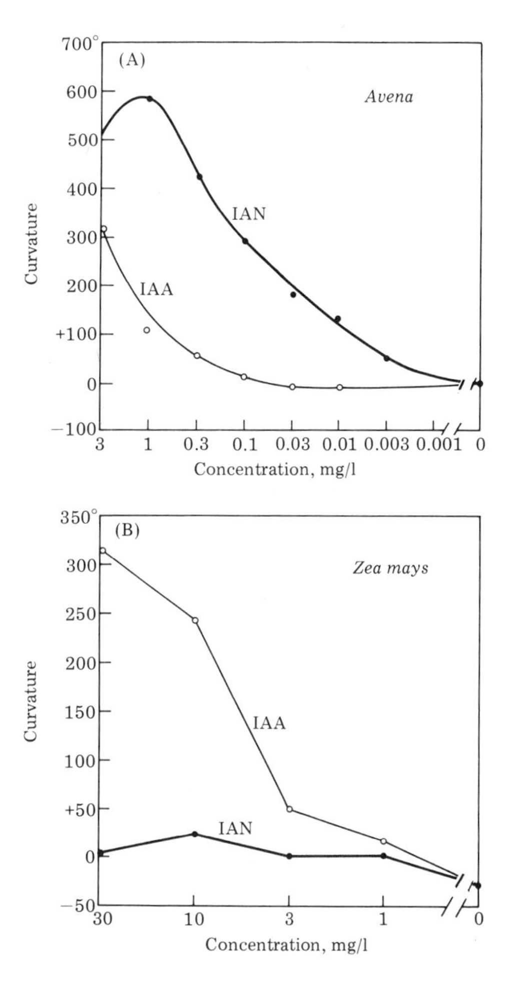

8-2 The curvature of slit coleoptiles of two plants in indoleacetonitrile (IAN) and indoleacetic acid (IAA). (A) oats, nitrile highly active; (B) corn, nitrile almost inactive. From Thimann, 1954.

doleacetic acid produces curvatures up to 300° at 3 mg per liter, and reaching 0° at 0.03 mg per liter. Indoleacetonitrile not only gives larger curvatures—up to 580°—but at 0.03 mg per liter, where IAA has no effect, it is still giving large curvatures. Thus the ratio of activities in the linear parts of the two curves is more than 10 times. This was the first time that an auxin of activity higher than IAA had been obtained from plants.

However, more careful studies showed that the activity of IAN, like that of indolethanol, varies with the species of plant. On pea stems it had no activity at all, on *Avena* it had 10 times the activity of IAA, while in some plants, like *Lupinus* or *Zea* (fig. 8-2), it had weak activity that plateaued at a very low value. Such behavior strongly suggested some kind of interconversion. It was soon shown that if one does this test with *Avena* coleoptiles, saves the solution and subsequently places slit pea stem sections in it, they do show auxin curvatures. Evidently the *Avena* segments bring about a change in the IAN which makes it active on pea stems. From that it was not difficult to show that what happens is that the nitrile group is in some plants, especially apparently in monocotyledons, hydrolyzed to the carboxylic acid.

Such a conversion is brought about when most organic nitriles are boiled with alkali for an hour or two, but here it was obviously taking place in the cold, in the presence of a small amount of plant tissue. Plant tissues were ground up and found to yield an enzyme which carries out the hydrolysis. Barley leaves were found to be the best source; they were better than *Avena* and the leaves were better than the coleoptiles. Thus from the barley leaves a crude enzyme was prepared which carries out the reaction shown in figure 8-1. At room temperature in M/20 pH 7.0 phosphate buffer the amount of indoleacetic acid liberated is close to the stochiometric value. The enzyme releases ammonia and IAA quantitatively without apparently much in the way of side reactions. We wanted to call the enzyme indoleacetonitrilase at first because it works on indoleacetonitrile, but it was soon evident that it hydrolyzes very many nitriles, in fact virtually all nitriles that have been studied up to now. In consequence it is named nitrilase.

The equation for the reaction ($ICH_2CN + 2H_2O \rightarrow ICH_2COOH + NH_3$ where I = indole) would suggest that it might proceed in two steps: insertion of one mole of water would lead to the amide; a second mole of water would lead to the acid plus ammonia. But there is actually no evidence that the reaction proceeds via the amide, and the enzyme does not attack it. Thus the two steps are probably both carried out while the material is bound to the enzyme. From a series of purifications we obtained a preparation which on electrophoresis on starch gel gave just four protein bands. Only one of these shows great activity and none of the others show any. Hence the enzyme is apparently a single protein, carrying out both steps. The reaction is almost certainly something like the formulae in 8-3.

The carbon of the nitrile has a fractional positive charge, and the enzyme, which must therefore be slightly negatively charged, attaches to that. A water molecule is introduced to form—not exactly the amide—but a hydroxy-amino compound which is still attached to the enzyme. While it is attached to the enzyme a second molecule of water is brought in, producing the dihydroxy compound which is unstable and at once rearranges to the carboxyl. It is interesting that the reaction is greatly facilitated by having a fractional negative charge on the R-group. In testing a number of nitriles we found that those R-groups which are electron-withdrawing greatly increase the rate at which hydrolysis takes place.

On the other hand hydroxyl groups which are electron-donating, tend to push the electron toward the carbon and correspondingly confer only low sensitivity to hydrolysis. There is an almost straight-line relationship between the electron donating or removing part and the rate of hydrolysis. Although there are some secondary spatial relationships, clearly the enzyme is strongly influenced by the nucleus and its tendency to give or take electrons away.

Another interesting case of an auxin-related enzyme which is very dependent on the nucleus was discovered by Wain in England in regard to indolebutyric acid. When indoleacetic acid was being tested as an

8-3 Scheme for the mechanism of hydrolysis of nitriles by nitrilase. E, enzyme. From Mahadevan and Thimann, 1964.

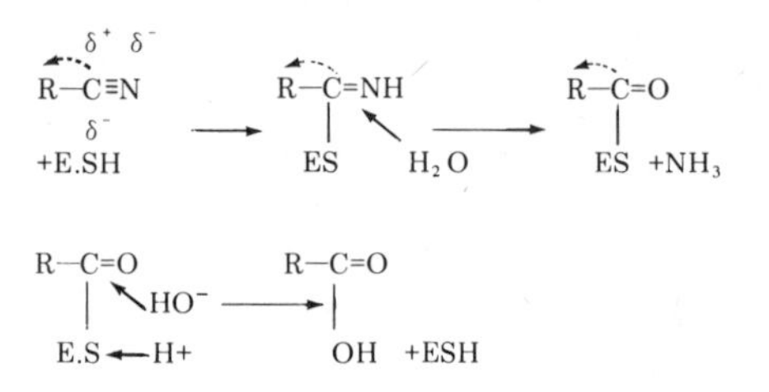

auxin in the field for various purposes, it became clear that for several responses indolebutyric acid is more active than indoleacetic. If the butyric side chain were being oxidized to acetic the indolebutyric acid would be much *less* active, and this was indeed found in some plants. The same thing was also shown by 2, 4-Dichlorophenoxy-butyric acid, as it is more active than the corresponding acetic acid in some plants and less in others. Although oxidation of the side chain with loss of two carbon atoms would represent the standard type of oxidation, Wain came to the conclusion that it depended on the aromatic nucleus whether or not the plant enzymes could oxidize the side chain. Evidently in different plants there are enzymes of rather different specificity.

To return to the nitrilase, there are several subsidiary points to be mentioned. In the first place, if one carefully measures the affinity between the substrate and the enzyme it is found to be very high, suggesting that if the enzyme were present in a given plant one would never find the nitrile there. We therefore deduced that the nitrile was present in the Brussels sprouts only because the nitrilase enzyme was not present. However, the enzyme *is* present; the explanation of this curious situation is given below. Secondly, among the nitriles that can be hydrolyzed is the purely synthetic one, of great interest, α-napthaleneacetonitrile. This nitrile is readily hydrolyzed to α-napthaleneacetic acid (fig. 8-4), which is a powerful auxin, stronger than indoleacetic acid on many plants. There may be a future horticultural application here, for the conversion goes with very high yield. Thirdly, it is curious that in some cases the activity of IAN is actually greater than that of indoleacetic acid. Even if the nitrile is converted in very high yield, as with napthaleneacetonitrile, still the activity would not be expected to be actually higher than that of IAA. The explanation finally turned out to be quite simple. It is that a neutral substance like the nitrile is likely to enter the tissues more rapidly than an acid. An acid would be extremely dependent upon its dissociation. In a neutral medium the acid will be largely dissociated, and it is well known that dissociated salts do not enter living cells as rapidly as free acids, so the entry will be dependent on pH. Nitriles are highly soluble in lipids and so would be soluble in membranes; thus they would enter rapidly, and the entry rate should be roughly independent of pH.

To confirm this deduction we used indoleacetonitrile labelled with carbon-14. Figure 8-5 shows the radioactivity in the sections as a function of time in hours. Since these are sections of *Avena* coleoptiles, which are able to convert the nitrile, it is relatively unfavorable material. Yet even so, at pH 4.5, which is near the pK of indoleacetic acid, so that it will be 50% undissociated, the indoleacetonitrile is entering a little more than twice as fast as IAA. In neutral buffer, pH 7.5, there is very slow entry of indoleacetic acid, and now the entry of indoleacetonitrile is about 10 times as fast as that of the acid. All these curves have the same general shape—a rapid period at first, which we think represents physical uptake into cellular spaces and cell walls, and a slower rate which represents the active process. Experiments in

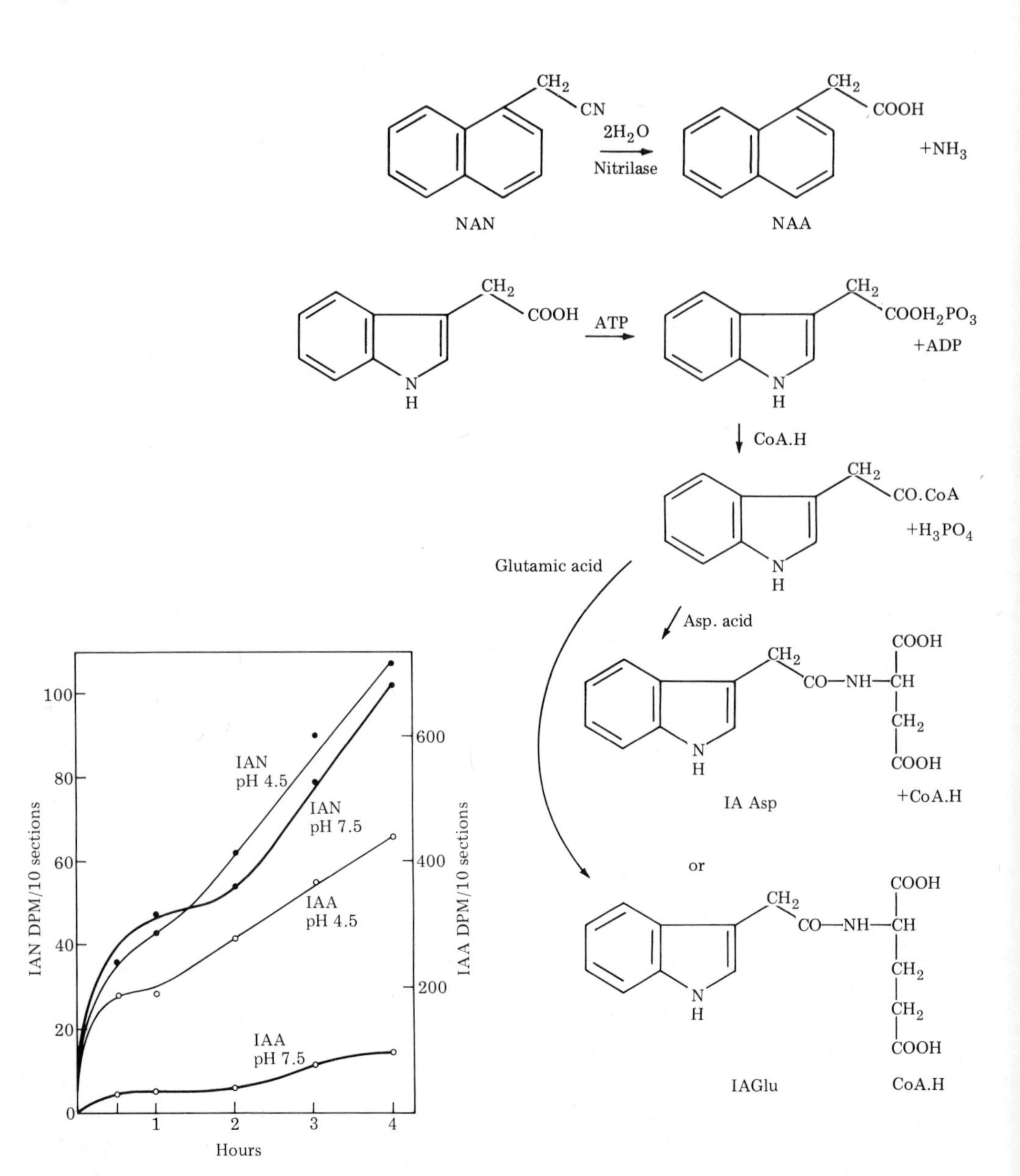

8-5 Entry of IAA and IAN into *Avena* coleoptile segments at two different pH's. IAA enters far faster as the acid than as the salt, while the entry of the nitrile is insensitive to pH. From Poole and Thimann, 1964.

8-4 Above, hydrolysis of 1-naphthylacetonitrile by nitrilase. Below, Zenk's scheme for the mechanism of formation of indoleacetyl peptides. From Zenk, 1960.

nitrogen give the rapid phase about as in air.

A fresh turn was given to the nitrile discovery by a subsequent discovery in the laboratory of Gmelin and Virtanen in Finland that there are present in nitrile-containing plants members of the so-called glucosinolate group. These are aliphatic or aromatic compounds which contain glucose and thiocyanate linked together. One is indoleglucosinate, glucobrassicin (see formula, fig. 8-1). These compounds, most of which occur in seeds, are responsible for the so-called mustard oils in plants. They decompose in the presence of plant enzymes—especially a preparation called myrosinase, which probably contains several enzymes—to yield glucose, sulfur, potassium hydrogen sulfate and an organic nitrile. In other decompositions there are formed instead thiocyanates, RCH_2CNS, and isothiocyanates, RCH_2NCS. In mustard the R-group is allyl, C_3H_5; and allyl isothiocyanate (mustard oil) happens to be very volatile and to have a sharp smell and taste. That has been known from very early days and people have always imagined that C_3H_5NCS occurred in onion and mustard, but in fact it does not occur free. What is present is a glucosinolate which hydrolyzes rapidly in the plant to free the mustard oil. When glucobrassicin was discovered it was quickly deduced that indoleacetonitrile was not occurring free in the plant but was being set free from this glucosinolate by the enzyme myrosinase. I mentioned that Jones's group at Oxford had to collect the Brussels sprouts over a period of days and therefore put them in chloroform and froze them. These are very unfortunate conditions. An enzyme which does not have a very strong Q_{10} will go on acting in the refrigerator; myrosinase, like most hydrolases, is such an enzyme, and the result was breakdown by myrosinase and production of the nitrile. Thus it has never been proved that free IAN was in the sprouts—the bulk of it was certainly being liberated from this precursor.

There are several interesting aspects to the glucosinolates. They yield not only organic thiocyanates and isothiocyanates, but also some free thiocyanate *ions*. No one seems to know exactly how all these reactions take place, especially how elemental sulfur could be liberated, but the thiocyanate ion is important because it reacts with iodide, and this is the reason why cattle sometimes get goiter. They eat cabbages, mustards, onions and turnips, and goiter may result. The goiter can often be cured by simply feeding them iodide, which competes with the thiocyanate, but, in some of these plants there is another goiter-producing compound which is not competed with by iodine, so that the disease does not always clear up.

Schraudolf, in Germany in 1965, and again recently Ettlinger and co-workers in Denmark have found that glucobrassicin, the indoleacetonitrile-yielding glucosinolate, is not limited to the cabbages but is found in three other small families: the *Resedaceae, Tovariaceae,* and *Capparidaceae.* Massachusetts taxonomist Dr. A. C. Smith tells me that these are in fact close to the *Cruciferae,* and sometimes linked with the *Cruciferae* in a group. Also there is one member of the *Euphorbiaceae, Drypetes,* that contains mustard oils. As it turns out

Ettlinger has been in correspondence with Dr. Smith about this plant because it is not a Crucifer. So the group is both taxonomically and chemically rather well defined.

Indoleacetic acid occurs in two other important forms. One form, which has been largely discovered by a group at Michigan State, is as compounds with inositol or glucose. These are biochemically interesting because they yield their IAA very easily; one has only to warm with acid or alkali. On some test plants, indeed, they give weak auxin effects directly, so we know they are readily hydrolyzed. In these compounds one (or rarely two or three) of the OH groups of inositol is substituted by IAA; one of them is simply indoleacetyl-inositol, another contains indoleacetyl on one OH and a molecule of arabinose substituted on another OH. Bandurski and his colleagues isolated four such compounds, two of which contain only indoleacetyl-inositol and two contain also a molecule of arabinose. Later a fifth one bearing a molecule of galactose, and a rhamnose ester found by Sircar's group in India, have been added. Trace amounts of inositol have also been found by Bandurski bearing two or even three indoleacetyl groups. There are several glucose esters also. Ehmann has isolated three from maize, bearing the indoleacetyl group at the 2-, 4-, and 6-positions respectively.

Now the molecular weight of indoleacetyl-inositol is 332 and it is water-soluble. The acid group is of course covered up so that it does not function as an acid until it is hydrolyzed. Apparently this ester is rather widespread. The actual isolation of it was from maize grains, but *if* it were also present in the maize coleoptiles, it could well explain the high molecular weights (around 330) originally found, and that would represent the last step in exorcizing the ghosts of auxin A and B.

The one other major form in which indoleacetic acid occurs in plants is as complexes with protein. About these we know very little. In transport experiments with labeled indoleacetic acid, auxin comes out into the receiver block; but at the end of the experiment there is always bound auxin left in the tissue. At least in *Avena,* this labeled auxin is no longer soluble in ether. It can be extracted with water to some extent, and usually it can be liberated by chymotrypsin or other protease; some can be liberated also by boiling with alkali. Evidently, therefore, it has become bound to a protein. Many years ago when numerous workers were engaged in extracting auxin it was found, both in my laboratory and elsewhere, that plant tissues usually yield auxin to ether only slowly over a period of days or even weeks. Acid did not help. When it was found that water must be present for release to occur, we hypothesized that something was being hydrolyzed, and so various hydrolytic enzymes were tested. With spinach leaves the fraction containing chloroplasts, the globulins, and the albumins were a source of slowly released auxin; the fibrous material did not yield any auxin. In experiments with *Lemna* plants, which are thin and have a minimum of fibers, we found that after incubation for 24 hours at 37°C more auxin was extractable than if the plants were extracted directly. If they were boiled only very small amounts were released. But figure 8-6 shows that when they were incu-

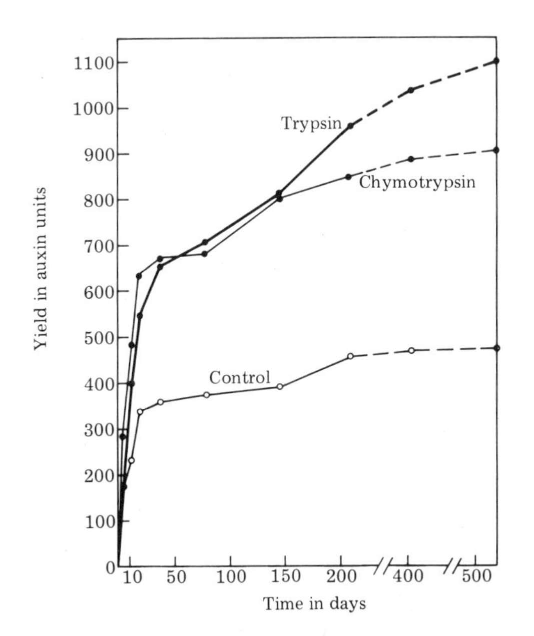

8-6 Liberation of IAA from dried *Lemna* plants in serial extractions with ether. Proteases evidently increase the liberation. From Thimann, Skoog, and Byer, 1942.

bated at 37° with chymotrypsin very large amounts of auxin were released. Thus, again, most of the auxin is bound to a protein, possibly a globulin; some is bound to an unknown fraction of the chloroplasts; and the remainder may be free. Such a distribution was found generally, especially in leaves.

The last group to be mentioned is an artificial group, namely, the compounds that are formed when plants are treated with exogenous indoleacetic acid. When stem segments, or whole plants, are grown in indoleacetic acid solution, after a few days—or even hours sometimes—no detectable IAA remains. It has not been destroyed, but all of it has been converted generally to indoleacetyl aspartic acid (IAAsp), a peptide. Like some of the other growth-promoting compounds it has a very weak auxin activity in some tests, so that it can evidently be hydrolyzed by peptidases, which are present, as far as we know, in all plants. If one does not allow the peptidases to act, by not crushing the plant material but leaving it intact, IAAsp will accumulate in large amounts. Formation of this compound is one reason why it is so difficult to work with auxin in legume roots. Most roots make indoleacetyl aspartate with great readiness: some years ago Dr. Andreae, in Canada, treated pea roots with IAA at higher than physiological levels and extracted them one hour later; there was not a trace of auxin there—all of it had been converted to indoleacetyl aspartate. The related compound indoleacetyl-glutamic acid (IAGlu) is not so well known but has been noted on more than one occasion. This formation of aspartate occurs also with synthetic auxins. M. H. Zenk, in Germany, one of the most careful workers in this field, has treated a large number of different plants with ^{14}C-labeled α-napthaleneacetic acid, and has shown that they all make α-napthylacetyl-aspartic acid in the same way. They also make α-napthylacetyl glucose and α-napthylacetyl-glutamic acid. In an interesting taxonomic study, the lower plants such as mosses tend to make glucose complexes, while the flowering plants tend to make aspartic or glutamic complexes. The figures are rather variable, but since the table shows some surprising zeros in some columns, and large numbers in other columns, it may indeed be partly an evolutionary phenomenon. These reactions may be of mainly agricultural interest, since they occur in general when one administers auxin at more than physiological levels. They are of biochemical interest, though, for Zenk has shown that the peptide is formed by way of indoleacetyl coenzyme A, and thus the reaction parallels the biogenesis of lipides and terpenoids in plants. The reactions are summarized in figure 8-4.

THE RELATION BETWEEN STRUCTURE AND ACTIVITY IN THE AUXINS

It is worthwhile to make this diversion from our main subject because so much work has been done in this area; and in comparison with some of the other structure-activity studies, those with the vitamins for instance, the work has been somewhat more successful. We have at least finished up with a model of what it takes for a substance to have auxin activity, though one can never be sure that a new compound will not turn up which defies the rules we have developed.

As soon as the structure of indoleacetic acid was known, 30 to 40 years ago, it seemed likely that one ought to be able to find synthetic substances that imitated the natural compound. At the time the idea was particularly strong because Kögl had proposed formulae for auxin A and B, which had 5-membered rings, but did not contain nitrogen. Hence it seemed likely that the nitrogen of indole was not a necessary atom in the structure of an auxin. My first step in those days was to consider the corresponding compound with a carbon atom instead of nitrogen in the ring. This would be indene-3-acetic acid, and I found that indene-3-acetic acid had been synthesized by Thiele in Switzerland back in 1912. So I wrote to the laboratory, from which he had long ago retired, and his successor was kind enough to look through the chemicals in storage and locate a sample of it. To my great pleasure it turned out to be active. It was not as active as indoleacetic acid, but it had about 20% of the activity. Another possible isostere (these compounds are isosteres if they occupy the same form in space) is coumaryl acetic acid, and this was kindly synthesized by Professor Reichstein, also in Switzerland; we found it to have moderate auxin activity. Thionaphtheneacetic acids, with an S atom in place of the O, also have weak activity, up to 3% of that of IAA.

An interesting property of these compounds is that although they have moderate activity when applied to test plants in solution, they have very little activity when placed asymmetrically on coleoptiles in which they have to be polarly transported. It was immediately clear that the ability to be polarly transported was separable from the ability to cause elongation. In fact, with indeneacetic acid this is evident because, for the same angle of curvature, the curved zone is very much shorter than with indoleacetic (fig. 8-7).

With this in mind, we paid more attention to tests in which the plant material—sections of coleoptiles, or slit pea stems—floated on the surface of the solution, so that the transport would not be limiting. In such tests the activity, relative to that of IAA, was always considerably higher.

Publication of the activity of indoleacetic acid led other laboratories to test related compounds, and among the first compounds tested by Zimmerman and Hitchcock in New York was naphthalene-1-acetic acid, which of course has a similar, though not isosteric, structure. It takes up about the same amount of space, it contains no nitrogen atom (which as we have seen is not necessary), and it turns out to be highly active. Napthaleneacetic acid introduced one new factor—the realization that the position of the side chain was of critical importance, because β-napthaleneacetic acid, with the side chain in the 2-position, has very little activity. Similarly coumaryl-2-acetic acid is less active than coumaryl-3-acetic acid.

When a larger number of compounds began to be available, the nature of the inquiry changed a little: it became, "what is the form or the properties of a structure

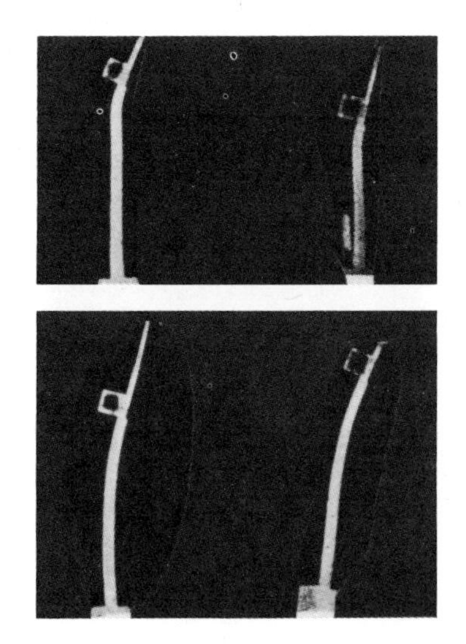

8-7 Demonstration of the limited transport of indeneacetic acid. Above, indeneacetic acid, curvatures 16° and 18°. Below, indoleacetic acid, curvatures 16° and 18°. Both photographed after 110 minutes. From Thimann, 1935.

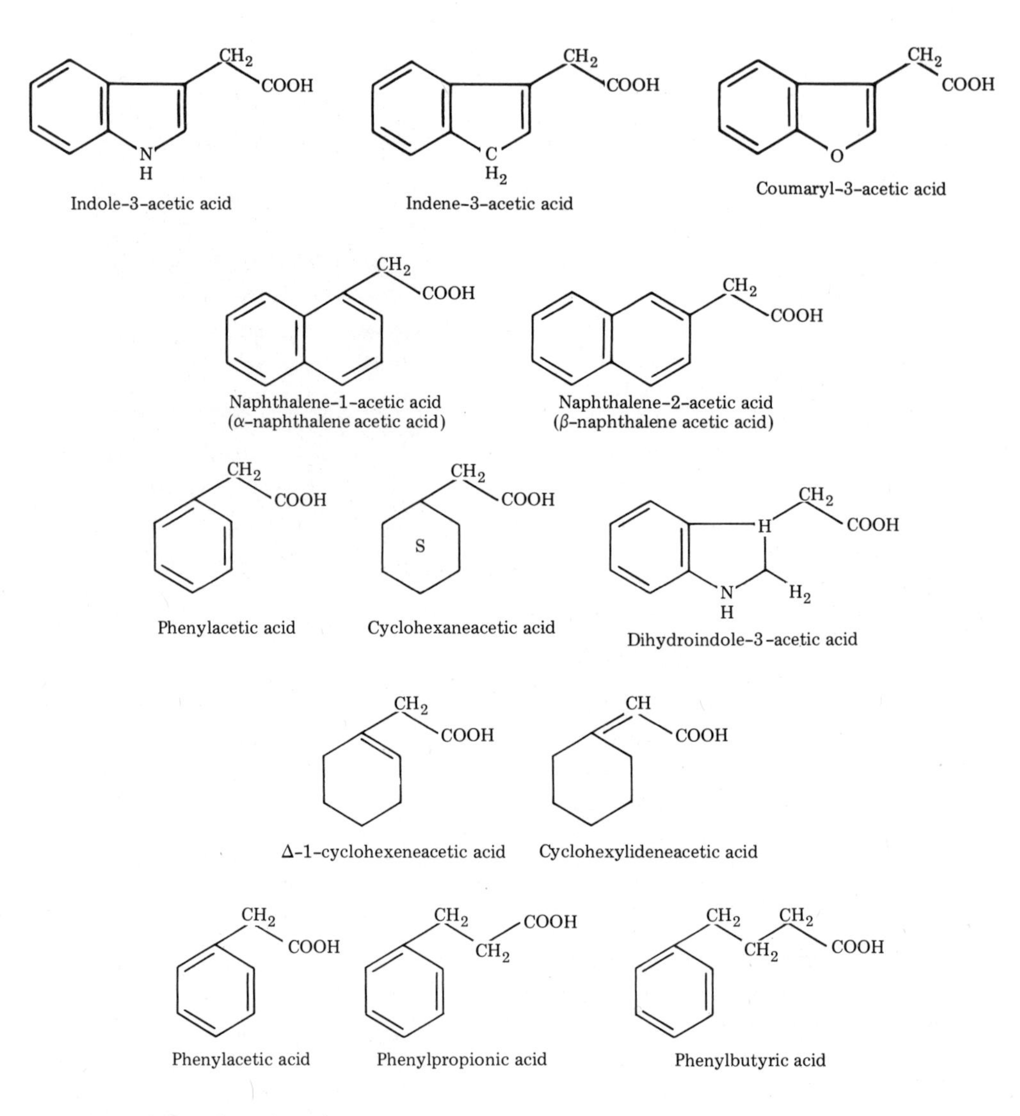

8-8 Some of the early auxin analogs.

such that it will have auxin activity?" For this, floating tests were used and a great number of synthetic compounds were tested, including compounds with a single ring, i.e., benzene derivatives. Not only is there some activity in these but one or two of them have become extremely important in later years.

Figure 8-8 shows some of the early studies. Dihydroindoleacetic acid, with the indole double bond reduced, has virtually no activity, perhaps a few tenths of one percent. Because most of these assays that depend upon floating material give results in proportion to the logarithm of the concentration, rather large differences are needed in order to be sure they are real. For this reason very low activities are hard to quantitate.

Phenylacetic acid, with a single ring and an acetic side chain, has activity varying in different tests from 2% to 10% of that of indoleacetic acid. This compound, by the way, is a natural auxin and has been shown to be present in many plants, especially by Frank Wightman of Ottawa. On the other hand cyclohexaneacetic acid, with no double bonds in the ring, is completely inactive. Thus we can draw one conclusion, namely that some property of double bonds is very important for activity. We see this again in a second group. Cyclohexeneacetic acid, strictly, δ-1-cyclohexeneacetic, with a single double bond has appreciable activity. But cyclohexylideneacetic acid, with a single double bond in the side chain, has none. We therefore conclude that the double linkage needs to be in the ring, to confer the potentiality of auxin activity; it is not the property of a double linkage itself (such as the potential of undergoing an addition reaction). We shall see that that conclusion is quite important.

Next we compare a group of simple aromatic acids. Phenylacetic acid again has 2% to 10% of the activity of IAA. If we shorten the chain to that of benzoic acid, we lose activity. If we lengthen the chain to that of phenylpropionic acid, we lower the activity, though we do not lose it completely. With a longer chain, namely with the butyric acid side chain, the activity is higher again. Thus, if we plot activity against number of methylene groups, we get an alternating series. This is of considerable interest, because one knows very well, from the work of Knoop in 1904 and many later studies, that ring compounds with long side chains are subject to β-oxidation—that is, oxidation at every other carbon atom. The carboxyl group and one CH_2 group will go off as CO_2, so that indolebutyric acid will in fact be converted, in both plants and animals to indoleacetic acid. The same is true of phenylbutyric acid. It seems likely, therefore, that the increase in activity with the longer chain is not due to the longer chain itself but is due to the fact that the compound can be converted to IAA. In itself indolebutyric acid probably has very low activity and, as we will see later, that is borne out in some work by Wain and his group in England.

The introduction of methyl groups in the side chain has instructive results. Phenylpropionic acid with the phenyl in the 1-position (i.e., 1-phenylpropionic acid) has about the same activity as phenylacetic acid. Thus, this methyl in the side chain does not change the activity much. But if

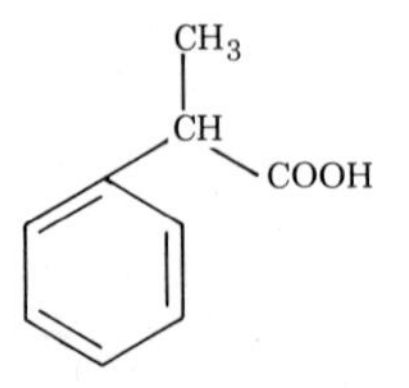

1-phenylpropionic acid

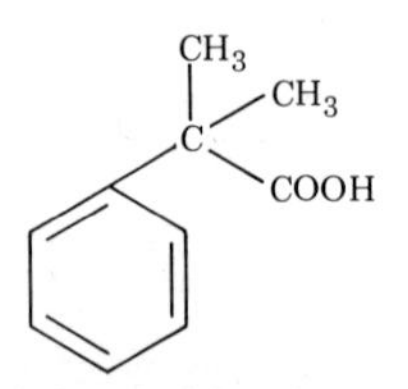

1.1-dimethylphenylacetic acid
(= 1-phenylisobutyric acid)

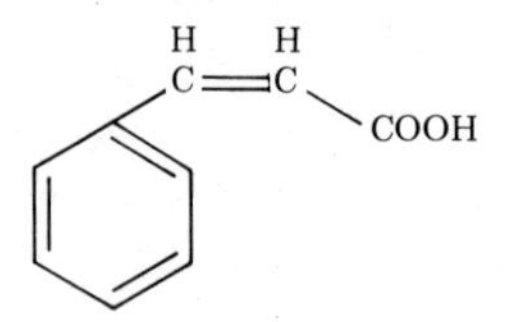

cis-cinnamic acid

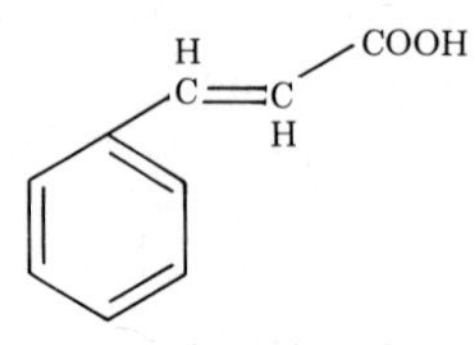

trans-cinnamic acid

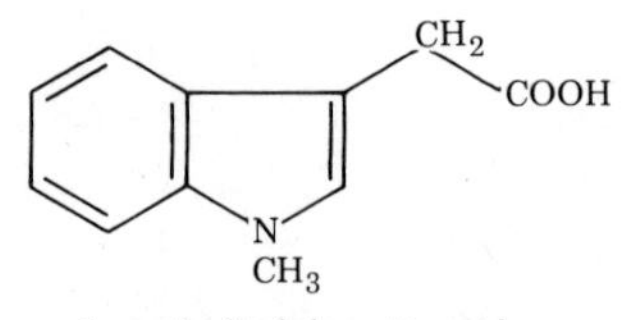

1-methylindoleacetic acid

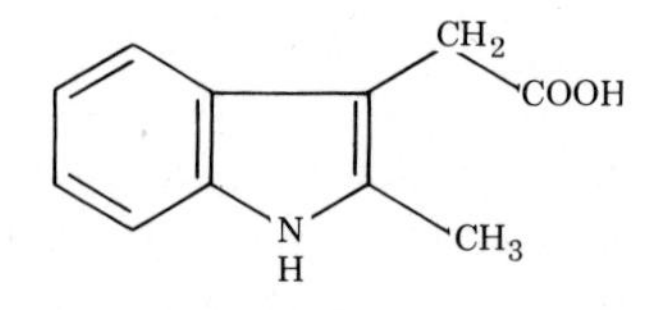

2-methylindoleacetic acid

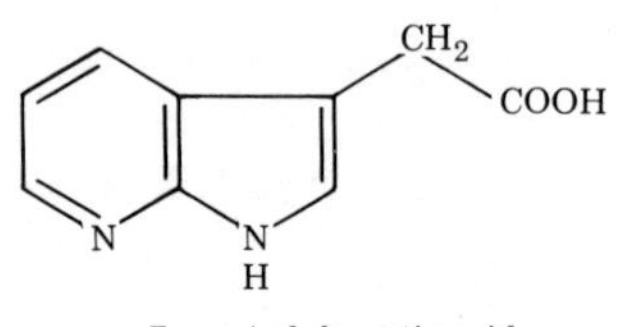

7-aza-indoleacetic acid

8-9 Compounds discussed in text.

we put 2 methyls in the side chain, the activity goes to zero. Two theories have been put up to account for this: one, that a free hydrogen in the side chain is essential; and the other, that there is a steric effect of those two methyl groups that somehow prevents the compound from reaching its substrate.

The latter view is supported by the case of the cinnamic acids, which exist in two forms, *cis*- and *trans*- (fig. 8-9). The *cis*- form has definite but small activity, about 3% of that of IAA; the *trans*- has none. So again we see that the configuration in space is of critical importance. If the carboxyl is held out, as it were, at arm's length, away from the ring, there is no activity; whereas if it can come around to reach a certain proximity to the ring, there is activity. This gave, I think, the first clear-cut indication of the necessity of a specific configuration in space. Some substituted cinnamic acids behave similarly. Several pairs of optical isomers also show a marked difference in activity.

If all bioassays are performed with the same test (usually the slit pea stem test), in which the response is in proportion to the logarithm of the concentration, then in a log plot (which is a "dose-response curve") one compound will give a certain straight line, and another compound will give another line roughly parallel to the first. Thus one can readily deduce the relative activity. However, for reasons that are not understood, all compounds do not give straight lines; some compounds yield very peculiar dose-response curves. Often low concentrations are quite ineffective and higher ones suddenly become quite effective. With such behavior it is hard to assign a definite activity as a fraction of the activity of indoleacetic acid.

The upper half of figure 8-10 shows the behavior of two compounds with a methyl group substituted in indoleacetic acid, namely, 2-methylindoleacetic and 1-methylindoleacetic (see formulae, fig. 8-9). As just mentioned, in the test using elongation of floating coleoptile segments most compounds give a roughly straight line on a log plot. The line for IAA is not very straight in this particular test, but look what the 2-methyl group does: it cuts out all activity in the lower levels and yet allows the compound at high concentrations to be as active as IAA. The 1-methyl substitution does drastically reduce the effectiveness by an order of 100 times or so. Indoleisobutyric acid is another compound which yields an S-shaped curve (see below). As yet no one has explained this behavior.

The lower half of figure 8-10 shows the exact opposite type of behavior. First we see two tests on different days done with indoleacetic acid; both give beautiful straight lines reaching saturation eventually. Next we see the substance to be tested, 7-azaindoleacetic acid, a substance which is closely related to IAA, but has a nitrogen atom in the aromatic ring. In view of the lack of need for the nitrogen atom in indole, we felt that this might not introduce a large difference, and in fact it does not. The activity is of the order of 10% (note that this figure shows a log scale with powers of 3). Thus, so long as the shape of the molecule is maintained, the presence or absence of a nitrogen atom in the benzene ring makes little difference.

One important conclusion, and it derives from tests of *many* compounds, is that the presence of a hydroxyl group has a drastic effect. To compare with phenylacetic acid, which as we saw has from 2% to 10% of the activity of IAA, mandelic acid (fig. 8-11) has very low activity indeed. Next we consider two indole compounds with hydroxyl groups, indolelactic acid and indoleglycolic acid. These have two different chain lengths, of course. Indoleglycolic acid, the parallel case to mandelic acid, carries a tiny amount of activity, about 1/1,000 of that of indoleacetic acid. Indolelactic acid, with an extra methylene group, compares with indolepropionic acid. Its activity is clearly less. Finally, we have the case of ring hydroxyl groups on the indole nucleus, namely 5- and 7-hydroxyindole acetic acid. Their activities are only 1% or 2% of that of IAA. Thus the introduction of hydroxyl groups, whether on the ring or on the side chain, definitely reduces the activity, generally rather strongly. It may be remembered that Kögl's auxin A had a cyclopentene ring and a series of 5 carbon atoms, with hydroxyl groups in 3 of those 5 positions. If we had known what we know now of synthetic compounds, we could have said at once that this compound cannot be active as an auxin, let alone have an activity comparable with that of IAA. But this is of course hindsight.

The benzene compound in which the methyl groups interfere spatially with the arrangement of the molecule has a parallel with indole, namely, indoleacetic acid substituted with two methyl groups; this is indoleisobutyric acid. Each hydrogen is replaced by a methyl group, and the rotation

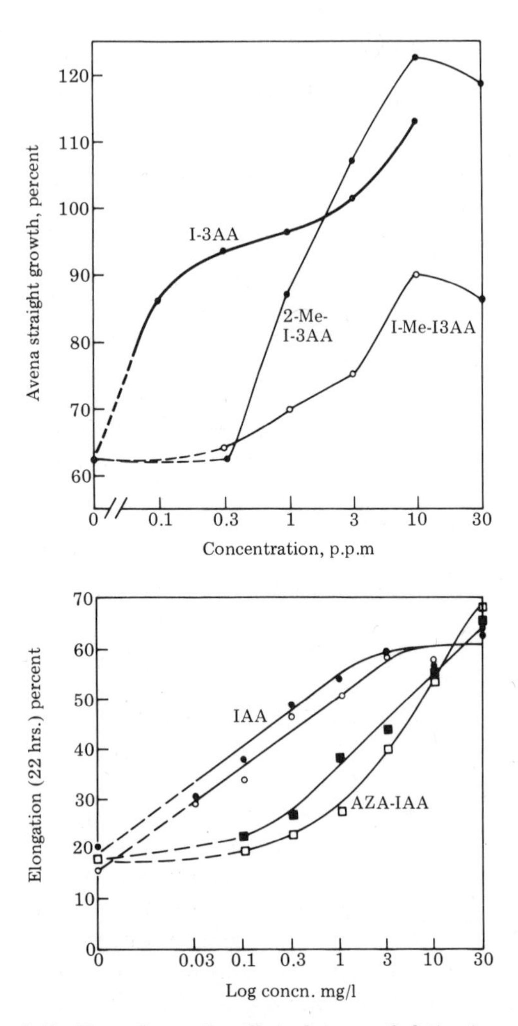

8-10 Upper figure, the effect of ring methylation in lowering activity of IAA on *Avena* coleoptiles. Lower figure, the effect of insertion of a ring N atom into IAA on the growth of pea stem segments. From Thimann, 1972 and 1958 respectively.

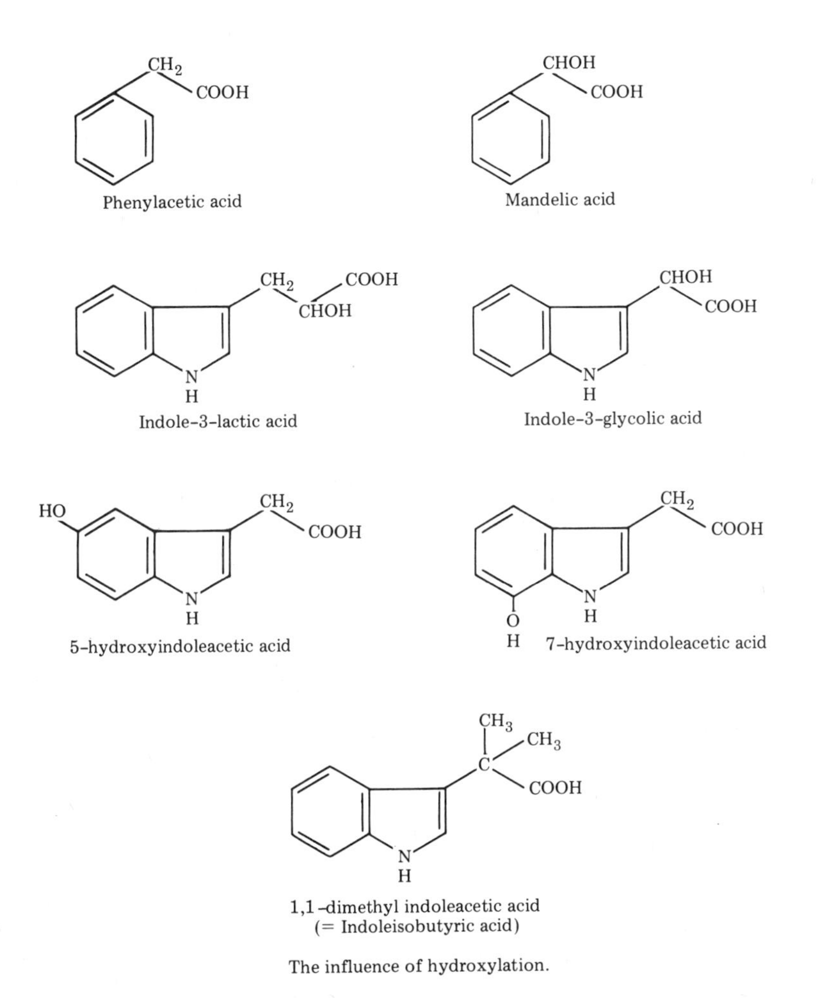

The influence of hydroxylation.

8-11 Above, cases where hydroxylation lowers activity. Below, one case where side-chain methylation lowers activity.

of the carboxyl is evidently constrained. Indoleisobutyric acid gives a plot unlike any other known compound. It was at first thought that it was an anti-auxin, a compound which would antagonize the effects of auxin and therefore be very useful in physiology. However, it is not. It has real activity, but with a very peculiar relationship between dose and response. Again qualitatively we see that a free hydrogen atom in the side chain is not essential.

At this time, historically, theories began to build up as to the critical nature of the structure that makes an auxin. One of these theories, that of Veldstra, was that what is important is the ratio between the lipid solubility and the water solubility of the compound. He felt that a number of the substitutions which confer activity, like chlorine atoms on the ring, are those which increase lipid solubility, while substitutions which decrease activity, like OH groups, confer water solubility. As we saw, a hydroxyl group, wherever it comes, always seems to lower or destroy activity, and of course it greatly increases the water solubility.

About this time, the substance 2,4-dichlorophenoxyacetic acid (2,4-D for short) was discovered. In the course of testing various aromatic compounds for insecticidal activity, not only phenylacetic acid and its congeners, but phenoxyacetic acid and its derivatives, with an extra oxygen atom, were tested by the Dupont organization. Of little use as insecticides, they were later tested as growth substances. Phenoxyacetic acid had no activity, but some of its relatives showed strong activity, particularly the compound with 2 chlorine atoms, in the 2- and 4-positions (fig. 8-12). There are three interesting things about its activity: (1) it is comparable with that of indoleacetic acid, and indeed higher than indoleacetic acid in some tests; (2) its lipid and water solubilities and the ratio between the two are very similar to those of indoleacetic acid; and (3) it is immune from the IAA-oxidizing system which, as we have seen, works with peroxidase as an oxidase. It is attacked only very slowly in plants, therefore it was obvious that this compound would be valuable for those applications where long times are involved. As you know, it has become the most used of all auxins in agriculture. We noted that phenoxyacetic acid unsubstituted has no activity, and since it has only low lipid solubility that fitted the theory. The two chlorine atoms greatly increase the lipid solubility and they very greatly increase activity. Dr. Veldstra in Holland was very happy about this at first, until, in testing related compounds, we found that 2,4,6-trichlorophenoxyacetic acid (fig. 8-12), which has a little greater lipid solubility and has the same great stability to enzymic destruction, has virtually *no* auxin activity whatever. This was a serious blow for the lipid solubility theory. However, as so often happens, it gave rise to another theory, namely, that an active auxin requires not only a ring, a side chain, and a special configuration of the side chain with relation to the ring, but also in the ring position adjacent to the side chain there must be a free hydrogen. This theory thus requires that the ortho position to the side chain must be unsubstituted. The inactivity of 2,4,6-trichlorophenoxyacetic acid supports

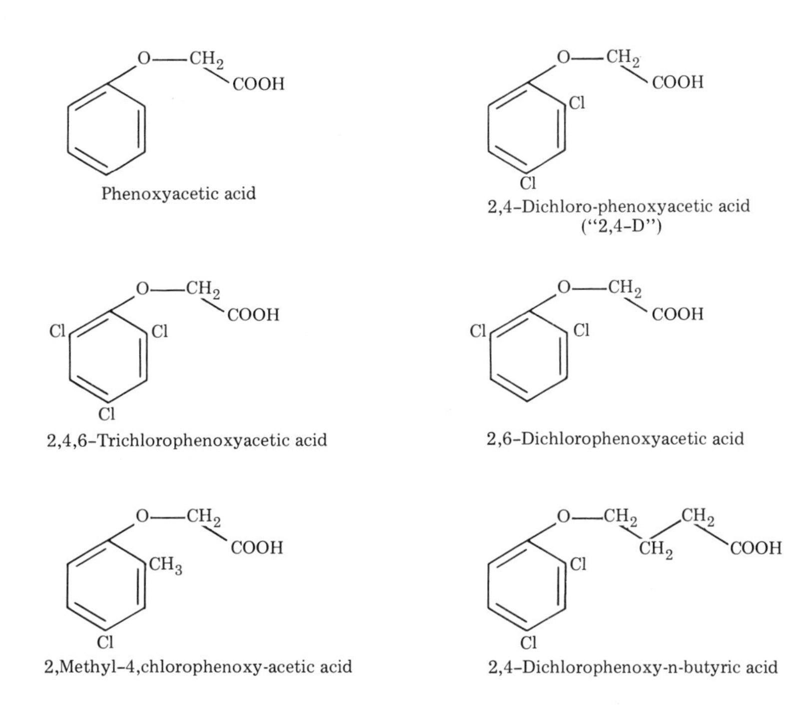

8-12 Phenoxyacetic acid and some chlorinated derivatives of special auxin interest.

that very well. But this could be tested: if we leave unsubstituted the 4-position, but substitute the 6-position, we have a compound (2,6-D) with the same number of chlorine atoms as 2,4-D, therefore a similar lipid solubility, and no ortho position free. Is this active? It was predicted that this compound would be inactive, but it turned out to be active, though not *as* active as 2,4-D. In general, 2,4-D measures out about 1 log unit more powerful than indoleacetic acid, though the curve curiously flattens off at the lower values. 2,6-D is clearly also an auxin, though it is about 1.5 log units less powerful then IAA. 2,4-D would have a relative activity of around 800% and 2,6-D would have a relative activity of 7%. Thus, if the ortho position is regarded as an absolute requirement, the theory cannot be maintained, but certainly it is *favorable* for the development of auxin activity.

Today one of the chief applications of 2,4-D is as a herbicide. High concentrations of auxin are always toxic to plants. Concentrations around 10 times those which are maximally effective, which probably means 100 times those which are physiologically present, are toxic with any auxin. In the case of dichlorophenoxyacetic acid the effect is enhanced because of its great stability to the oxidase system. The first use of auxins as herbicides, in Britain, was with napthaleneacetic acid, but when 2,4-D was discovered everyone went over to it, because it was both more active and cheaper to make. Curiously enough, in England patents were not available on 2,4-D. The British workers had made 2-methyl-4-chlorophenoxyacetic acid quite independently of the Americans and had their patents on that; so for many years that compound was used as a herbicide in England and in Western Europe, while the 2,4-dichloro derivative was used in America. Their activities are comparable.

In connection with the β-oxidation, Wain, at Wye College in England, made the interesting discovery (1955) that the butyric acid analog (2,4-dichlorophenoxybutyric acid) which should be subject to β-oxidation as before, is herbicidal to some plants and not to others. (This specificity is within the dicotyledons, because none of these auxins is very herbicidal to monocots.) With 2,4-dichlorophenoxybutyric acid certain families, for instance legumes, are susceptible and others are not. Wain and his group followed this up very carefully. They treated the susceptible plants and subsequently extracted them and showed that the butyric acid side chain had in fact been broken down to acetic acid; thus there had been β-oxidation and CO_2 had come off. Insensitive plants did not do this. The sensitivity of the plants, then, paralleled their ability to break down the side chain. The theoretical significance of this is that it means the ability to oxidize the side chain does not depend only on the nature of the side chain; it also depends on the nucleus. The presence of the chlorine atoms there makes some plants unable to break down the side chain. In other cases like indolebutyric acid, there is no such specificity. Hence the enzyme has not only to attack at the β-position, but it has to have some connection with the rest of the molecule in order for it to act.

From all this work and thought we can derive several properties of a molecule which seem to be needed for auxin activity: (1) the optimum chain length is clearly established, and the apparent exceptions are due to β-oxidation; (2) there is a particular (as yet undetermined) orientation of the carboxyl to the ring; (3) there is the necessity for some unsaturation in the ring; and (4) we have to remember that the naturally occurring substance of high activity is an indole derivative.

Some years ago, with the help of William Porter, I tried to put all these ideas together. It seemed evident that the unsaturation in the ring may be necessary to maintain the ring in the form of a plane. A cyclohexane ring is not planar, it is boat-shaped, and other saturated rings are far from planar, while benzenoid rings are very flat. The idea that the ring must be flat had been noted earlier by Veldstra, who had proposed that the side chain must be nearly at right angles to the plane of the ring.

Secondly, we see that all these compounds are acids—that is, they are able to dissociate. Where there is activity in neutral compounds, as with indoleacetonenitrile, it turns out to be due to hydrolysis by nitrilase to the acid. Thus the neutral com-

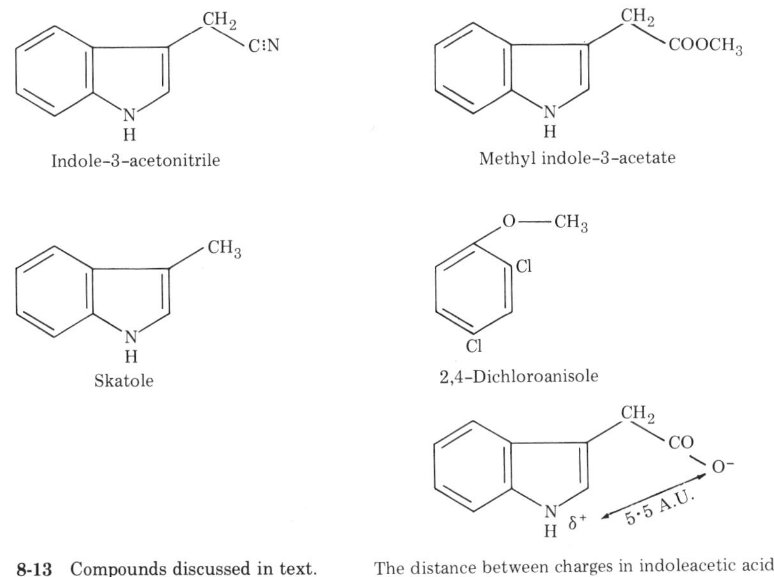

8-13 Compounds discussed in text. The distance between charges in indoleacetic acid

pound does not have activity of itself, and plants which cannot hydrolyze it do not recognize it as an auxin. Another example is furnished by the indoleacetic acid esters. These are sometimes active; and when they are active, Kögl and Kostermans, back in 1937, showed that the *lowest* esters are the most active; the methyl ester is far more active than the butyl or hexyl ester, and this is in accord with the well-known ability of esterases to hydrolyze esters, that is, the ability to hydrolyze esters is roughly inversely proportional to the number of carbon atoms in the esterifying group. Thus the ability of esters to show activity also evidently rests on their conversion to the free acid. Not only are two apparent exceptions, the nitrile and the esters, both satisfactorily explained; but when there is no carboxyl group at all nor anything convertible to it, as with the relative of IAA, skatole, or the relative of dichlorophenoxyacetic acid, 2,4-dichloroanisole (fig. 8-13), there is no activity. In fact, they inhibit activity of the others, though they are not exactly anti-auxins because their inhibition cannot be removed by adding more of the auxin, as the inhibition due to a true anti-auxin should be. In any case it is enough to say that they have no activity at all. Hence a clear-cut conclusion is that the dissociable acid group is an essential part of the activity. So we have our planar ring with a dissociable group.

What does the peculiar orientation in space mean? A single-bonded side chain is free to move about, when in solution, so evidently certain positions are more favorable than others, and what matters is the location of the center of the orbital rotation. In the case of indoleacetic acid, the *mean* distance from the dissociable oxygen of the carboxyl to the nitrogen atom comes out to about 5.5 Angstrom units.

Now we noted above that the natural compound is an indole derivative. What is the special property of indole? One of the most characteristic properties of the indole nucleus is the charge distribution on it. It belongs to the so-called π-excessive group, in which the carbons of the 5-membered ring have a fraction of a negative charge, while the nitrogen of the adjacent carbon atom no. 8 has a fraction of a positive charge. In the case of indole, its value is about 0.3 of a positive charge, and carbon atoms 1 and 2 share a comparable negative charge about equally.

We began to think of 2,4-dichlorophenoxyacetic acid in this connection, because chlorine atoms, with 7 electrons, have the well-known ability to draw electrons towards them. As a result the adjacent positions are distinctly negative, and the other carbon atoms then become positive; particularly the one meta to the chlorine atom develops a considerable positive charge. In the case of 2,4-D this charge would be mainly located at the 6-position.

We begin to see some unity in this idea. We have a positive charge, a particular orientation of the substitution, a dissociable carboxyl, and some range of the distance between the carboxyl and the positive charge. Two other discoveries were made at this point. The first was the revival of an old stager, namely, the benzoic acid derivatives. We saw that benzoic acid has no activity and the substituted derivatives that were tested also had very little, but it turned out, much later, that there is one substitution that *does* produce very good activity—2,3,6 trichlorobenzoic acid. Now why should that have considerable activity? We noted earlier that benzoic acid is inactive because, as with indolecarboxylic acid, the chain length is too short. But if this optimum chain length were not really a *chain* length, but the distance between the carboxyl and some particular point in the molecule, it might be satisfied in a benzoic acid derivative, if that point in the molecule is not too close. Let us suppose as before that this distance is around 5.5 Angstroms: the chlorines in the 2- and 6-positions would tend to orient a fractional positive charge to the 4-position, and in fact that position is at about the right distance (cf. fig. 8-13). Also 2,4,6-trichlorobenzoic acid is inactive. So that begins to fit.

The second discovery was an active compound without a ring, which came about in the following way. For many years there was much interest in the derivatives of carbamic acid, esters like ethyl carbamate or phenyl carbamate; they are called urethanes and have been well known as respiratory poisons since the classical experiments of Warburg with tissue slices which are taking up oxygen. Phenylurethane and ethylurethane are both very effective in preventing oxygen uptake. These compounds are being investigated by a group in Holland for their function as fungicides, and they found that if the 2 hydrogens on the nitrogen atom are substituted by methyl groups, the fungal toxicity is increased. Furthermore, if one substitutes

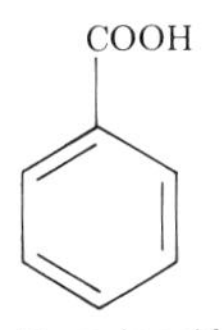

Benzoic acid

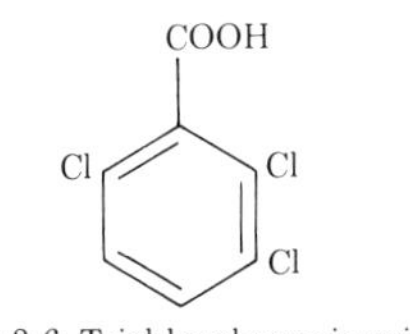

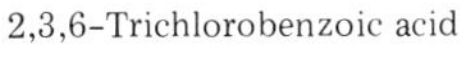
2,3,6-Trichlorobenzoic acid

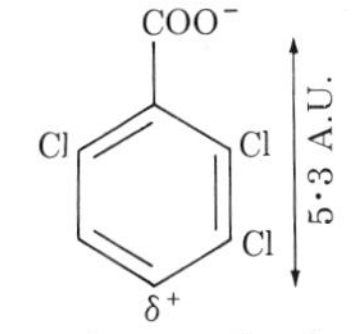

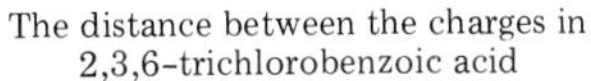
The distance between the charges in 2,3,6-trichlorobenzoic acid

$NH_2 — COOC_2H_5$
Ethyl urethane

$(CH_3)_2N—C(=S)—OC_2H_5$
N,N-Dimethyl ethylthiocarbamate

$(CH_3)_2N—C(=S)—OCH_2COO^-$
δ^+
5·5 A.U.
(*ca.*)
N,N-Dimethyl carboxymethylthiocarbamate

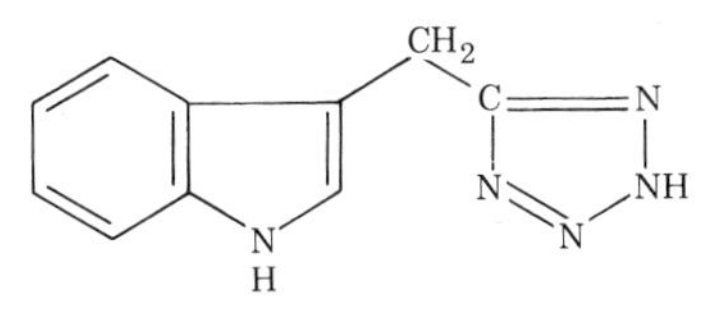

Indolemethyltetrazole

8-14 Compounds discussed in text.

one of the oxygens by sulphur to form a thiocarbamate, it becomes still more potent. As a result they prepared a number of derivatives, diethyl and dimethyl thiocarbamates. Among these were a few compounds having a dissociable acid group, i.e., carboxymethyl derivatives. In testing the N,N-dimethyl compound with the carboxymethyl group (fig. 8-14), they found that the host plants, on which they were testing it against a fungus infection, reacted oddly. In short, it turned out that this is an auxin. It caused curvatures and elongations and had all the typical auxin effects. The activity is not very high—about 10% of that of IAA—but fully significant. At this point some workers concluded that all the theories are wrong: a flat ring was not needed; no ring is needed.

However, as it turns out, this nonring compound fits very well into the preceding scheme because, although it is not a ring, the two double bonds are so arranged that they hold the central part of the molecule flat. Not only is there really a flat molecule, but the presence of two methyl groups confers a marked positive charge on the nitrogen and again the distance to the carboxyl oxygen is just about 5.5 Angstroms. We have thus a unitary framework for viewing the activity of all the compounds known to be active. They have to have a flat structure, a fractionally positively charged atom situated at a distance of 5.5 Angstroms from the dissociable oxygen of a carboxyl group. Actually it need not be a carboxyl group, but it has to be dissociable, and hence negatively charged. That is an important detail which had been cleared up by Veldstra in Holland in two distinct ways: (1) Napthalene-1-nitromethane occurs in two isomeric forms:

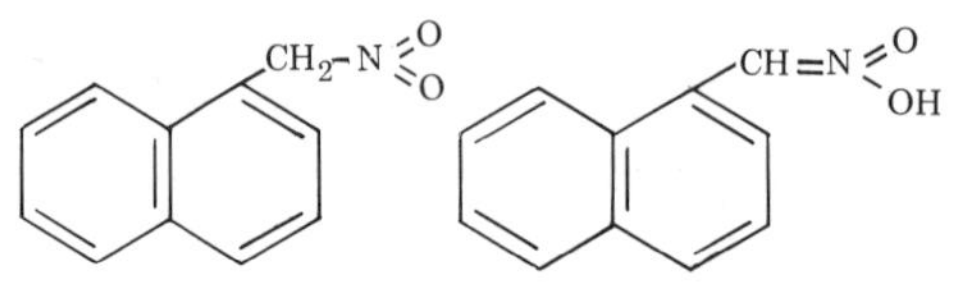

the second, called the *aci*-form, bears a group having the same structure as a carboxyl. Veldstra showed it to have some auxin activity. (2) The other way involves a group of compounds called tetrazoles. In these tetrazoles, which are moderately stable (much more so than azides), the 4 nitrogens have such a tendency to divert electrons toward them that the hydrogen atom becomes dissociable; and in fact many organically substituted tetrazoles are acids. Dr. Veldstra had a co-worker who synthesized indolemethyltetrazole, which is the exact analog of indoleacetic acid but with the dissociable proton coming from the tetrazole instead of from the carboxyl group. And it had real activity; it is not a highly active compound but the fact of its activity shows that a carboxyl is not the only source of a dissociable hydrogen. It is also important as showing that auxin activity cannot depend on forming an acyl derivative of the type of indoleacetyl CoA, as had earlier been suggested.

To get evidence to support the above concept we have tried to measure the fractional positive charge which was optimal for the job. In the infrared spectrum of indoles, the NH bond, which is vibrating back and forth, shows up as a clearly recognizable band, and in different compounds the NH frequency appears at slightly different positions. The tighter the linkage, the more rapid the vibration; the looser the linkage, the less rapid the vibration. Hence if we measure the NH stretching and relaxing frequency it will be a measure of the amount of fractional positive charge. The differences are very small, but my colleague William Porter was able to make such measurements, and he found that the NH stretch frequencies clearly go parallel with the dissociability of the carboxyl group. The dissociability of the carboxyl group is measured by Hammett's sigma constant, which is the difference between the log dissociation constant of a given acid (or base) and of the same acid (or base) substituted in some way. For example, suppose benzoic acid has a pK of 4.2: we substitute a chlorine in it, and the resulting 2-chlorobenzoic acid is a stronger acid with a pK of 3.7; the difference, sigma for 2-chlorobenzoic acid, is 0.5. Thus for each configuration there is a certain value which is changed by substitution in this way. To compare these numbers, Hammett makes use of a sigma multiplied by a *rho,* where the rho represents the basic type of molecule and the sigma represents the substituting group. According to Hammett, once rho is determined for a given molecule like benzoic acid, 1 chlorine will always have the same sigma, namely 0.5. Figure 8-15 shows that for a series of indole derivatives, if we plot the NH stretch frequencies against the sigma values, we obtain a beautiful straight line. The significance of that is, essentially, that substituents which increase the strength of an acid increase the stretching frequency of the vibration at the NH bond and therefore increase its fractional positive charge. We synthesized two new compounds shown in the figure—IAA substituted with chlorine and bromine in the 2-position. We had predicted that they should be more active than indoleacetic acid, because the halogen would further withdraw electrons and thus increase the positivity of the N atom; and so they are, by a considerable amount.

In the figure below we have plotted the change in the stretch frequency (that measures the tendency of the nitrogen to be positively charged) against the auxin activity in a given test. These data refer to the slit pea-stem test. They do give satisfactory straight lines. We can note, as was pointed out earlier, that hydroxyl substitution drastically lowers activity. The value of 7-hydroxy-IAA is so low that it does not even come on the plot, whereas halogen substitution always increases it in different amounts according to the position of the halogen atom. There is one exception here—the 5,7-dichloroindoleacetic acid, which does not lie on any of the lines—but it was the only exception at the time this work was done.

Now, with some evidence to support a consistent theory, the next question is, "what does it tell us?" "What does an auxin have to do that it needs this particular structure and charge distribution?" Since it is a flat molecule with an acidic group protruding, it probably has to associate itself with a flat surface. The most attractive concept, though pure speculation at present, is that this fractional positive charge and molecular shape is highly suggestive of adenine or another purine base, a compound with the same basic flat structure, though with several nitrogen atoms in the rings (we saw that the one extra nitrogen in 7-aza-IAA did not affect activity much). Unlike indole and its derivatives, the ni-

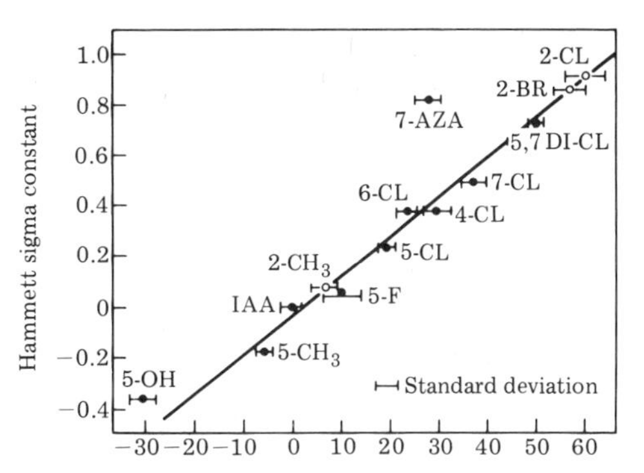

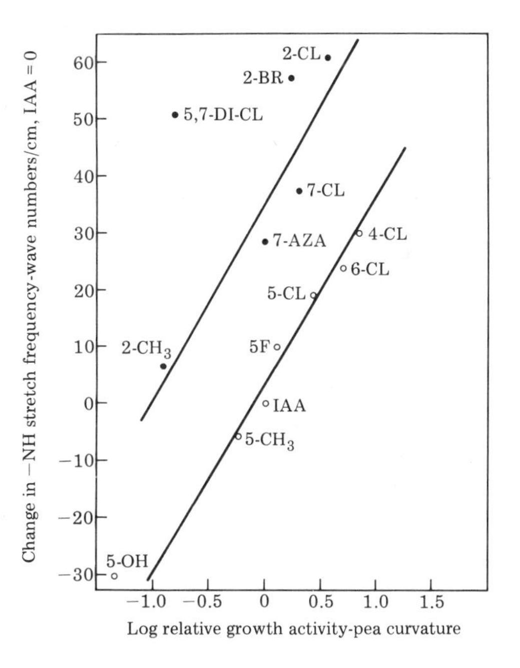

8-15 Upper figure, relation of the hydrogen-bonded −NH stretch frequency in the infrared to the Hammett Sigma constant in substituted indoleacetic acids. Lower figure, log of the auxin activity on slit pea stems plotted against the −NH stretch frequency for the same acids. From Porter and Thimann, 1965.

trogens in adenine derivatives have fractional negative charges and one of them usually bears a sugar molecule together with a phosphate group, which is highly charged, at a little distance away. There is obviously a certain affinity between the two molecular types, though I repeat that this is only speculation. However, the idea that auxin has to enter into close relationship with one or the other of the bases of a nucleic acid has received a good deal of support from physiological experiments that will be presented in chapter 14.

As soon as one begins to get a consistent theory one begins to use other compounds to test this theory. Two compounds are worth special mention: one of them does not fit very well with the theory, but the other does, and reinforces it. The first is a compound in which, instead of an acetic acid side chain being on the 3-position, it is on the nitrogen, the 1-position. There are two related substances, the N-indoleacetic acid (fig. 8-16) and N-indolepropionic acid. Figure 8-17 shows that in tests with *Avena* coleoptile segments, the former is a good deal more active than indole-3-acetic acid. (The odd shapes of these dose-response curves may not be significant, because every test does not produce as straight a line as it should, though the mean of many different observations would almost certainly give a straight line.) Thus the substitution of the carboxymethyl on the N atom instead of on carbon atom no. 3 confers greater activity, and by a considerable margin, than that of IAA—perhaps by a factor of 5 or so. As yet it has been difficult to decide what would happen to the charge distribution in the 5-membered ring by the substitution of a carboxy-methyl group on

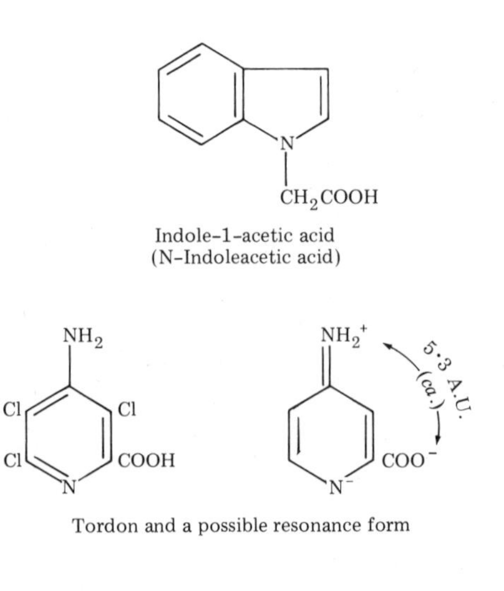

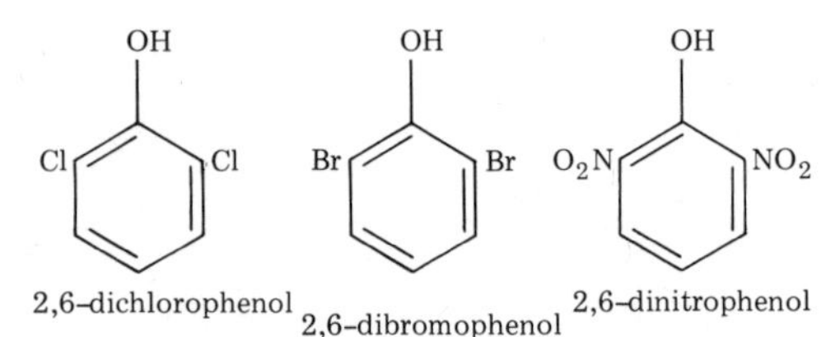

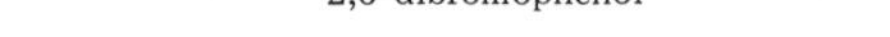

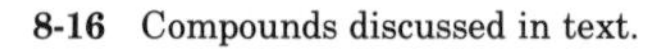
8-16 Compounds discussed in text.

the nitrogen. At present, therefore, this stands out as an unexplained exception. By contrast, in the *Avena* elongation test, the 1-propionic acid side chain confers no activity whatever, but even shows a little toxicity at the top level.

The second compound turns out to offer unexpected support to the theory. Tordon is a substance which was developed purely as a herbicide without any thought of its being an auxin. It is a derivative of pyrimidine. This compound, 2,3,5-trichloro-4-aminopicolinic acid (fig. 8-16) is a very powerful herbicide. It is an auxin too, which raises the question of the activity of pyrimidine rings instead of the pyrrole ring which occurs in indole. Pyrroleacetic acid was too unstable to test; it decomposes in the solution. Pyrimidine compounds have been tested, however, and since phenylacetic acid has some weak activity, we naturally thought at one time that pyrimidineacetic acid ought to have some activity. It has none, and for a long while this seemed rather curious, until later it was realized that in a pyrimidine ring the situation is not the same as in a 5-membered ring. The five carbon atoms each contribute 1 electron to the *pi* system, so there is only room for one from the nitrogen. But its lone pair like to stay together (so to speak), so the *pi* system is somewhat deficient in electrons, and as a result the nitrogen is *not* positively charged—it has a fractional negative charge. Thus the absence of activity fits with the absence of a fractional positive charge. However, its substituted derivative, Tordon, was found by Kefford, in Hawaii, to be an auxin, which was unexpected, because pyrimidine derivatives, as just mentioned, have so far not shown activity. But in this compound we have *three* chlorine atoms, all drawing electrons away from the ring, and furthermore, a compound like that would be in resonance with the second form shown. The result would certainly confer a strongly positive charge on the nitrogen atom. The extent to which the different forms would be present is not for a mere biologist to determine, but qualitatively there would be a strong tendency to develop a positive charge. The location of this positive charge, as it turns out, is just 5.5 A.U. from the dissociable carboxyl. Thus it fits perfectly with the scheme. In fact it has somewhat higher activity than indoleacetic acid itself.

8-17 Dose-response curves for indole-3-acetic acid (I-3-AA) and indole-1-acetic acid (I-1-AA) on *Avena* coleoptiles. From Thimann, 1972.

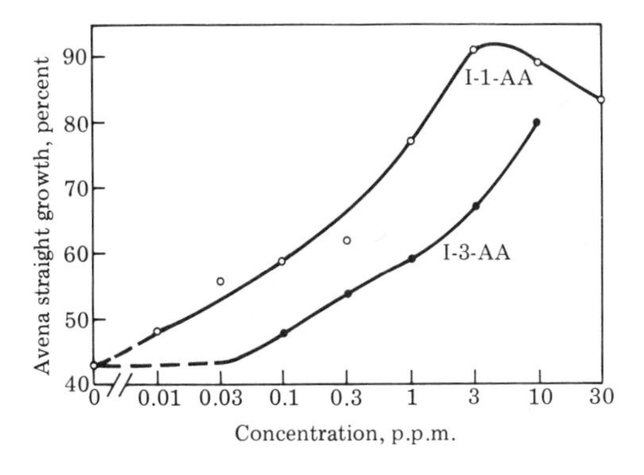

A further support to the idea is given by the recent discovery of Harper and Wain that some di-substituted phenols are (weakly) active. This at first seems surprising, since we saw above that hydroxy-groups greatly decrease activity. But in the case of 2,6-dichlorophenol, or the 2,6 dibromo-, iodo-, nitro-, or cyano-substituted analogs, electrons are withdrawn by all these substituents, so that the phenolic -OH becomes a fairly strong acid. The electron-withdrawing groups also induce a fractional positive charge at the 4-position, and this is at about 4.3 A.U. from the dissociable acid -OH. In these phenols, as in the benzoic acids, substitution at the 4-position kills all activity, as the theory would predict. Since the activity is at best only a few percent of that of IAA, the activities of this group of compounds fit well into the scheme proposed.

In conclusion, we have here a consistent theory which seems to apply to compounds of many different chemical types. So far it is the *only* theory of the relation between structure and auxin activity which has not been disproved. It will be most interesting to see how it stands up to the discovery of new auxins in the future.

THE CYTOKININS

At this point it will be worthwhile to go back over the whole basis of the work on cytokinins, and say a little of the bioassays and somewhat more in detail about the relation between structure and activity. As we saw above, this is a field that with auxins has been extensively explored, and we have come out with some concept of what it takes for a molecule to have auxin activity. To some degree this is true also with the cytokinins, and in both groups there are certain rather peculiar exceptions.

A. THE PATH TO THE DISCOVERY OF THE CYTOKININS

The history starts really from Philip White's studies on that hybrid tobacco—*Nicotiana tabacum* hybridized with *N. Langsdorffiana*—in which any wound, however slight, leads to the production of a relatively large tumor. This had been observed by tobacco growers—it was not discovered by White—and was fairly well known, being regarded as an oddity. White picked it up and succeeded in growing the tumor tissue in culture. What he found was that it grows well on agar, and the culture becomes a mass of callus, looking very much like the gall or tumor growing on the intact plant. When he tried to grow it in liquid culture, however, its character changed: there were produced small nests of dividing cells, and some of these grew out into buds. White reported this and did nothing more about it at that time, but it was the beginning of a whole series of interesting studies. For some reason, the tissue was obviously delicately poised on the edge of producing buds, but in air it never did. Putting it in submerged culture in solution made it produce buds. The role of the submergence has not been explained, and as far as I know no one has worked on it further.

But the fact that here was a culture which could produce organs with a relatively slight stimulus interested Skoog very much, and when he was with us at Harvard I got White to give me a culture and Skoog began growing it. He very soon found that while the growth is adequate on ordinary sugar-salts-auxin medium, it is increased by adding adenine or AMP. Later (1948) Skoog and Tsui published a very important paper in which they explored this ability of adenine and other purines to promote rapid growth and the initiation of buds on a medium that was adequate for ordinary growth as callus.

This work led not only to the study of specific addenda to the growth medium but also to a series of studies of different kinds of tissue. Among other results may be mentioned the growth of pea stem segments, which do not produce a transferable callus in culture, but which produce a callus on the cut end of the segment. In these, adenine promoted growth only under one special condition—a condition which takes the mind back 40 years—namely, that vascular tissue must be present. You will recall that Haberlandt, back in 1910 or so, in his attempts to get plant tissue cultures, had studied various kinds of pithy material, potato tuber, and the spongy parenchyma of leaves, and found with several of them that there were practically no cell divisions (he had no auxin then of course) unless vascular tissue was present. It was this that led Haberlandt to postulate the "Zellteilungshormon" which I have mentioned on previous occasions. The Haberlandt phenomenon is as clear in potato tuber slices as it is in pea stem segments. These tissue slices do not grow easily; for one thing they cannot use indoleacetic acid because they destroy it—so that synthetic auxins are necessary. The presence of internal phloem, even in quite small amounts—just a few strands—will often lead to very vigorous development of cell division around them, while the rest of the tissue shows only cell enlargement.

Prominent among the tissues studied by Skoog and his group were those of normal tobacco plants—not the special hybrid, but pith tissue from normal tobacco stems. This material showed active cell enlargement in the presence of auxin, as does artichoke tuber tissue, but almost no cell division. To obtain cell division one had either to include vascular tissue, which had a weak effect, or to add various tissue extracts, some of which had a strong effect. The occasional appearance of excellent cell division and the formation of little masses of tissue led the Wisconsin group to wonder whether, since cell division was involved, perhaps the effective extracts contained a nucleic acid. As a result they started testing nucleic acid preparations. One particular preparation from yeast was very powerful, and led to the great development of tobacco pith cultures which have been under intensive study ever since.

This was a yeast DNA preparation, and it was an old one; the old one was a good deal better than several new ones, which seemed very strange. Furthermore, when they sterilized it by filtration, the potency was much less than when they autoclaved it. This indicated that the activity was not due to the DNA itself, but to some breakdown product of it. Accordingly some extensive chemical work was undertaken, in association with some of the chemists at Wisconsin, who finally found that there was a chloroform-soluble purine there. Since purines are rarely soluble in nonpolar solvents like chloroform, this was a highly characteristic property. Making use of this and many other properties, they finally isolated the kinetin that we have discussed earlier. It is extremely potent in causing tissue cultures of numerous materials to grow and form shoots and leaves. As it turns out it is not the most potent member of the cytokinin group. The name cytokinin was afterwards coined for the group because they seemed particularly to promote cell division (cytokinesis), whereas the DNA synthesis which precedes cell division could be promoted perfectly well by auxin. Later, studies were made of cytokinin activity in plant tissues, particularly in immature fruits; and Letham, working in New Zealand, finally isolated the first higher plant cytokinin from unripe corn (*Zea mays*) grains in the milk stage. Since it was from *Zea* it was christened zeatin, and had the formula shown. It was the great activity of coconut milk, the liquid endosperm of unripe coconuts, which had directed attention to unripe fruits and especially liquid endosperm in general as a source of cell-division promoting activity. Historically, indeed, coconut milk had been used by Blakeslee and van Overbeek as early as 1941 for the culture of immature ovules, and it had brought them to maturation in sterile culture. This particular wheel came full circle in 1975 when Van Staden and Drewes, in South Africa, isolated both zeatin and zeatin riboside (in about equal quantities) from coconut milk.

Immediately after the establishment of the structural formula of kinetin, in fact within three days, the first synthetic imitation had been produced by the very assiduous Wisconsin chemistry group. The compound they synthesized was benzylaminopurine, and it turned out to be even more active than kinetin. Still later a fourth compound, which I mentioned earlier, the so-called 2iP (the convenient short name of γ,γ-dimethylallylpurine, or $\gamma\gamma$ DAAP) came on to the scene, first synthetically, and then naturally.

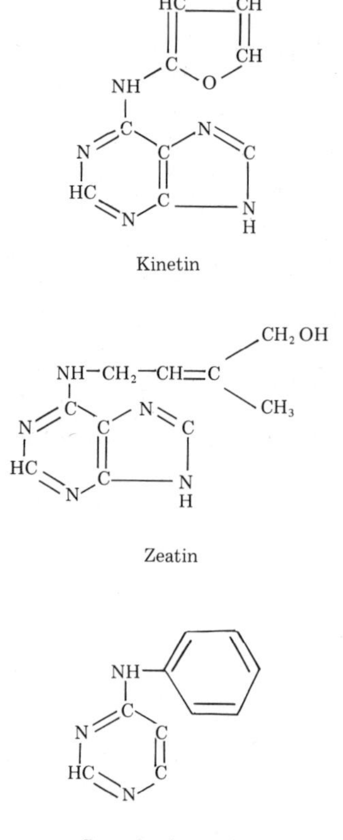
Kinetin

Zeatin

Benzylaminopurine

B. BIOASSAY

Figure 8-18 shows one of the early studies by Skoog and Miller on bud formation in

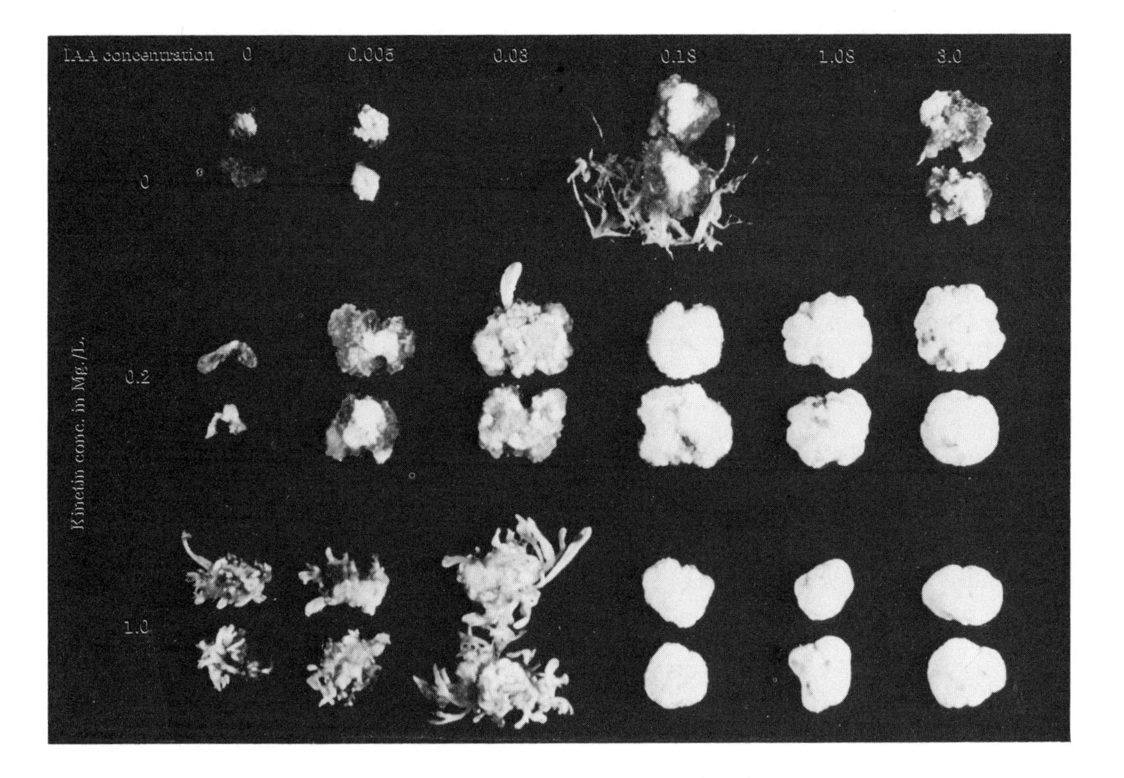

8-18 Tobacco pith cultures. Without added kinetin, IAA produces many roots; with high kinetin and low IAA, masses of buds develop. From Skoog and Miller, 1957.

tobacco pith cultures. The process is inhibited by auxin, very much as it is in whole plants, except that here we are dealing with the *formation* of buds and not with their growth. Thus this is not bud inhibition in the ordinary sense, but the inhibition of bud formation, and what is plotted here is the *number* of buds per segment. Adenine has quite a marked effect in increasing the number of buds and in removing the inhibition exerted by auxin. Notice that with tissue cultures the concentrations are usually very low; the concentrations we are dealing with here are micrograms per liter, parts per 10^9. We see here that for low auxin, low adenine will suffice to remove the inhibition; if the auxin level is higher, it takes more adenine to reverse it. In other words, there is a true balance between the two factors. The point of balance, or ratio of concentrations, is not necessarily the same in different tissues. On the other hand, in most of these assays, it turns out that at the optimum concentrations of auxin and cytokinin there is a synergism, i.e., more total growth occurs than with either one alone. We shall see in chapter 10 in regard to bud inhibition that the curves all go to an optimum point well above 100%. The same is true in these experiments.

These definite quantitative relations between the concentrations of cytokinin and of auxin certainly suggest (although of course they do not prove) that they are both acting in some way on the same basic system. In order to develop good bioassays, the Wisconsin group made their tissue culture system extremely quantitative, so that they could simply set up a logarithmic series of concentrations and after a standardized growth period on standard medium could read them off by eye. The subsequent figures present some striking examples of how well this sort of thing can be done, and how effectively structure and biological activity can thus be correlated.

C. CHEMICAL STRUCTURE AND ITS RELATION TO BIOLOGICAL ACTIVITY

Using these methods, Dr. Skoog's group at the University of Wisconsin have determined the growth activity of a large number of compounds. These have been synthesized by Prof. Nelson Leonard's group at Illinois, who are experts in the chemistry of purines.

Figure 8-19 is typical of the work of the combined Illinois and Wisconsin groups. Here they have compared a series of adenine compounds substituted in the 6-position with alkyl groups of differing chain lengths. Figure 8-20 shows how the results can be made quantitative by simply weighing the tissues. The same series of increasing chain lengths is shown, namely, methyl, ethyl, propyl, butyl, pentyl, hexyl, heptyl, and finally decyl. We see very little activity with the methyl, then activity increasing to a peak with around 4, 5, or 6 carbon atoms in the side chain. After that with more carbon atoms it decreases again. It is interesting that the most active of the synthetic straight-chain compounds compare closely with the natural compounds in that they also have 5 carbon atoms, though they are not in a straight chain. It suggests the behavior of the auxins, implying that what is needed is not necessarily a special chemical structure so much as a particular configuration in space. The molecule has to fit into some particular shape of receptor. Incidentally, none of these compounds is anywhere near as active as kinetin or 2iP; even the optimum compound of this series is certainly 300 times less active.

For comparison with the straight-chain compounds many ring substitutions were also made, as for instance in benzyladenine. The different rings show some interesting relative activities. Benzyl is the most active, furfuryl (as in kinetin) less so, cyclohexyl still less, and phenyl is about equally low. Also with zeatin, Dr. Leonard and colleagues have synthesized the two possible configurations, the *cis* and *trans* configurations about the double bond. The *trans* would have the hydroxyl group away from the ring. The case is of some interest because with the auxins we have the cinnamic acid pair, *cis*- and *trans*-cinnamic acids. In those, the *cis* acid is active, showing about 3% of the activity of IAA, while the *trans* acid is virtually inactive. With the cytokinins, the reverse is true. The *cis* compound has very low activity, only 1% or 2% as active as the *trans,* while the *trans* has, in some bioassays, the highest activity of any of them.

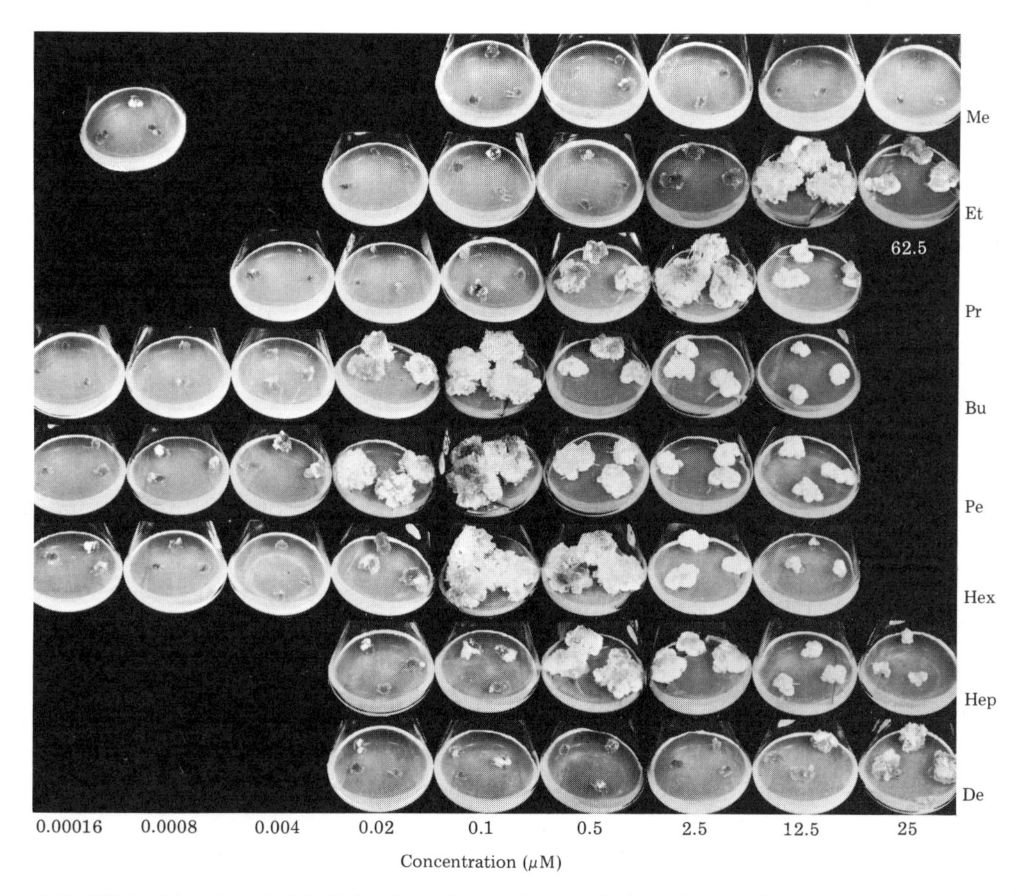

8-19 Effect of chain length of 6-alkyl-aminopurines on their cytokinin activity on tobacco pith cultures. Me=methyl; Et=ethyl; Pr=propyl; Bu=butyl; Pe=pentyl; Hex=hexyl; Hep=heptyl, and De=decyl. From Skoog et al. 1967.

Lest one conclude that the structure needed for a cytokinin is now understood, one needs to be reminded of the peculiar exception that was found in studying the activity of coconut milk. Coconut milk is, of course, immature endosperm. It happens to be from the biggest seed and thus represents the largest amount of immature endosperm that can be found in the vegetable kingdom, and it was used very early by Blakeslee, Conklin, and van Overbeek in studies on tissue culture, subsequently extended by Steward's group at Cornell. In isolating one of the active components, this group did not obtain an adenine derivative (although we know now that zeatin is present), but instead found that much of the activity was in symmetrical diphenylurea. One does not see much resemblance there to the type of substituted adenines, but recently Kefford and co-workers in Australia have tested out a large series of these, because they are easily synthesized, and have confirmed the activity. They have shown that there are certain rules: one needs the unbroken NH-CO-NH structure, and one cannot substitute both of the hydrogens, but more than that it is hard to generalize from the enormous list of different compounds having different activities. None of them is nearly as active as the natural cytokinins, but one or two come up to about ¼ of the activity of kinetin. These findings make it difficult to conclude that we know exactly what is needed for cytokinin activity. Of course one can draw out formulae and twist them around to make them look like substituted adenines, but I do not find this very convincing. On the other hand it would be unwise to conclude that the substituted ureas and the substituted adenines are acting in different ways.

8-20 Data from weighing the cultures of the preceding figure. From Skoog et al. 1967.

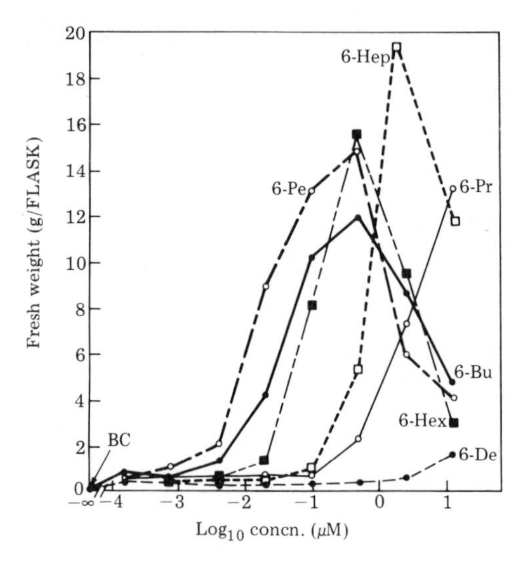

An interesting development took place in 1972. The Wisconsin group began looking for closely related substances that were not active, in the thought that they might function as antagonists to the cytokinins. They developed a group of substances based on the isopyrimidine formula (fig. 8-21), but the activities turned out to be about the same as kinetin. However, when they took a step further, substituting a methyl group in the 9-position, the activity was much less, and when they went still further and introduced a saturated side chain, the resulting compound had no activity and is now believed to be a cytokinin antagonist. It is of interest that two changes in the

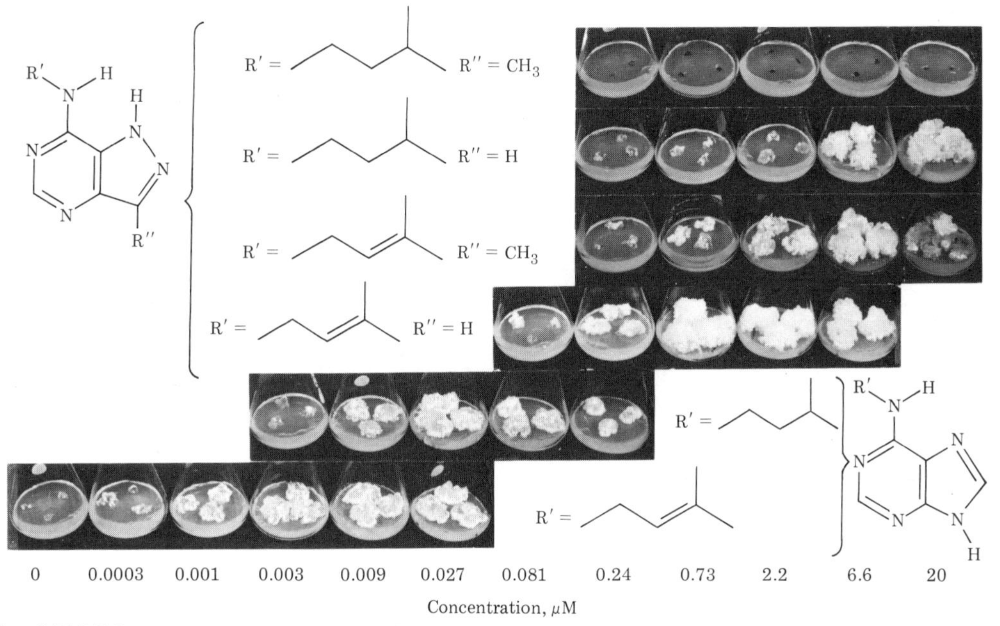

8-21 Activities, reading from bottom to top, of a series of compounds, showing how 3 steps convert a highly active compound into an antagonist. Bottom, 2′-isopentenyl purine; next, same with the side chain saturated; next, 2′-isopentenyl derivative with the N atom in the 5-membered ring displaced; next, the same methylated in the 9-position; next, the same with the side chain saturated; last (top) the same methylated. Concentrations are on a log scale to the base 3. From Skoog et al. 1973.

molecule still left some activity, but it is only with three differences from the active molecule that antagonism appears. Figure 8-21 shows the series of cultures; the substance with peak activity at .009 micromolar is 2iP.

Naturally there is great interest in the possibility of an anti-cytokinin, just as there was in the possibility of an anti-auxin. But in the case of anti-auxins, there were a great many false leads. Many people thought that quite modest changes in the auxin molecule would make anti-auxins. They merely made inactive substances, or substances with very weak activity. One of the few synthetic relatives of auxin that is really an auxin antagonist is the 2,4D derivative, 2,4-dichlorophenoxy-isobutyric acid. But the corresponding indole derivative, indoleisobutyric acid, is not an antagonist; it has very weak auxin activity. It is evident that the requirements for antagonism are not as simple as they look. It is not a simple matter of competing at the same site for molecules of slightly different structure; it needs a great deal more specialty in the structure of the molecule. Some compounds that are not quite antagonists but have comparable structures exert very powerful synergism. This is a phenomenon which is worth much more study than it is getting. A classical example is triiodobenzoic acid synergizing with indoleacetic acid. Triiodobenzoic acid inhibits the transport of auxin, and can be used as a transport inhibitor, but in some tests it will synergize with very low auxin concentrations to make them act like very high ones.

D. SOME VARIED BIOLOGICAL ACTIVITIES

As to the activities of the cytokinins, some of them we have seen already. They include synergism and antagonism with auxin, the removal of bud inhibition (chap. 10), including both the inhibition of bud growth and the inhibition of bud formation, and also the removal of inhibition due to abscisic acid. This does not take place in all systems, but abscisic acid is a rather generalized plant growth inhibitor, and in several systems cytokinin removes that inhibition pretty completely. I think the clearest case is van Overbeek's, with *Lemna* cultures. As figure 8-22 shows, *Lemna* cultures are inhibited in their growth by abscisic acid, and if a cytokinin (in this case benzyladenine) is added, the inhibition (if it is not too great) can be completely removed. If the inhibition is very powerful, it can be only partially relieved, but the effect is there.

Then there is the interesting behavior of the *Convolvulus* roots studied by Torrey and his colleagues. *Convolvulus* roots are rather like the tobacco pith culture in that they can form both buds and roots. And they provide a nice example of an organ culture which responds in the same way as the tobacco tissue culture does. With a high ratio of auxin to kinetin in the medium, *Convolvulus* roots form lateral roots—just as the roots of many other species form laterals when auxin is provided. If the ratio of auxin to kinetin is low, the cultured roots make buds, and can make quite large numbers of them.

Torrey and Bonnett noted that in the

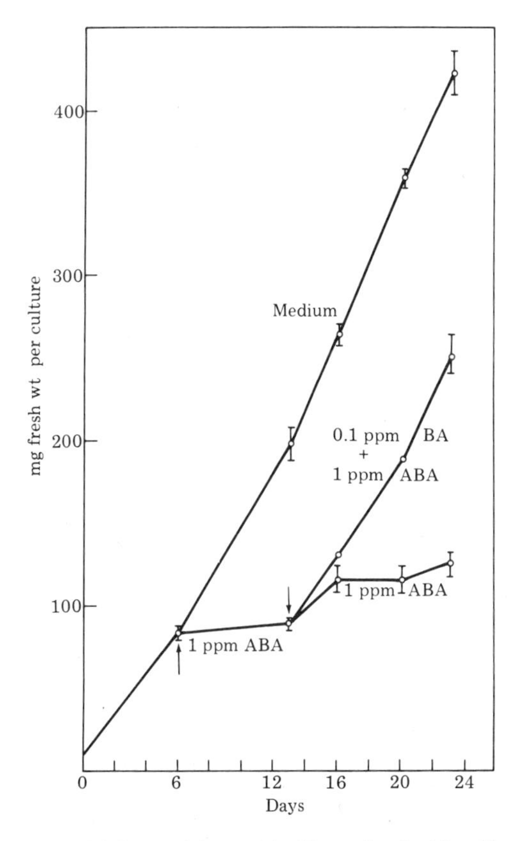

8-22 Inhibition of the growth of *Lemna* by abscisic acid (ABA) and its reversal by 6-benzylaminopurine (BA), (added at second arrow). From van Overbeek et al. 1968.

early stages of the growth of these lateral roots one sees a little nest of cells, just as White saw with his immersed tumor cultures, and one cannot decide whether they are going to be buds or roots. There is at this stage no distinction between root initials and bud initials, unlike the situation in intact plants, wherein one can usually tell the difference immediately.

Another peculiar effect, discovered by Japanese workers, is that if plants are treated with a cytokinin and then exposed to moderate cold, they are protected against freezing. Usually protection against freezing, to a small degree, involves an increase in the solute concentration, so that the freezing point is lowered. But one thing we know about the effects of cytokinin is that it tends to promote the synthesis of polymers, so that its action would tend to work the other way—it would decrease the osmotic content, because it would be producing polymeric substances. There are several cases where cytokinin treatments have been shown to promote the synthesis of starch. A *Lemna* culture grown with cytokinin becomes very much richer in starch than normal *Lemna* cultures. Thus the reported protection against freezing is hard to understand.

A number of years ago Drs. Richmond and Lang at Michigan, and later Prof. Kurt Mothes in East Germany, discovered an important effect of cytokinin. What they discovered is this: if one treats aging, yellowing leaves with a cytokinin, the part treated remains green when the rest of the leaf turns yellow; in effect, senescence is arrested. Dr. Mothes felt that this must mean that protein is being synthesized in the green zone (since senescence is usually accompanied by proteolysis); and therefore he tried applying amino acids. As the result of a series of such experiments, he found that if the cytokinin is applied on one part of the leaf and the amino acid on another part of the leaf (using ^{14}C-labeled amino acids), the amino acid migrates to where the cytokinin is. In other words, the cytokinin "exercises an attraction" for amino acids. Mothes' first interpretation of that was that the cytokinin produces protein synthesis and that lowers the concentration of amino acids in the cells to which it was applied; therefore there is a diffusion gradient, and the amino acids move in along the gradient. But then this was made less probable by his testing amino acids like γ-aminobutyric acid, which are not incorporated into protein but behaves in the same way. Figure 8-23 (left) shows a zone treated with kinetin and a zone treated with α-aminobutyric acid, which is not a constituent of protein. After the leaf has become yellow, a radio-autograph (right) shows activity where the cytokinetin was, so the amino acid has migrated there under the influence of the cytokinin. It may be, of course, that the nonprotein amino acid has been converted to a protein constituent, but there may be some quite other explanation not yet known.

The greening part of this phenomenon has gradually become adopted as a bioassay for cytokinins, based on the maintenance of chlorophyll in aging or senescing leaves (chap. 12). It can be used easily as a bioassay, for all one has to do is cut out the green part, extract it in hot acetone, and read off

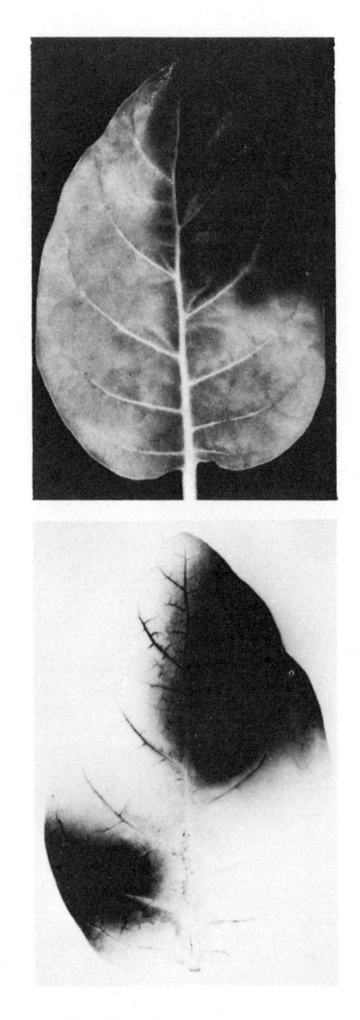

8-23 Example of the experiments of Mothes et al. with isolated tobacco leaves. Both have been sprayed with kinetin on the top right half,, and with ^{14}C-α-aminobutyric acid on the lower left half. On the left is a photo, after storing in the dark, showing that the kinetin-sprayed area remains green. On the right is a radio-autogram, showing that much of the label migrates to the area treated with kinetin. From Mothes et al. 1961. (Original photo supplied by Prof. Mothes.)

the chlorophyll content. Figure 8-24 shows the chlorophyll content, as measured by the absorption at the peak of chlorophyll A, 665 nm, with a logarithmic series of concentrations of 3 different cytokinins. Tests on 3 different days are shown to indicate how much variation there is. Evidently the slopes are extremely reproducible. The ratio of activities of phenylaminopurine and kinetin is about 10,000 times. This topic is treated in more detail in chapter 12.

Since the cytokinins are derivatives of adenine, and since adenine characteristically operates as a ribosephosphate derivative, many people thought no doubt that the first thing that happens is that the cytokinins are combined with ribose and phosphate to produce a relative of AMP. This question could not be dealt with until suitable synthetic methods were developed, but when they were a number of cytokinin-ribose-phosphates were synthesized and tested. In every case their activity is only weak, and clearly less than that of the uncombined compound. One could ascribe it to hydrolysis or breakdown, rather than to their action as ribosephosphates. Nevertheless it is a curious fact that in some cases where known cytokinins have been applied to plant tissue and were subsequently extracted, cytokinin-ribose-phosphates have been identified. Thus the plant readily makes the conversion, but it does not follow that it is significant for the biological activity. Such compounds would be just like AMP in that they would enter cells only with difficulty, which might help to explain their low relative activity. But the ratio of activity of the free purine to that of the purine-ribose-phosphate is too great for

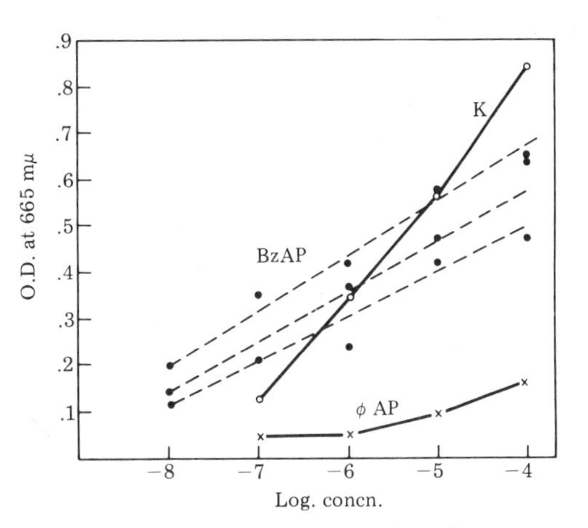

8-24 Chlorophyll content (measured by the optical density at 665 nm) of oat leaves treated with a series of concentrations of three cytokinins and then left for five days in the dark. BzAP, 6-benzylaminopurine, three separate experiments; ØAP, 6-phenylaminopurine; K, 6-furfurylaminopurine (kinetin). From Thimann and Sachs, 1966.

that excuse to be satisfying. A complication is that extracted cytokinin activity seems to be exchanged between butanol-soluble, water-soluble and petroleum ether-soluble forms following various light and dark treatments. This happens in seeds (Van Staden and Wareing, 1972) and in other tissues, and it may mean the formation of a number of different complexes, from which liberation of the free base would be the physiologically important step.

When cytokinins are fed to seedlings they are rapidly metabolized. Letham and his colleagues (1975) have identified three different glucosides, bearing the glucose molecule in the 3-, 7- and 9-positions, as metabolites of zeatin or of benzylaminopurine. The 3-glucoside still has high cytokinin activity, but the others are very low. Lupin seedlings also make zeatin into a compound with alanine at the 9-position. The functional significance of these derivatives is not yet known.

OTHER GROWTH-CONTROLLING COMPOUNDS OF CHEMICAL INTEREST

A. THE PHENOLIC ACIDS

A large group of naturally occurring compounds is primarily of interest for the way in which its members *interact* with the hormones, rather than for their direct actions. These are the phenolic acids. They are perhaps more important than is usually believed, if only because they occur so very widely in plants and at concentrations often 100 times higher than hormonal levels. Virtually all plants contain phenolic compounds of one sort or another, contrasting with animals in this respect. In several cases a specific compound has been looked for and nearly always found; parahydroxybenzoic acid has been found in 97% of the several hundred species in which it was looked for, and caffeic acid similarly was found in 95% of the many species studied.

There are two series that are the most widespread in higher plants, the benzoic series and the cinnamic series (fig. 8-25). Parahydroxybenzoic and protocatechuic acids are probably the most widespread in the benzoic series, protocatechuic being probably a breakdown product of many larger molecules. Also common is gentisic (2,5-dihydroxybenzoic); vanillic, which has one hydroxyl group methylated; and syringic acid, which is 3,4,5-trihydroxybenzoic with two of the hydroxy groups methylated. This last is important because it is related to a lignin polymer and is set free on the breakdown of some lignins. Of the cinnamic series, cinnamic itself and paracoumaric are the most common. Orthocoumaric scarcely exists, because the hydroxyl adjacent to the 3-carbon side chain causes a ring to form. The resulting product, coumarin, is responsible for the smell of freshly cut hay, which certainly suggests rapid breakdown of some phenolic precursor during the first stages of drying. It is present in many other plants too. Caffeic is dihydroxycinnamic; ferulic acid is methylcaffeic, and sinapic

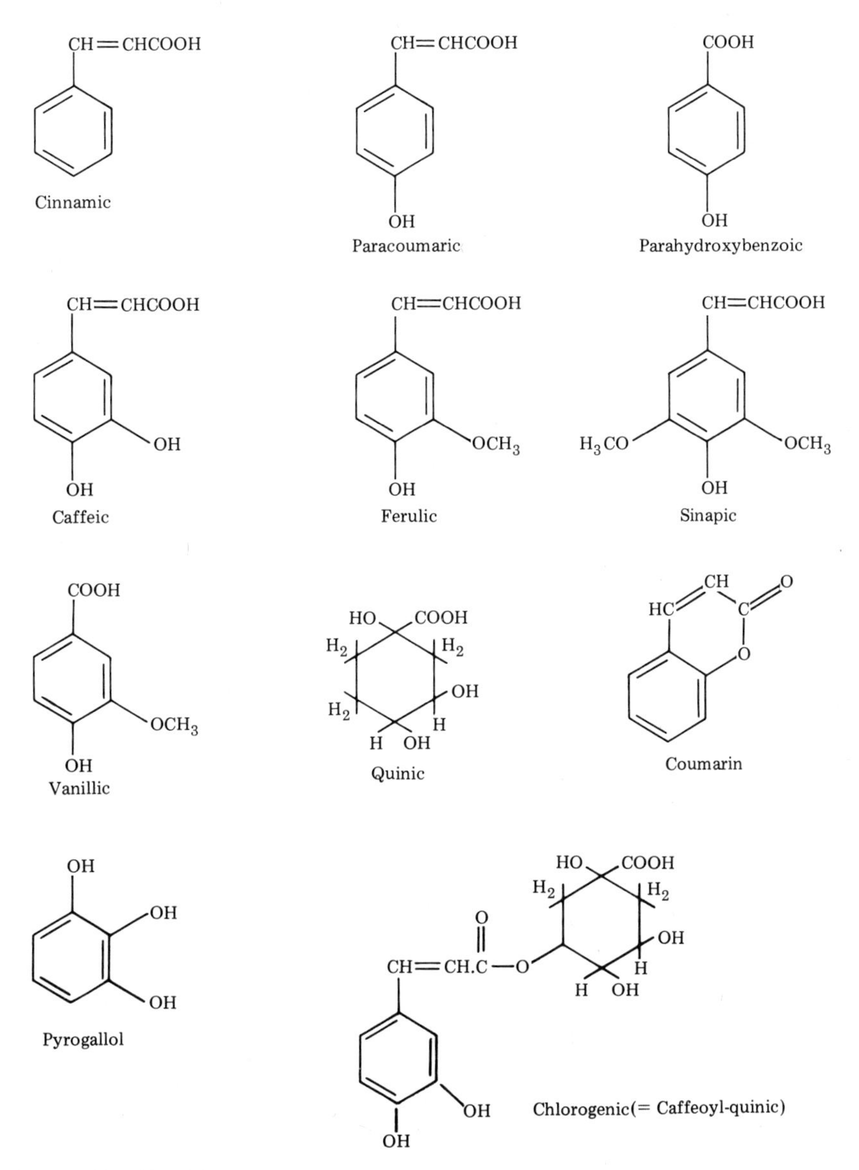

8-25 The major phenolic acids of plants.

acid is the dimethyl derivative, which corresponds in structure to the syringin which is so prominent a component of lignin. Shown here also is quinic acid, which has a saturated ring; it is not a benzene derivative but related to cyclohexane. Last comes chlorogenic acid, a compound of quinic and caffeic acids. The name literally means green-making—not because it makes chlorophyll, but because it gives a green color with ferric chloride. This compound is of importance not only because it occurs widely and in good amounts, at least in many leaves, but because it is very interactive with auxins.

All these phenolic acids, and indeed all phenol derivatives, are characteristic of plants and bacteria. They are not as a rule formed in animals nor in animal tissues; indeed the phenolic amino acids are essential food elements for mammals just because they are not synthesized. Insects, however, are like plants in that they form relatively large amounts of phenols, and indeed the brown color of so many pupae is due to oxidized polymers of phenols.

To give some idea of the amounts of these acids that are found in plants, we give here some figures from a recent French analysis of corn leaves. In parts per million of fresh weight we have:

Coumaric	50
Vanillic	35
Sinapic	25
Ferulic	20
Parahydroxybenzoic	15

For comparison, the amount of indoleacetic acid in the same units would be less than 0.1. Thus there are 100 or more times as much of *each* of these phenolic acids as there is of auxin, and in most tissues there will be at least 6 to 8 such compounds.

The reason why these compounds are interesting in the present context is because they greatly modify the effects of auxins. We know this because when indoleacetic acid is applied to different plants in physiological concentrations the quantity and type of phenolic acid have a great influence on the resulting growth. For this reason, and especially because of the relatively large amounts commonly present, we cannot doubt that the same interaction must occur *in vivo*.

In general, monophenols like parahydroxybenzoic or paracoumaric acid promote the oxidation of indoleacetic acid, so that they *reduce* its apparent activity under natural conditions. Diphenols, whether they are ortho- or para-, have the opposite effect; they inhibit the oxidation of indoleacetic acid. When Peter Ray, at Harvard, was studying the oxidation of IAA by plant peroxidase he was estimating the peroxidase activity by a standard procedure using pyrogallol as substrate, and he found that the pyrogallol interferes with the oxidation of the IAA. The formula in figure 8-25 shows that it can indeed be considered as a diphenol. But what was interesting and unexpected was that IAA, in turn, prevented the oxidation of the pyrogallol. The competition is mutual, so that when both are present in sufficient concentration neither is oxidized.

A fascinating consideration arises from the presence in most plants of the enzyme polyphenol oxidase, which can introduce a second hydroxyl into the ring of a

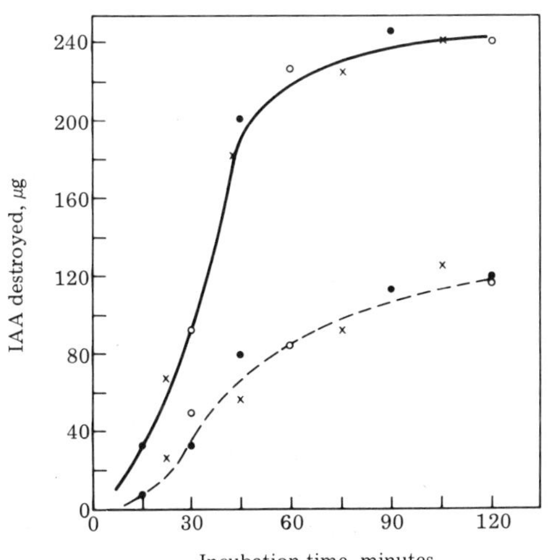

8-26 The oxidation of IAA by an enzyme system from lentil roots, showing that parahydroxybenzoic acid (solid line) doubles the rate. Dashed line, controls. From Pilet, 1966.

monophenol. This second hydroxyl, as a result of the electron transfer mechanism, comes in ortho- or para- to the first. (The enzyme also further oxidizes the resulting diphenol to a quinone, which is of less direct concern here.) The point is that if this enzyme becomes active, a phenolic compound which promotes auxin oxidation will be converted to one having the *opposite* effect. The ramifications involved here have not yet been much explored, but it is evident that one needs to know more about the native concentrations of phenolic acids before one attempts to quantitate auxin effects on the growth of whole plants.

Figure 8-26 gives an example of an auxin-phenol interaction; the amount of IAA destroyed in this particular system is obviously greatly increased by parahydroxybenzoic acid. This experiment is due to Prof. Pilet of Lausanne, but the effect was first shown by Zenk and Muller in 1963; and we have also done many comparable experiments using ordinary growth tests, with segments of pea stem or coleoptile, in auxin solution with and without phenols. By using ^{14}C-indoleacetic acid we can measure the amount of $^{14}CO_2$ given off into a KOH receiver during the growth and compare with the amount of growth resulting. Table 8-1 shows that with the addition of monophenols the CO_2 evolved is larger, and the elongation is on the whole smaller, than in the controls. While there are some variations, there is a particularly clear correlation between the amount of C^{14} remaining in the tissue and the amount of elongation resulting. The effect is decreased because much of the elongation takes place in the first hours before the auxin destruction has proceeded very far. Similarly in presence of diphenols the CO_2 production is less, and the elongation is greater, than in the controls. In such growth tests the phenol exerts its greatest effect, as might be expected, when the auxin concentration is low. Thus at 10^{-7} molar IAA chlorogenic acid causes a very large growth promotion, but as the auxin concentration increases the effect of the chlorogenic acid decreases. However, it is the low auxin level which best approximates the situation in the intact plant, and for this reason I am inclined to ascribe considerable importance to the interactions of the phenols with the auxin system. We shall see more examples in connection with tissue culture (chap. 9).

A special case is presented by coumarin. This compound is not really a phenol because the phenolic oxygen atom has been incorporated into the ring (fig. 8-25). Correspondingly its effects are not typically phenolic. Coumarin has a characteristic effect on elongation in the presence of auxin; if we hold the auxin constant and vary the coumarin we find that while it does inhibit the growth at fairly high concentrations (as a monophenol would do); yet at lower concentrations it clearly promotes the growth (fig. 3-21). In other systems, too, it exerts effects which do not seem to relate readily to an interaction with auxin. The prime example is its action on the germination of light-sensitive lettuce seed. We saw in chapter 2 that Grand Rapids lettuce seed is caused to germinate by red light and that this is reversed by far-red (within certain time limits). Coumarin inhibits the red light effect, making the lettuce seed behave as though in darkness; put in another way, it acts like far-red light. To some extent the coumarin action can be antagonized by increasing the intensity of the red light, so that there is a balance between a chemical reaction and a photochemical reaction, acting in opposite directions. Since the phytochrome system is sensitive to sulfhydryl reagents, and the action of coumarin is

Table 8-1 The fate of IAA$-1-{}^{14}C$ in green epicotyl segments of light-grown Alaska peas. Radioactivity of supplied IAA=8525 cpm. Concentrations of phenolics Nos. 1 to 4, 5×10^{-5}M; nos. 5 and 6, 1×10^{-4}M. Time, 24 hours.

Solution	Radioactivity: cpm in $^{14}CO_2$	cpm remaining in tissue	Elongation: as percent of initial length minus elongation of controls in water
IAA alone	282	291	14.0
IAA plus *Polyphenols*			
1. Chlorogenic acid	145	391	22.0
2. Caffeic acid	255	324	16.4
3. Sinapic acid	153	356	20.2
IAA plus *Monophenols*			
4. Vanillic acid	449	307	15.1
5. Parahydroxybenzoic acid	916	220	13.2
6. 2,4-dichlorophenol	590	233	14.2

From Tomaszewski and Thimann, 1966.

often ascribed to its reacting with -SH groups of enzymes, there may be a *modus operandi* here. In any case, there is more to the action of coumarin that can be explained by its semiphenolic character.

B. ANTHOCYANINS AND FLAVONOIDS

Another group of widely occurring compounds has a good deal in common with the phenols. Many plants contain anthocyanins, the red and purple pigments of leaves and flowers, and also flavones and flavonols, pale yellow pigments of similar composition. These substances contain a sugar (mono- or di-saccharide, glucose being the commonest) and a three-ringed molecule, of which two of the rings are clearly phenolic (fig. 8-27). The left-hand ring, ring A, is a derivative of phloroglucinol; it has the three hydroxyls in meta-positions. This results from its synthesis by combination of three molecules of malonyl coenzyme A, which lose both the coenzyme and one carboxyl group in polymerizing. The three -C-OH groups of the ring are formed from the C-O groups of the other carboxyls. Ring B, containing an oxygen atom, has in the anthocyanins peculiar properties which give a basic character to the molecule; the anthocyanins are usually crystallized as chlorides. Ring C is clearly a phenol and is known to be synthesized by the same mechanism as is responsible for forming the other phenols, both in plants and in bacteria. Both anthocyanins and flavonols have one or more hydroxyl groups on this ring, and these may sometimes be methylated. Thus in all plants possessing these pigments—and that includes the

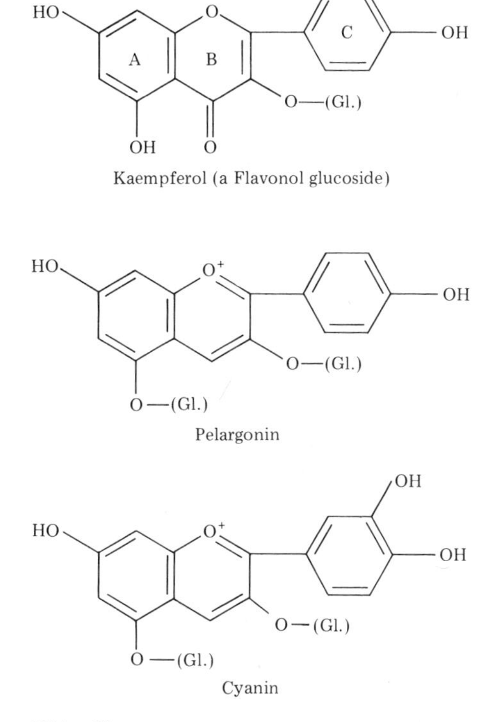

8-27 A typical flavonol and two typical anthocyanins.

majority of flowering plants—there are relatively large amounts of combined but reactive phenols present. The pelargonin of our common *Pelargoniums,* with one -OH on ring C, act as monophenols in the reaction with IAA, while anthocyanins like cyanin of the bachelor's button (*Centaurea cyanus*), with two -OH groups ortho to one another, act as diphenols in this reaction. The extent to which they do react *in vivo* may perhaps be minimized by the limited contact they can make with the auxin, for these pigments are commonly limited to the vacuole, while auxin must presumably be located in the cytoplasm in order to exert its many effects. A possible hint of the natural interaction is seen in those orchids whose flowers have a very dark purple spot: here the diphenolic anthocyanin is clearly associated with a marked over-growth, sometimes almost tumorous in appearance, in the pigmented area.

Perhaps other common plant materials exert effects as yet unrecognized on the action of the growth-controlling hormones. The observations here described serve only to give a hint of how little we yet understand about hormone action in the whole plant.

C. DIHYDROCONIFEROL

This phenolic alcohol occupies a special place among natural phenols. It has come into prominence only very recently. Seedlings of lettuce and cucumber are stimulated to elongate very strongly by gibberellins (cf. fig. 3-32), but if the cotyledons are removed their response to the gibberellin is greatly reduced. Extracts of the cotyledons will restore most of the response. Now Shibata and Kamisaka, in Osaka (1974, 1975), have identified the active "cotyledon factor" as dihydroconiferol. Coniferol itself, which is a constituent of the giant molecule of lignin, has a double bond in the 3-carbon side chain and this makes it inactive as a cotyledon factor.

Because of its diphenolic character (for the $-OCH_3$ group acts like -OH in this regard), dihydroconiferol inhibits the enzymatic oxidation of IAA. However, this is apparently not the basis for potentiating the action of gibberellin, since neither caffeic nor ferulic acid shows any such effect. The compound which has -OH instead of $-OCH_3$ acts about as well as dihydroconiferol, but other phenolic compounds with different substituents are all either less active or quite inactive.

There is some evidence that dihydroconiferol occurs rather widely in cotyledons, so that its role as a synergist for gibberellins may prove to be quite general. It is too soon to know whether it has any other effects.

D. THE GIBBERELLINS

The chemical study of the gibberellins, begun by Yabuta and Sumiki in Tokyo, was continued in a most thorough manner in England. Much of the structural work has been done by MacMillan and his collaborators, first at Imperial Chemical Industries and then at Bristol University. The structures of the known gibberellins were presented *in extenso* by Paleg and West in 1972 and have appeared more recently in several places. The new book,

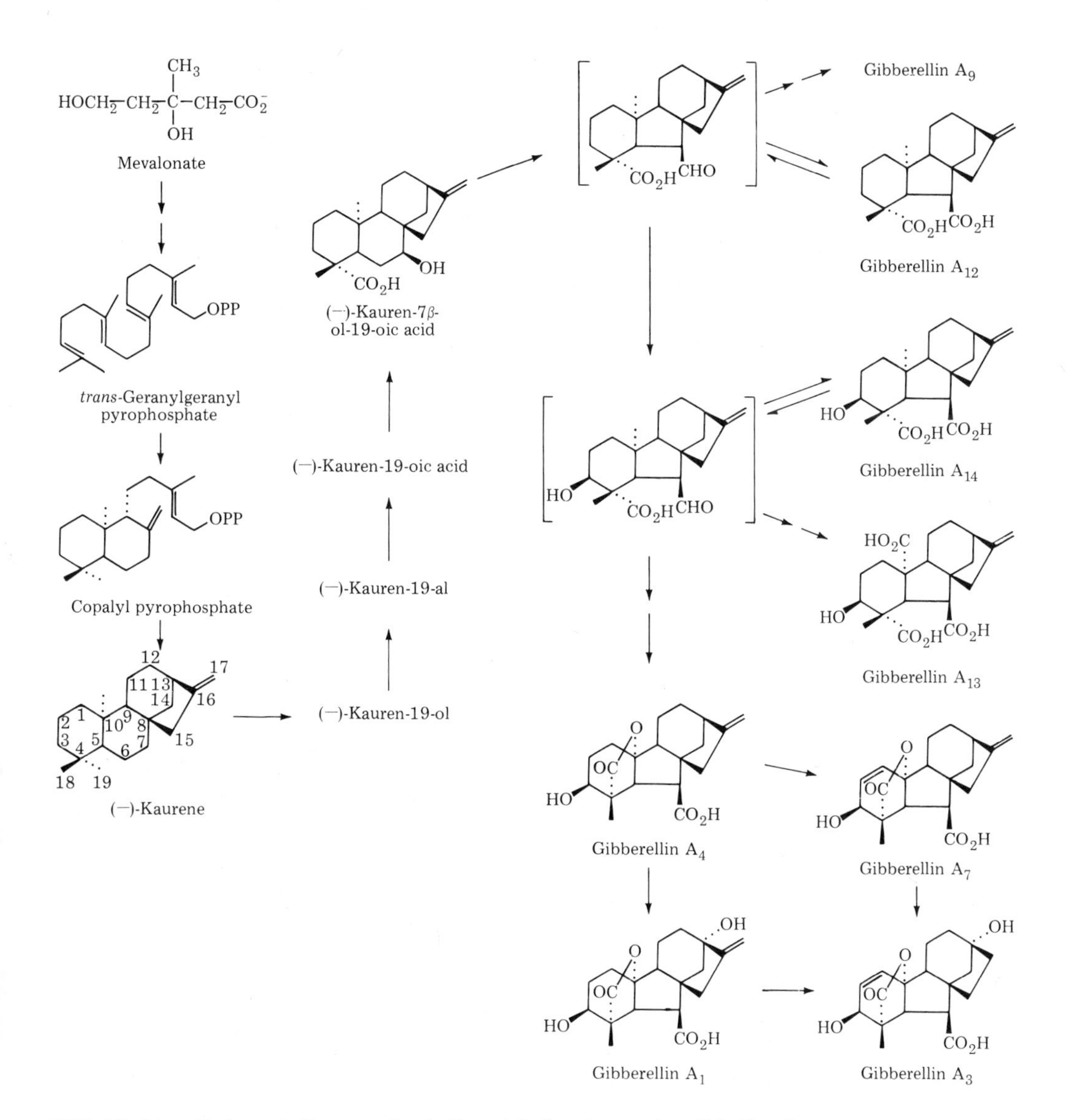

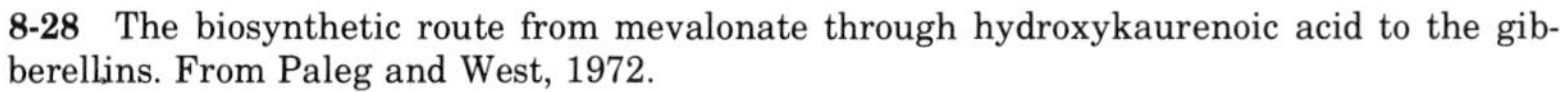

8-28 The biosynthetic route from mevalonate through hydroxykaurenoic acid to the gibberellins. From Paleg and West, 1972.

Gibberellins and Plant Growth (edited by Krishnamoorthy, 1975) brings all the information up to date and also includes discussions of the biosynthetic pathways from mevalonic acid to the C-19 and C-20 gibberellins. West has summarized the stages, leading via the steroid relative hydroxykaurenoic acid to many of the key gibberellins, in figure 8-28. As yet it has not proved possible to deduce much as to the relations between structure and activity in these molecules; therefore I shall not treat the chemical aspects of the gibberellins again here.

REFERENCES

Bandurski, R. S., and Schultze, A., Concentrations of Indole-3-acetic acid and its esters in *Avena* and *Zea*. *Plant Physiol.* 54, 257–262, 1974.

Gmelin, R. Occurrence, isolation and properties of glucobrassicin and neoglucobrassicin. In *Regulateurs naturels de la croissance végétale,* pp. 159–167. Paris: C. N. R. S., 1964.

Harper, D. B., and Wain, R. L. Studies on plant growth regulating substances. 30. The plant growth regulating activity of substituted phenols. *Ann. Applied. Biol.* 64, 395–407, 1969.

Kefford, N. P., Bruce, M. I., and Zwar, J. A. Retardation of leaf senescence by urea cytokinins in *Raphanus sativus*. *Phytochem.* 12, 995–1003, 1973.

———, and Kelso, J. M. Sulphur analogues of IAA. Synthesis and biological activity of some thionaphthenacetic acids. *Australian J. Biol. Sci.* 10, 80–84, 1957.

Krishnamoorthy, H. N., ed. *Gibberellins and plant growth*. New Delhi: Wiley Eastern Ltd., 1975.

Larsen, P. Indole-acetaldehyde as a growth hormone in higher plants. *Dansk Bot. Arkiv* 11, 11–132, 1964.

Letham, D. S. Chemistry and physiology of kinetin-like compounds. *Annu. Rev. Plant Physiol.* 18, 349–364, 1967.

———, Wilson, M. M., Parker, C. W., Jenkins, I. D., Macleod, J. K., and Summons, R. E. Regulators of cell division in plant tissues. 23. The identity of an unusual metabolite of 6-benzylaminopurine. *Biochem. Biophys. Acta* 399, 61–70, 1975; also MacLeod, J. K., Summons, R. E., Parker, C. W., and Letham, D. S. Lupinic acid, a purinyl amino acid and a novel metabolite of zeatin. *J. Chem. Soc.,* Chem. Comm., 809–810, 1975.

Mahadevan, S., and Thimann, K. V. Nitrilase 2. Substrate specificity and possible mode of action. *Arch. Biochem. Biophys.* 107, 62–68, 1964.

Mothes, K., Engelbrecht, L., and Schütte, H. R. Uber die Akkumulation von α-aminoisobuttersäure im Blattgewebe unter dem Einfluss von Kinetin. *Physiol. Plantarum* 14, 72–75, 1961.

Ohwaki, Y. Thin layer chromatography of diffusible auxin of corn coleoptiles. *Botan. Mag.* (Tokyo) 79, 200–201, 1966.

Overbeek, J. van, Loeffler, J. E., and Mason, M. I. R. Mode of action of Abscisic acid. In *Biochemistry and physiology of plant growth substances,* ed. F. Wightman and G. Setterfield, pp. 1593–1607. Ottawa: Runge Press, 1968.

Pilet, P. E. Effect of parahydroxy-benzoic acid on growth, auxin content and auxin catabolism. *Phytochem.* 5, 77–82, 1966.

Poole, R. J., and Thimann, K. V. Uptake of

indole-3-acetic acid and indole-3-acetonitrile by *Avena* coleoptile sections. *Plant Physiol.* 39, 98–103, 1964.

Porter, W. L., and Thimann, K. V. Molecular requirements for auxin action. Halogenated indoles and indoleacetic acid. *Phytochem.* 4, 229–243, 1965.

Shibata, K., Kubota, T., and Kamisaka, S. Isolation and chemical identification of a lettuce cotyledon factor, a synergist of the gibberellin action in inducing hypocotyl elongation. *Plant and Cell Physiol.* 15, 461–469, 1974.

———. Dihydroconiferyl alcohol as a gibberellin synergist in inducing lettuce hypocotyl elongation. An assessment of structure-activity relationships. *Plant and Cell Physiol.* 16, 871–877, 1975.

Skoog, F., Hamzi, H. Q., Sweykowska, A. M., Leonard, N. J., Carraway, K. L., Fujii, T., Helgeson, J. P., and Loeppky, R. N. Cytokinins; structure/activity relationships. *Phytochem.* 6, 1169–1192, 1967.

———, and Miller, C. O. Chemical regulation of growth and organ formation in plant tissues cultured *in vitro.* In *Symp. Soc. Exptl. Biol.* 11, 118–131, New York, Acad. Press, 1957.

———, Schmidt, R. Y., Bock, R. M., and Hecht, S. M. Cytokinin antagonists; synthesis and physiological effects of 7-substituted 3-methylpyrazolo (4,3-d) pyrimidines. *Phytochem.* 12, 25–37, 1973.

Thimann, K. V. On an analysis of the activity of two growth-promoting substances on plant tissues. *Proc. Kon. Akad. Wetensch. Amsterdam* 38, 896–912, 1935.

———. The role of ortho substitution in the synthetic auxins. *Plant Physiol.* 27, 392–404, 1952.

———. The physiology of growth in plant tissues. *Amer. Scientist* 42, 589–606, 1954.

———. Auxin activity of some indole derivatives. *Plant Physiol.* 33, 311–321, 1958.

———. Auxins, an informal summary of some recent work. In *Hormonal regulation in plant growth and development,* ed. H. Kaldewey and Y. Vardar (Proc. Adv. Study Inst, Izmir, 1971), pp. 155–170. Weinheim: Verlag Chemie, 1972.

———, and Sachs, T. The role of cytokinins in the "fasciation" disease caused by *Corynebacterium fascians. Amer. J. Bot.* 53, 731–739, 1966.

———, Skoog, F., and Byer, A. C. The extraction of auxin from plant tissues. 2. *Amer. J. Bot.* 29, 598–606, 1942.

Tomaszewski, M., and Thimann, K. V. Interactions of phenolic acids, metallic ions, and chelating agents on auxin-induced growth. *Plant Physiol.* 41, 1443–1454, 1966.

Van Staden, J., and Wareing, P. F. The effect of light on endogenous cytokinin levels in seeds of *Rumex obtusifolius. Planta* 104, 126–133, 1972.

Veldstra, H. The relation of chemical structure to biological activity in growth substances. *Annu. Rev. Plant Physiol.* 4, 151–198, 1953.

Wain, R. L. A new approach to selective weed control. *Ann. Applied Biol.* 42, 151–157, 1955.

Wightman, F. Biosynthesis of auxins in tomato shoots. *Biochem. Soc. Symp.* 38, 247–275, 1973.

Zenk, M. H. Enzymatische Aktivierung von Auxinen und ihre Konjugierung mit Glycin. *Zeit. für Naturforsch.* 15 b, 436–441, 1960.

9 Differentiation and Plant Tissue Cultures

A. HISTORICAL AND GENERAL

The development of the methods of tissue culture with plant material has enabled fresh approaches to be made to many problems of growth and morphogenesis, and it has also opened up some new problems of great importance. One of its most interesting aspects is the way in which the whole development has been inextricably intertwined with that of the plant hormones. For it was the discovery and availability of auxin that made tissue cultures possible, and it was the working out of the procedure with different types of plant material that led, in turn, quite directly to the discovery of another whole class of plant hormones, the cytokinins. Then use of the two hormones together, largely but not wholly in tissue cultures, taught us a great deal about how the formation of plant organs is controlled. Study of the auxin relations of tissue cultures brought us what seemed at one time to be a new approach to the problem of the physiology of tumors, but this, while it has led to the useful concept of a "tumor-inducing principle," has so far yielded more theory than fact, in an admittedly difficult area.

Early attempts to grow plant tissues in culture, beginning with Haberlandt, failed, though they did inspire Ross Harrison to try the same procedure with animal tissues. These succeeded, with the result that the methods of animal tissue culture quickly became of central importance in several branches of medicine. The key to success with plants came only in the 1930s after many years had been spent in improving the mineral medium and optimizing the sugar concentration. In 1934, we saw that both cell elongation and the formation of roots on stems were controlled by one hormone, indole-3-acetic acid, and the very next year Robin Snow at Oxford found that the same substance activated the cambium of seedlings, i.e., stimulated cells to *divide*. At about that time Roger Gautheret in Paris was studying a culture made from the cambium of willow; he established sterility, and observed that the cells did divide once or twice, but then they stopped. It would have seemed natural for the next step to be to add a little of the substance which had just been shown to initiate and stimulate cell division.

Actually that test was performed, but with a different tissue culture—that from carrot roots—and the same experiment was carried out simultaneously by two different workers in France—Gautheret in Paris and Pierre Nobécourt in Grenoble. The simultaneity was remarkable, though the use of the same tissue was actually not such a coincidence, because this was popular tissue. They both used carrot tissue, they both added indoleacetic acid (at about 10^{-7}M) to what otherwise had seemed like a good medium. In both cases the cells divided and *continued to divide,* and so tissue cultures were born. This was in 1939, and Snow's experiment was done in 1935, so actually not a lot of time was wasted.

They both published at almost the same moment little papers, one in the Compts-Rendus of the Société de Biologie and the other in the Compts-Rendus of the Academy; and as it turned out Gautheret, the one who published with the Academy, later got the Academy award. There was some to-do about that: I was at the Botanical Congress in Stockholm in 1950 when Gautheret gave a paper on his tissue cultures. Of course the intervening war had cut down most laboratory activities so that in 1950 it was still very interesting news, and there had been some new developments. When Gautheret gave his paper Nobécourt was in the audience; so of course he got up and began to explain how he had done all this before and that Gautheret had done nothing new. Poor Gregory, the professor at Imperial College in London, was chairman of the meeting, and he had, as can be imagined, an extremely difficult time of it (an Englishman trying to separate two voluble Frenchmen in a Swedish ambiance).

This was, then, the beginning of plant tissue cultures. The material happened to be carrot tissue. It might as well have been several other kinds, but I will not say it might have been anything, because many other tissues would not have responded as clearly as these did. They grew and produced *callus,* which as you know, is simply a nondescript name for nondescript cells. They are essentially parenchyma cells, thin-walled, usually not containing chloroplasts, except in very exceptional conditions; and they grow without producing any organs. However, in a good medium they do produce some differentiation in the form of vascular bundles. While this was not really seen in Gautheret's earliest experiments, it became clear later, and especially with *Helianthus* tissue cultures.

Slices of artichoke tubers, *Helianthus tuberosus,* are very reactive. Figure 9-1 shows low magnification pictures of a whole tuber slice. The control shows only parenchyma tissue, and indeed, since there is no auxin here, there is very little cell expansion and it soon stops. With the auxin there are nests of vascular tissue, an occasional evident xylem strand and some probable phloem tissue. So not only are all the cells dividing, but many are differentiating.

In all cases the xylem and the phloem are formed beneath the surface, and in some sense the structures in the figures are suggestive of what is seen in the natural stem. However, there are no organs—no roots, buds or anything else other than this cellular differentiation.

A great many species were grown in tissue culture like this, and most of them eventually formed vascular tissue. Camus,

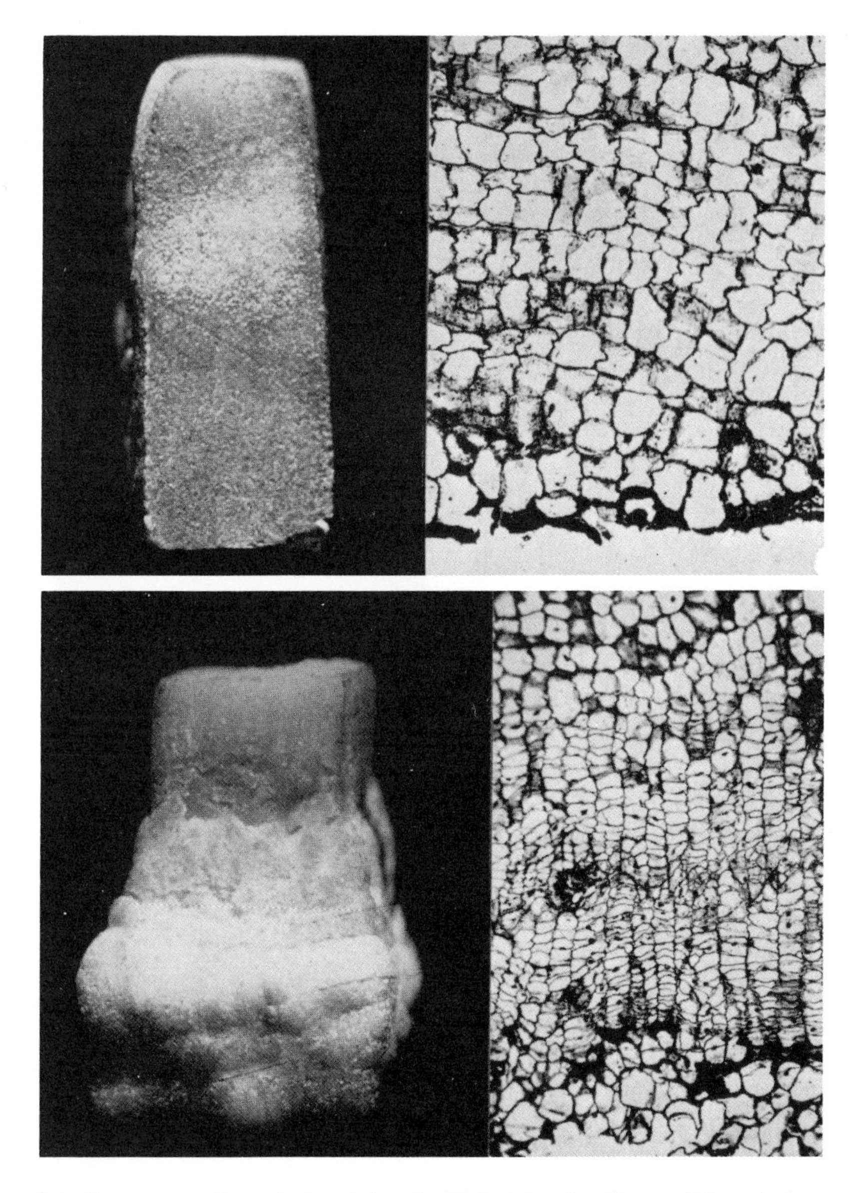

9-1 Above, cross section and external view of artichoke tuber slice after growth in auxin-free medium. Below, similar slice after growth in same medium but with 0.1 ppm IAA added. From Gautheret, 1946.

in particular, working in the same institute as Gautheret, showed not only that vascular bundles were formed, but that they had a tendency to join to one another. Figure 9-2 is a reconstruction of a tissue culture of *Syringa* (lilac), into which a bud has been grafted. The bud has initiated a small amount of vascular tissue (mainly xylem). On the right is a similar culture to which a little auxin has been added in the graft notch; far more xylem has been formed. Also Camus and others have noted that in tissue which already has vascular bundles in it, the bundles formed in the newly grown tissue have a tendency to join up with the existing bundles. Thus one begins to see that there is some parallel between the happenings in a tissue culture and those in a normal stem. We shall encounter some other parallels below.

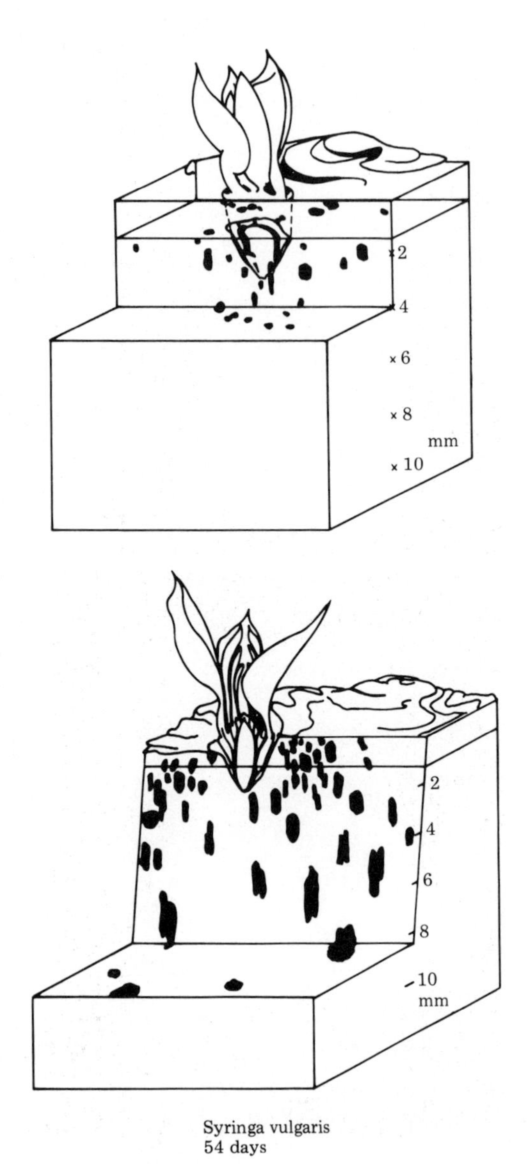

9-2 Vascularization (shown in black) of a callus tissue of *Syringa vulgaris* (lilac) 54 days after a bud had been grafted into it. Above, bud alone. Below, bud with 0.5 ppm naphthaleneacetic acid added. From Wetmore and Sorokin, 1955.

B. FROM AUXIN TO CYTOKININ

The experiments of the French workers were done from 1937 to about 1950; and towards the end of that time Folke Skoog became interested in the growth of the peculiar tissue which is formed by the hybrid tobacco discussed in chapter 8. It will be remembered that this hybrid has the odd property of making a gall-like outgrowth—a sort of tumor—when wounded. The outgrowth is not due to infection with a tumor-producing bacterium like *Agrobacterium tumefaciens,* but it is due to the nature of the hybrid, for neither of the two parents reacts in this way. In any event this rapid growth of tumorous tissue suggested that it might be grown in tissue culture, and Philip White, at the Rockefeller Institute, did in fact succeed in getting it into culture. Later he gave a culture to Skoog. The route from here to the discovery of kinetin, and thence to the other cytokinins, was described in part 3 of chapter 8. As we have seen, kinetin proved to be only the first example of the group of compounds called cytokinins (from their action on cytokinesis, or cell division). The isolation of naturally occurring cytokinins soon followed, beginning with Letham's preparation from immature maize grains called zeatin. *Corynebacterium fascians,* which causes a disease of seedlings characterized by formation of masses of little buds, produces a similar compound, but without the oxygen atom of zeatin, thus having two methyl groups; it is 6 (2′-isopentenylamino) purine, abbreviated 2iP. Many of the related synthetic compounds were discussed in chapter 8. As we have seen, all are purine derivatives, and most are in fact substituted adenines. They generally combine with ribose at the 9-position to form ribosides, and these can also form phosphates, i.e., ribotides like AMP.

That the cytokinins should have turned out to be purine derivatives was, in a way, not too great a surprise, for it had already been deduced that this property of producing buds was common to several purine derivatives. Even adenine, in the work of Skoog and Tsui, had the property of producing a few buds on tissue cultures. The first question to be asked in this connection was, why is kinetin the product of autoclaving the nucleic acid? We know that furane rings are often produced when 5-carbon sugars are heated, indeed qualitative tests for pentoses rest on this reaction. Thus it has now been made pretty clear that the precursor is adenosine or a closely related compound, and at autoclaving a migration takes place, the ribose being converted to a furane and becoming attached at the 6-position. There is a model for this migration in the behavior of triacanthine, which undergoes a similar change when heated. As we saw in chapter 8 also, the ribosides of kinetin and zeatin have been prepared but do not seem to be any more active than the free bases themselves.

The action of cytokinins in promoting cell division in tissue cultures is dramatic. It was early shown that the cells of tissue cultures when treated with auxin alone will divide, but after a while the nuclei tend to react only faintly with the Feulgen stain. When a little kinetin is added to the medium, not only is the division greatly stimulated, but the nuclei retain their normal dark staining, which means they retain their DNA content. Later it was shown by Skoog and colleagues that the total RNA, though not necessarily the total DNA, was clearly increased. Another characteristic thing about cell division in presence of kinetin is that not only are there cambium-like divisions in which small cells divide, but quite large cells, which seem mature, being to undergo divisions and thus become subdivided into many small cells. Figure 9-3 is from a preparation of Gautheret's, made when he was using coconut milk as a source of cytokinin. Auxin was present as well, and almost every one of the originally large cells is seen to have been activated to divide. Figure 9-4 shows a similar phenomenon in Skoog and Miller's work; and figure 9-5

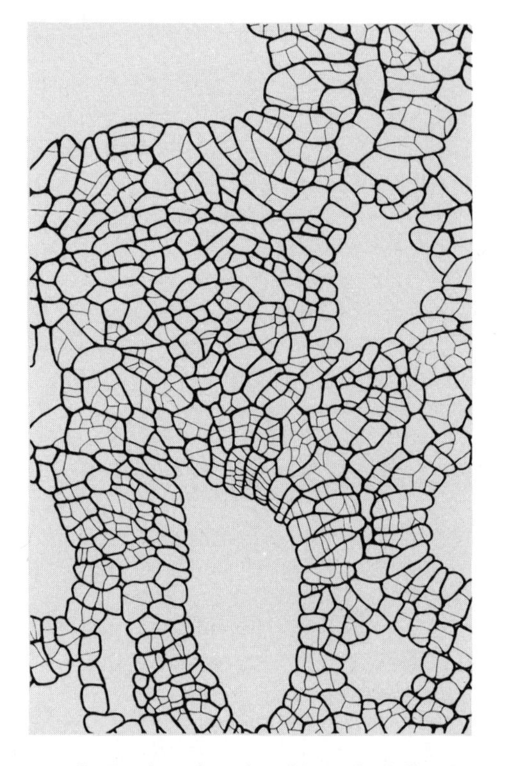

9-3 Section through a tuber of *Amorphophallus rivieri* treated with naphthaleneacetic acid 0.5 ppm plus 15% coconut milk, showing multiple subdivision of the original cells. From Gautheret, 1959.

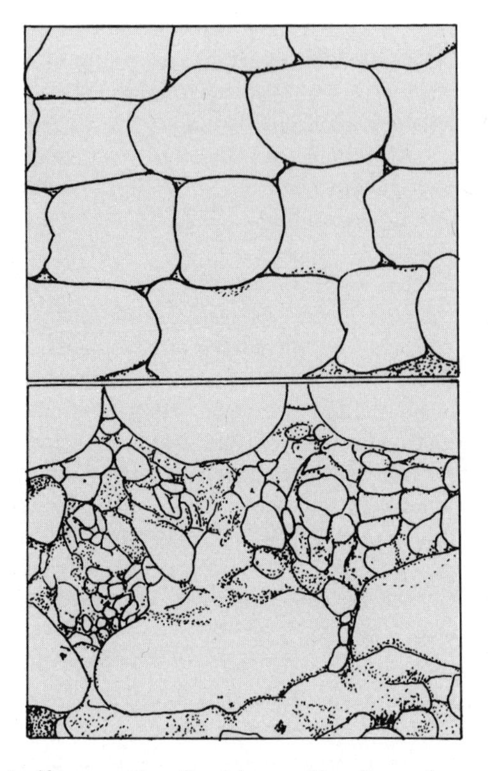

9-4 Above, section of a tobacco pith culture after growth on a basal medium. Below, section of a similar culture after growth on same medium with IAA 2 ppm and kinetin 0.2 ppm added. From Skoog and Miller, 1957.

shows a comparable response in the large pea root cells studied by Torrey and coworkers. The response is almost like creating embryonic initials in the middle of what seemed like fully mature parenchymatous tissue. Such cells would not be expected under normal conditions to revert to the meristematic state. Obviously only a very powerful stimulus for cell division could have this effect.

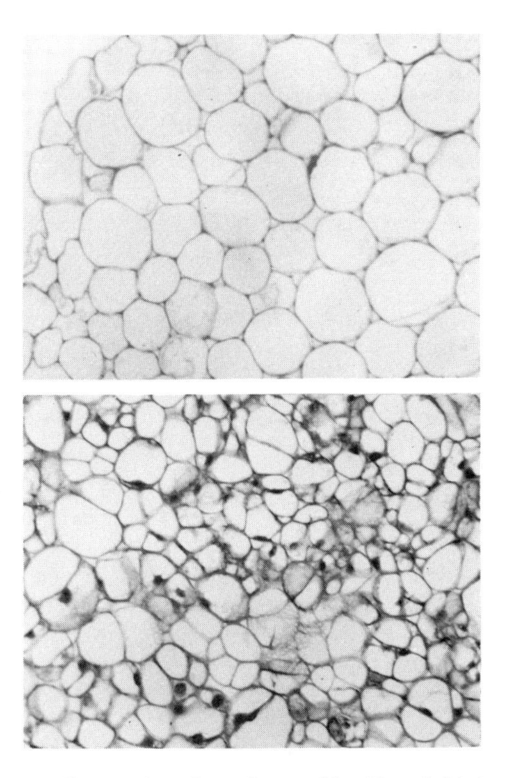

9-5 Cross sections of root of pea seedling 10 mm behind apex. Above, at start. Below, after 5 days on medium containing 0.5 ppm kinetin, 10^{-6}M 2,4-D and 10^{-6}M IAA. From Torrey and Fosket, 1970.

Thus it began to appear that in a tissue culture, in order to maintain continued cell division and growth, not only auxin but also a cytokinin is needed. All members of the group have this effect to widely varying degrees. However, they are not assayed by counting divisions, which would be both slow and difficult, but by the weight of tissue formed under standard conditions. This means, of course, that the newly divided cells must *enlarge* also. Indeed, enlargement following division is so universal a character of plant cells that botanists often thoughtlessly speak of "growth by cell division," although division in itself would not constitute that increase in volume which is the criterion of *growth*.

Besides growth, cytokinins sometimes cause the formation of chloroplasts in tissue cultures; that is, they convert ordinary parenchyma cells to something like leaf tissue. This occurred spontaneously in one of Gautheret's early carrot cultures and has been occasionally noted since. With the cultures of cassava (*Manihot*) tubers grown by Bruinsma's group at Wageningen, Holland, zeatin or 2iP produced dark green tissue; other cytokinins were not so effective. Evidently rudimentary chloroplasts are normally carried, though not developed, in many (perhaps most) parenchymatous cells. Since cytokinins powerfully prevent yellowing in senescent leaves (chap. 12), it seems they may exert some particular action on the chlorophyll-synthesizing system and its associated proteins.

Most plant tissues normally contain a little cytokinin and a little auxin at the time when isolated from the parent plant. Thus it is nearly impossible to say with surety what one of these hormones alone will do; one is nearly always working in the presence of traces of an endogenous hormone of another type. The isolated tuber tissues are perhaps the nearest to being hormone-free; for instance, potato tuber slices will produce few if any adventitious buds unless they are first rooted. But one cannot draw too simple a conclusion as to the formation of cytokinin in roots here, for Miedema has found that application of cytokinins to unrooted slices will not give rise to buds; in other words, roots contribute some influence for which cytokinins will not substitute. Thus the interpretation of experiments in this field is still somewhat hazardous. We can, I think, conclude that mitosis is stimulated by both auxin and cytokinin, but with auxin "alone" mitosis and DNA synthesis are the main results, whereas with auxin plus cytokinin actual cell division (cytokinesis) can follow and continue. When kinetin "alone" is added to a tissue that contains a mixture of diploid and tetraploid cells, then commonly it is the tetraploid cells that divide, not the diploid, suggesting that the synthesis of DNA may be the limiting factor and that this may require added auxin.

In addition to auxin and cytokinin, phenols can be important in differentiation. Most tissue cultures form phenolic acids during their growth, and some, especially if from plants that form red leaves in spring or fall, form anthocyanins as well. That these substances certainly influence growth and organ formation can often be shown by adding them to the medium. Generally diphenols increase the action of auxin, as would be predicted from their action on the IAA-oxidase system. In artichoke tuber and willow cambium cultures they promote root formation, a typical response to high auxin concentrations. In tobacco pith, by contrast, tyrosine, which is a monophenol, promotes bud development, which is characteristic of a high ratio of cytokinin to auxin. By themselves the phenolic compounds seem to exert little real effect (Paupardin, 1972).

The synthesis of phenolics is often very strong in tissue cultures. For instance, artichoke tuber tissue produces seven phenolic acids, of which parahydroxybenzoate is formed in largest amount. Potato tissues produce caffeic acid and quercetin, both diphenols.

C. WHOLE PLANTS FROM TISSUE CULTURES

We return now to the tissue cultures themselves. A most important finding has been that if, instead of being grown on agar, they are grown in a liquid medium which is constantly shaken, what results is not the usual mass of callus but instead a kind of suspension which contains some single cells and some small clumps of half a dozen or so cells joined together. The great protagonist of this work was F. C. Steward, who grew these shaken cultures as "free cells" from carrot phloem tissue for years, obtaining unlimited growth and multiplication. Steward did not use cytokinin, he used coconut milk, which contains not only cytokinins but other active materials as well (cf. chap. 8, section 3).

The interesting thing about these "free cells" is that after long study Steward and his colleagues at Cornell found that some of them grew into clumps that were almost indistinguishable in form from plant embryos (fig. 9-6). These little "pseudoembryos" were cultured separately, and to everyone's delight they grew into entire plants. They produced rootlets and an apex, which soon gave rise to leaf primordia. Steward, carefully using the term "free cells," would never commit himself as to whether single cells would grow into whole plants, or whether small clumps were needed. It was 10 years later that Murashige, now at Riverside, carefully separated out single cells and showed that in an enriched medium they can in fact grow into whole plants.

One of Dr. Torrey's students, Elizabeth Earle, made a study of this in 1965 using essentially bacteriological methods. If one spreads bacteria on an agar plate and counts the resulting colonies a day or so later, by knowing how many bacteria were applied and how many colonies develop one can calculate what is called the *plating efficiency,* the percentage of the original cells which will grow on the plate to produce a colony. When this is done with the free cells of liquid cultures the plating efficiency is pretty low. But addition of several compounds not normally included in the medium, especially glutamine, can greatly increase the plating efficiency. Thus the fact that only small numbers of single cells grow is almost certainly a matter of their being provided with compounds that they may synthesize but at a minimal rate. The richer the medium, the larger fraction of cells can grow into whole plants.

When I go into an orchid nursery and mention that Georges Morel was my friend, people almost fall down and worship. The reason is that Morel, in making so many tissue cultures, developed good techniques for getting very small fragments of tissue to grow into whole plants (1963). They were not isolated apices, since they had leaf primordia on them, but they were fragments, and they grew rather rapidly. For a plant that is difficult to propagate—and the prize example is orchids—one can break up the meristematic tissue into perhaps a thousand fragments. Each of these can be cultured and the majority will grow. From one plant you can thus produce a thousand plants without going through the business of waiting 6 years for it to flower, and then getting the minute orchid seeds to grow on a nutrient medium. The seeds grow poorly and very slowly. They need complicated nutrients and then one must still wait quite a few years before the seedlings reach flowering age. But these meristem plants can be produced in two or three years from fragments. In the case of a new hybrid (which is the usual cause of excitement in the orchid-growing fraternity), one can have many hundreds of them very quickly.

It is curious, nevertheless, that these little clumps, or even single cells, will make whole plants when it is so difficult to get isolated apices to grow into whole plants.

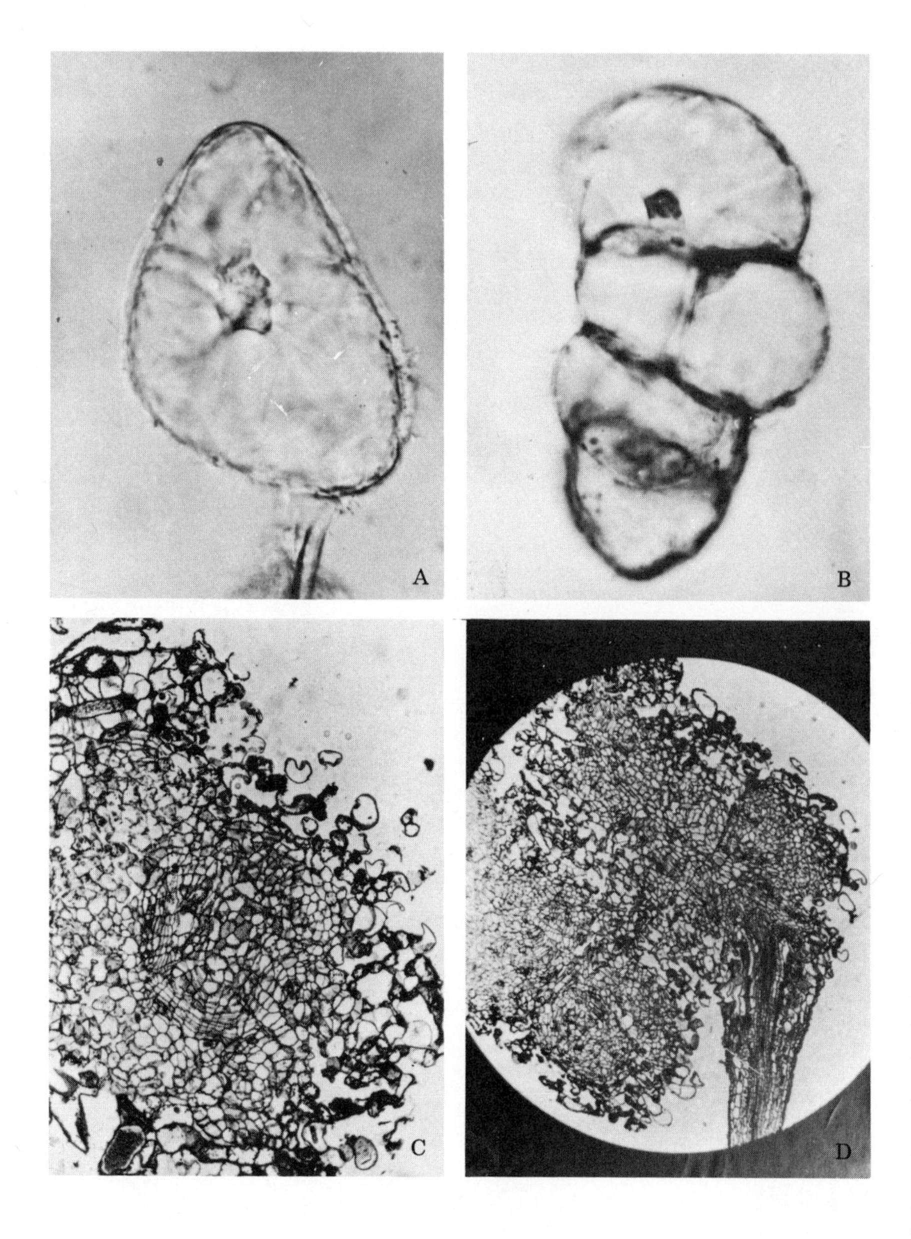
A
B
C
D

9-6 Successive stages in the development of free carrot cells into a plant, in liquid culture. (A) individual cell; (B) clump of cells; (C) cell colony with organized cambium-like activity; (D) formation of a root; (E) whole plant; (F) and (G) whole plant with (F) normal carrots; (H) cell division in types A and B showing tendency to aneuploidy. From Steward, 1964.

Many workers have struggled for years to get apices to grow; in the large majority of cases, they do not succeed in growing much at all, although they will form whole plants if leaf primordia are present. Recently, however, Murashige and his colleague Smith did succeed in getting apices to grow into whole plants, and Gamborg and co-workers have done the same with cassava (*Manihot*). Again this is largely a matter of the medium, for the apex is normally dependent on the rest of the plant. It must have a carefully planned medium, rich in most of the compounds it needs but devoid of what it does not need. One of the materials it evidently does not need is gibberellic acid, which Ernest Ball tested thoroughly and found to have a bad effect on the apices; it made the cells in them vacuolate and not divide. The nuclei stain faintly; there are still a few divisions, but no organ formation occurs, and after a while these cultures die out.

9-7 Flasks with lateral bulbs attached so that liquid culture will flow in and out on rotation. From Steward, 1964b. Note small growing tissue clumps in each bulb.

Figure 9-7 shows Steward's culture flask, which has lateral bulbs in which the cells lie. The flask (along with many others) is mounted on a rotating drum. As it rotates the cells fall out into the main body of the medium and settle again in another bulb, so that they are constantly gently stirred. The figure shows a little cluster of cells, which is typical. These multiply to produce an embryoid or pseudoembryo. It clearly has a cotyledon-like outgrowth; and what turns out to be a root soon develops supplementary roots, then leaf primordia, and finally becomes a whole plant. Unexpectedly, auxin now appears to inhibit embryoid formation.

The most used tissue, next to carrot, is

probably the pith of the stem of tobacco plants. These cells first form callus, and the callus then forms organs as a function of what is added to the medium. As we saw in chapter 8, if IAA alone is added, with only traces of cytokinin, they will produce roots, just as stems will produce roots in response to an auxin. If indoleacetic acid and kinetin are both added, and in the right proportions, then they not only produce roots but will form buds, often in very large numbers, crowded together (fig. 8-18). With still higher kinetin levels we get back to callus again; so buds are formed in only a fairly narrow range of concentrations. Finally, at the highest kinetin level the culture does not grow at all. So, although kinetin is not generally a particularly toxic substance, it can be present in excessive amounts. This presents a nice case where organ formation is wholly controlled by the hormonal composition of the medium.

Even more spectacular than growing whole plants from single cells is growing them from protoplasts—cells without their cell walls. By treating cultures (or sometimes intact plant tissues) with cellulase from certain fungi, the cell walls can be completely dissolved without killing the cells. The medium must have a high enough osmotic pressure to keep the cells from bursting, for the plasmalemma unsupported by the polysaccharide wall is very weak. In presence of sugar in the medium many protoplasts can regenerate their cell walls and can then be transferred to a normal medium and cultured to produce a normal callus. Protoplasts from tobacco leaves and from petunia stems have recently been coaxed into giving rise to whole plants in this way, by Vasil and Vasil (1974).

D. CULTURES OF ROOTS

A contrast to all these properties of shoot tissue is given by root tissue. All roots can now be cultured without major difficulty. The first successes were with the roots of dicotyledons, which were grown by Philip White as long ago as 1937, at the same time as Gautheret started tissue cultures. Isolated roots growing in a nutrient medium are not really tissue cultures; they are organ cultures, in that they grow only as roots. White's great discovery was that they will grow on a simple medium without auxin, but with a little yeast extract. One can well understand this since auxin tends to inhibit the elongation of roots. The yeast extract of course covers a multitude of sins, and in this case it was supplying amino acids in traces and, more importantly, supplying thiamine. It was not realized at that time that thiamine is essential for the growth of roots. One could dispense with yeast, in fact, and just add thiamine, sugars, and appropriate mineral salts, and some roots will grow well. Many roots need additional vitamins; legume roots need nicotinic acid and grow poorly without, but basically it is thiamine, salts and sugars that are their main materials. As the culture media improved, roots of more and more species were brought into culture, including a number of monocotyledons, such as wheat. Actually if one adds IAA to these root cultures, which grow without auxin, there are some interesting results. While the auxin in all but the very lowest concentrations inhibits elongation (probably via its stimulation of ethylene production), as we have seen in preceding chapters, yet it can have a powerful effect on differentiation. John Torrey found some years ago that when roots have been decapitated and are regenerating new tips, auxin increases the number of vascular bundles. He used pea roots, which have triarch structure, and found that the regenerated tips in those that were treated with auxin had hexarch structure. Controls did not change their vascular structure on regeneration, and elongation was not immediately affected by the auxin (fig. 9-8).

Addition of kinetin to the medium generally inhibits the growth of roots, and it inhibits the growth of roots both on cuttings (chap. 7) and on tissue cultures. It strongly inhibits the formation of laterals on the main root also. On the callus from artichoke tubers, kinetin prevents root formation at concentrations around one millionth molar; its action on cuttings is not exerted at such low concentrations. In some tissues kinetin actually makes the roots form buds; they have even produced small buds right down within a millimeter of the root tip. It must be remembered, too, that inhibition of root formation is not highly specific, since gibberellic acid has the same effect, as was seen in chapter 7.

So we come back to Julius Sachs and his idea of a rhizocaline and a caulocaline. In some respects, enough to make one feel uncomfortable in writing off the crude concepts of our ancestors, Sachs was rather close to the mark, because auxin in tissue cultures is in fact acting as a rhizocaline, in that it produces roots. It stimulates the cul-

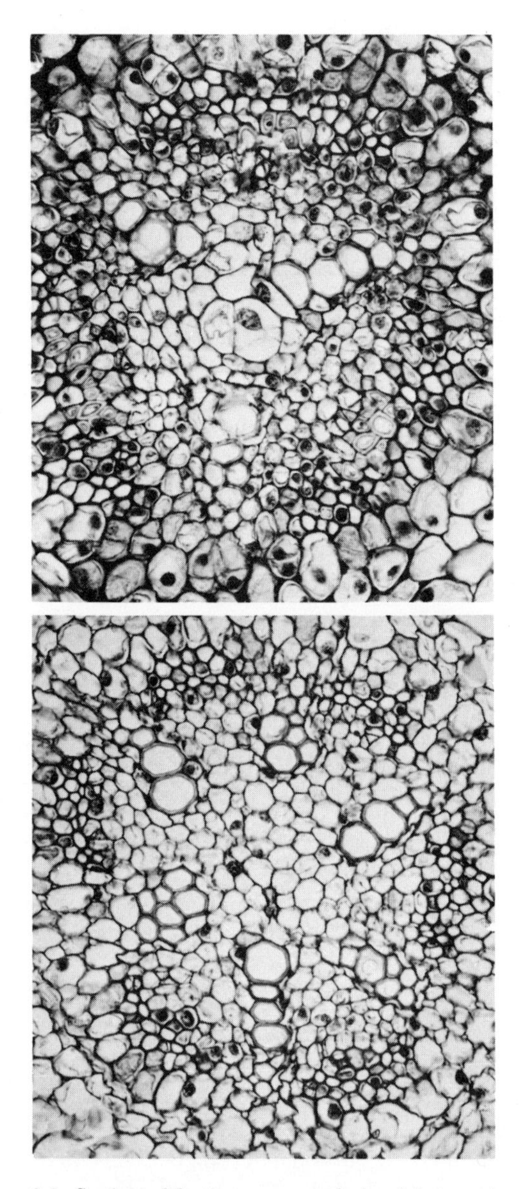

9-8 Sections of the tips regenerated when 0.5 mm root tips of Alaska peas were removed. Above, in control medium, showing triarch organization; below, medium with IAA added, showing hexarch organization. From Torrey, 1957.

tures to grow, but it also makes them form roots; whereas cytokinins stimulate the cultures to grow and also make them produce shoots. Thus the actions of auxin and cytokinin are close to the old ideas of rhizocaline and caulocaline respectively. We have no phyllocaline, at least as yet; for the stated function of phyllocaline, in Sachs's ideas, was to produce leaves, and of course a growing shoot apex always produces leaf primordia anyway, which grow into leaves. Some leafless cactaceae produce spines, but the spine is believed to represent a reduced leaf.

One other interesting effect on root cultures is that if cultures of pine tree roots are grown with dilute solutions of auxin for a very long time, they slowly produce the distorted forms, called *mycorrhizae,* that these tree roots, in association with a fungus, normally produce in the soil. It has been shown, especially by Visvaldis Slankis in Canada, that either the fungus itself (*Boletus luteus*) can be inoculated with the roots or a sterile extract of the fungus culture can be added to the roots; either one will make the roots grow as mycorrhizae. Now instead of either of those, simply IAA or NAA, at the right concentrations, will have the same effect (fig. 9-9). Evidently what is needed is a moderate inhibition to the growth, allowing the growth to continue over very long times, usually a month or more, perhaps with some stimulation of branching, to produce perfect imitations of a mycorrhiza.

E. THE PHENOMENON OF ACCOUTUMANCE

As this subject developed, Morel and Gautheret, at Versailles, made the rather peculiar discovery that if one cultures a callus-type tissue with added auxin, after a number of transfers the tissue changes its character. Instead of being tough and opaque it becomes friable and watery. Later analyses showed that the dry matter content shrinks greatly; instead of constituting about 10% of the fresh weight like most plant tissue, it is only 3% or 4% of the fresh weight. These cultures now *no longer require auxin for growth.* If transferred to an auxin-free medium, not only will they grow without auxin, but with any more than a trace of auxin added they are actually inhibited. Thus the level of auxin that normally is physiologically just right is now in excess, and is inhibiting.

In figure 9-10 the thin solid line represents the growth of a normal tissue culture; as the auxin is increased, there results a very nice growth up to the order of 1000 mg of tissue (under Morel's conditions) and higher concentrations up to at least 10^{-5} molar do not decrease this. The dotted line shows the growth of the tissue which has been cultured for several transfers. What happens is that in the absence of auxin it grows almost as well as the control tissue does *with* auxin. Furthermore its growth is drastically inhibited at high concentrations. Morel was also culturing a tumor somewhat similar to that of the *Nicotiana* hybrid described above, but actually derived from a crown gall. Crown gall is a bacterially produced tumor which can be subsequently cultured bacteria-free, when it grows like a callus, and characteristically grows quite well without auxin. There is a small increase due to auxin but it is

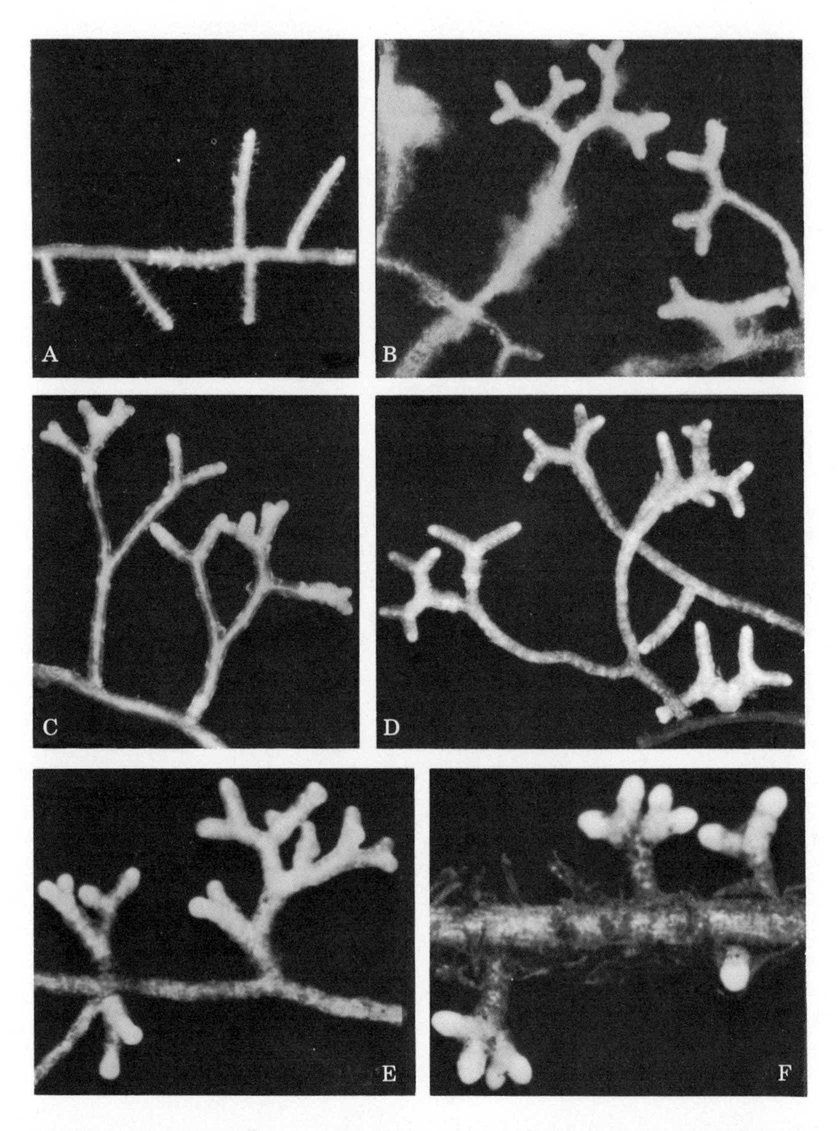

9-9 Production of mycorrhiza on roots of *Pinus sylvestris* (Scotch pine). (A) in basal sucrose-salts medium; (B) same with *Boletus luteus* mycelium; (C) with aqueous exudate of same mycelium; (D) basal medium with indolebutyric acid 1 ppm; (E) with naphthaleneacetic acid 2.5 ppm; (F) is root of intact seedling grown in basal medium with NAA 9 ppm. From Slankis, 1951.

strongly inhibited at concentrations above the optimum. Evidently the normal tissue which has grown through several transfers has changed, both in appearance and in auxin requirements, to become like tumor tissue. Gautheret calls this material accoutumé, or habituated, tissue. Since it is now growing without added auxin and responds only negatively to auxin applied externally, the natural interpretation is that it has now developed a taste for making its own auxin. That turns out to be the case.

Some quite striking data on auxin content of such tissues were accumulated by Mlle. Kulescha, working in Gautheret's laboratory. As one example, grape stem tissue when it was first cultured contained 41 μg of auxin per kilogram and then after it had been grown in culture for 45 days it contained only 11 μg. Apparently, therefore, it had been using up its auxin faster than the tiny amount in the medium could supply it. In another case Kulescha analysed a strain of accoutumé tissue, which had become able to grow without auxin, and its analysis showed a content of 66 micrograms per kilo, which was more even than the initial material. Since this tissue was growing without any added auxin, the increase must have actually been synthesized. For comparison, she cultured a crown gall of the same plant, which after culture for a while had 50 μg of auxin per kilo. Similar results were obtained with normal, accoutumé, and crown gall tissue cultures of three other plants. They leave little doubt that the accoutumé material has, by continued growth in a low auxin medium, favored the emergence of mutant cells which have the enzymes necessary for auxin synthesis.

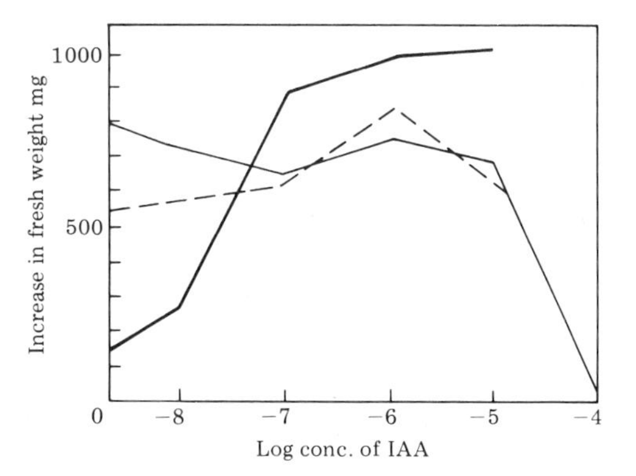

9-10 Responses of three kinds of *Scorzonera* tissue to IAA. Solid line, normal tissue; dashed line, accoutumé tissue; dotted line, crown gall tissue. Modified from Gautheret, 1950.

The significance of the discovery that cells can develop the synthesis of an important growth factor was not lost on the animal cancer fraternity, and from time to time this aspect of their problems has been revived, but without success. Evidently life within the whole plant somehow results in suppressing, in these cells, the ability to synthesize auxin. Whether this is due just to the continued auxin supply from the apex, or to some more complex situation, one cannot now say. Skoog has subsequently found that callus tissue grown on high cytokinin can synthesize their own thiamine—apparently a parallel case (see below).

F. THE FORMATION AND ORIENTATION OF VASCULAR BUNDLES

We come back now to the formation of xylem and phloem, which is really the heart of our topic. We begin with two relatively small points.

In several tissue cultures it is now clear that the formation of xylem, stimulated by auxin, is inhibited by ethylene. In other plant responses we shall see that very often auxin induces the formation of ethylene. Thus it seems a curious contradiction that a product which may be the result of auxin treatment should actually inhibit a characteristic auxin effect. Figure 9-11 shows one of Stanley Burg's pictures of the 4 vascular bundles of a pea seedling treated with ethylene. The pith cells have enormously enlarged, and though the number of vascular bundles remains the same the number of lignified cells is greatly decreased. Some of the anomalies of the tissue type produced by very high auxin concentrations will doubtless find their explanation in this response.

Secondly, it is clear now (as it was not for a long time) that not only xylem but also phloem can be induced in tissue cultures by auxin. It seems that cytokinin is not needed; what is needed in order to produce phloem is a high level of sugar. This is rather suggestive, because sieve tubes are mainly engaged (*in vivo*) in transporting sucrose, and sometimes other sugars as well. Apparently this kind of tissue is not only generated by sugar but has a sort of special affinity for sugar. With more sugar than one normally adds to such cultures, e.g., 4% or 5% sucrose, beautiful phloem results. Thus the whole vascular bundle can be produced, though it is sometimes irregularly oriented, compared to the usual arrangement in the stem.

In tissue cultures the formation of vascular bundles, especially of xylem, is promoted by gibberellin. Indeed, Wareing reported this earlier (1964) in the twigs of trees which form limited xylem with auxin alone, and which form more layers, with more lignification, when GA was added as well. Light, temperature, gibberellin and auxin all interact in this process. The interaction of chemical and physical factors in xylem formation is neatly shown in table 9-1. All the artichoke cultures received auxin (naphthaleneacetic acid, 1 ppm) in the nutrient medium. Gibberellin also tends to regularize the arrangement of the vascular bundles in this material. From the table one could conclude that high temperature or high GA can substitute for light.

It has often been noted that where cambium-like activity occurs in tissue cultures, whether induced by added growth substances or by endogenous causes, phloem tends to differentiate towards the outer side and xylem towards the inner side. Such development in cultures thus parallels what happens in the normal stem. Perhaps an oxygen gradient is interacting here with the obvious gradients of water and nutrients.

Table 9-1 Xylem formation by artichoke tissue cultures

		Mean no. of lignified layers of xylem formed:	
Conditions	GA_3 concn. (ppm)	at 13°	at 26°
Light	0	14	20
Dark	0	2	16
Light	0.001	15	31
Dark	0.001	4	15
Light	0.01	18	27
Dark	0.01	11	29
Light	0.1	27	33
Dark	0.1	18	31

Data of Saussay and Gautheret, 1974.

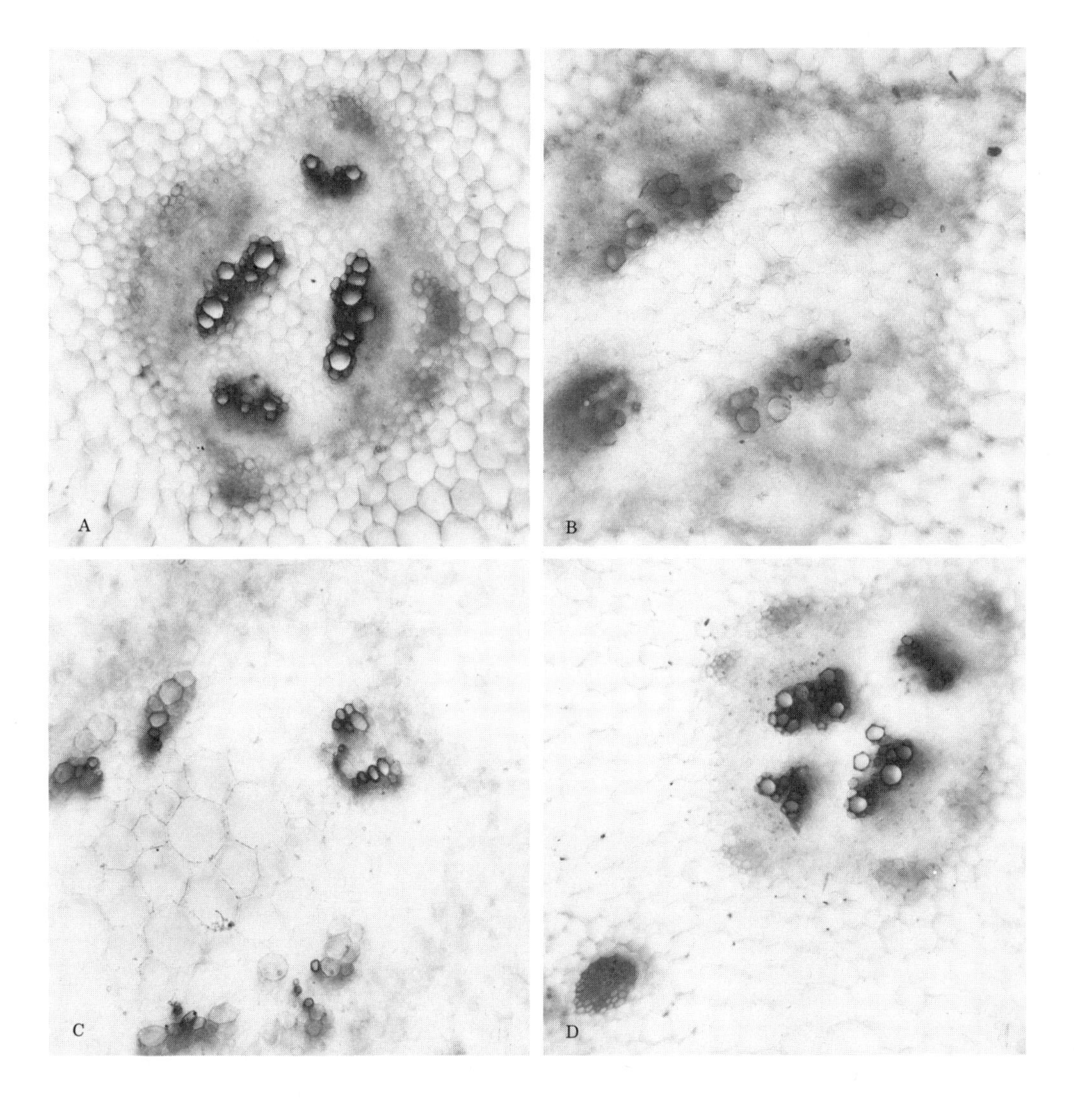

9-11 Effects of ethylene and auxin on lignification in the vascular bundles of etiolated Alaska pea plants. All sections stained for lignin with phloroglucinol. (A) control plant; (B) plant treated with ethylene; (C) plant sprayed with 2,4-D 1 mM; (D) plant sprayed with 2,4-D 0.05 mM. All after 72 hours. From Apelbaum et al. 1972.

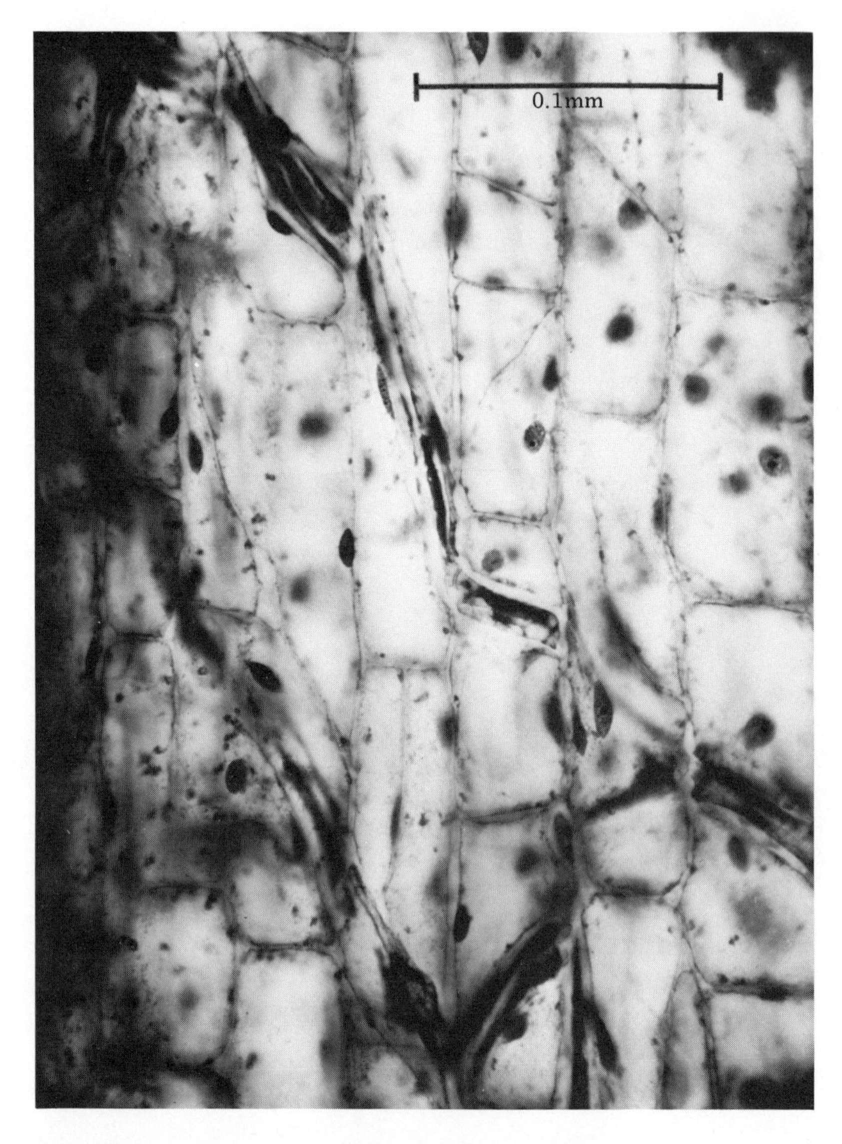

9-12 Regeneration of phloem elements in the 5th internode of *Coleus,* 5 days after severing the vascular bundle and applying IAA. The angle of the regenerating phloem strands enables it to make connection with the phloem below the wound. From La Motte and Jacobs, 1963.

The formation of xylem and phloem occurs in isolated segments of some plants, much as it does in tissue cultures. For instance, Jost long ago observed that when a vascular bundle in a young stem of *Coleus* was cut, new vascular tissues would develop and join on to the existing vascular bundles below the cut. More recently, Jacobs has brought this observation up to date with auxin treatments. He wounds the square stem of *Coleus* in one corner so that the bundle there is cut. The amount of xylem that grows, to make the connection around the cut, is found to be a function of the amount of auxin applied to the upper cut surface. Auxin on the surface below the wound is ineffective because of its polar transport. If the segments are provided with added sugar, phloem is also formed, and very characteristic sieve tubes can be observed. Figure 9-12, from Jacobs's *Coleus* experiments, shows the newly formed sieve tube cutting across the files of parenchyma cells on its way back to join the original severed vascular bundle, which is over to the right. A second strand is making its way parallel to the first. The sieve tubes are produced by local subdivisions which immediately differentiate. Thus there is more to this than a simple stimulus to divide. Figure 9-13 shows how quantitative the response is, for both the xylem strands and the sieve tubes are under the direct influence of the applied auxin, their number being proportional to the logarithm of the auxin concentration. One can even deduce the concentration of auxin present in the intact plant, namely, the equivalent of 0.03% IAA.

This result suggests that in the normal plant the vascular tissue, being under the control of auxin which undergoes such strong polar transport, must be differentiated from the apex down. The flow of cambial activation from apex to base in trees in the spring has of course long been known, and has been instanced more than once in earlier chapters. This deduction has been controversial in the past, perhaps because in many plants the lateral organs are active and they are below the apex; thus in sectioning the apex, the early workers may not have taken enough account of the ability of organs just below to be producing some vascular tissue.

The flow of cambial activation is of course in the main the flow of auxin, and Tsvi Sachs (1974) has come to the interesting

9-13 Effects of apically applied IAA in lanolin on xylem and phloem regeneration around a wound in *Coleus* 5th internode. Asterisks show the cell numbers in the intact plant. From Jacobs, 1967.

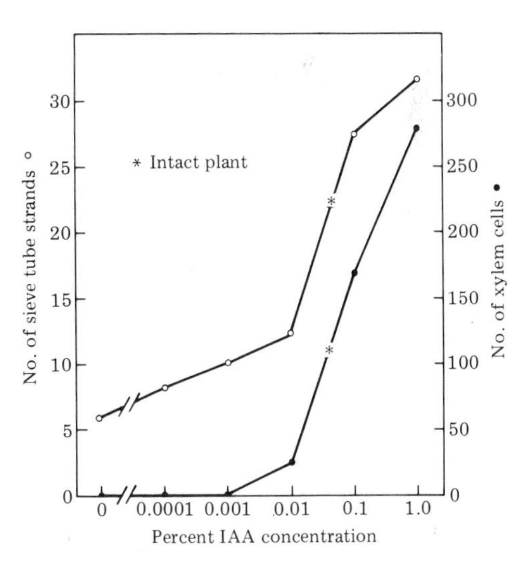

conclusion that it is the transport of auxin *through* the cells, rather than just its presence, which because of its directional aspect causes the formation of longitudinal xylem elements. In very young coleoptiles it appears that the first strand of protoxylem may actually differentiate acropetally from the seed. Figure 9-14 suggests strongly that in the youngest coleoptile the single strand is going upwards and ending short of the tip. As the coleoptile extends, this original strand starts to multiply, and one now sees quite clearly the new strands coming down. They look almost as if they were dripping. This could be considered as an anatomical picture of the polar transport of the differentiating auxin. In the tip of *Avena* the two opposite bundles approach one another but do not meet; in the tip of *Triticum,* however, they do actually meet and fuse (fig. 9-14).

The subject of wood formation is more difficult, because there is involved not only the formation of the vascular strands, but also the several different types of lignification: helical, annular, and scalariform reinforcement, fiber cells, gelatinous cells, and so on. In general there is a powerful influence of auxin not only on the formation of the xylem strands and therefore of the wood as a whole, but also in part on the type of xylem produced. Auxin and gibberellin, in particular, interact in the formation of "normal" wood, as Wareing and his coworkers have clearly shown. Auxin influences can be especially clearly seen in those special modifications of wood types which develop when branches or stems grow at an angle. When conifers grow at an angle there is a special "red wood" produced on the lower side. This can be induced in vertical plants by application, on one side, of auxin. When dicotyledonous trees grow at an angle, special so-called "tension wood" develops on the upper side. This can be imitated by *lowering* the auxin content, for example by preventing the polar transport of auxin into the tissue. Triiodobenzoic acid induces such tension wood, through its known ability to inhibit auxin transport. In normal stems gibberellin synergizes with auxin to increase the amount of xylem formed.

G. DIFFERENTIATION IN GENERAL

I want to finish up with an overall view of differentiation. It is not easy to realize the complexity that differentiation involves. For instance, to make the smallest bacterial cell it has been calculated that there are required something like a thousand enzymes. Now, suppose that in differentiating a plant cell we are going to make chloroplasts, i.e., to make chlorophyllous tissue out of colorless parenchyma cells. Think of what it takes to make chloroplasts: the outer membrane, the various lipids and proteins, the layers of thylakoids, all the enzymes needed in the biosynthesis of chlorophyll and of the carotenoids, and the enzymes needed for photosynthesis—one can see that that would involve at least several hundred more enzymes. And cell division, the process of mitosis, is not a simple matter either. The chromosomal material has to be taken out of the nucleus and assembled into little bodies; then we need a set of enzymes to make the spindle material and the microfibril to cause the two chromatids to separate; and then we

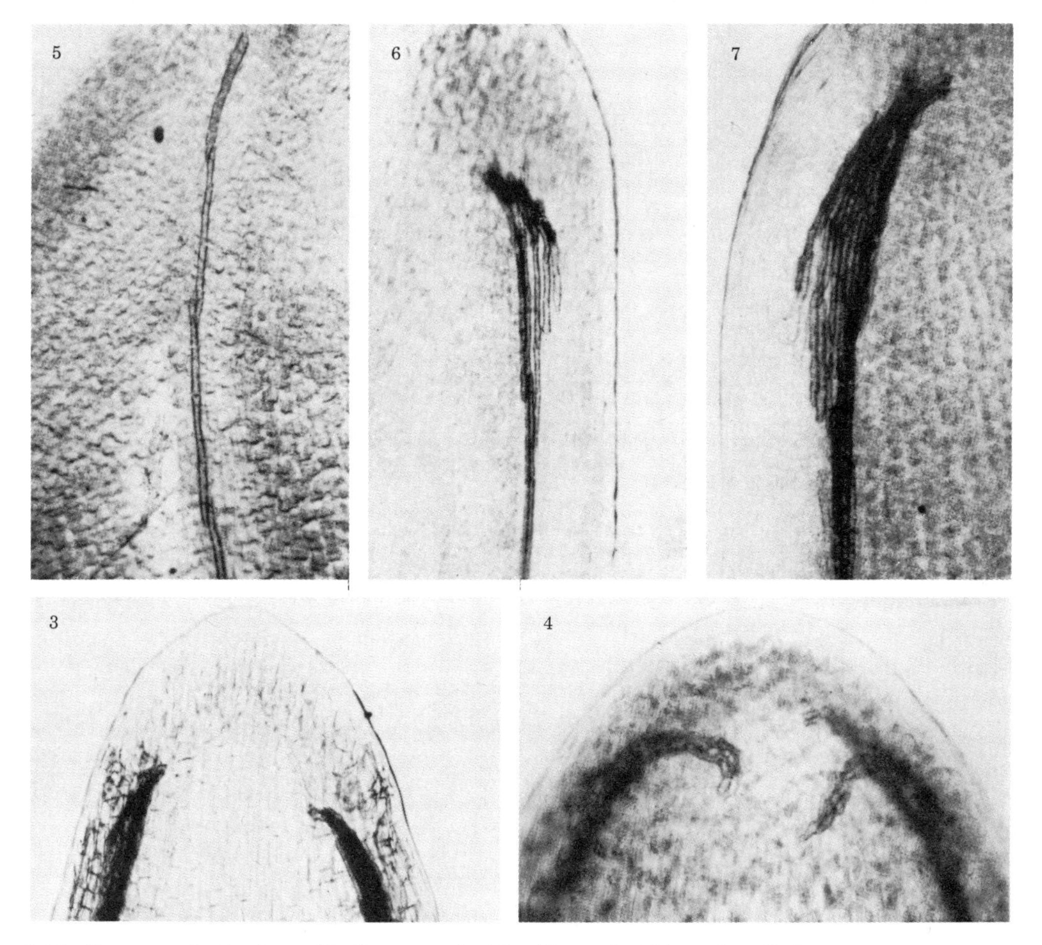

9-14 Above, vascular development in the young coleoptile of *Avena:* 5, at 1.5 mm long; 6, at 3.5 mm long; 7, at 4.0 mm long. Below, vascularization in the tips of mature (30 mm) coleoptiles; 3, *Avena,* X 85; 4, *Triticum,* X 110. From Thimann and O'Brien, 1965.

need to bring it all back into functional condition inside the nucleus, which probably requires quite a number of enzymes in addition.

Yet in all cases the differentiation process goes through intact. Only in the rare exception is some kind of imperfect chloroplast formed, only in the rare exception does mitosis fail. One can make it fail with certain drugs, but basically these things go to completion time after time. It is clear that what is needed in all cases of differentiation is to produce at one blow, coordinated, a group of enzymes which work together to carry out the normal process smoothly and to completion.

We have seen in the case of the auxins and of the cytokinins that there are a number of cases where an enzymatic change is produced and that change becomes permanent. The accoutumé change is the most extreme, but Skoog and Miller have found several others. For instance, their cultures with high cytokinin level turn out, after a few transfers, not to require thiamine (fig. 9-15). Apparently the tissue richly supplied with kinetin has learned to synthesize thiamine. Since changes of this kind remain during continued subculturing, we deduce that they are genetic. Thus what we deal with in these differentiation processes is the activation, or the release, or the evocation, of a whole group of genetic factors which are to produce a group of enzymes which bring about the differentiation. In the phenomenon of operons, described in *E. coli* and other bacteria, a gene is situated, so to speak, to the left of all the other genes in that area of the chromosome, and when it is activated all the genes on its right are also activated, giving rise to a closely related group of enzymes. The classical case is in the synthesis of histidine, where the group of enzymes that synthesize histidine are all activated by the histidine operon. A single gene thus activates a series of genes and so the whole process goes through. There is little or no accumulation of intermediates, because the rate of each process is modified by the rate of the one preceding it, so that the final product is also the only product.

If we accept this as a view of differentiation we get a somewhat different impression of the role of a hormone. We see that a hormone may call forth a group of genetically controlled activities or enzymes. In principle it calls forth something already there, but inactive. And I think we can well understand that in a complex process like cell division, one hormone may well call forth one operon-like group, and another hormone may call forth another operon-like group—not in sequence, but perhaps overlapping. Thus in Skoog's data on cell division the cytokinin apparently activates especially cytokinesis proper, the auxin activates especially DNA synthesis, and as a result, the two together activate normal mitosis. It follows that plants do not have special differentiation hormones, cell division hormones, rhizocalines, etc. Instead the existing abilities in the tissue, which in the normal event come into play one after the other, can be stimulated to do so by the semispecific action of individual hormones. And in different cell types the same hormone may activate different operons, giving rise to the multifarious effects recorded for the same auxin or cytokinin.

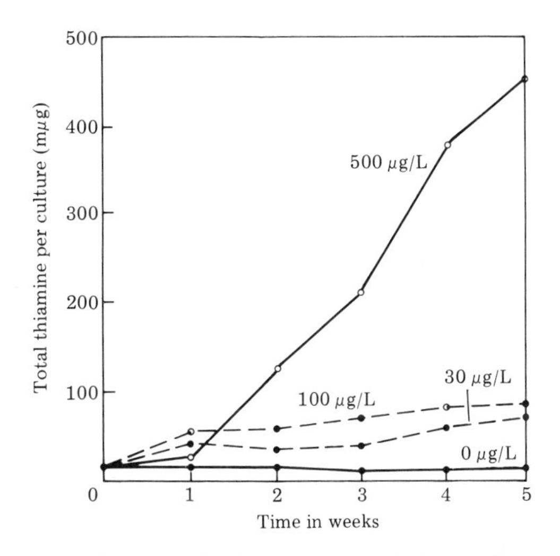

9-15 Changes in the thiamine content of tobacco callus cultures after growing for 25 days on media containing, respectively, 0, 30, 100, and 500 μg per liter kinetin. From Digby and Skoog, 1966.

REFERENCES

Apelbaum, A., Fisher, J. B., and Burg, S. P. Effect of ethylene on cellular differentiation in etiolated pea seedlings. *Amer. J. Bot.* 59, 697-705, 1972.

Camus, G. Recherches sur le role des bourgeons dans les phénomènes de morphogénèse. *Rev. Cytol. Biol. Végétale.* 11, 1-199, 1949.

Digby, J., and Skoog, F. Cytokinin activation of thiamine biosynthesis in tobacco callus cultures. *Plant Physiol.* 41, 647-652, 1966.

Earle, E. D., and Torrey, J. G. Colony formation by isolated *Convolvulus* cells plated on defined media. *Plant Physiol.* 40, 520-528, 1965.

Gautheret, R. J. Sur la possibilité de réaliser la culture indéfinie des tissus de tubercules de Carotte. *Compt. Rend. Acad. Sci.* 208, 118-120, 1938 (cf. *Compt. Rend. Soc. Biol.* 127, 259-262, 1938).

———. Plant tissue culture. *Growth Suppl.* (6th Symposium on Development and Growth) 21-43, 1946.

———. Vues nouvelles sur le cancer végétal. *Vierteljahreschr. Naturforsch. Ges. Zurich* 95, 73-88, 1950.

———. *La culture des tissus végétaux.* Paris, Masson, 1959, pp. 863 + xviii.

Jacobs, W. P. Comparison of the movement and vascular differentiation effects of the endogenous auxin and of phenoxyacetic weedkillers in stems and petioles of *Coleus* and *Phaseolus. Ann. N. Y. Acad. Sci.* 144, 102-117, 1967.

La Motte, C. E., and Jacobs, W. P. Quantitative estimation of phloem regeneration in *Coleus* internodes. *Stain Tech.* 37, 63-73, 1962.

———. A role of auxin in phloem regeneration in *Coleus* internodes. *Devel. Biol.* 8, 80–98, 1963.

Morel, G. La culture in vitro du méristème de certaines Orchidées. *Compt. Rend. Acad. Sci.* 256, 4955–4957, 1963.

Nobécourt, P. Sur la perennité et l'augmentation de volume des cultures de tissus végétaux. *Compt. Rend. Soc. Biol.* 130, 1270–1274, 1939 (cf. *Compt. Rend. Acad. Sci.* 205, 521–523, 1937).

Paupardin, C. Contribution a l'étude du role des composés phenoliques dans les phenomènes de morphogenèse manifestes par quelques tissue végétaux. *Ann. Sci. nat. botan.* (Ser. 12) 13, 141–210, 1972.

Sachs, T. The induction of vessel differentiation by auxin. In *Plant Growth Substances* 1973, 900–906, Tokyo: Hirokawa Pub. Co., 1974.

Saussay, R., and Gautheret, R. J. Action de la lumière, de la temperature et de l'acide gibbérellique sur la production de formations cribrovasculaires par les tissus de rhizomes de Topinambour cultivés in vitro. *Compt. Rend Acad. Sci.* 279 D, 1871–1875, 1974.

Skoog, F., and Miller, C. O. Chemical regulation of growth and organ formation in plant tissues. *Symp. Soc. Exptl. Biol.* 11, 118–131, 1957.

Slankis, V. Uber den Einfluss von Indolessigsäure und anderen Wuchsstoffe auf das Wachstum von Kiefern. *Symbolae Botan. Upsalienses* 11, 1–63, 1951.

Steward, F. C. Totipotency and variation in cultured cells. In *Plant tissue and organ culture,* ed. P. Maheshwari and N. S. Ranga Swamy, pp. 1–25. Delhi: Int. Soc. Plant Morphologists, 1964.

———. Carrots and coconuts; some investigations on growth. In ibid, pp. 178–197.

Thimann, K. V., and O'Brien, T. P. Histological studies on the coleoptile. 2. Comparative vascular anatomy of coleoptiles of *Avena* and *Triticum. Amer. J. Bot.* 52, 918–923, 1965.

Torrey, J. G. Auxin control of vascular pattern formation in regenerating pea root meristems grown in vitro. *Amer. J. Bot.* 44, 859–870, 1957.

———, and Fosket, D. E. Cell division in relation to cytodifferentiation in cultured pea root segments. *Amer. J. Bot.* 57, 1072–1080, 1970.

Vasil, V., and Vasil, I. K. Regeneration of tobacco and petunia plants from protoplasts and culture of corn protoplasts. In *Vitro* 10, 83–96, 1974.

Verhille, A-M., Saussay, R., and Gautheret, R. J. Nouvelles expériences sur le néoformation de racines par les tissus de Topinambour (var. Violet de Rennes) cultivés in vitro; persistance des actions exercées par quelques facteurs de rhizogénèse. *Compt. Rend. Acad. Sci.* 278 D, 1199–1204, 1974.

Wareing, P. F., Hannay, C. E. A., and Digby, J. The role of endogenous hormones in cambial activity and xylem differentiation. In *The formation of wood in forest trees* (2nd. Cabot Symp.), ed. M. Zimmermann, pp. 323–344. New York: Academic Press, 1964.

Wetmore, R. H., and Sorokin, S. On the differentiation of xylem. *J. Arnold Arboretum* 36, 305–317, 1955.

White, P. R. Potentially unlimited growth of excised tomato root tips in a liquid medium. *Plant Physiol.* 9, 585–600, 1934.

Apical Dominance 10

A. THE FORMATION OF LATERAL MERISTEMS

As the dicotyledonous plant develops, the apex is pushed upward by elongation of the cells below, and the leaf primordia up on the flanks of the apex enlarge and arch over it, as in figure 7-1. The growth of the apex itself and the elongation of the internodes just below are almost certainly controlled by different factors. For when apices are grown in culture media, as we noted in the previous lecture, success can be had only if auxin is added to the medium. Thus it seems certain that the apex does not secrete much auxin, and the leaf primordia are the main sources of the considerable amounts of auxin which so readily diffuse out of the tips of terminal buds. The elongation of the zones below is therefore to be ascribed either to auxin or to some other hormonal influence coming together with it out of the young and developing leaves.

In the past we have always thought of the influence coming out of these young and developing leaves as being primarily auxin, but I think that conclusion is now open to some slight doubt, especially because of the work of William Jacobs at Princeton (which, however, was unfortunately not extended to the youngest internodes). This work seems to show that if one analyzes a series of internodes of a growing dicot, from the youngest to the oldest, measuring the rate at which each internode elongates and the amount of gibberellin present in it, there is a rather remarkable parallel. It may not be any more than a mere parallel, but it is close, and since the corresponding parallel does not work out very well with auxin, I am inclined to think that the elongation of these first internodes may be somehow controlled by both elongation substances. This is nevertheless a point that is very hard to establish.

The production of gibberellins by young leaves has not been clearly worked out, largely because it is very difficult to make gibberellin determinations on small quantities of tissue, except by extraction. And as we have seen, the determination of auxin *by extraction* is not always reliable as a guide to auxin effects. One really needs to measure the actively polar transported hormones, which, with auxin, can be done by transport experiments, but with gibberellin the yields are so small that the results are almost meaningless. I would only say that if one were to try to establish

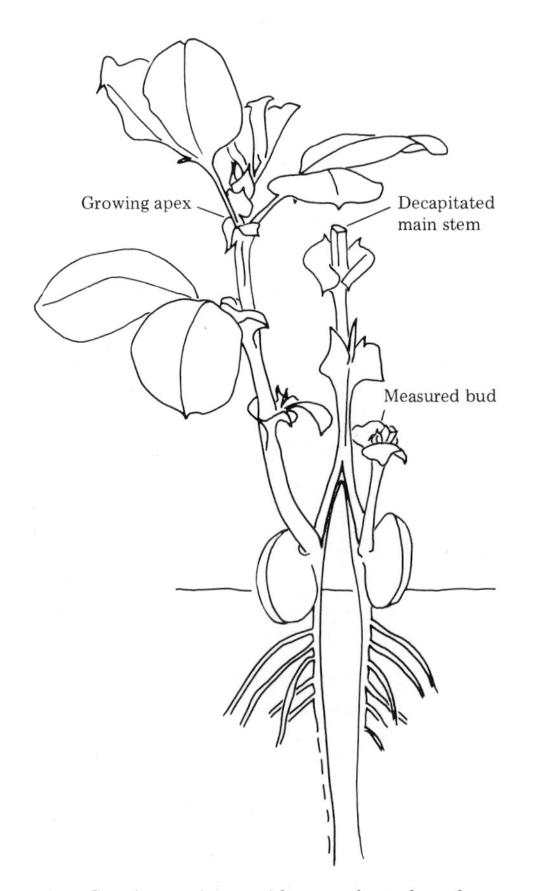

10-1 Snow's experiment with a two-shoot plant showing that the bud next to the nutrient-rich cotyledon, and without a growing center above it, is nevertheless inhibited by the apex on an adjacent shoot. From Snow (personal comm., 1937).

the same parallelism with auxin, one would find that the relative internode auxin content would be different in different plants. The distribution of elongation rates, on the other hand, might be very similar. For instance, in the *Ginkgo* branch, the internode auxin content shows a maximum about two-thirds of the way down the stem (fig. 7-7). The value is rather low just below the apex, suggesting that neither the apex nor the young leaves makes auxin in nearly the amount that the medium-sized leaves do, or else that the rate of production is less than the rate of transport. In the latter case the auxin is drawn away from its point of maximum accumulation, so that the apparent auxin maximum comes far down the stem. With gibberellins, since they are transported relatively freely in all directions, and not polarly, this parallel between accumulation and growth response is perhaps easier to establish.

Leaving aside this issue of the exact nature of the hormonal influence, it is clear that *some* influence reaches the subapical tissue from the young leaves, whether it be auxin, gibberellin or both. This influence causes the cells to elongate axially and so to push the apex and the newly forming internodes continually upward, and, as it is lifted upward, the lateral organs enlarge and develop. Thus there is a certain zone in which cells are dividing at varying rates (depending on their distribution in the layers) and then a zone just below that in which elongation supersedes division.

Most of the cell division takes place in the apex itself and on the flanks where the lateral organs are being produced. As the number of cells in the lateral organs increases, they begin to elongate too. As a result the relatively massive meristems of the leaves are pushed up, and there is left behind, at the base of each of these laterally developing leaves, a small mass of meristem. This small meristem left behind develops into a bud in the axil of the leaf. Thus in the developed series of internodes, each leaf bears in its axil a small meristem which differentiates into a bud. It may increase in size as time goes on—sometimes it does and sometimes it never does. In some plants it is no more than two or three cells of meristem; in some of the *Nyctaginaceae* one can see no more than one pair of divided cells at this point—there is no mass of meristematic tissue at all. And the subsequent fate of that little bit of meristem also varies widely with different plants. We will consider mainly the situation in which it does grow to some extent and produces a small bud.

B. THE NATURE OF APICAL DOMINANCE

So long as the apex is growing rapidly, auxin and gibberellin are being fed in, and the stem is maturing, the small bud in the leaf axil generally does not develop. The reason is not because it is dormant in any sense or unable to develop; it can develop the moment the apex is removed. For when the apex is removed one of two things can happen. Either the most basal buds develop into shoots, as in such plants as *Vicia faba*, or in many tree branches, or else the basal buds remain undeveloped and one or more of the most apical ones develop into shoots. Whether there is basal or apical bud development again depends upon the plant (see below), but that is not the immediate problem. The immediate problem is that they do develop and therefore they demonstrate that this is active tissue which, in the absence of the apex and young leaf primordia, can grow perfectly well.

So the basic problem of apical dominance is: why do axillary buds normally not develop? This is a problem that has exercised botanists for a long time, and in the 1890s Karl Goebel, a great observer of plants, wrote about it in his voluminous publications on more than one occasion. His idea was that the rate of development of any bud is a function of the rate of supply of nutrients. Somehow the terminal bud constitutes what we should now call a "sink" into which the supply of nutrients becomes attracted, so that the poor little lateral buds are sentenced to seeing the good nutrient sweeping by all the time without its coming to them, because it is all destined for the growing apex. But why it should go by the lateral buds and never reach them, he was not able to explain. Nevertheless this kind of idea was prevalent for a long time, that the growth of a bud is a function of the supply of nutrient, which in turn is somehow polarized towards the bud which is using it, so that other buds are unable to receive it.

There were two classical experiments which cast doubt on Goebel's idea; one was carried out in the 1920s by Reed and Halma at Riverside, California. Dr. Reed was a great student of the growth of orange and lemon trees and he had been particularly interested in the growth of their buds. In this experiment they selected young, developing shoots with well-developed lateral buds which, however, were not growing, and they made a cut just above each of the lateral buds. Especially if the shoot was horizontal, this small cut was found to be sufficient to allow the bud to grow out. Note that these cuts made the bud *below* them grow, not the bud above them—providing clear evidence that the Goebel idea of a flow of nutrient was probably wrong, because what was interfered with was a flow of something from *above*. Stopping such a flow would allow the bud *below* the cut to grow.

The second experiment was done on a different tissue by Havránek, a student of Dostál in Czechoslovakia. This concerned the buds on potatoes. A potato usually bears a largish bud at the terminal end and others in various places on the tuber. These are essentially lateral buds, although they do not occur in the axils of leaves. What Havránek found was that if he made a sort of ditch around the upper side of one of these buds by cutting through the periderm and removing a little of the surface tissue, then this bud tended to develop. Other buds remained unaffected. Thus again some influence coming from the main bud end of the potato seemed to be *preventing* the growth of these lateral buds. So evidence began to point away from a basal supply of nutrients which *promotes* growth to an apical supply of some agent which *inhibits* growth. The two viewpoints are almost exactly opposite.

It is doubtful that these experiments had much influence, but looking back on them, one sees how clearly they foretold the future. For in 1927 Robin Snow, a careful observer of the growth of young leaves, and a student of phyllotaxis, made a most interesting group of experiments at Oxford. He saw clearly that this idea of an influence coming from the apical end to inhibit growth was much nearer the truth than the idea of a growth-promoting influence coming from the base. He defoliated seedlings in various ways, sometimes causing growth of the lateral buds, and decided that the youngest developing leaves in the light (this was green tissue) were the prime influence. Only if these leaves were taken off did the axillary buds develop.

Then Snow carried out the logical but rather difficult experiment which is shown in the next figure. It is somewhat analogous to Boysen Jensen's well-known experiment on the movement of the phototropic stimulus in coleoptiles, which showed that the influence could pass across a cut surface, and must therefore be either an electric stimulus or a diffusible substance (Boysen Jensen seems if anything to have favored the former). Snow's experiment is more complicated (fig. 10-1). Broad bean *(Vicia faba)* seedlings, with the cotyledons still on them, were severed as shown, so that the lateral bud to be observed was isolated from the apex. The two halves of the stem were bound together tightly. Then in half of the plants he decapitated the main stem, and as expected, the lateral buds in the axils of the cotyledons at once began to grow out. But the point is that so long as the main stem was not decapitated these buds did not grow out, or at least their growth was retarded. There were several variations on this experimental theme. Here, then, we have a growing apex with young leaves (as we have just seen, he concluded that young leaves are the main in-

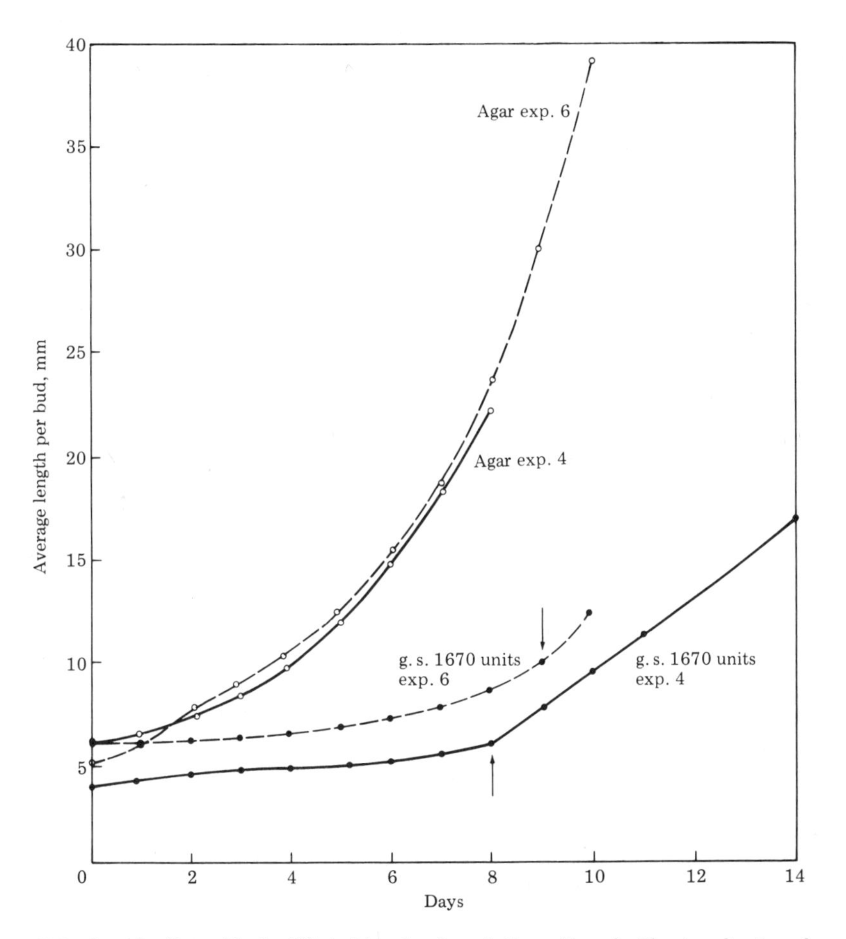

10-2 Growth of lateral buds of *Vicia faba* after decapitation with and without application of crude IAA (g.s.) to the cut surface. Application of auxin stopped at arrow. From Thimann and Skoog, 1934.

hibiting organs) which not only inhibit the growth of the buds below, but can exert this inhibition across a cut surface; the inhibited bud was still in direct connection with a cotyledon and hence had a normal supply of nutrient. There could be no remaining doubt that the growth of the lateral bud is not controlled primarily by the supply of nutrient from below, but by an inhibiting influence coming from the apex and young leaves above.

C. AUXIN AS THE INHIBITING AGENT

Soon afterwards Folke Skoog and I were studying the production of auxin by different plant organs, and we had observed that young leaves produced relatively large amounts of auxin for their weight (cf. table 7-1). And so the suggestion was obvious that the inhibiting influence that came from the young leaves might be the auxin which they were shown to secrete. In that case, one should be able to remove the terminal bud with its young leaves attached and maintain the inhibition by supplying auxin in its place. We did this, but at first the results were rather unsatisfactory. We used *Vicia faba*, in which it is the basal buds which primarily develop into shoots; we would decapitate the plants and apply auxin in small agar blocks, in the same way as we were applying it asymmetrically to coleoptiles. There did indeed seem to be some delay in the development of the buds but the effect was not very strong. Fortunately, at the same time, Bonner and I were studying the kinetics of the curvature tests with agar blocks applied on one side of a coleoptile. We would put blocks on, allow curvature to occur, and then after various times take the blocks off and estimate their residual auxin content by applying them to other plants. We soon found that the auxin leaves the agar block rather rapidly; about 15% of it goes into the test coleoptile in 90 minutes, so that in a few hours most of it would be exhausted. Now in the bud inhibition experiments we were applying these blocks for 24 hours, and of course in that 24 hours the auxin would long since have been used up, so that the plant would be for a good part of the time essentially untreated. The conclusion was very simple: take off the blocks and apply fresh ones every few hours before each of them gets exhausted. Then, as figure 10-2 shows, the result is very clear. The plot shows the length of the basal axillary buds, which are the ones that develop when the plant is decapitated. In two experiments, the growth of the buds on the decapitated controls follows clearly parallel lines. The amount of auxin applied is given in units, which were decided upon from measurements of the number of units that the young leaves or terminal buds can diffuse out into agar. When this amount of auxin replaces the terminal bud, the lateral buds are nicely inhibited. At the arrows the application of auxin was stopped, and immediately the buds grew out, and their rate of growth was very much like the rate of growth of controls on decapitated plants. Thus the buds have only been inhibited and not damaged; they are still able to grow very well.

Still more striking effects are obtained with the buds in the cotyledonary axils that Snow used, especially in peas. The cotyledonary axils grow very rapidly when decapitated pea seedlings are grown in the dark. When, after peas have grown for about 7 days in the dark, the main stem is cut off, this axillary bud at once develops. (There are actually two axillary buds, but only one grows.) The photo in figure 10-3 was taken 14 days later. These buds are close to the supply of water from the roots and the supply of nutrients from the cotyledons, thus they are ideally fed and can elongate normally in the dark.

Now if to the cut surface of the stump of the main stem we apply a little auxin, the results are as shown in the figure. Inhibition is virtually 100% (although one of the buds does show a *tiny* amount of enlargement). Thus this system is extremely sensitive—complete inhibition can result with a very tiny amount of auxin applied. These data leave no doubt that auxin coming from the young leaves or the terminal bud is a powerful inhibitor of the growth of lateral and axillary buds.

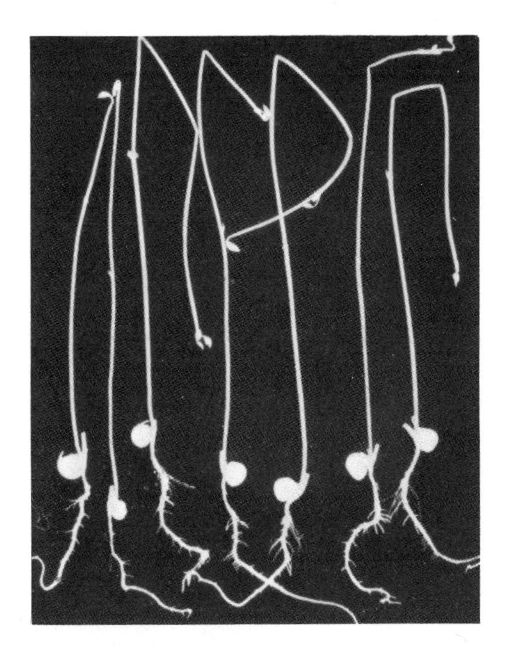

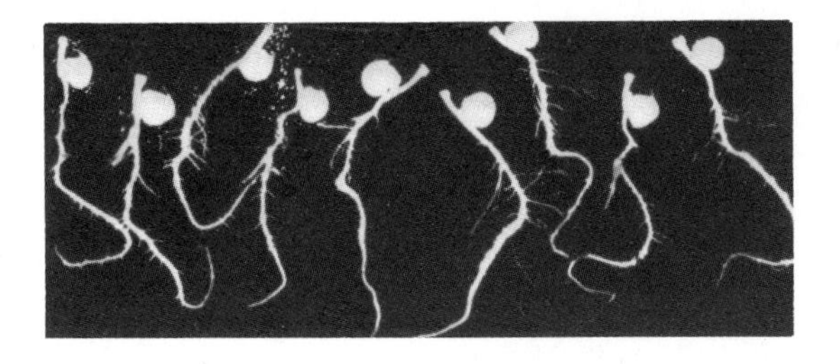

10-3 Growth of buds in the cotyledonary axil of pea seedlings decapitated two weeks earlier. Above, controls. Below, IAA applied to stump. From Thimann, 1937.

D. THE ROLE OF THE LEAF BLADE

There are some second order complications. Many years ago Dostál* found that the growth of axillary buds in decapitated plants of *Scrophularia nodosa* could be hastened by cutting off the leaves in whose axils they rested. So, back in 1922, before even the work of Snow, Dostál had deduced that leaves exercise an inhibiting action on the buds in their axils. This phenomenon was lost sight of for many years, but some 10 years ago it was reinvestigated, particularly by Champagnat in France. Champagnat showed that the influence of the leaves, in plants with an active terminal bud, is not readily detectable because the inhibiting influence of the apex is overwhelming. But if the shoots are decapitated, then the elongation of the buds, and sometimes the number of such buds which grow, is a simple function of the presence or absence of the leaf blades. Table 10-1 shows an experiment with long shoots of lilac, *Syringa vulgaris*. The shoots are decapitated and then varying numbers of leaves are removed. In this case the shoots bore 16 pairs of leaves. If no leaves were removed, 23 lateral buds developed. If all the leaves were removed, in spite of the fact that the shoot is defoliated and therefore deprived of nutrition except any that may come from other parts of the plant, twice as many lateral buds develop. Also their growth rate increases, that is, the number of days required for them to reach a certain size steadily decreases as the number of leaf blades removed increases. It is clear from data of this sort, and other similar results, that the leaves do exercise an inhibiting influence on their axillary buds. The extent of this varies with the species and the vigor of the plant, but it is obviously a second factor contributing to the general inhibition of lateral buds on the stem of a dicotyledonous plant. My colleague Dr. Skoog has produced a little diagram which puts these facts together (fig. 10-4). On the left we have the intact plant—apex, young leaves, mature leaves, the base and the roots. In (2), we decapitate, the buds in the axils develop. If we decapitate and apply auxin (3), the buds in the axils are inhibited. Incidentally, you will notice that the auxin applied produces additional roots. However, if we apply the auxin basally, there is no effect—the buds grow just as in the plain decapitated plant. In other words, in young seedlings polarity is strict enough so that little or no auxin reaches those buds. If this experiment had been done with older plants, there might have been a little hesitancy about (4).

In (5), we decapitate and also defoliate, à la Dostál; we get excellent growth of buds. But if we leave one blade on and defoliate the other (6), then there is a marked differ-

* Dostál, who died only recently in his nineties, worked in the Agricultural University in Brno, Czechoslovakia.

Table 10-1 Effect of defoliation on the growth of upper lateral buds after decapitation

Number of leaves removed	Number of lateral buds developed	Days required for upper buds to reach 8–10 mm
0	23	30
8	26	28
12	30	18
14	39	13
16	45	9

Data of Champagnat (1950) with Lilac (*Syringa vulgaris*).

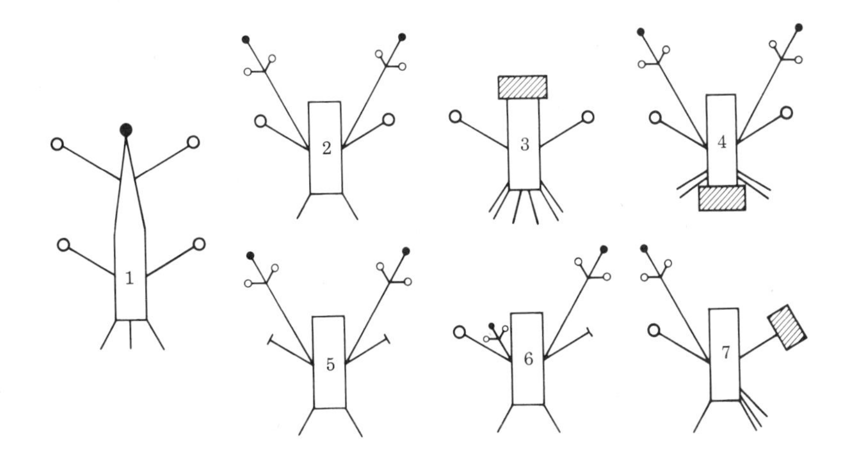

10-4 Diagram of the growth of lateral buds in an opposite-leaved plant. (1) intact; (2) decapitated; (3) decapitated, auxin on apical cut surface; (4) decapitated, auxin on basal cut surface; (5) decapitated and defoliated; (6) as 5, but one leaf left on; (7) as 6, but auxin applied in place of removed blade.

ence in the rate of growth of the buds; one bud will be at least partially inhibited by the leaf in whose axil it resides. We can imitate that effect by applying auxin in place of the blade (7); then the opposite one will grow more. In other words, the inhibition by the leaf blade is a subsidiary one, easily overcome.

E. COMPETITION AND BALANCE AMONG BUDS

A lot of work has been based on this competition or balance between pairs of buds in opposite-leaved plants. Mostly it shows this kind of subsidiary influence. In the case of cotyledonary buds, it is particularly interesting because these are leaves but are also sources of nutrient. In some conditions, in a seedling with 2 cotyledons, the bud in the axil of the cotyledon which is left on will actually grow faster than the one whose cotyledon has been removed. This means that the nutritional effect of the cotyledon is greater than its auxin-supplying effect.

There are a large number of variations on this theme. We noted earlier that, on decapitation, in some plants it is the apical buds which grow and in others it is the basal buds. The French call the latter habit *basitonie*, and some recent French studies show that basitonie is a function of the environment, the species, and even the temperature. *Sambucus* shoots, for instance (in France, that is; it would not necessarily be true here), develop the most basal buds when the temperature is near 15°C; if they are decapitated in the summer when the temperature is nearer 25°, the most apical buds are the ones which develop. The

Gingko described in chapter 9, on the other hand, is strongly *acrotonal,* and when it is decapitated only the buds of the most apical short shoots grow out into long shoots. Usually 2 or even 3 such buds develop, and the environment does not seem to control.

These effects are probably due to differences in the transport of auxin. In a basitonal plant like *Vicia faba*, relatively little auxin must reach the basal buds compared to that which reaches the apical buds; most of the auxin remains in the upper part of the stem. On the other hand, at higher temperatures, where transport goes more rapidly, transport can be faster than production; so that the more apical buds run out of auxin first. Correspondingly in *Gingko* the auxin maximum is near the base. This explanation is by no means proven but seems likely. If so, with 2,3,5-triiodobenzoic acid, which powerfully inhibits auxin transport, one might perhaps be able to change the habit of plants from basitonal to acrotonal.

There are also cases where lateral buds can have alternative fates: they can be flowering or vegetative; they can grow vertically, be plagiotropic, or even (as in the peanut) positively geotropic; or they can become stolons or rhizomes. In the last case recent work of Fisher and his colleagues (1974) on *Cordyline* offers a remarkably simple explanation. For in horizontal stems of this woody monocot, buds on the upper side develop into leafy shoots, those on the lower side into rhizomes. ^{14}C—IAA applied to the apical end of a stem segment accumulated on the lower side to a ratio of 7 : 1 (not 3 : 1 as in coleoptiles). Girdling lowers the auxin content (isotope accumulates above the girdle) and buds that would have become rhizomes now become leafy shoots. Thus the rhizome seems to result from an increased auxin level, the leafy shoot from a decreased auxin level. Fisher predicted this earlier on morphological grounds.

We saw in chapter 7 (section B) that the buds of Ginkgo trees also have alternative fates, some growing out into normal branches while others produce only the leaves already present as initials in the bud, and remain short. This differentiation into "long shoots" and "short shoots" is apparently controlled by hormone relations, and the short shoots have much in common with inhibited lateral buds. Decapitation causes the first two or three laterals to grow out as long shoots, but if auxin is then applied to the cut surface the buds develop only as short shoots.

This type of behavior is of special interest in fruit trees, for in these (particularly apple trees) it is the short shoots that flower. The hormonal relationships seem, from largely indirect evidence, to be about the same in apple as in Gingko. Placing the young trees in a horizontal position, so that auxin would move to the lower side, promotes the development of short shoots on the lower side and long shoots on the upper side. Recently Plich and co-workers in Poland have made further interesting findings with apple buds. Cutting out every other bud on young trees increases the incidence of long shoots, which could well mean that buds are tending either to inhibit one another through their auxin production, or else to compete with one another for cytokinin. Both may be happening. Direct treatment of every other bud with IAA increased the percentage of short shoots from 50% in controls to 81% in treated shoots. Conversely, cytokinin (benzyladenine was used) decreased the percentage of short shoots to 25%. Gibberellin had no clear effect but somewhat antagonized the auxin. These results and others both from the Polish group and from Wareing's laboratory all support the general idea of a balance between auxin and cytokinin (perhaps modified by gibberellin) as controlling not only bud development itself (as shown below) but also some aspects of the fate of the buds when they develop.

F. DOMINANCE IN LOWER PLANTS

Apical dominance is a very generalized phenomenon throughout the plant kingdom and is certainly not limited to higher plants. Back in 1938 Harry Albaum in New York demonstrated apical dominance with fern prothallia (fig. 10-5). The prothallium is roughly heart shaped, and the growth center, analogous to an apical meristem, lies in the indentation of the heart. Albaum found that when prothallia were growing actively on a good nutrient medium, if he cut out the central growth zone a new growth zone would be produced. The phenomenon could even be repeated, giving rise to a series of little heart-shaped zones one above the other. If now the growth zone were cut out and auxin applied in its place, the outgrowth did not occur. Thus in these little plants there is auxin inhibition of the development of a secondary growth zone comparable to the inhibition of the growth of a lateral bud (which can also be thought of as a secondary growth zone).

The fungus *Phycomyces* shows apical dominance too, but it has not been shown to be controlled by auxin. The sporangiophore, a long rapidly growing hyphal branch which bears the sporangium at its tip, is the locus of this dominance. For if the sporangium is removed, or if the tip is lightly damaged in any of various ways, then sur-

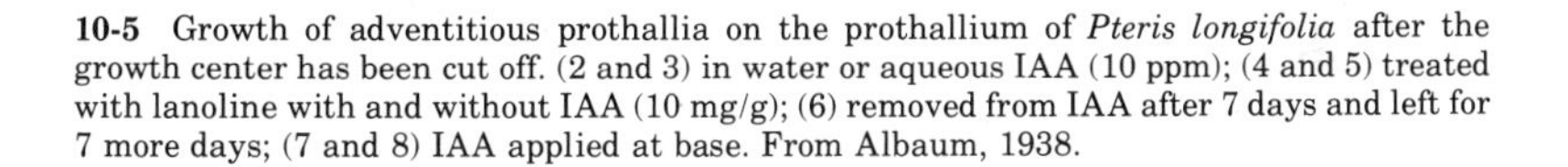

10-5 Growth of adventitious prothallia on the prothallium of *Pteris longifolia* after the growth center has been cut off. (2 and 3) in water or aqueous IAA (10 ppm); (4 and 5) treated with lanoline with and without IAA (10 mg/g); (6) removed from IAA after 7 days and left for 7 more days; (7 and 8) IAA applied at base. From Albaum, 1938.

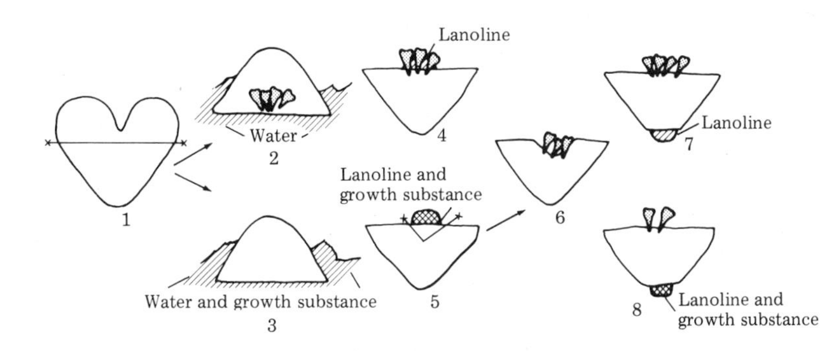

prisingly, lateral branches develop. This is the more remarkable because normally there is no branching at all in these sporangiophores. Thus, even in an organ that normally never branches, a part of the cell retains the capability of producing such branches. Also, since the sporangiophore is a single cell, we see here that apical dominance is not limited to multicellular tissues. Dostál has observed somewhat comparable reactions in the alga *Codium* which, since it is coenocytic, can be regarded as a sort of special single cell.

G. THEORIES OF AUXIN ACTION

The bud-inhibiting action of auxin and the dominance of the apex in higher plants gave rise to a great number of theories in the late 1930s. At one time there were nine such theories. I will not present all of them but will mention a few that are of some interest. One is that the lateral buds do not grow because there is *compensating growth* elsewhere. The growth of the new tissue just below the apex, which is lost on decapitation, is the growth which the lateral buds compensate for. The idea is based on a sort of constancy of the growing tendency—when one part is growing rapidly, other parts will not grow, and if the rapidly growing part is removed, then compensating growth will take place.

Snow was puzzled by another aspect of apical dominance: the downward movement of auxin in the seedling stems is very polar; if that is so, how can it go up, at least for a millimeter or so, into a lateral apex? To explain this he proposed the theory that auxin travels down the stem and in the basal part of the stem is converted to, or gives rise to, an inhibiting substance which travels upwards, and it is that which comes up into the lateral buds to inhibit their growth.

Libbert, in East Germany, another student of apical dominance, extended Snow's concept and proposed that the auxin goes down to the roots, and in the roots produces an inhibitor which travels upward into the stem. He knew that the extracts of roots are rather inhibitory, as indeed are most extracts of plants.

Van Overbeek proposed an ingenious theory, to the effect that auxin, in travelling down, clogs the channels of transport which could supply materials to the buds. This could only happen if the molecules of auxin were of a size such that in lining the transport system they essentially filled part of the lumen of the transporting cells. At one time I went so far as to compare this with slowing down the flow of the Mississippi River by placing a line of pebbles along the shore.

But there was one theory that survived for a number of years, and this is the one shown in figure 10-6. It was based on the idea that different organs respond to auxin in different ways. It was well known that roots have their growth increased by auxin only to a very small degree, and only at exceptionally low concentrations—around 10^{-10} or 10^{-11} molar—while at most ordinary, even physiological concentrations, they are inhibited. In contrast, the elongation of stems is promoted by all physiological auxin concentrations and only inhibited by unphysiological, exceptionally high ones. So it was suggested that buds fall in

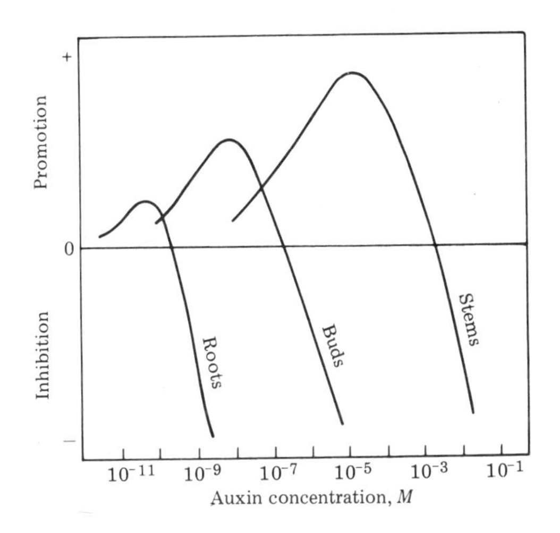

10-6 Ranges of growth response to IAA for different plant organs. Schematic. Modified from Thimann, 1937 by Leopold, 1955.

between these two, having a range in which they are promoted and a range in which they are inhibited, but that the range in which they are inhibited happens to coincide with the range in which stems are promoted. Thus in a normal growing plant the stem would be elongating under the influence of auxin while its buds are being inhibited by the same auxin concentration. The exact choice of abscissae, of course, would depend upon the plant.

This theory did survive for some time, but it began to be clear that it was impossible to maintain because of careful measurements (made mostly by van Overbeek) of the auxin content of inhibited buds, as determined by extraction into ether. In brief, it was found that after decapitation, when the buds start developing, the amount of auxin in them increases. Thus, instead of developing they should become more inhibited, so the change in the auxin on decapitation is in the wrong direction. However, extraction of buds or young leaves yields only a fraction of the amount of auxin they would produce by diffusion into agar, as Wilkins' group recently found.

H. DOMINANCE AND INHIBITION WITH SINGLE BUDS

In all of this kind of work, the complexities of the whole plant present serious limiting factors, and it seemed that no real progress could be made until we were able to avoid these complexities and operate a much simpler system. The simpler system that we finally adopted was to use just a single internode with one bud attached. Because

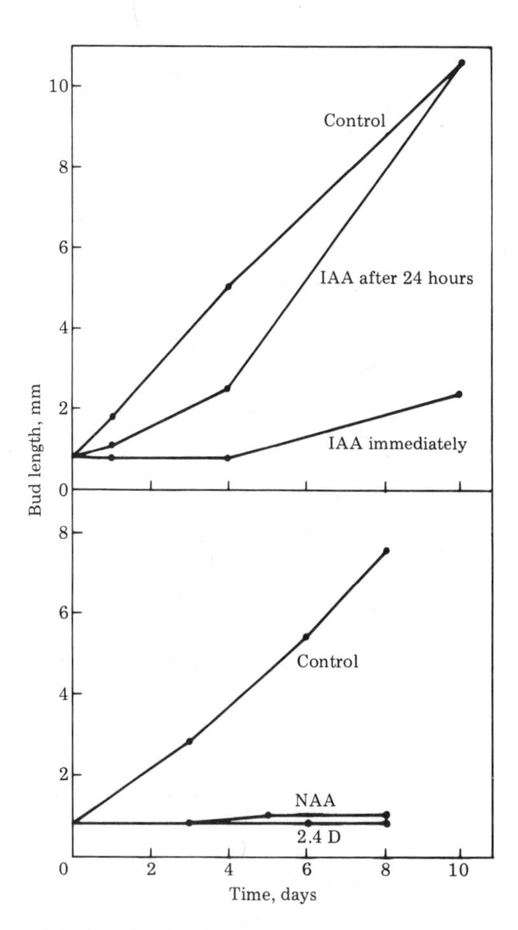

10-7 Growth of buds on isolated node-internode segments of etiolated Alaska pea plants in light in 1% sucrose. Above, IAA supplied immediately or delayed for 24 hours. Below, with synthetic auxin (1 ppm). From Wickson and Thimann, 1958.

of their great length, the node plus internode of an etiolated pea seedling was found convenient. Furthermore, the internode tissue can supply the bud with a limited amount of nutrient material. If one floats such isolated nodes and internodes in a nutrient solution (which as it turns out need be nothing but sugar solution, because there are amino acids in the stem) and places it in light, then the bud turns green (thus providing a still richer supply of nutrients), and it begins to develop. In this simple system there is enough nutrient to allow a modest amount of development and the response takes place in a few days. To this system auxin can be added in varying concentrations, and figure 10-7 shows some of the results, in terms of bud lengths. The control buds grow 11 mm in 10 days, while if auxin is supplied, 10 parts per million gives almost complete inhibition. Auxins other than IAA are even more potent; figure 10-7 shows that one part per million of either napthaleneacetic acid or 2, 4-D produces complete inhibition—an inhibition that shows no sign of weakening up to 8 days.

This system brings out another limitation that we mentioned earlier, that it is necessary to apply the auxin immediately. If the application of auxin is delayed, then the buds will generally start growing, and once they have started they are more difficult to inhibit. Where auxin is in the solution from the start, the buds show normal inhibition, but where nothing is applied for the first 24 hours and then auxin is introduced, the buds, after a little hesitancy, grow out and catch up to the controls. This tendency for the buds to *break away* from the inhibition applies in all experiments involving physiological auxin levels.

These experiments rule out several theories in one blow. All theories connected with the idea that the auxin has to go into the roots can be abandoned. All theories based on compensating growth in another part of the plant are removed because there is no place for compensation to take place. Furthermore, theories which depend upon diversion of nutrient by another growing organ can also be forgotten because there is no other growing organ present. Thus the problem has been simplified to a large extent.

One problem that is not removed is Snow's difficulty about the importance of polarity. Snow did not see how the downward-moving auxin could ever travel up into the lateral buds to reach their apices. For, in general, polarity in these seedlings is very strong. This problem led Margaret Wickson and me to restudy the movement of auxin in isolated segments of such plants floating in solution. In order to do that one had to use ^{14}C-indoleacetic acid labeled in the carboxyl group, as has been mentioned on several occasions before.

Unfortunately, though the auxin in the solution is now labeled, there is one thing that is not controlled, namely the point of application of the auxin. It is true that the internode is covered with cuticle, and we know that cuticles interfere very powerfully with the entry of auxin; nevertheless there are two cut surfaces, one at each end, which provide for rapid auxin entry. Sealing one end with paraffin has been tried, but the results are very inconclusive and one is not sure whether films of solution do not enter under the seal.

Some results are clear-cut with this system, and an example is shown in figure 10-8. Here we compare the amounts of radioactivity (previously shown to be actually due to unchanged auxin) in five segments, each 7 mm long; the whole internode is 35 mm long. We see that the amount of auxin is high in the most apical segment, decreases as we go down the internode, and finally increases steeply in the basal segment. Preculture in auxin makes little difference, though it does somewhat increase the uptake all round, as we saw in section C of chapter 4. The accumulation at the base is partly the result of transport from above, since most of the entry is certainly through the apical cut surface. It may also be, since the cuts were made while the segments were in the solution, that there is some passive entry at the basal cut surface as well. (In that connection, note that preculture in IAA influences this basal uptake relatively less than the transportable uptake.) The polarity is clearly expressed but may be by no means 100%.

If the plants are grown in darkness or with red light, the amounts in the apical part are always less and the accumulation in the basal part is always steeper. This suggests to me, as many other experiments do, that the transport in etiolated or red-light plants is more rapid than the transport in green plants, and this conclusion is borne out by diffusion experiments. For green pea seedlings, cut into segments, yield measurable amounts of auxin by diffusion into agar, as Scott and Wilkins have shown; but in etiolated seedlings only traces of auxin diffuse out of stem segments. In any event, apical dominance is usually much stronger in etiolated plants than in light-grown plants, for, as we shall see in section I below, white light decreases dominance. It could be, too, that the red light used with etiolated seedlings is rich in far-red, which greatly increases the inhibition through perhaps a quite different mechanism (see below).

In figure 10-9 we see how the distribution of the labeled IAA in the different segments builds up with time. Here the whole internode is divided into five segments, of which the bud is attached to no. 2; the internode floats on the solution, and we take samples every few hours to measure the radioactivity in the segments. In the most apical segment, which is morphologically above the bud, the rate of entry is very rapid, as might be expected, and it plateaus after about 40 hours. In the second segment, the C^{14} comes up more slowly, not quite reaching the same levels; the third and fourth segments show still less; and then again there is a buildup in the fifth. This behavior agrees exactly with that shown in figure 10-8. The data for the fifth segment are plotted on the same curve as those for the most apical segment, for there is not much difference in their isotope content.

What is of importance is the entry of the ^{14}C-auxin into the bud. This entry into the bud is sluggish at first, and the auxin level is actually below that in the lowest segment. Later, however, it increases, and by about 24 hours the level of auxin in the bud is about equal to that in segment no. 2,

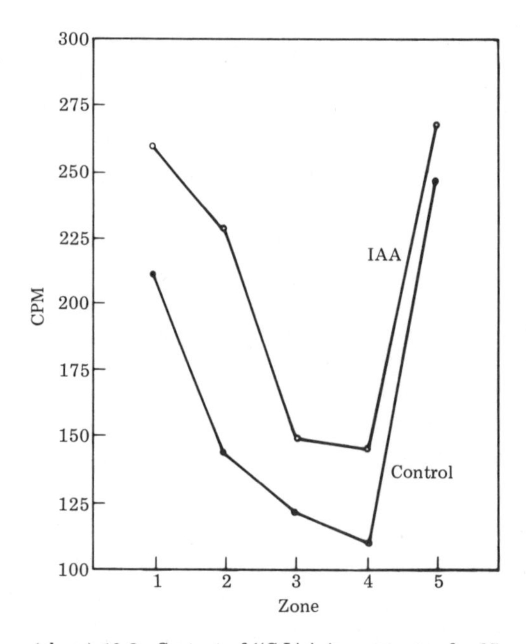

(above) **10-8** Content of ^{14}C-IAA in segments of a 35 mm pea stem internode after 5 days with (upper curve) or without (lower curve) ^{12}C-IAA (3 ppm), then 24 hours in ^{14}C-IAA. From Wickson and Thimann, 1960.

(below) **10-9** Increase of the ^{14}C-IAA content in the five 7 mm zones of the pea stem internode and (dashed lines) in the attached bud. Times are from placing in the ^{14}C-IAA (3 ppm). From Wickson and Thimann, 1960.

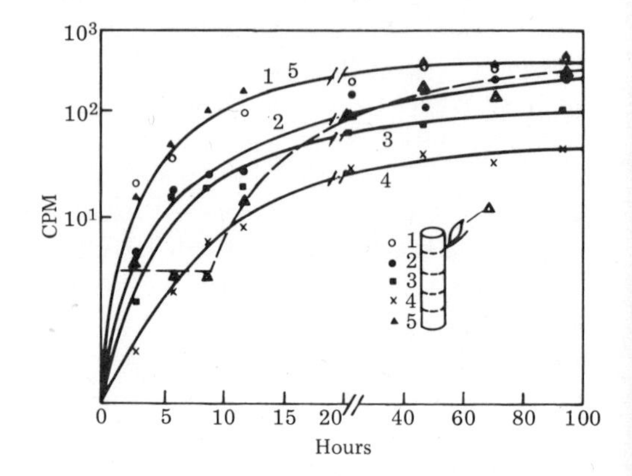

which is the segment on which the bud rests. Thus the polarity is not extremely strict, although it is strict for short times; transport experiments in the first 3 hours, the period wherein transport experiments are usually done, would indicate that the polarity is very strong.

A major step in these experiments was to make a comparison between the amount of radioactivity in the buds and the extent to which they were being inhibited. Some buds are inhibited more than others, and by analyzing the buds at different times one could find a series of buds which showed different degrees of inhibition. In figure 10-10 we see that plot. The degree of inhibition of the bud (compared to that of the control) is a very direct function of the amount of auxin that has reached the bud at that time. This result goes at least a long way towards dealing with Snow's reservation.

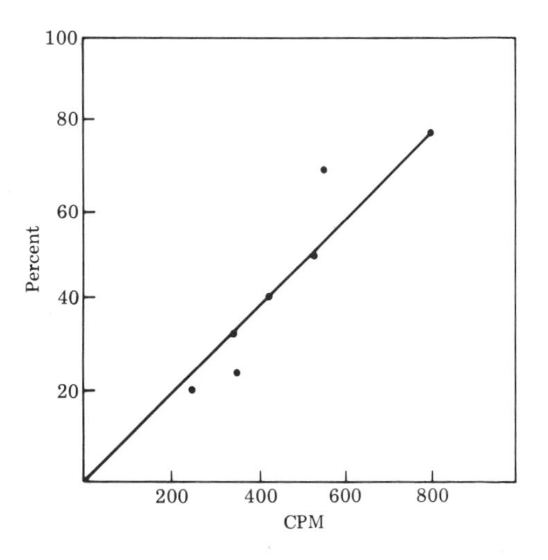

10-10 Correlation between the radioactivity in the lateral bud and the degree of inhibition under different conditions. From Wickson and Thimann, 1960.

I. INFLUENCE OF ENVIRONMENTAL FACTORS

One other important generalization from these and previous experiments is that the extent of inhibition exerted either by auxin or by the apex varies with the plant and with the environmental conditions. In some plants apical dominance is extreme. In the sunflower, for instance, a stem can be 4 feet high—in fact we treat sunflowers in the lab by applying gibberellin to get stems 7 feet high—and not a single lateral bud develops. Thus apical dominance here is extremely strong. On the other hand in some bushes, and especially in dwarf plants which (almost by definition) have a low auxin supply in the stem, many lateral buds will develop and extra buds may develop from the base. The extent of inhibition is a function of the species.

Secondly, dominance, or inhibition, is a function of the nitrogen supply. This has been shown very clearly by Gregory in his studies on flax, *Linum usitatissimum*. When flax is grown under conditions of relatively high nitrogen supply, numerous lateral buds will develop, even though the terminal bud is still in place. On the other hand, when grown with low nitrogen supply it shows the classical picture—very strong apical dominance.

Thirdly, the process is very sensitive to light. Jacobs at Princeton with his *Coleus* plants, grown in the greenhouse in full light, found in many experiments very little apical dominance. At almost every node lateral buds would develop, and decapitation promoted these, yet when he applied auxin in place of the apex, he could get very little control. Jacobs at one point even speculated that the inhibiting action of auxin in apical dominance may be exerted only in leguminous plants. However, Sachs and I were able to show that inhibition in *Coleus* is primarily a function, not only of nitrogen nutrition (his plants were richly nourished with nitrogen) but of light intensity. Table 10-2 compares greenhouse plants with laboratory plants. Using the same *Coleus* variety as Jacobs and studying intact plants, we chose a bud in a particular internode. Grown in weak light the buds are very undeveloped. In white light of 16 hours day-length there are quite long buds—11 mm in this particular group. In intermediate light we see an intermediate bud length. If we decapitate and apply water, all buds grow at about the same rate. In fact in the brightest light the growth rate is actually a little less, while the other two are the same. If we decapitate and apply auxin (we have to use high auxin concentrations because the amount is very small), we get excellent inhibition in the low light intensity; but in high light intensity the bud elongation is about the same as in the intact plants. So there is no control with high light intensity, as Jacobs found. Intermediate light intensity gives intermediate results, i.e., partial inhibition. But it is clear that in these plants the extent of the dominance is a function of light intensity, being weaker in high light than in low light. If one is seeking for an explanation, the most logical one is that light produces something which antagonizes the inhibiting effect. We shall return to that later.

Table 10-2 Effect of light intensity on growth of *Coleus* lateral buds. Mean bud lengths in millimeters.

Treatment	1800 ft.c. (16 hrs)	300 ft.c. (continuous)	50 ft.c. (continuous)
Plants intact	11.4	6.5	1.0
Decapitated, water applied	24.2	30.2	30.0
Decapitated, IAA (100 ppm) applied in droplet	13.9	8.5	3.3
Inhibition due to IAA, as %	42	72	89

Data of Sachs and Thimann, 1967.

J. SOME SPECIAL CASES

There are some curious experiments by Champagnat with different plants, especially with shrubs that have a number of shoots from the base. Champagnat shows that if, of such a group of shoots, one is more vigorous than the others, i.e., its growth rate is higher, then on this more vigorous shoot lateral buds will tend to grow out. On any ordinary basis of simple auxin control one would expect the opposite, for the most vigorous grower is probably the highest auxin producer and therefore its lateral buds would not be expected to grow, while on the slower growing ones lateral buds might grow, as they tend to do on the dwarf varieties of many plants. Champagnat's observation suggests that a vigorously growing plant, like those in sunlight, may have available to it some influence which antagonizes the inhibition.

A variation on the theme is shown by plants with prostrate shoots. Here some of the lateral buds are no longer symmetrically placed but are located on the upper and lower sides. In this case, when the terminal bud gets far away, or begins to age, lateral buds will begin to develop, and there we see that, in general, the lateral buds on the upper side will develop while those on the lower side will remain inhibited. This constitutes a confirmation for whole plants of the gravitational movement of auxin to the lower side, which was established for the coleoptile in chapter 5.

K. THE ROLE OF CYTOKININS

Summarizing the discussion up to this point, it is clear that auxin does in fact inhibit lateral buds under a variety of physiological conditions and at reasonably low physiological levels. On the other hand, there were several reasons to suspect that this could not be the only factor involved. Events have shown how true this deduction was. The idea was brought to a head by the work of Skoog and Miller on tissue cultures. As we saw in chapter 8, in both tobacco callus and tobacco pith cells, not only was growth promoted by adding a cytokinin, but in some cultures buds were formed. When they held the auxin level constant and varied the level of added kinetin (kinetin being the first member of the cytokinin group to be isolated) they came to an intermediate value in which the cultures produced buds. As figure 8-18 showed, there were a large number of buds, and though they developed very slowly (for callus is rather slow-growing material) they grew side by side. With almost no cytokinin but with the auxin present, on the other hand, large numbers of roots are produced. The roots form when the ratio of auxin to the kinetin is around 100.

It struck me, although Skoog and Miller did not make a point of it, that the fact that the buds were developing so close to one another suggested that under these conditions they were not inhibiting one another's growth. This led us to restudy the whole matter of apical dominance with the aid of the cytokinins. (Dr. Skoog was good enough to supply some kinetin for the first experiments.)

After many trials we returned to that single node-internode system described above, for it gives remarkably clear-cut results. Using etiolated pea internodes bear-

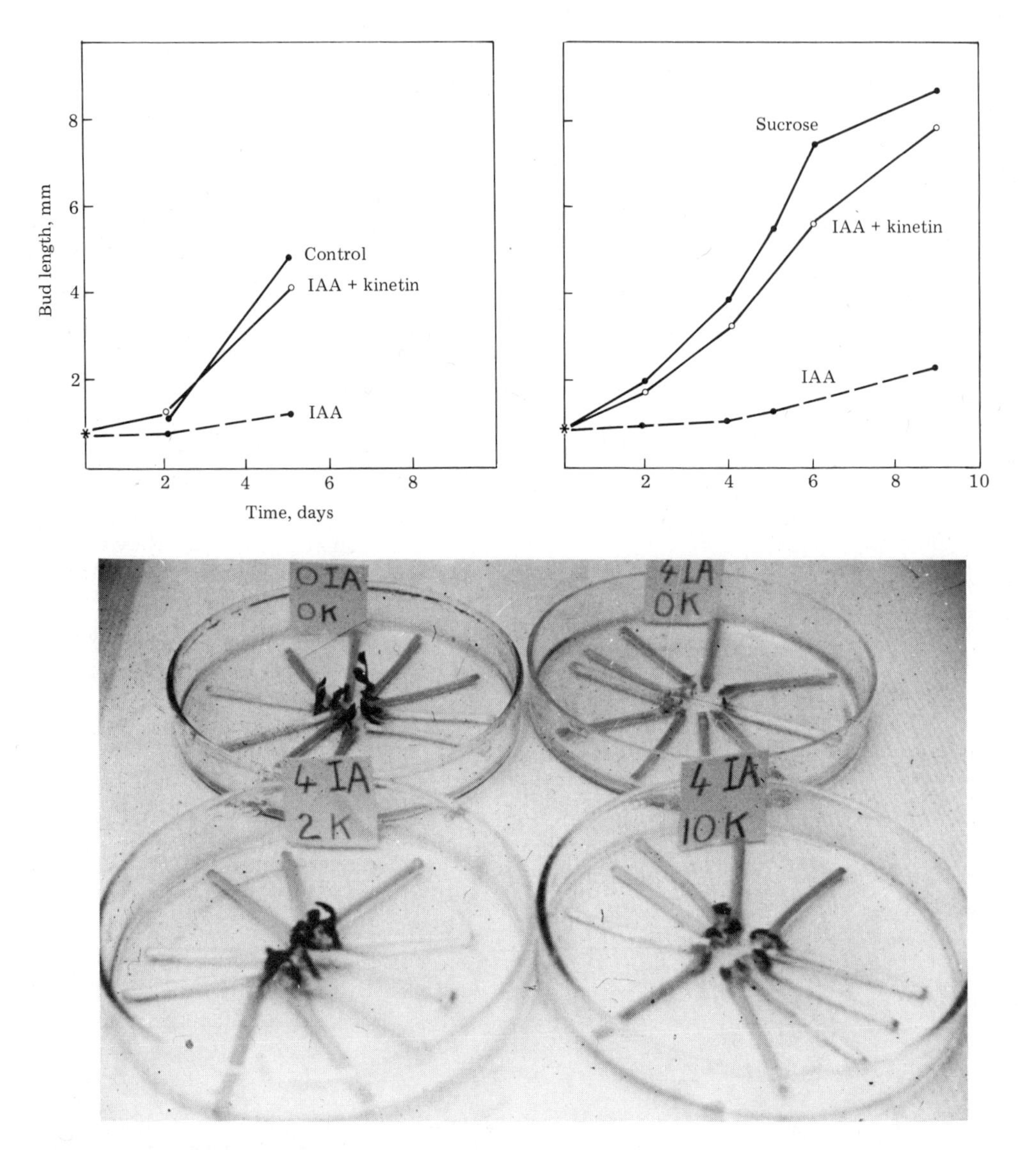

10-11 Above, release of the inhibition of lateral buds on isolated pea stem segments by kinetin; left, IAA 1 ppm; right, 3 ppm. From Wickson, 1959. Below, photograph of a typical experiment like the above; concentrations in ppm.

ing one node and a single bud, cultured in sugar solution in light over a period of 6 days, we found that auxin inhibits the growth of the bud completely. But figure 10-11 shows that with the same auxin concentration with the addition of kinetin, there can be a complete reversal of the inhibition, and the buds grow just like the control buds.

The more one studies this sytem, the clearer it is that it is an optimum system, exactly like the tissue cultures. On the controls, with nothing added, except of course the sucrose, the single bud has developed on every one. With the addition of auxin (4 ppm) the development is virtually 100% inhibited (though there is a tiny growth on one of them). With 4 ppm of auxin and 2 ppm of kinetin (it so happens that their molecular weights are nearly the same) the buds are growing exactly like the controls. Finally, however, with 5 times the amount of kinetin the effect is already less. Thus we have an optimum point at which the amount of the cytokinin bears a certain relationship to the amount of the auxin; this relationship may well depend upon the conditions, and it depends too on the specific auxin and cytokinin used, but each set of conditions seems to determine a point of balance between the two hormones.

Figure 10-12 summarizes a series of such experiments, each carried out for 6 days. Growth is plotted as percentage of the growth of controls without auxin. Auxin (IAA) at 1 ppm gives the curve at the left; auxin at 5 ppm, the next, and at 15 ppm, the curve on the right. The concentration of kinetin needed to bring the growth to 100% of that of the controls can be read off these curves. In another series, the 100% growth was reached with 1 ppm of auxin and about 1 ppm of kinetin. Five ppm of auxin takes 4 to 5 ppm of kinetin to reverse it, and 15 ppm of auxin takes about 9 ppm of kinetin. The ratio is not very far from 1 : 1, and the molecular weights are nearly the same. Thus we evidently have to do with an auxin: kinetin balance. However, the ratio does vary with the conditions, and is different in light and in dark.

There is another point to be made from this figure. In every case, the optimum combination of auxin and kinetin gives *more* growth than the controls. The increase is of the order of 25% over the growth of the controls. A similar increase can be seen in table 10-3, where I have put together experiments with two different auxins, indoleacetic acid and napthalene-1-acetic acid. With indoleacetic acid by itself the buds are inhibited to 27% of the control; then as we add kinetin we reach an optimum level and finally a decrease again. I have underlined the figure which comes nearest to overcoming the inhibition and bringing the growth back to 100%. In every case, whether the auxin is IAA or NAA, the optimum combination gives more than 100% growth, so that there is not only a balance but at the optimum levels an *increase* in the growth. And generally, as we increase the auxin by 10 times, from 0.33 to 3 ppm, we increase the kinetin needed to balance it by 10 times, from 0.5 to 5 ppm. Minor variations in the growth both of the control and of the experimental buds make the constancy of the ratio far from perfect, but it is about as good as one can reasonably expect. When the conditions are

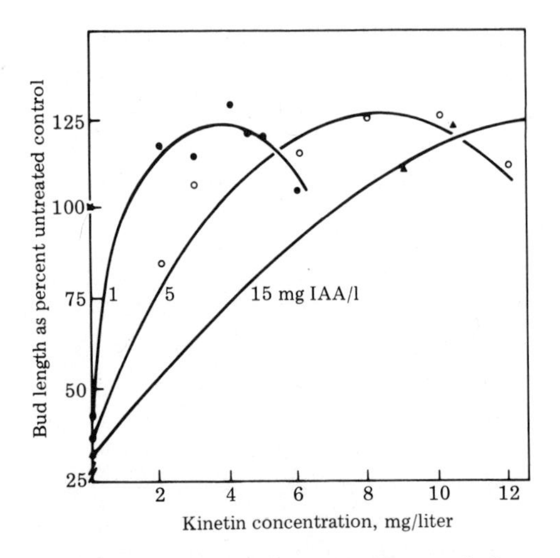

10-12 Growth of lateral buds as in fig. 11 with three concentrations of IAA and serial concentrations of kinetin. From Wickson and Thimann, 1958.

10-13 Release of dominance of the apex over lateral buds on de-rooted pea seedlings with bases in sucrose 1%. Kinetin for etiolated plants was 6 ppm; for green plants, 2 ppm. From Wickson and Thimann, 1958.

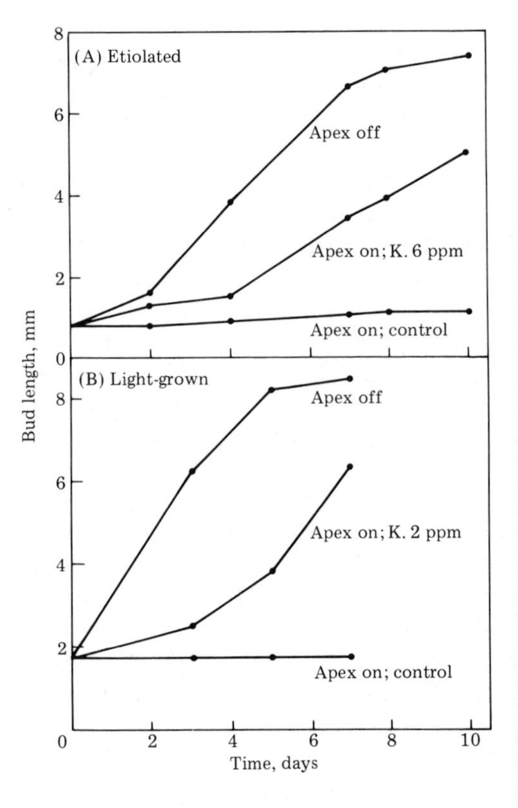

closely controlled, the growth of these buds evidently constitutes a fairly reliable, repeatable system.

I mentioned that the position of the balance point, the amount of kinetin needed for a given amount of auxin, does vary with conditions; and at Amherst, Rubinstein and Nagao have shown that the point of kinetin excess appears to vary with the age and development of the bud. Thus the balance is determined for any given system, and as the bud changes its age or size the point of balance may shift. In isolated tillers (lateral buds in the crown) of wheat, Langer and colleagues in England have found, somewhat similarly, that only when the bud is small (less than 3 mm) is its growth being inhibited by auxin. But kinetin promotes its growth whether it is young or old.

Do these experiments, with exogenous materials added, really apply to natural apical dominance? It is pertinent to try a few experiments in which the apical dominance is not provided by auxin but is exerted in the normal way by the bud itself. Figure 10-13 shows two sets of plants, an etiolated group at the top and a light-grown group below. In both cases, with the apex on there is total dominance; with the apex on and kinetin applied, the buds grow. In the light-grown plants the same thing is true. At that time we had not learned the technique of applying kinetin most effectively, but about 1 or 1½ more days would have brought the growth up to that of controls. In this case the plants were derooted and had their bases in the test solution, which, it afterwards turned out, is not the optimum means of supplying kinetin.

While these experiments were going on, much interest was developing as to the effects of gibberellin on the elongation of stems, and it was a natural question to ask, wouldn't gibberellin do the same thing? The answer is, of course, that it would not. For although gibberellin, when applied to a growing bud, will cause it to elongate very rapidly, when applied to an inhibited bud, it will in general not release the inhibition. It will only produce its growth effect after the inhibition has been released, which can be done either by removing the source of auxin or by supplying a balancing amount of kinetin. The addition of gibberellin does make a very tiny difference on the 11th day, but at that time the inhibitions are beginning to wear off. Thus the release from dominance is in fact specific for the cytokinins. Reports that gibberellin participates in apical dominance are probably

Table 10-3 The auxin-kinetin balance with IAA and NAA. Bud elongation as % of control after 8 or 10 days.

		In presence of kinetin (mg/liter):							
Auxin (mg/liter)	Alone	0.3	0.5	1.0	1.5	2.0	3.0	5.0	9.0
IAA (0.3)	27	95	147	126	93	—	—	—	—
IAA (3.0)	10	—	—	—	21	—	58	127	—
NAA (0.15)	66	—	—	—	111	—	137	140	90
NAA (0.30)	10	—	—	11	—	—	58	105	71

The numbers underlined are the elongations at the lowest kinetin concentration capable of reversing the inhibition.

to be ascribed to its effects on transport or supply of auxin (chaps. 3 and 4).

L. INFLUENCE OF THE MODE OF APPLICATION OF CYTOKININ

As it turns out, experiments on buds are very sensitive to the means of application, because if the bud once begins to develop and to elongate, it rapidly becomes harder and harder to inhibit. In this particular plant, growth to a length of 3 mm is about the limit; beyond that, auxin can slow down its growth, but it is very difficult to stop it altogether. For this reason, the means of application is extremely important. Lanolin is almost useless, as Kuse found in 1961. Of the many experiments on this, one of the first was the so-called downward-pointing strip method, beloved of the German morphogenesists and also of Robin Snow. The idea is to apply the experimental material (in our case, cytokinin) locally, but below the bud, so that translocation will bring it in. A slit is made in the stem for several centimeters and a cut halfway through the stem is made at the lower end of the slit. The resulting downward-pointing strip can then be immersed in a little vial with a small volume of solution. In the light, transpiration will occur and the material will be drawn up.

This system does not work for kinetin application, however, for the use of kinetin in this way caused the development of relatively large nodules, so that the downward-pointing strip almost looked as if it were producing tumors. In these growths there have been large numbers of cell divisions in nearly every kind of tissue. Table 10-4 summarizes a series of sections. In pith parenchyma there are about twice as many cell layers as in controls, and the cambial zones are very greatly activated. Only the cortical parenchyma is unresponsive. This observation supports the idea that in the development of cambial, and in general of mitotic, activity, cytokinins have a very large stimulating part to play. IAA has some effect alone but the total control of cell division is more complex.

Other procedures such as the spraying of the bud with aqueous cytokinin solution were equally ineffective. Instead we worked out a procedure in which one dissolves the cytokinin in an alcoholic solution and then mixes it with a suitable polymer. Carbowax (polyethyleneglycol) is very good, and this material prevents evaporation, holds the solution close to the plant tissue, and also, being aqueous, contacts the plant tissue completely so that the uptake is very rapid. For in order to reverse the dominance exerted either by the terminal bud or by auxin, the cytokinin must be morphologically extremely close to the bud. The

Table 10-4 Cell division in downward-pointing stem strips of *Vicia faba*

Tissue type	Number of cell layers: In water	Number of cell layers: In kinetin
Pith parenchyma	8	16
Secondary xylem	8	19
Cambial zone	2	6
Phloem	6	10
Tissue derived from nodules	—	21
Cortex	11	12
Total	35	84

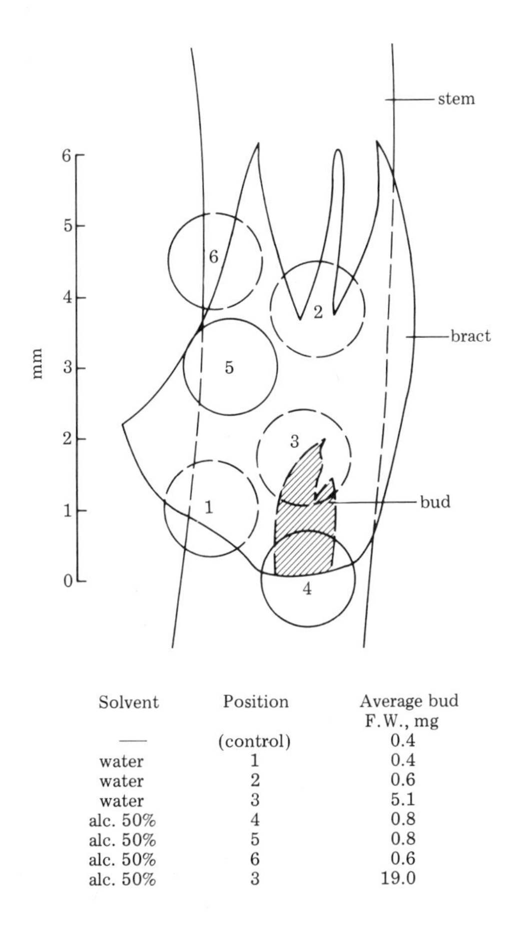

Solvent	Position	Average bud F.W., mg
—	(control)	0.4
water	1	0.4
water	2	0.6
water	3	5.1
alc. 50%	4	0.8
alc. 50%	5	0.8
alc. 50%	6	0.6
alc. 50%	3	19.0

10-14 Importance of the point of application of kinetin on lateral buds of etiolated pea seedlings in light. Bud lengths after 3 days. From Sachs, 1965.

cytokinin that is taken up is not transported either up or down the stem very well, as we saw in the case of the downward-pointing strip, and so only those arrangements in which it is supplied directly to the bud are effective. Figure 10-14 shows an intact pea plant with the inhibited lateral bud in the axil of the large stipule. The positions of droplets applied are shown on the diagram, together with the fresh weight (in milligrams) of the bud after a week. Position 2, directly above the bud, which if there were polar transport, as with auxin, would reach the bud rapidly, is no better than the control. Position 3, however, when the droplet actually rests on the young leaf primordia, gives an excellent effect. When the solvent is improved, with 50% alcohol, almost the same position (position 3) gives an excellent effect, while none of the others shows any result. Evidently there is extremely little free movement or transport of the cytokinin, which is perhaps not surprising, for there is no great reason to think that it would be readily transported. Dr. Osborne at Oxford, who has studied these matters extensively, claimed at one time that the cytokinins were very readily transported in petioles, and it is true that in her experiments, about 2% of the material was transported in about 24 hours. Veen and Jacobs at Princeton found from 3% to 7% in 24 hours. These figures show only a very small amount of transport compared to the transport of auxin, which would probably reach 60% in 3 hours. Thus the movement of cytokinins in tissue is actually very slight. Such movement as there is takes place in conducting tissue and not from cell to cell.

M. INFLUENCE OF EXTERNAL CONDITIONS ON INHIBITION AND ITS RELEASE

In discussing the apparently weak apical dominance in *Coleus*, we saw that when the experiments are repeated in a lower light intensity, there is a normal, fairly complete inhibition by auxin. Indeed, this inhibition also shows normal reversal by cytokinin. If we use a system exactly like the single node of *Pisum* but with *Coleus*, i.e., *Coleus* stem segments containing only one or two buds, the system reacts exactly like that in the pea. Light-grown *Coleus* has extremely short internodes, so that actually one has to remove a few buds, but the scars do not seem to exercise any particular influence, and with this system indoleacetic acid and other auxins all inhibit and cytokinins all release, just as in any other plant. So the suggestion that the auxin–kinetin balance holds only for leguminous plants is disproved.

In experiments of this sort we found, curiously enough, that the effect of light is independent of day length. Both short-day and long-day conditions give very different results from continuous light (fig. 10-15). In continuous light the kinetin concentration needed for release is far less than with interrupted light. It is as though the 1:1 ratio could be reduced by adding continuous light, and that again suggests that the light synthesizes a cytokinin.

Recently, in a more elaborate study of light effects, Tucker and Mansfield in England have grown *Xanthium* plants on long days under fluorescent light and then at the end of the day have given them half an hour of incandescent light. In all, they had 16½ hours of illumination. The added half-hour

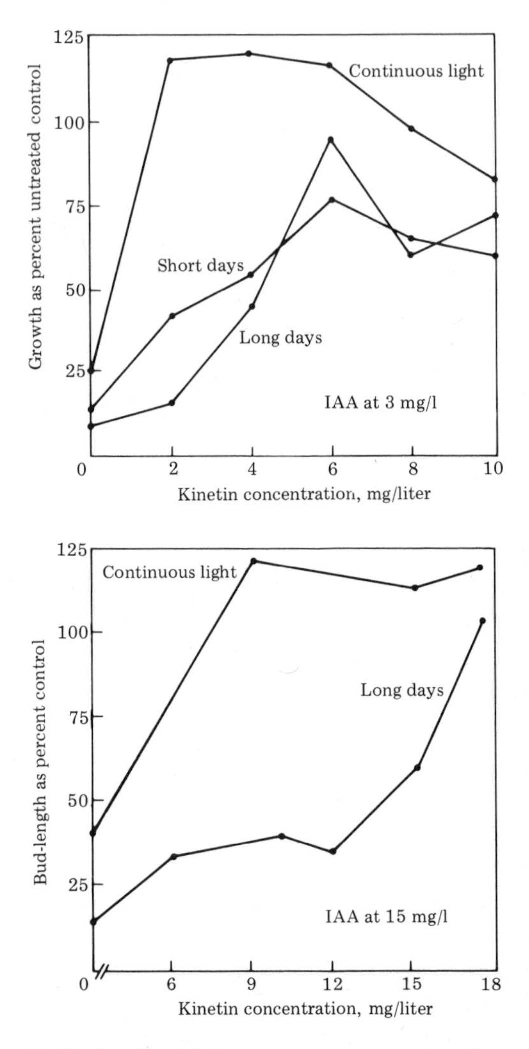

10-15 Growth of buds on pea stem segments under varied light conditions. Continuous light sensitizes the IAA inhibition to release by kinetin. Above, low auxin. Below, high auxin. From Wickson, 1959.

of incandescent light, which they regard as providing far-red light (since the fluorescent contains none) greatly strengthens the apical dominance.

The buds do grow out very slowly. The experiments extend over *weeks*, not days, in comparison with our experiments which would have been all over in the first 7 or 10 days. Only after about 2 weeks do the buds *begin* to grow out. But if the additional half-hour of incandescent light is provided, the apical dominance is fully maintained. Thus these are plants in which the dominance is evidently weak unless it is reinforced. In a series of experiments designed to find out what could produce this reinforcement, these workers have traced it to the formation of very large amounts of abscisic acid in the bud. This suggestion brings us to a third hormone in this system. Table 10-5 shows that really startling quantities of abscisic acid are produced, compared to those in the controls given 16½ hours of fluorescent light. In the apical bud and in the youngest leaves the light regime has no effect. It is in the axillary buds on the intact plant that the amount of abcisic acid is so greatly increased. After the three youngest leaves have been removed, it becomes apparent in this plant (somewhat as in our old experiments on *Ginkgo*) that the inhibition due to the leaves is greater than the inhibition due to the young apex. The amount of abscisic acid decreases in the other parts of the plant. These experiments raise an interesting set of questions which we are not in a position to answer. That is, does abscisic acid regularly participate in the dominance exerted by the apex? And in those cases where the dominance is weak, as in *Coleus* plants in bright light, is this due to the removal of abscisic acid which otherwise somehow supplements the inhibiting action of the auxin coming from the young leaves or the tip? It could be thought that high light, especially high incandescent light, exerts a powerful effect in this way, but in *Coleus* the best apical dominance occurs in weak light, not in strong light, while in Tucker and Mansfield's experiments it occurs with added (incandescent) light. So the effect would have to be exerted in opposite directions in the two species, which is improbable. Some indication of the generality of the far-red effect is given by Tucker's recent results with the tomato, which are very striking (fig. 10-16).

The large amount of AbA formed, and its very rapid fall, just in 48 hours, reminds us that the one thing we do know about AbA is the rapidity of its formation and disappearance. This was discussed in chapter 7 and will be taken up again in connection with abscission. Since AbA is an acid there is a

Table 10-5 Influence of far-red illumination at the end of the day on formation of an abscisic acid-like material in *Xanthium strumarium*. Data are ppm of dry wt.

Tissue extracted	With far-red	Without far-red
Apical bud and 3 youngest leaves	247	242
Axillary buds on intact plants	12,220	193
Axillary buds on plants whose 3 youngest leaves have been removed (measured 48 hours after leaf removal)	855	—
Mature leaves	57	66
Internodes	250	255

From Tucker and Mansfield, 1973.

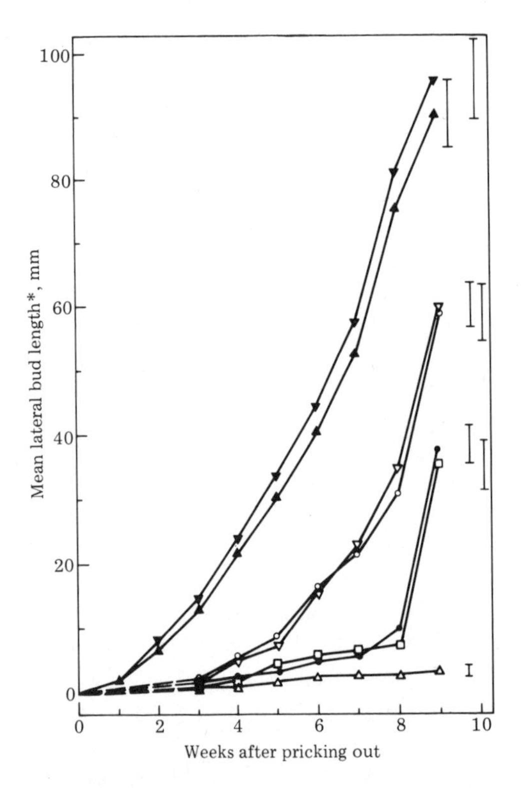

10-16 Mean lengths of lateral buds on tomato plants given 5 minutes far-red light at the end of the 16-hour day, then transferred to greenhouse after: ▽ (2 weeks); ○ (3 weeks); ● (4 weeks); or □ (5 weeks). ▼ = control in greenhouse throughout. △ = far-red throughout. From Tucker, 1975.

possibility that it is liberated by hydrolysis, and the effect of drying or wilting may have something in common with the effect of the far-red light. Tucker and Mansfield also have some data on the variation of the amounts of auxin and cytokinin in the buds; but, although the auxin content appears to be tripled, these are relatively small changes compared to the changes in AbA, and they are determined by an assay which depends upon a logarithmic scale.

We saw in table 8-1 how a number of diphenols strongly inhibit the decarboxylation of IAA and correspondingly promote elongation when IAA is the auxin used. Monophenols act in the opposite direction, promoting decarboxylation and decreasing the elongation. These effects are seen also in bud inhibition. They are at their strongest when the concentration of IAA is low, and thus would be maximal under physiological conditions in the stem. For example, paracoumaric acid, a monophenol, completely prevents the partial inhibition exerted by IAA at 0.05 ppm; conversely, chlorogenic acid, a diphenol, intensifies the inhibition caused by IAA at 0.5 ppm and even prevents the "breakaway" which would normally ensue at this auxin level after about 3 days. There is scattered information to the effect that phenolic derivatives can be both synthesized and modified by light. The formation of anthocyanins and flavonols has been studied a good deal, and these represent the best understood classes of phenolic derivatives. The formation is mainly a function of high light intensity but can be modified by the red-far-red system. Thus the role of the natural phenols in phenomena of this sort may well prove to be a major one.

N. POSSIBLE INTERACTION WITH ETHYLENE

There is one other hormone that may play a part. Our knowledge of it results from a discovery made only a few years ago, whose significance cannot yet be fully evaluated. Ethylene, although a gas, is to be regarded as a hormone in every respect. It is produced in one part of the plant, exercises its actions in other parts, and is operative at concentrations around 10^{-6} molar or less. Those properties seem to be approximately the necessary requirements for designating the substance a hormone. Since the hormonal definition does not specify that it be a solid or a liquid or anything else, there is no real reason why a gas cannot be a hormone. The reason for bringing it into this subject is that Burg has shown that when auxin is applied, or when apices are present, the amounts of ethylene normally produced in the plant are greater at the nodes than in the internodes. Figure 10-17 shows an etiolated pea plant in which the amounts of ethylene produced are given in nanoliters per gram per hour, i.e., parts in 10^9. Although the amounts are thus very small, the substance is very potent. The apical bud itself and the hook tissue produce large amounts, and the nodes produce relatively large amounts compared to the internode tissue. We know from the work of Abeles, the Burgs and others, that when somewhat high physiological concentrations of auxins are applied to plants the ethylene production is increased. In general, concentrations of auxin greater than one or two parts per million will elicit a considerable increase in the amounts of ethylene. Thus one possibility would be that when auxin is applied, or is coming from the apical bud, it stimulates the production of ethylene at the node, and it is this ethylene that inhibits the growth of the buds. In other words ethylene would be the equivalent of cyclic-AMP in animal tissues, the "second messenger," an intermediary hormone. If this were true, it should be enough to place the buds under reduced pressure, e.g., ½ or ¼ atmosphere pressure (suitably adjusting the oxygen tension), to draw off the ethylene, since both Burg and Abeles have shown that reduced pressure draws ethylene out of the tissue very quickly. Then of course all the inhibited

10-17 Distribution of ethylene production rates in 7-day-old etiolated pea seedlings. Rates measured from 6th to 30th hour. From Burg and Burg, 1968.

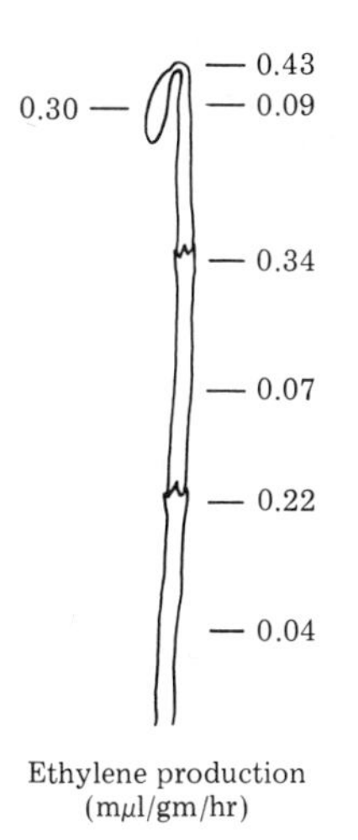

buds should burst out. Dr. Burg has done this and you can guess the result: it has no effect. Lateral buds do not develop in intact plants under what is called "hypobaric" (lowered pressure) conditions. If the inhibition is due to auxin in lanoline applied to decapitated plants, lowered pressure does cause some release, because these are conditions where the auxin level is high enough to induce ethylene formation. Thus ethylene is apparently not the determining cause in natural apical dominance.

O. ROLE OF THE CONDUCTING SYSTEM

I will take up now a set of observations which at one time looked like explaining the morphological basis of the inhibition of buds; indeed it may still do so. These observations concern the development of the vascular system leading to the axillary bud. A bud is unlikely to develop to any extent unless it is well supplied with phloem and xylem, since it must have supplies of sugar, nitrogen and water. Axillary buds, we know, are poorly supplied with vascular bundles; and it is of great interest that when one removes the auxin source, or balances it with a cytokinin, the vascular system to the bud develops very rapidly.

Figure 10-18 shows, at left, an inhibited bud on an isolated stem preparation after 120 hours in auxin. The bud traces (arrows) show no xylem whatever, although the stem below has strong bundles. After 72 hours some buds, especially on the intact plant, may show modest enlargement of the leaf primordia, but they show virtually no development of a vascular strand. The xylem normally develops in two directions: acropetally, towards the bud from below; and basipetally, growing downwards from the base of the bud. Buds on intact pea plants at this age show little or none of the latter.

The right-hand figure is a section through a bud which has been released with kinetin and is rapidly elongating. It has developed excellent xylem strands and probably phloem too, though this is hard to see. The basipetal and acropetal strands have joined and make contact directly with the main xylem in the stem. Such contact takes in general 48 to 72 hours to develop. When cytokinin is supplied, basipetal xylem begins to differentiate, acropetal xylem comes up to meet it, and as far as we could determine they make contact about at the time that appreciable bud elongation takes place. A view of the newly formed xylem at higher magnification is shown below on the left. The question is, are these differentiations causative or are they only an accompanying phenomenon? There are a number of complications. The first is that sometimes xylem makes a very weak connection, perhaps a single strand, and still the buds do not grow out. This may be because the xylem formed under the influence of auxin is at first not functional; it appears to be clogged with colloid. Also it is not normally lignified, but when cytokinin is added, it changes its quality, the lignification becoming scalariform, and we have deduced that it then becomes much more functional. The situation is complicated by another factor. We have seen that ethylene is probably produced close to the inhibited bud, and one of the effects of ethylene is to inhibit the differentiation of xylem (fig.

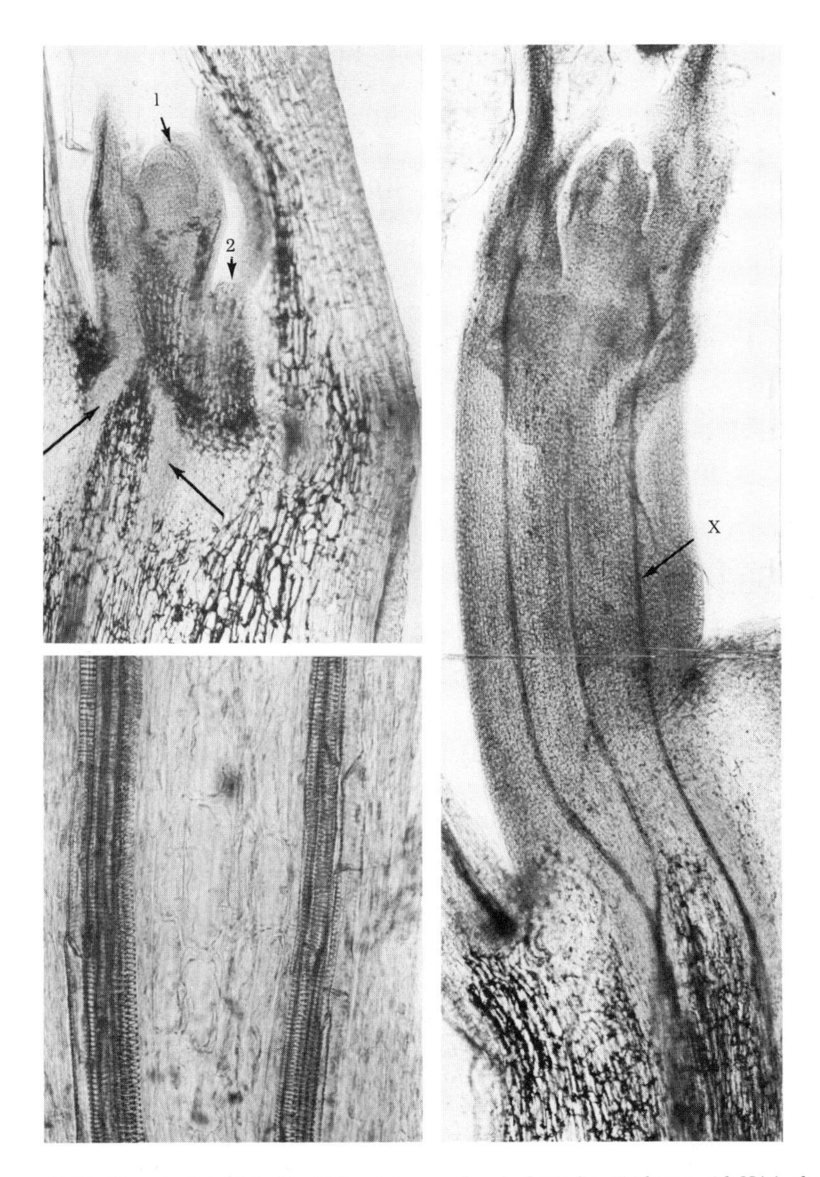

10-18 Above, left: inhibited bud (1) and its secondary bud (2) after 120 hours with NAA; the bud traces (arrows) are devoid of xylem; neither bud has enlarged. Above, right: released bud after 72 hours with same NAA plus kinetin 5 ppm; complete xylem connections (X) joining the vascular bundles in the stem. Below, left: close-up of the vascular strands on the right, showing helical thickening on inner side and scalariform thickening on outer side; all buds given sucrose. From Sorokin and Thimann, 1964.

9-11). It greatly decreases the size of vascular strands in growing plants, and it largely inhibits lignification. These phenomena may be the normal accompaniment of bud inhibition and outgrowth, but, if they *are* the normal accompaniment, they may still not be a *necessary* part of it, but merely the indications of the presence of ethylene.

When the bud of the Alaska pea has been inhibited for several days with auxin, although it does not grow, hairs (technically, *trichomes*) develop. But hairs are normally never seen either on the buds or on the young leaves of Alaska peas. They are typically glabrous at all stages, young or old. The development of these trichomes thus seemed to us rather unexpected until Burg showed that when plants are treated with ethylene, both trichomes on the shoot and root hairs on the root develop strongly. Hence the appearances of these trichomes on the bud may support the idea that perhaps some of the phenomena of bud inhibition are merely secondary changes brought about by the ethylene liberated by the auxin treatment.

The role of the vascular strands, however, does appear to be of prime importance, and some most interesting and suggestive results have been obtained by my former colleague Tsvi Sachs, now at the Hebrew University in Jerusalem. In general Sachs has found that as soon as inhibition is to be released, whether by cytokinin or by other methods, the development of vascular strands is extremely rapid. And in following up the problem of the connection of vascular bundles between the bud and the main stem he has made a strange discovery.

Figure 10-19 shows the stem of a pea plant which has been decapitated and to which auxin has been applied on the cut surface a little to the right of the main bundle. Here, after one week, auxin has induced the development of a new vascular bundle, which cuts across the parenchyma and may or may not make contact with the existing main bundle. The finding is this: if no auxin, or only a relatively low auxin concentration, is applied to the main bundle, then the induced strand will make contact (A and C in the figure). If the auxin on the main bundle is equal to, or not too much lower than, the concentration of auxin inducing the new bundle, then the strands of the induced bundle will be "reluctant" to make contact with the main bundle (B and D in the figure). It is as though around a strand which is engaged in transporting auxin there is an atmosphere of repression which prevents the junction from taking place. Again, that strongly suggests the evolution of a small amount of ethylene, perhaps only enough to diffuse 4 or 5 cell-widths away, but enough to inhibit the last stages of differentiation of xylem so that it does not make vascular contact with the main bundle. One can see how the stream of auxin coming from the apex might thus prevent the connection to the bud and thus prevent its growth. This is a very interesting system and a good example of how, when we start to study the action of one hormone, we find ourselves saddled with an interaction between several different hormones. No one hormone operates alone.

At one time we thought it of interest to see whether in the development of buds the supply of materials is a controlling factor,

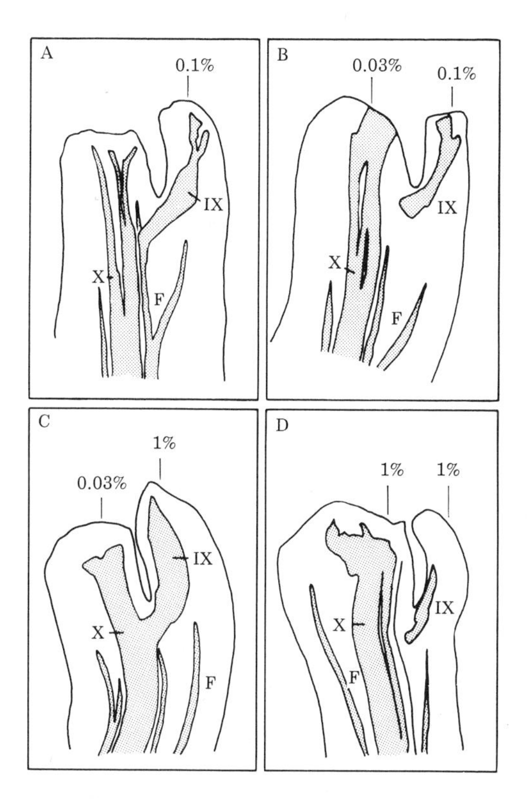

10-19 Experiment described in text. X, main xylem bundle; IX, induced xylem; F, leaf traces. IAA as percent in lanoline. From Sachs, 1969.

since, if the vascularization is of critical importance, one ought to be able to show that ordinary nutrients (sugars, amino acids, etc.) are going into the buds. One set of such experiments is illustrated in figure 10-20. The growth of the buds, measured in fresh weight rather than in length, is shown for three classes of buds: the uninhibited controls, the buds inhibited by auxin application, and the buds whose inhibition is reversed by cytokinin. As far as fresh weight is concerned, the controls and the cytokinin treatment go together, while the inhibited buds naturally have less weight—although after some days they "break away" (as they often do if the auxin concentration applied is a little low), and fresh weight begins to increase.

In dry weight the kinetin-treated buds did not reach the value of the controls. In protein-nitrogen there is rather little difference; they both go up rapidly together and again, when the breakaway occurs, the protein-nitrogen falls off. And the same is true roughly for the amino acid-nitrogen: the controls and the kinetin-released buds move in parallel—with a little variation, but they go well together—and the inhibited buds stay behind. So, roughly one can say that the supply of what the Germans call plastic substances, the materials for growth, more or less goes along as one would expect. There is little basis for assigning cause and effect to the nutrients.

The last consideration to be brought up here is the role of cytokinin in the normal plant. It is generally believed, following the demonstration of the presence of a cytokinin in guttation fluid, that in the normal plant cytokinin is synthesized in the roots,

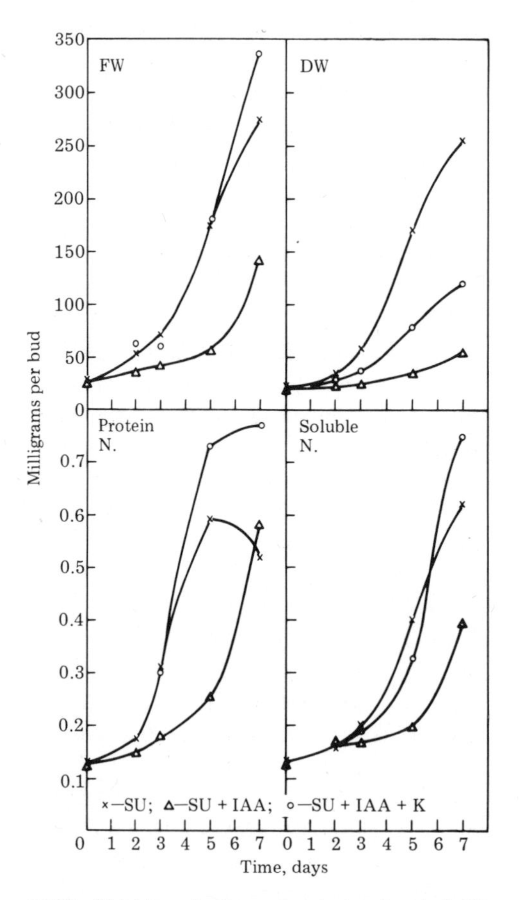

10-20 Weights and nitrogen contents of control (X), inhibited (△), and released buds (○) on pea stem segments in sucrose solution. Concentrations: IAA, 1 ppm, kinetin 4 ppm. From Thimann and Laloraya, 1960.

along with a number of amino acids. These would all be carried up to the shoot part of the plant through the xylem. Analysis of the xylem sap of woody plants does show amino acids in it, as Ballard, working in New Zealand, first showed. The amounts are not large, but xylem sap represents a large volume, so that even if the concentration of amino-nitrogen is low the xylem sap may constitute an important source. Also we know that the nitrate reductase system is very active in roots, so that as nitrate is taken into the root tissue a great deal of it would be reduced therein. Free nitrate occurs in leaves only under exceptional conditions, because most of the reduction takes place before the nitrogen reaches the leaves. This being so, one might expect that the roots would be of great importance in apical dominance. Specifically one would expect it would be much more difficult to release the buds in derooted plants than in rooted plants, because rooted plants would have a native supply of cytokinin coming from the roots, which derooted plants would not have. Actually, the amount of cytokinin needed to reverse the inhibition, as we saw in the little experiment with application to the stem, is not especially high, provided application is made directly to the bud; the presence of roots seem to exert only a very minor influence, if any. Indeed, the success of this single node system, in part, was that it showed that the roots could be dispensed with and a very normal-looking apical dominance situation remains.

However, we have just seen that light exerts a considerable influence on inhibition and its reversal, the presence of light essentially acting as though it was supplying part of the cytokinin. This indicates very strongly that cytokinin may be produced in light; if so, it would not be produced in the roots but in the buds themselves. A few years ago an interesting antibiotic called Hadacidin was discovered; it never came into use as a medical antibiotic for some reason, but it was produced in small amounts at Merck's and the details published in the mid-1960s. This very simple molecule, N-formyl-N-hydroxyglycine, occupies positions in space rather similar to those of aspartic acid (see formulae). Thus one might suspect that it would be a competitor for aspartic acid in some system.

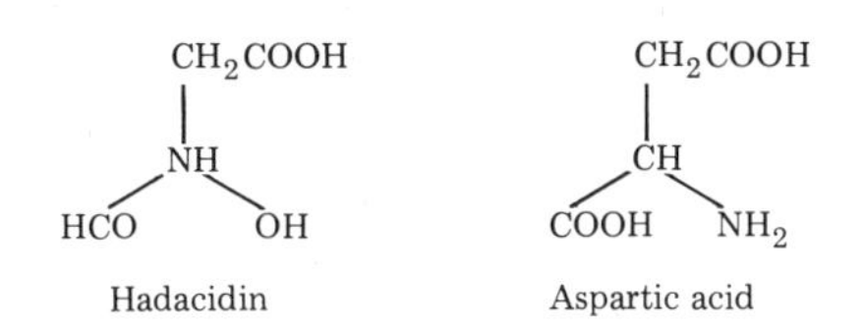

And so it turns out to be. The reason why it acts as an antibiotic is that it interferes with a particular enzyme, adenylosuccinic-synthetase, which takes a molecule of succinic acid and applies it to a molecule of inosinate, as an intermediate step in the synthesis of the purine ring. It was not until some years after this had been published that I realized that if cytokinin is to be produced locally, it is presumably produced by the same pathway as other purines. Hence it would be interesting to see if Hadacidin exerted any influence on apical dominance.

To test this idea, whole pea plants were decapitated so that the lateral buds would grow out. Roots were present as usual. Figure 10-21 (above side) shows the growth of bud 1 for about 10 days, and also the growth of bud 2 which usually is slower, since peas are not basitonal generally. On adding Hadacidin, buds 1 and 2 are both totally inhibited; there is no sign of release of inhibition in 10 days. If we then cut off the grown buds and now apply Hadacidin plus kinetin to buds 1 and 2, they begin to grow, and grow as fast as the previously inhibited bud 3 on controls. Thus these buds are not killed by Hadacidin, but truly inhibited, and they remain able to grow. Thus the inhibition is exerted by Hadacidin and can be released by cytokinin, very much like the inhibition due to auxin. Indeed, when the Hadacidin is mixed with a cytokinin from the start, the inhibition is largely prevented (fig. 10-21 below).

Now in the adenylosuccinic synthetase system, aspartic acid, given in sufficient quantities, will partially reverse the inhibition exerted by Hadacidin. And we see in figure 10-22 that the inhibition of bud growth due to Hadacidin can also be partially reversed by aspartic acid. Unfortunately, it takes rather high concentrations to do it. A relatively large amount of aspartic acid produces a *small* reversal, but larger amounts of aspartic acid produce a very satisfactory reversal. Aspartic acid by itself has no promotive effect, and the aspartic acid points simply fall along the control curve, so that I plotted the same curve for both. Thus we are dealing with a genuine partial reversal. However, the obvious question that comes to one's mind is, are we interfering with the synthesis of *all*

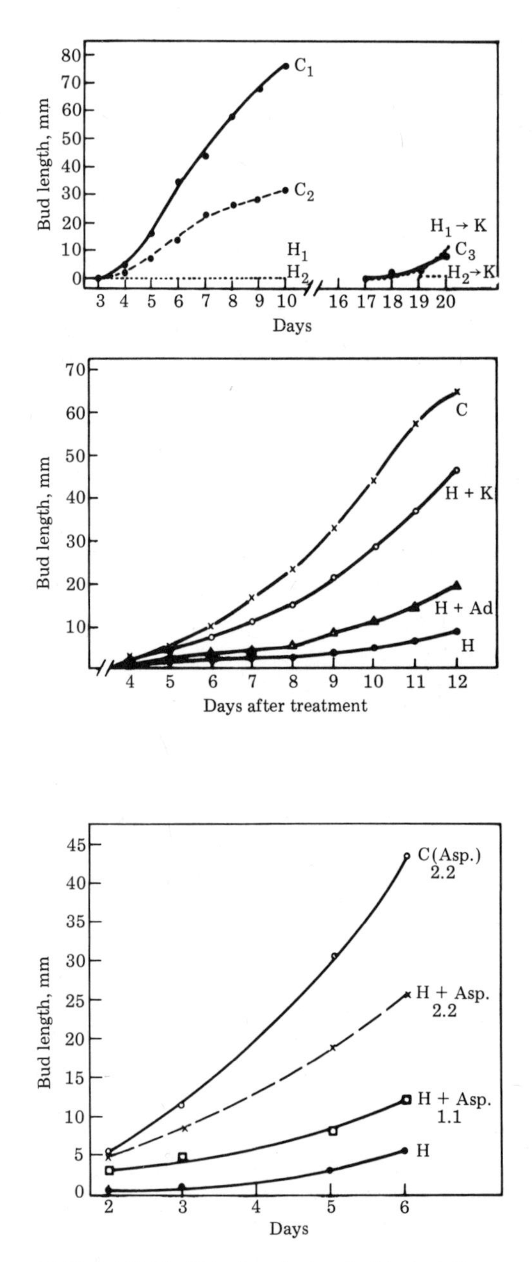

10-21 and **10-22** Experiments described in text. From Lee, Kessler, and Thimann, 1974.

purines, or is the inhibition of synthesis specific for cytokinins? Growing buds naturally need ATP and other purines. However, adenine by itself produces very little reversing effect. In a few experiments we have obtained a barely significant reversal with the maximum amount of adenine. Thus Hadacidin is not primarily exerting a general effect on purines, but a rather specific effect on cytokinins. But the most important deduction is that Hadacidin is obviously not being transported down to the roots and there influencing the formation of cytokinin, which then comes back up, for several reasons. First, the experiment works equally well without roots at all. Second, when the Hadacidin is applied to only one bud, that bud is inhibited and the bud below is not. (The method of application was again critical; the use of carbowax and alcohol was found to be essential in the cytokinin mixture.) For if the Hadacidin were being transported rapidly down the plant, then the lower bud should have been inhibited. Such local inhibition, not exerted on the second bud, is a result quite opposite to what would be seen if, for example, IAA were the inhibitor. From that, Lee and I deduced that Hadacidin is only very weakly, if at all, transported in tissue, and that it exerts its effect locally. If that is true, it means that the cytokinin must be produced locally. These buds are in light. Buds in darkness show more extreme apical dominance, and it was noted above that there was reason to believe that light acts to produce cytokinins; if so it must be acting *in the lighted part* of the plant, not in the roots.

A similar conclusion can be drawn

(though the authors do not draw it) from some recent experiments by Hewitt and Wareing (1974) on buds of woody plants. When poplar or willow buds are about to burst in the spring, their cytokinin content, and that of the sap, reaches a peak a week or two before bursting. The point is that this peak occurs also in *excised* twigs, so that the cytokinin increase can hardly be coming up from the roots. So both from the Hadacidin experiments and these studies of spring opening, we have to conclude that whether or not some cytokinin comes from the roots, a considerable supply—perhaps in the case of apical dominance the critical supply—is made directly in the bud and is made under the influence of light.

REFERENCES

Albaum, H. Inhibitions due to growth hormones in fern prothallia and sporophytes. *Amer. J. Bot.* 25, 124-133, 1938.

Burg, S. P., and Burg, E. A. Auxin-stimulated ethylene formation; its relationship to auxin-inhibited growth, root geotropism and other plant processes. In *Biochemistry and physiology of plant growth substances*, ed. F. Wightman and G. Setterfield, pp. 1275-1294. Ottawa: Runge Press, 1968.

Champagnat, P. Les corrélations d'inhibition sur la pousse herbacée du Lilas. *Bull. Assoc. Phil. d'Alsace et de Lorraine* 9, 36-38, 54-56, 1950-51.

———. Recherches sur les "rameaux anticipés" des végétaux ligneux. *Rev. Cyt. Biol. Veg.* 15, 1-51, 1954.

Fisher, J. B., Burg, S. P., and Kang, B. G. Relationship of auxin transport to branch dimorphism in Cordyline, a woody monocotyledon. *Physiol. Plantarum* 31, 284-287, 1974.

Gregory, F. C., and Veale, J. R. A reassessment of the problem of apical dominance. *Symp. Soc. Exp. Biol.* 11, 1-20, 1957.

Havránek, P. Dissertation, Brno, 1931; cited by Dostál, R., Korrelationswirkung der Speicherorgane und Wuchsstoff. *Ber. Deut. Bot. Ges.* 54, 418-429, 1936.

Hewitt, E. W., and Wareing, P. F. Cytokinin changes during chilling and bud burst in woody plants. In *Mechanisms of regulation of plant growth,* ed. R. L. Bieleski, R. Ferguson, and M. M. Cresswell, pp. 693-701. Wellington, New Zealand: The Royal Society, 1974.

Jacobs, W. P. Comparison of the movement and vascular differentiation effects of the endogenous auxin and of phenoxyacetic acid weed killers in stems and petioles of Coleus and Phaseolus. *Ann. N.Y. Acad. Sci.* 144, 102-117, 1967.

Lee, P. K.-W., Kessler, B., and Thimann, K. V. The effect of Hadacidin on bud development and its implications for apical dominance. *Physiol. Plantarum* 31, 11-14, 1974.

Leopold, A. C. *Auxins and plant growth.* Berkeley: Univ. of Calif. Press, 1955.

Overbeek, J. van. Auxin distribution in seedlings and its bearing on the problem of bud inhibition. *Bot. Gaz.* 100, 133-166, 1938.

Plich, H., Jankiewicz, L. S., Borkowska, B., and Moraszczyk, A. Correlations among lateral shoots in young apple trees. *Acta Agrobotanica* 28, 131-149, 1975.

Reed, H. S., and Halma, F. F. On the existence of a growth-inhibiting substance in the Chinese lemon. *Univ. Calif. Publ. Agric. Sci.* 4, no. 3, 99–112, 1919.

Sachs, T. On the mechanism of apical dominance. Dissertation, Harvard University, 1965.

———. Polarity and the induction of organized vascular tissues. *Ann. Bot.* 33, 263–275, 1969.

———, and Thimann, K. V. The role of auxins and cytokinins in the release of buds from dominance. *Amer. J. Bot.* 54, 136–144, 1967.

Snow, R. The correlative inhibition of the growth of axillary buds. *Ann. Bot.* 39, 841–859, 1925.

———. The young leaf as the inhibiting organ. *New Phytol.* 28, 345–358, 1929.

Sorokin, H., and Thimann, K. V. The histological basis for inhibition of axillary buds in *Pisum sativum*, and the effects of auxins and kinetin on xylem development. *Protoplasma* 59, 326–350, 1964.

Stoddart, J. L., and Lang, A. The effect of daylength on gibberellin synthesis in leaves of red clover *(Trifolium pratense L.)* In *Biochemistry and physiology of plant growth substances,* ed. F. Wightman and G. Setterfield, pp. 1371–1383. Ottawa: Runge Press, 1968.

Thimann, K. V. On the nature of inhibitions caused by auxin. *Amer. J. Bot.* 24, 407–412, 1937.

———, and Laloraya, M. M. Changes in nitrogen in pea stem sections under the action of kinetin. *Physiol. Plantarum* 13, 165–178, 1960.

———, and Skoog, F. On the inhibition of bud development and other functions of growth substance in *Vicia faba. Proc. Roy. Soc. B.* 114, 317–339, 1934.

Tucker, D. J. Far red light as a suppressor of side shoot growth in the tomato. *Plant Sci. Letters* 5, 127–130, 1975.

Tucker, D. J., and Mansfield, T. A. Apical dominance in *Xanthium strumarium. J. Exptl. Bot.* 24, 731–740, 1973.

Wickson, M. Control of the growth of lateral buds in the pea. Dissertation, Harvard University, 1959.

Wickson, M., and Thimann, K. V. The antagonism of auxin and kinetin in apical dominance. 1. *Physiol. Plantarum* 11, 62–74, 1958; 2, ibid 13, 539-544, 1960.

The Role of Hormones in Flowering and Fruiting 11

FLOWERING

The subject of flowering is very large and full of complications; it is curiously overexplored in many areas and underexplored in others. All we shall deal with here is the role of hormones in flowering. But in order to do that we first have to make a brief survey of the entire subject.

Unlike other areas of hormonal action, which are full of facts—sometimes too many facts—and elaborate experiments, this field has been curiously dominated by a sort of religion. This religion consists of dogmatic belief on the part of most botanists and plant physiologists, in spite of the lack of evidence, in a flowering hormone. We shall see that the situation is really very complex and is not to be solved by oversimplifications of this kind. For the flowering hormone or "florigen," in spite of our 40 years of stubborn clinging to the idea, has not in fact appeared.

A. GENERAL FACTORS CONTROLLING FLOWERING

Flowering seems to be controlled in the majority of plants by one of two factors. The first is a very simple one on which we need not spend much time—age. The age factor is especially noticeable in trees. Fruit trees such as pears and apples, although they may be very vigorous vegetatively, do not flower for two or three years or sometimes more. *Pinus strobus,* our local white pine, takes around ten years before it forms cones, and several other pines are comparable. So here we have a simple phenomenon of age, about which no one has really done anything. Nor is it obvious what could be done.

It is well known also that in general there is a rough antithesis between vegetative growth vigor and flowering. Dwarf fruit trees are well known to be earlier flowering than their tall growing sibs. Thus one of the great advantages of the dwarf apples developed at the East Malling Research Station in England is not only that the trees are dwarfed, but that they flower and set fruit several years earlier than the normals. These trees are not genetically dwarfed themselves, but are grafted on to dwarfing root stocks, which change the whole subsequent vegetative growth and promote reproduction.

The second important factor is light.

About light, of course, one can do a great deal. By flowering we mean the formation of flower primordia, not, as a rule, the opening of flowers. In some cases there are special factors which influence the opening of flower buds, and these are mainly environmental. Some flower buds open as a function of temperature or moisture or under the control of an inherent biological rhythm.

The control of flower formation by light is mainly a function of the duration, and not the intensity, of the light. The duration of light means the length of day, and all plants are divided into three main classes in this respect: (1) plants that flower when the day is longer than a certain minimum; (2) plants that flower when the day is shorter than a certain maximum (that maximum is different for different plants); and (3) plants whose flowering is not dependent on the length of day, but which flower equally well on a 1-hour day or a 24-hour day.

The situation is easily shown by a simple graph in which we plot the number of days required for flowering against day length, or hours of light. The short-day plants will in general use a minimum number of days to flower when the day length is around 6 to 8 hours (fig. 11-1). When the day length is still less they may suffer from shortage of light. This last is not primarily the effect of duration, but simply that with very short

11-1 Number of days required for flowering. Solid lines, short-day plants; dashed lines, long-day plants. From Melchers, 1952.

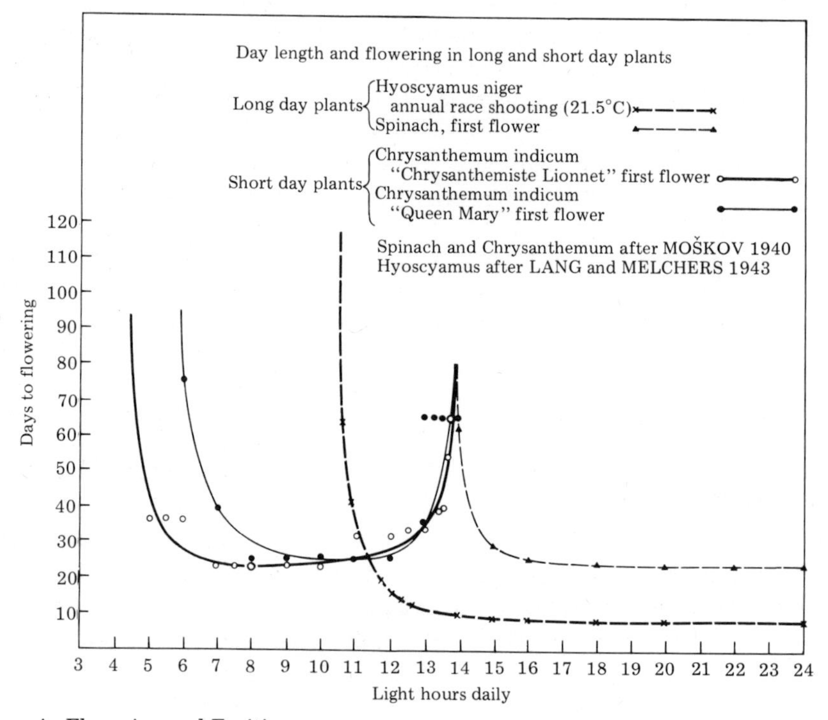

day length over a period of time it is difficult to provide adequate light; thus with very short days the curve goes off into space and the plants are virtually etiolated. The interesting thing is the other end of the curve: if the hours of light exceed a certain maximum, the number of days to flowering goes up extremely rapidly and finally the curve goes to infinity. Such plants will essentially never flower. The particular plant shown here will practically never flower on a day length of 11 hours, (though it does depend to some extent on temperature, so that sometimes this upper part can be modified).

On the other hand another type of plant requires a minimum time to flower when on long days—12, 14 or 16 hours. This type will show the opposite behavior, in that as the day length gets shorter they take longer to flower, and finally the curve may or may not go quite straight up, so that some plants may flower after *very* long times in a short day length. The exact position of these limiting values will vary with the particular plant; some plants that flower best on short days will tolerate surprisingly long periods, up to 13 hours or so, so that we cannot define short days by 8 hours or any such fixed value. We can say merely that short-day plants flower on less than a certain maximum, and that long-day plants will flower on more than a certain minimum. It is understood that there are modifications of various kinds, but because duration is the main factor, the response is called photoperiodism.

One such modification is that of plants whose response to day length is quantitative rather than qualitative. Thus Lona, at Parma, showed as early as 1946 that of two *Xanthium* (cocklebur) species, one has a qualitative response and the other a quantitative one. *Xanthium italicum* will not flower on day lengths greater than 16½ hours; it is a typical short-day plant. But *Xanthium strumarium* has an early-flowering variety which will flower on all day lengths from 6 to 24 hours; its flowering is merely *accelerated* by short days.

Other modifications are of different kinds. *Ipomoea,* which flowers on short days, develops tendrils on long days, changing its habit of vegetative growth slightly. Some varieties of dwarf beans behave similarly and even develop the climbing habit. The change from a rosette to a vertical type of growth is the most common example of this category. Winter cereals, wheat, barley, and oats, are special cases of another sort, in that they do not flower well (or usually do not flower at all) unless they have experienced a cold spell during the early period of growth. In general there is no defined relationship between day length and the form of vegetative growth, for some long-day plants remain as rosettes on short days and others do not. Thus we are not dealing here with the classical antithesis between vegetative and reproductive growth.

The most important experimental work on the hormonal control of flowering began about the middle of the 1930s with two Russians, Chailakhian and Moškov. Curiously enough, working in different institutions and with different plants, they actually made very similar discoveries. The essential experiment was based on this question: must the whole plant receive the short days? Therefore in Chailakhian's first experiments he covered up the bud whose conversion to the flowering state was being studied and simply illuminated the leaves. He found that if the leaves were given short days the plants would flower even though the bud itself that was going to produce flower primordia had not been exposed to short days. This led to a whole series of more elaborate experiments of which the neatest is this one: from a (short-day) plant with a terminal bud all the leaves except one are removed. This one leaf—the rest of the plant can be covered up—is exposed to 8-hour days, 16-hour nights, and then the plant is returned to long days. After a while the terminal bud becomes a flower. So only one leaf needed to be exposed to the specific day length for flowering to be induced.

However, Chailakhian found that if he exposed this leaf to an eight-hour day, but other leaves on the plant, along with the bud, were in 16-hour days (or even if the bud was in darkness), then the effect might be neutralized. In other words, a group of leaves receiving the noninductive day length could antagonize the leaf that received the inductive day length. Thus not only must the inductive day length be received by the leaves, but the result depends on a *balance* between leaves receiving different amounts of light. Another perhaps more important conclusion that we can draw from this type of experiment is that the stimulus is *transported.* The stimulus is received in a leaf and the response is at the apex. So just as in Darwin's classical experiment where the stimulus of phototropism is received in the tip and responded to lower down, here we have the same thing, except that, as it happens, the opposite polarity is involved. We derive several concepts from these relatively simple experiments: the idea of a receptor, the idea of transport, and the idea of a balance within the plant.

B. BIRTH OF THE FLORIGEN IDEA

The occurrence of transport seems to point so directly to a flowering hormone that Chailakhian as early as 1936 actually wrote a book (in Russian) called *The Hormonal Theory of Plant Development,* which was devoted to the thesis that since growth in plants is controlled by a growth hormone, so flowering must be controlled by a flowering hormone.

Moškov did another interesting experiment. He detached one leaf and then reattached it to the petiole with a glass vial filled with liquid, which was essentially the equivalent of Boysen Jensen's 1911 experiment (in which he cut off the tip of the coleoptile and replaced it in contact, and the phototropic stimulus was transmitted). Moškov claimed that in this way the stimulus could be transmitted to the bud through a discontinuity, hence that there was no need for a tissue connection. But although most of his other experiments were very good, this particular one could never be confirmed. Several people tried, both in Russia and the West, without success; Leopold particularly made a number of good trials, and we have to conclude that a flowering stimulus is not transmitted across a discontinuity. The many experiments in which the stimulus evidently crossed a graft depend for their success on the tissues' actually growing together, i.e., the graft must "take." Hamner and Bonner's experiment in which the graft partners were separated by lens paper (see fig. 11-2 below) was afterwards shown to work only when cells grew through the apertures in the lens paper to effect true junction.

It must be said at once that while many of these ideas certainly point very strongly to a hormone, this last negative tends to point against it. It must also be noted that when we classify a plant as a short-day plant we often find that the induction of flowering by short days can be modified by other circumstances, and especially by temperature. For instance the tobacco cultivar Maryland Mammoth, which is listed as a short-day plant, is a short-day plant at the normal temperatures at which tobacco plants grow (which are fairly high, around 24°–30°), but if one persists in growing it at low temperatures, at 12° or 13°, it does not show any short-day characters, and while it flowers only slowly it flowers equally in short and long days. It essentially becomes a day-neutral plant. This is not the only example. *Xanthium* is one of the most classical short-day plants, yet *Xanthium pennsylvanicum* can actually flower on a 16-hour day—a typical long day—provided part of the light time is given at low temperatures and only part is given at normal growth temperature. Thus if *Xanthium* is grown on 16 hours of light and 8 hours of darkness the controls would not flower at all, but if 8 of the lighted hours are given in the cold room and the other 8 hours are given in the greehouse, then it flowers perfectly well. It is as though the plant did not *notice* the light if the temperature were low enough. Thus the day length must be provided under the "normal growing conditions" for that type of plant.

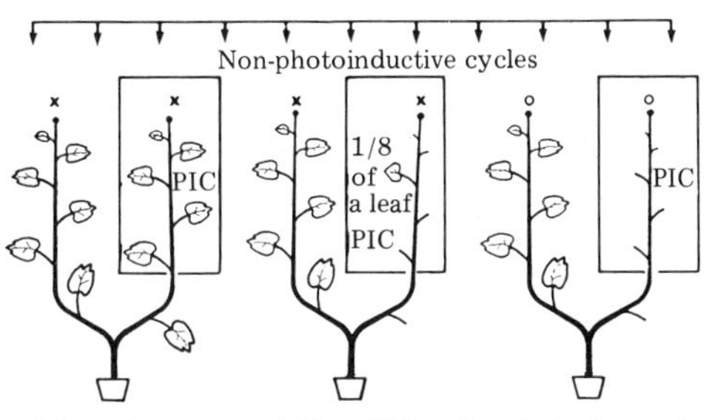

11-2 Responses of two-shoot plants of *Xanthium pennsylvanicum* (cocklebur), one being on long days and the other on short (i.e., photo-inductive cycles, PIC). With no leaf blades the photo-induction is not perceived and therefore nothing is transmitted to the other shoot; with only one-eighth of a leaf blade it is both received and transmitted. From Hamner, 1942, after Hamner and Bonner, 1938.

The experiments of Chailakhian and

Moškov were greatly extended by Hamner and Bonner, working with *Xanthium* (fig. 11-2). Among the things they showed was that one does not need even one leaf to get the flowering response. They found that one could take off half that leaf and still place the remaining attached half in an 8-hour day and the rest of the plant (with no other leaves) in 16 hours, and flowering would still take place. So then, having halved the leaf, they quartered the leaf, and still it was able to detect. Even one eighth of the leaf was sufficient to detect the stimulus and induce flowering in the bud. These experiments certainly strengthen the concept that there can be a photoreceptor area and a responding area, thoroughly well separated from one another.

C. TIMING, SPECTRAL SENSITIVITY, AND RHYTHMS

The other point that Hamner and Bonner established concerned the 16 hours of night—must this be continuous? They tried breaking the night with a brief illumination in the middle of the night, or at various times in the night. They found that if, for instance, the plants receive 8 hours of light and then 1 hour of dark, and then again another hour of light, the plants still flowered. But if this second patch of light were given later, say after 6 hours of darkness, then the plants refused to flower. It is as though they did not notice the intervening dark period and considered that because it was still light after 6 hours it must have been light all the intervening time; as a result they regarded it as a 14-hour day, which is not the condition under which they flower. The plant, in other words, behaves as though it *integrates* the illumination over an intervening dark period. In the first case, it integrates the 8 hours of light and 1 hour of dark and 1 hour of light, which still means only a 10-hour day, on which it will still flower nicely. This is of course an oversimplification, and there are many more complex experiments; in fact one of the disadvantages of this whole field is that many of the experiments are extremely complex.

In the early period there came also a remarkable discovery of Flint and McAllister with lettuce seeds. Directly it had nothing to do with flowering, but indirectly it had a great deal to do with it. As mentioned in chapter 2, Flint and McAllister showed that in a particular cultivar of lettuce, Grand Rapids, the seeds will not germinate in total darkness but will germinate after very brief exposure to red light. And if a sample of seeds would partially germinate in darkness, then they were inhibited from germinating if the red light was too far on the long wavelength side, in the far-red. In other words the red and the far-red have opposite effects, and the optimum wavelength of the far-red for inhibiting germination was around 730 nm. The optimum wavelength of red for promotion would be close to 660 nm, which as it happens is about the peak of chlorophyll *a* absorption in extracts, though that can hardly be of any importance, since there is precious little chlorophyll *a* in a lettuce seed. The two opposite effects were traced to the two forms of a single pigment. Figure 2-3 shows the absorption spectra of the two forms of this activating pigment, phytochrome.

In any case this observation with lettuce seeds was very soon integrated into the flowering problem by the work of Hendricks and Borthwick in the U.S. Department of Agriculture. It was highly appropriate that it be done there, because it was in the U.S. Department of Agriculture, oddly enough, that photoperiodism was discovered by Garner and Allard in 1925. This is one of the major contributions that the staff of the Department has made to *pure* plant science. What Hendricks and Borthwick did was to determine exactly what wavelengths of light are active in inducing flowering. They reduced each of their plants to a single leaf so that they could expose the leaf to a broad spectrum projected down a long hallway; and they soon found that the same phenomenon occurred as with the lettuce seeds, namely, that red and far-red act in opposite directions. When one is interrupting the night of short-day plants with a short light exposure, red acts very much like white light, while far-red will antagonize this and return the plants to the flowering state. The same kind of action spectrum, red acting in one way and far-red in the opposite way, was found whether the light was used to *inhibit* the flowering of short-day plants by turning it on in the darkness, or to *promote* the flowering of long-day plants by using it to supplement their day length. If long-day plants, for instance, are held on a subthreshold day length, say 8 hours, and then given a couple more hours of light, they will now flower. This extra light could also be given at different spectral wavelengths and just as with the preventing of flowering of short-day plants, the effectiveness of those wavelengths in promoting the flowering of long-day plants can be compared. The identity of specific actions of red and far-red thus provides a suggestion that the flowering of long-day plants and of short-day plants is basically subject to the same internal process but simply inverted in the two cases.

More recently Erwin Bünning, in Germany, who was interested in the movements of leaves, came to the conclusion that those plants in which there are clear-cut day and night leaf movements are subject to a sort of internal rhythm or internal "clock," according to which the leaves move from a daytime position to a nighttime position at specific periods. The movement is really the movement of petioles and depends on changes in cell size at the junction of petiole and stem. The timing of the movements is dependent on the day length in which the plants have grown, and one can find daily cycles which correspond nearly to a long day or to a short day. Hence Bünning felt that the influence of day length in causing flowering was somehow connected with this internal rhythm, which continues in the plant even when you place it in constant conditions. This last point is important. If a plant which has been growing on 12-hour days and thus has a leaf movement on a 12-hour basis is put in darkness for 2 or 3 days (not indefinitely in darkness), the movements will continue on exactly the same 12-hour rhythm. In other words, it is not immediately dependent on the external stimulus, but something endogenous continues to function. If the night is interrupted with a brief light period the rhythm of the leaf movements can be prevented, and the most effective time to do that is in the middle of the dark period. Bünning interpreted that as meaning that the endogenous rhythm was at that moment at the bottom of the dark part of the cycle.

The idea that the length of day is important because it interacts with an endogenous rhythm has never been wholly accepted. It is clear that the endogenous rhythm plays some modifying role, but not that it is the determining factor. Many new experiments have been carried out, interrupting days and nights at different times. For instance, plants were grown in 72-hour cycles, with 8 hours of light and 64 hours of darkness. Then if the plants were operating according to their internal rhythm, which is basically a 24-hour one, there would be certain times in the long dark period at which a flash of light or a short exposure to light would be most effective, because it came in the middle of the dark part of the rhythm, and other times at which it would be less effective because it occurred in what corresponds to the light part (fig. 11-3). In the plant shown (soybean) the endogenous rhythm was derived from growth on 8 hours light, 16 hours dark. Therefore we should normally prevent flowering most effectively with a light break in the middle of the first 16 hours of the dark period. But when the plant enters the next 24-hour period, we could prevent flowering equally with light at the 40th hour, or later at the 64th hour, which would be in the center of the dark spell of the third 24 hours. In other words, there should be three time zones at which light is at its most effective, with intermediate periods of little or no effectiveness. That seems like a simple thing to test, and yet it is remarkable that the results are conflicting. Several workers, especially Takimoto in Japan, and Hamner and Coulter in the United States, have obtained three clear periods like those just described (fig. 11-3), while others have been unable to find much difference; one person has even found a clear single 64-hour dark period. One can only conclude that the matter is far from proven.

If, instead of measuring the percentage of flowering, we measure the *stage of the development* of the flower buds, scoring it by dissecting the buds and noting the percentage of flowers, we see again how very effective a short light period is at the eighth hour and how little effect it has on either side of that time. There have been many experiments of this kind, and the results are essentially the same as with the other method.

So these are basically the facts. Details could be multiplied but they would add little to the general picture. The key question is: what is the nature of the flowering stimulus? What can we say about it?

D. SOME PROPERTIES OF THE PROPOSED FLORIGEN

Some of the things we have to say about the flowering stimulus are contradictory. Perhaps the most interesting is the case of plants that require several light or dark periods. The majority of plants do not respond to a single short day or a single long day; they respond to 4, 5, or 6, or sometimes even more; *Bryophyllum* requires 10 or 12 such days. What can we deduce? If a short-day plant requires a series of unbroken long

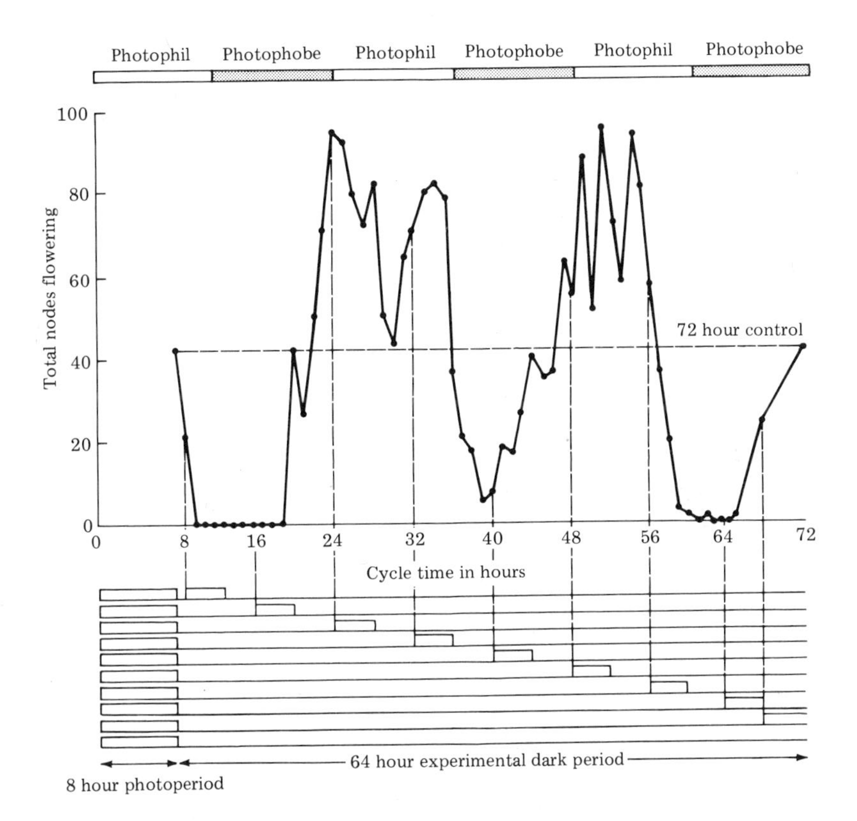

11-3 Soybean plants on a 72-hour cycle, with 8 hours of light and then 4-hour light breaks at different times during the long dark period (as shown under the curve). Flowering shows a diurnal cycle with the most complete inhibitions 24 hours apart. From Coulter and Hamner, 1964.

nights, we must suppose that after the first long night some stimulus is produced, but it is not enough to induce flowering. After the second long night a little more of it is produced, and so on. It would follow that the flowering hormone or stimulus is stable; it lasts over through the next light period and accumulates until finally enough is present to cause flowering. In fact Schwabe, at Rothamsted, working mostly with *Bryophyllum,* has essentially measured this stability by giving two such inducing periods and then waiting several days, thus giving a noninducing period. Afterwards the plants were given the rest of the needed long days. If the period intervening is not too long, the plants "remember" that they had those two long days and will add them to the subsequent long days and will flower. But if the intervening period is too long, they "forget" the previous period and must be given the full number of long days again. This result is interpreted to mean that the flowering substance is stable, but not fully stable. It gradually decomposes over a period of several days. Now consider the 8-hour-day plant that requires only one day and one long night, and suppose we give it a short spell of light during that long night. Why does that spell of light prevent flowering? Because, we understand, it destroys or leads to the destruction of the flowering hormone. Therefore we conclude that it takes 16 hours to make the stimulus and a single hour of light in the middle to destroy it, or to destroy enough of it to prevent flowering. These two arguments lead to opposite conceptions: one, that the hormone is stable enough to survive and accumulate *in the light* over a period of days; the other, that it is unstable enough to be destroyed by very brief exposure to light. Also, in the case of long-day plants, they can flower on a 24-hour day length, which means that the material in them is totally stable to light, since it is not destroyed by indefinitely long light periods. Some may even need several such 24-hour periods in order to flower. Day-neutral plants offer still more extreme cases. We must deduce either that the substance is entirely different in the two cases, or that the substance is the same and the associated phenomena, like enzymes to destroy it, are entirely different, and indeed opposite, in long-day and short-day plants.

A very different set of conclusions is derived from grafting experiments. These give equally clear-cut results. Perhaps the simplest case is *Hyoscyamus niger,* the black henbane.* This plant comes in two forms, a biennial and an annual form. The biennial form flowers only after a cold spell (biennials are essentially annual plants which require a cold spell). If both young plants grow side by side in a day length which is also appropriate, the biennial will not be flowering and the annual will be flowering. Now if we take a branch from this annual, using one which has not yet flowered but is going to flower, and graft it on to the biennial, which would not normally flower until next year, the biennial will flower this year (fig. 11-4). Thus the influence or stimulus which is already present in the annual can be transmitted to the biennial and make it flower in the first year. This fits well with the old Chailakhian experiment; obviously a stimulus is

* The many plants mentioned in this chapter are listed in table 11-1.

Table 11-1 Plants discussed in this chapter

SHORT DAY (SD)
Bryophyllum (requires long days first)
Chrysanthemum, many cultivars
Glycine max cv. Biloxi (soybean)
Nicotiana tabacum cv. Maryland Mammoth (tobacco)
Perilla nankinensis
Pharbitis nil (Ipomea)
Xanthium pennsylvanicum (cocklebur)
Xanthium strumarium

LONG DAY (LD)
Brassica napus (mustard)
Hordeum (barley) winter cultivars
Lolium perenne (perennial rye-grass)
Mimulus spp. (monkey-flower)
Nicotiana sylvestris
Raphanus sativus (radish)
Rudbeckia (black-eyed Susan)

DAY-NEUTRAL (DN)
Gomphronia
Lycopersicum spp. (tomato)
Nicotiana tabacum (tobacco) several cultivars
Phaseolus spp. (bean)
Zinnia and other garden plants

COLD-REQUIRING
Avena sativa (winter oats), then requires LD
Chrysanthemum cv. Shuokan
Daucus carota (carrot)
Hyoscyamus niger, biennial strain (henbane)
Triticum vulgare (winter wheat), then requires LD

present in one part of the graft and is transferred to the other. Such grafting experiments—of which there are many—are perhaps most interesting when long-day and short-day plants are grafted together. Tobacco offers a convenient pair because *Nicotiana sylvestris* is a long-day plant and *Nicotiana tabacum,* var. Maryland Mammoth (MM), is a short-day plant. There are not many genera in which two closely related species have opposite day length requirements. Suppose we graft a Maryland Mammoth on a *sylvestris,* to produce a two-branched plant: once we have them growing satisfactorily together, we put up a light-barrier and give the two branches different light treatments. If MM is given long-day treatment and *sylvestris* short-day treatment, both will continue to grow vigorously but vegetatively. If we take away the opaque barrier and give both sides long days, then the *sylvestris* finds itself in flower-inducing conditions and the MM does not. What happens is that the *sylvestris* flowers, and, a little later, so does the MM (fig. 11-5). So in the MM we have a plant flowering in conditions under which normally (at the right temperature) it never would flower. This has been widely held to show that the flowering hormone is *the same in long-day plants as in short-day plants.* By whatever means it takes, we have formed some flowering "hormone" in the *sylvestris* and it has been transferred to the MM, which by definition does not have any. There are many other such experiments, e.g., figure 11-4, photograph.

Comparable experiments have also been done, not with short days, or with biennials, but with cold treatment. These suggest that

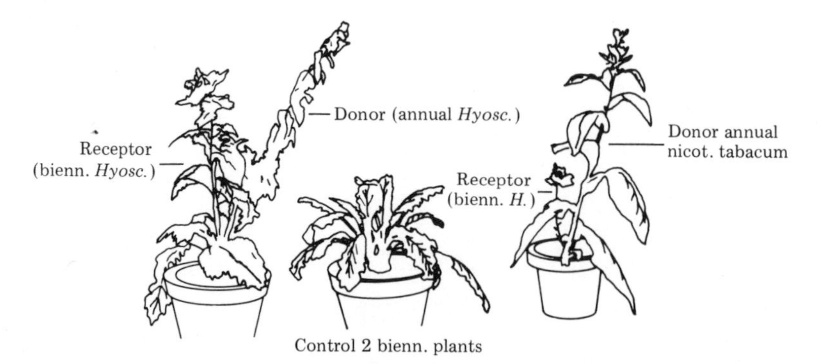

11-4 (Above) Flower initiation in the first year of biennial *Hyoscyamus niger* (black henbane) by grafting either an annual strain of *Hyoscyamus* or an (annual) species of *Nicotiana* on it. From Melchers, 1952. Original drawing by Erich Freiberg of the Max Planck Institute. (Below) When one-year *Hyoscyamus niger* (No. 3) is grafted on *N. tabacum* Maryland Mammoth in long days (No. 1), neither plant (separately) flowers but the graft pair (No. 2) forms a long stem and flowers, supporting the 2-factor concept of GA in *Nicotiana* and anthesin in *Hyoscyamus*. Original photograph supplied by Prof. Chailakhyan.

cold treatment and day length have the same hormonal basis. Chailakhian used a winter cereal, like winter barley, which had had a cold treatment and was being grown under long days so that it would eventually flower; he made an aqueous extract from this and supplied it to a short-day *Rudbeckia* plant growing in long day. The *Rudbeckia* flowered. Thus extracts are effective and cold treatment apparently engenders the hormone in much the same way as day length does. These effects were probably due to gibberellin (see section F). In Chailakhian's lab, extracts from *Perilla,* a short-day plant, were also applied to *Rudbeckia* and made it flower.

Thus there is evidence that there is a flowering hormone which is the same in different plants; it is produced under different conditions but the actual substance is the same.

What can we say about the chemical nature of this hormone? It is transportable, as we saw, but it does not cross a discontinuity. Why not? It seems that there are two possible reasons. One is that it is rapidly destroyed at cut surfaces. That is not likely for the following reason: when Withrow repeated the discontinuity experiment he found that if the tissues are left in place, so that they begin to graft, and if the graft has *just* taken, so that some tissues have just barely joined together, then the influence can be transmitted. It only takes a few days with very rapidly growing, well-nourished greenhouse plants till a minimal amount of graft has taken, and no sooner has a graft union formed than the flowering stimulus can move. Yet until it does move, it has certainly been in contact

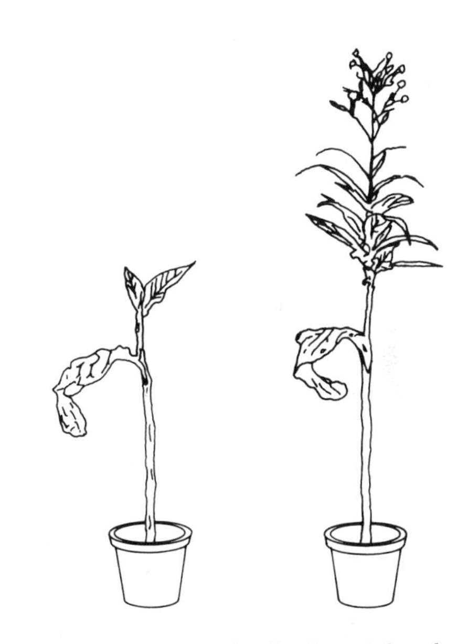

11-5 Short-day plant induced to flower in long days by grafting one leaf from a long-day plant into it. Left, *Nicotiana tabacum* Maryland Mammoth (MM) on a stock of MM (the so-called "autograft"); right, *Nicotiana sylvestris* on MM. Both plants on long days. From Melchers, 1952. Original drawing by Erich Freiberg of the Max Planck Institute.

with the cut surface. The second possible explanation of the failure to cross a discontinuity is that it is a very large molecule and therefore does not diffuse readily. That is an attractive explanation, because one would like to think that the substance could perhaps be a nucleic acid and therefore would indeed be a large molecule. Following that line of evidence several workers have tested antagonists of nucleic acid formation (see chapter 14, section E) and found that they do inhibit flowering. Basically the experiment is this: a plant is reduced to a single leaf, which is kept in flower-inducing conditions and at the same time is treated with fluorodeoxyuridine, 5-fluorouracil, or actinomycin D. Under some conditions this will inhibit flowering, though not under all. In tobacco stem segments grown *in vitro,* such antagonists actually promote flowering. Good inhibition has been reported in *Xanthium, Pharbitis* and *Lolium;* in the last-named case it was reversed by orotic acid. Generally, if the treated leaf is a long distance from the bud that is being studied, flowering is not inhibited, while if the leaf is close to the bud, inhibition does result. This result has been thought to show that the antagonist does not influence photoreception as such but influences some intermediate stage between that and the appearance of visible flower buds.

Another most important piece of evidence comes from the extraordinary experiment of Lona, at Parma, who in 1946 made grafting experiments of the type already mentioned. His plant was reduced to a single leaf, which was then exposed to flower-inducing conditions—short day, for instance; then this leaf was cut off and grafted on to another plant. The rest of the receiver plant was defoliated to avoid the balance problem, and after a fairly long time the terminal bud of the receiver plant, which had been grown in noninducing long-day conditions, flowered. Thus the leaf which had received the stimulus had transmitted it to the receiver plant. That is in itself not too extreme a development of the experiments we previously described. However, Lona brought in a third plant, also grown in long day so that its terminal bud remained vegetative. The second plant, on which the graft just described was made, was allowed to develop some new leaves. These leaves, which came from axillary buds, were thus leaves from a flowering plant, but they themselves were never exposed to the flowering stimulus. Now he took off one of these leaves and grafted it onto the third plant; after again a long time, this third plant now flowered. Thus we have transmitted the stimulus by way of a piece of the plant that *never actually received* the stimulus. And as Lona points out, this means that the amount of stimulating substance, whatever it is, must have increased greatly. There must have been so much that when he made the first graft it permeated the whole plant and flooded over into the leaves, which could then themselves transfer the stimulus across the graft. Thus we now can deduce these facts about the chemical nature of the substance: that it is apparently a large molecule, its formation or its action is inhibited by RNA inhibitors, and it *increases in amount.*

We can measure the time of translocation by cutting off leaves at certain times after causing induction, and seeing how far the stimulus has traveled. It turns out to move at 2 to 3 mm per hour, which is about 1/5 to 1/6 of the rate of auxin movement in stems (commonly 10 to 15 mm per hour) but nevertheless it is a rate which could be associated with a hormone.

E. THE CHANGE FROM THE VEGETATIVE TO THE FLOWERING STATE

We can deduce a few things about the nature of the change to flowering. This is usually envisioned as a change in the terminal bud, i.e., in the apical meristem. The typically vegetative apex, as shown in figure 10-1, has a tunica of epidermal cells and more or less isodiametric, nondividing, cells within. A little further down the cells are rapidly dividing, and around the flanks other rapidly dividing cells are producing leaf primordia. Now when the apex changes over to flowering, the distribution of cell division is nearly the opposite. The apex commonly expands somewhat in width, and the central cells, which were rather inactive, at least as far as division is concerned, now begin to divide rapidly, so that very soon the whole apex is covered with small primordia. If the flower is a composite these give rise to a mass of little flower primordia. Thus when the area of activity changes from the vegetative to the flowering stage, the changes are concentrated near the apex. We deduce that the flowering "hormone," if there is one, flows up to the apex and there causes its cells to change their pattern of division.

Chouard and Aghion, working at the Phytotron at Gif in France, made an interesting discovery (1961). They made tissue cultures from different parts of the stem of tobacco plants and found that they behaved differently. Cultures from the pith near the apex developed flower buds, while those from near the base formed mainly vegetative buds. (In tobacco plants, flowering normally takes place from the most apical nodes, while the buds in the axils of the lower nodes are usually vegetative.) If the plant is grown under such conditions that it will flower, it follows that there must be what Chouard and Aghion call a "flowering gradient" from apex to base. Skoog and his group have extended these experiments, and have pictured remarkable little tissue cultures bearing, in some cases, absolutely mature normal flowers. Tissue cultures taken from the pith of the first 3 internodes yielded 89% flower buds; those from the 9th and 10th internodes yielded only 37%. There is some hormonal control here too, because the formation of flower buds is in part a function of the auxin-kinetin balance, much as we saw in chapter 10 with the formation of buds in general. At very low cytokinin concentrations relative to the IAA in the medium, one or two flowers are formed per culture, while at ratios nearer to 1 : 1, the cultures are much more active and produce more flower buds, sometimes branched heads. However, the further development of the flowers may be delayed. Flower buds formed in the dark never gave rise to mature flowers, so that there may be a nutritional or hormonal control of the final stages of development.

Figure 11-6 shows one of Wardell's and Skoog's tissue cultures. A tiny piece of pith has produced several leaves, a mass of roots and a single flower. Although the tissue was not taken from near the apex yet the flowering stimulus is present in it. Adding auxin to the cultures inhibits flower bud formation, and curiously, this inhibition is reversed by several base analogues, like thiouracil or azaguanine.

In somewhat parallel work at about the same time, Colette Nitsch found that internode segments from *Plumbago indica* (a South African bush widely used as a hedge plant in California) when planted on nutrient medium give rise to a callus which produces many flower buds. Adenine promotes the process, yielding about 10% of flower buds, but the combination of adenine and zeatin produces 92% flower buds. Unfortunately only about a quarter of these go on to form mature flowers. Evidently the cytokinins act in the same way in both of these plants.

Thus we have here all the essential facts. Needless to say, there are far more than I have presented, but these are most of the critical ones. Many of the experiments described could have provided excellent bioassays for a florigen, by grafting, by extracting, or by tissue cultures. Yet none of them has yielded florigen. There has been ample opportunity, with 30 years of experience in bioassay. In the same period of time, or less, the red–far-red system has been discovered, isolated and chemically purified. In that case we have learned what the photoreceptor is—it is a bile pigment—we know a good deal about its properties and that its molecular weight is around 120,000; good bioassays are available for it. Furthermore it is bound to a membrane—a very suggestive fact. All of these properties

11-6 Tissue culture from pith of upper part of tobacco stem. IAA:Kinetin ratio = 17 : 1. From Wardell and Skoog, 1969.

could have been shown using bioassays for the flowering hormone, but so far they have not been.

F. ROLES OF KNOWN HORMONES

In contrast to all these negatives, a group of known factors does strongly influence flowering. Of these gibberellin is by far the most prominent, and some of its effects are dramatic. Soon after GA became available it was found that it influences many long-day plants to flower in short days. If a long-day plant on an 8-hour day is treated with GA it is as though the gibberellin were the equivalent of 8 more hours of light. This led to the view that gibberellin, for many plants, acts like light, as indeed it does in Grand Rapids lettuce seeds (chap. 2). Then it was found that some seeds which are inhibited by light are also caused to germinate by gibberellin. In those cases GA acts like darkness. Thus one cannot make simple generalizations of this kind.

In some cases, GA induces long-day plants to bolt, that is, to change from the rosette to the tall scape form, but the bolted plants do not always flower, although in some cases they do. *Brassica,* for instance, flowers, while *Centaurea* and *Mimulus* only bolt. On short days, *Mimulus* forms a rosette, and on long days it flowers, even if part of the 16 hours of light is given as far-red. When the rosette was treated with GA it bolted beautifully, looking exactly like the flowering plant, but the terminal bud was still a vegetative bud. *Brassica napus,* another long-day plant, in Lona's hands both bolted and flowered after being given GA on short days. Controls remained

11-7 *Rudbeckia bicolor* (black-eyed Susan) a long-day plant, grown in short days for 9 months. Left, control plant still in rosette form; right, plant given one drop of gibberellin solution (100 mg/liter) weekly for 4 weeks. From Bünsow and Harder, 1957.

11-8 *Bryophyllum crenatum* grown in short days for 4 years; three plants per pot, of which only one is treated in each. Details in text. Photographed after 2 months. From Bünsow and Harder, 1956.

in the rosette form. Incidentally, these are among the first cases in history where flowers were produced by purely chemical means. L. T. Evans of Canberra, in his 1971 review of flowering, concluded that GA is required for all flowering, but the Russian experiments below make that most unlikely. Nevertheless a nice example is given by the *Rudbeckia* (black-eyed Susan) in figure 11-7 which has been grown for nearly a year in short days. The control plants are still in perfect rosette form. The plant treated with gibberellin—a few drops of solution spread over a few weeks—is in the full flowering state.

Bryophyllum is a special case because it is a so-called long-short-day plant—it requires short days for flowering but has to be exposed to long days first. Exposure to long days for a week or two satisfies the long-day requirement; thereafter it needs short days in order to flower. Figure 11-8 shows a group of plants which have been kept for 4 years on short days, so that they never have seen long days and should therefore not be ready to respond to short days. Plant a has received a drop of gibberellin solution daily for 2 months, plant b the same for 6 weeks, plants c and d for 4 weeks. The controls in the same pots are fully vegetative, yet plants a and b have both bolted and flowered. Plants c and d are obviously preparing to do the same, but the treatments have been shorter; they did flower later. Here the gibberellin is not really substituting for something, because one can hardly substitute for the long days that it should have had long ago, nor is it substituting for light. It is acting simply as a *flower-forming stimulus;* it has made plants flower that were not in a flowering condition. Chailakhian later made the complementary experiment, keeping the plants in long days throughout; if they were then given short days they flowered, but if they were given gibberellin they did not; they only grew very tall.

Plants requiring a cold treatment, like the henbane of figure 11-4, show a clear gibberellin response. They are often classed as biennials. The "French forcing" carrot, for example, if grown in continuous warm temperatures remains vegetative, while plants treated with gibberellin are indistinguishable from those given a cold spell, in that they all flower. One could say that gibberellin is here substituting for cold, but it is simpler to say that it provides the flowering stimulus.

Although many long-day plants will flower in short days if given gibberellin, yet short-day plants will not flower in long days; this reverse change is never obtained with gibberellin. This has led to the common statement that gibberellin may be a flower-inducing substance, but it cannot be *the flowering hormone* since that must act in the opposite direction also. Nevertheless, we have seen now that gibberellin treatment will, in different plants, substitute for cold treatment, for day length, and for previous exposure to long days.

Furthermore the experiments with extracts which I mentioned above (in which extracts from flowers in the flowering condition caused plants in the vegetative condition to flower) were very likely due to gibberellin. Some gibberellin is usually

Table 11-2

A. Estimation of the amounts of GA_3 and presence of other gibberellin-like substances in leaves of long- and short-day tobacco plants

Species	Type	Grown on:	Free GA_3 (μg/kg fr.wt.)	Other GA-like substances
N. sylvestris	LD	LD	12	I,II,III and IV
N. sylvestris	LD	SD	4	I only
N. tabacum, MM	SD	LD	1.4	—
N. tabacum, MM	SD	SD	0	I only

From Grigorieva et al., 1971.

B. Estimation of the amounts of extractable and diffusible gibberellins in the leaves of red clover, a long-day plant

	Gibberellin in ng* per 100 leaves	
Plants grown on:	Extractable	Diffusible
Short days	25	1042
3 long days (following the short)	96	463
7 long days (following the short)	31	115
12 long days (following the short)	84	58
20 long days (following the short)	—	71
28 long days (following the short)	19	33

* ng = nanograms = 10^{-9} grams.
From Stoddart and Lang, 1968.

present and it is extractable with ethyl acetate or other common solvents. The crude extracts, however, are frequently somewhat toxic.

In 1971 Chailakhian and some chemist colleagues in Moscow extracted, purified, and determined the gibberellins in long- and short-day tobacco cultivars, namely *N. sylvestris* and *N. tabacum* Maryland Mammoth. Their results (table 11-2) show that the long-day plant had some GA in short days but tripled its GA content when given long days. The short-day plant yielded no GA at all on short days (when it would have flowered) and very little on long days. They conclude that the long-day plant flowers on long days because it produces GA, so that the long day exerts an *indirect* effect via GA. We have seen that many long-day plants do flower when given GA. The short-day plant, however, *neither forms nor responds to* GA; hence its flowering must be under an entirely different control. While such a conclusion may not be quite universally true, it does provide a welcome division of the problem into separate parts. Chailakhian has long considered that two substances control flowering, one causing the special growth involved, the other changing the buds to flower buds ("anthesin"). Thus the proposed florigen may, in long-day plants, have two components of which one is merely gibberellin. But whether the other has real existence as a *single substance* remains highly doubtful. In any case clover behaves in the opposite way, making more GA on short days (table 11-2, part B), so we cannot conclude much.

There is one more chemical flower-inducing substance to be mentioned: ethylene. It was discovered by Clark and Carns in Hawaii, some years ago, that if one adds to growing pineapple plants some auxin, especially naphthaleneacetic acid, NAA (indoleacetic acid is rather rapidly destroyed), they will come into flower earlier. The amount of auxin is fairly large (100 mg per plant) but NAA is relatively inexpensive, and when pineapple growers are faced with acres of pineapples which are not yet ready for flowering and fruiting, the prospect is obviously attractive, especially as it means that the fruiting time can be made uniform. And every treated pineapple flowers, the flowers are all normal, and they all set fruit. However, reports soon came in that Cooper in Florida was getting the same thing with acetylene water, i.e., water saturated with acetylene. Acetylene is chemically very close to ethylene, and the upshot is that it is *ethylene* that is responsible, for (as we have seen) high auxin concentrations cause plants to produce ethylene. Thus although ethylene inhibits a number of vegetative processes it actually causes flowering in a small number of plants. This is not by any means a widespread phenomenon. Pineapples belong to the *Bromeliaceae,* and other members of the family have been found to respond. Ethylene causes flowering in the sweet potato, *Ipomea batatas,* a response which is important for breeders because many sweet potato varieties are hard to bring into flower. The only other case so far known is the lichee nut, *Nethelium litchi,* which flowers, although less than 100%, with ethylene. There are also a number of cases in the literature where treatment with rather high auxin concentrations has inhibited flowering, and probably this was an ethylene effect, so that in general it tends to inhibit rather than to promote. Why just a few plants, in different families, should give the positive response is totally unexplained.

G. CONCLUSIONS: POSSIBLE DEATH OF THE FLORIGEN IDEA

What does all this mean? It certainly gives the impression that flowering is a complex phenomenon as is, indeed, the formation of gonads in the animal. The parallel would not just be the determination of sex, but the formation of the whole reproductive system. It is somewhat naïve to believe that a single hormone controls this and causes the change from the vegetative to the reproductive state. Probably there are a series of reactions and one can intervene in this series in different ways, in some plants with gibberellin, in others with ethylene, perhaps in one or two with auxin. For auxin sometimes seems to shift the balance towards flowering in plants that are near the balance between vegetative and reproductive states. Also there are one or two cases of active extracts whose action could not have been due to gibberellin, since they made short-day plants flower in long days. These were not very conclusive and will not be discussed in detail. But it seems that a flowering hormone is really any endogenous substance that causes flowering, and it need not be a universal florigen. What is a flowering hormone for one plant may not be a flowering hormone for another. We need a fresh look at this problem.

Let us try to visualize the control of flowering in the context of the several well-explored control processes in bacteria. There it is known that a specific *repressor* molecule binds to a region of the nuclear DNA, preventing that part of it from being duplicated and therefore preventing a whole group of proteins (enzymes) from being synthesized. An *inducer* molecule, which may be the substrate of one of these enzymes or it can be another, related molecule, combines with the repressor, either directly or more probably via a combination of inducer and RNA. This combination removes the repressor from the DNA and allows it to duplicate that special area which leads to the synthesis of the group of enzymes concerned.

Suppose that the group of enzymes necessary for flower formation are coded side by side in a specific area of the plant's DNA. So long as the repressor is bound to that area, the plant remains vegetative. When phytochrome in its active form (presumably P_{fr} in this case) combines with the repressor it removes the repressor from the DNA, and the enzymes for flowering start to be coded and synthesized. The complex needs, for specific day lengths and specific timing of light pulses, then boil down to means of holding the repressor off for long enough to enable the various stages of synthesis to go on: DNA to m-RNA, accumulation of the t-RNA compounds of the needed amino acids, assemblage at the ribosome, etc. All we need to hypothesize here is that, if the repressor is allowed to fall back into place on the DNA so that the message stops coming, these several stages will reverse, because the equilibrium states are not in the direction of flower formation and the reaction must be continually driven.

Then why the nonspecific effects—gibberellin, auxin, ethylene, temperature change, even inorganic ions in the case of Hillman's work on *Lemna?* Simply that the repressor can be removed from the DNA by a variety of events, by a group of different RNA-hormone complexes, by conditions which increase the instability of the bonding to DNA or, lastly, by increasing the rate of DNA synthesis so that it "outgrows" the repressor (as a fast-growing root outgrows its virus infection). Auxin, cytokinin and gibberellin are all known to increase the rate of DNA synthesis in certain instances. Some additional assumptions may prove to be necessary, but I believe that if we bear this generalized picture in mind we shall find that most of the phenomena of the control of flowering can be interpreted in its light.

THE GROWTH OF FRUITS

A. THE VARIETIES OF FRUITS

Pollination and the formation of the seed have been touched on in chapter 2, and it only remains here to consider the action of hormones in initiating and maintaining the growth of the fruit. Fruits are of so many kinds that it is difficult to generalize about them. In grasses and cereals the fruit is no more than a single seed with dry adhering chaff which is mainly the ovary wall; in legumes it is a long pod whose growth merely keeps pace with the growth of the seeds; while in many families, of which the Rosaceae is the most notable, the ovary tissue thickens and multiplies into a soft mass of parenchyma to form the many rosaceous fruits that we eat. In citrus fruits the same result is achieved by the outgrowth of multicellular hairs, which thicken and interdigitate like fingers to form the characteristic soft orange or lemon tissue.

Even in their tissue of origin, fruits are varied; they may consist mainly of the ovary wall, as in the apple or squash; the ovary wall may be supplemented by gross thickening of the placenta as in the tomato and pepper; or the fruit may consist largely of the receptacle fused to the ovary as in the strawberry. The variations in the final form are in every textbook—the drupe, berry, pod, pome, samara, etc. The action of hormones must therefore be expected to vary widely with the species. In what follows we shall be necessarily limited to the few which have been subjected to study, these being mainly the edible types.

B. SEEDLESS (PARTHENOCARPIC) FRUITS AND NORMAL FRUIT GROWTH

Many fruits, for example the pineapple, the banana, the navel orange, are naturally seedless. This is probably due to the sterility of the pollen. Bananas pollinated with a fertile species, such as *Musa rosacea,* can often set seed. Millardet, the inventor of Bordeaux mixture, pollinated grape flowers with the pollen from *Parthenocissus,* another genus in the same family (*Vitaceae*), and obtained seedless grapes. It was a short step in principle, but a long one in time, until Hans Fitting in 1910 showed that the first stages in formation of an orchid fruit—the falling of the petals and swelling up of the gynostemium—could be set off by an aqueous extract of the orchid pollinia. Laibach in 1932 showed that the active material in these pollinia could be extracted with ether, and a few years later I showed that the ether extract contained auxin. Thus Fitting's "pollen hormone" was an early example of auxin action. But to form a whole fruit by hormone action was first achieved by Yasuda in 1935 with the fruits of tobacco and cucumber, by treating the styles with aqueous extracts of the pollen. Such seedless fruits are termed *parthenocarpic* from the Greek *parthenos,* a virgin.

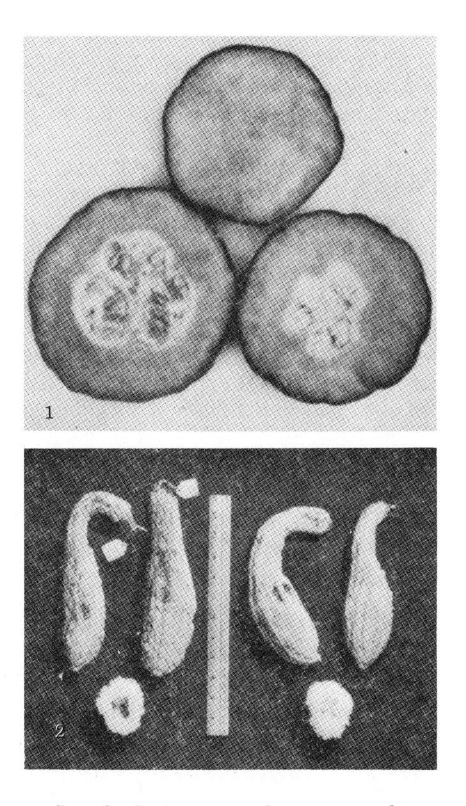

11-9 Gustafson's pioneer experiments on parthenocarpic fruit formation. Above (1), buttercup squash: on the left a normally pollinated and seeded fruit; top and right, fruits produced by treating the pistil with NAA. Below (2), crookneck squash: the tagged fruits on the left produced by treating the pistils with NAA; those on the right normally pollinated and seeded. From Gustafson, 1940.

In the 1930s F. G. Gustafson, at Michigan, was the first to use pure IAA to produce parthenocarpic fruits. He removed the immature stamens and sometimes also the style, so as to be sure no pollination occurred, and placed the auxin on the style. Tomatoes and squash of nearly normal size were obtained in this way. Later seedless peppers, figs, mulberries, grapes, and watermelons were obtained (but the watermelons contained empty seed coats) (fig. 11-9). A stamen-less variety of strawberry, in the hands of Nitsch in Paris, not only gave seedless fruits, but enabled an interesting study to be made (fig. 11-10). Here the "seeds" are really one-seeded fruits (achenes), and the "fruit" is the swollen receptacle. The achenes, being on the

11-10 Parthenocarpy in strawberries. Left, normal fruit; center, fruit from which all the achenes were removed and plain lanoline applied; right, same treatment but followed by application of auxin in lanoline. From Nitsch, 1953.

outside, are fully accessible to simple operations. By picking off the achenes and applying auxin to the receptacle Nitsch obtained good achene-less fruits. He further showed that if he left on one or two achenes the region just around these swelled up; hence the auxin normally comes from the achenes. Direct extraction confirmed this, for the achenes contain much more auxin than the receptacle tissue, and their auxin content parallels the growth rate of the fruit (fig. 11-11).

The testing of all available fruit types quickly brought out the new point that a number of fruits showed no response to the applied auxin, apples, pears and the stone fruits being notable examples. Some tomato cultivars set seedless fruits but these did not enlarge, indicating that other factors besides auxin were needed. The discovery of the gibberellic acids led to the testing of some of these on unpollinated ovaries. On the tomato, fig, and eggplant, GA_3 produced excellent parthenocarpic fruit. This led to much wider testing, and up to this point currants, several varieties of grapes, apples, pears, strawberries, *Solanum,* a few citrus varieties, and even several stone fruits have been grown to somewhere near normal size with gibberellins. Among the compounds tested, GA_5 seems to be the most active in this action.

The fact that some fruits can be set by *either* IAA *or* GA suggests some sort of interaction. Wittwer and Tolbert, at Michigan, showed this directly with a tomato cultivar. The unpollinated controls reached a mean diameter (in 12 days) of 3.0 mm; treated with IAA they reached only 4.2 mm, and with GA_3 only 4.8 mm; but with both IAA and GA_3 together they reached 12.3 mm. Thus it is indicated that *both* auxin and gibberellin are needed for normal fruit growth. Those fruits that respond to one hormone alone doubtless have endogenous supplies of the other.

A few fruits, including most cherries and the plum, are still resistant, although parthenocarpic Bing cherries have been obtained with a mixture of IAA and GA. The resistant types show us that there must be a still-undiscovered third factor, not necessarily hormonal. If the mature ovary is detached before pollination and placed on nutrient agar containing auxin, it can often develop into a (small) fruit. The nutrients here are critical, asparagine appearing to play a major part. This procedure, developed by Nitsch, deserves further investigation and might lead to identification of the missing third factor.

We can conclude that the normal enlargement of the fruit is caused and controlled by hormone supply coming mainly from the seeds. We have seen in chapter 2 that in cereal seed germination the gibberellin apparently comes from the *embryo*. Long ago Cholodny observed that in oat seeds, after 30 minutes' imbibition, auxin could already be detected in the endosperm. Recently (1974) Varga and Bruinsma in Holland have shown that the extractable cytokinin of parthenocarpic tomatoes is no more in amount than one-tenth of that in the seeded fruits. Removal of two-thirds of the leaves further increased the cytokinin content of the seeded fruits (though not of parthenocarpic ones), suggesting that some of the cytokinin normally flows from the seeds to the leaves. Thus we have already a few leads to the sites of formation of the necessary hormones, but to work out the actual localization will require a combination of new techniques on a microscale.

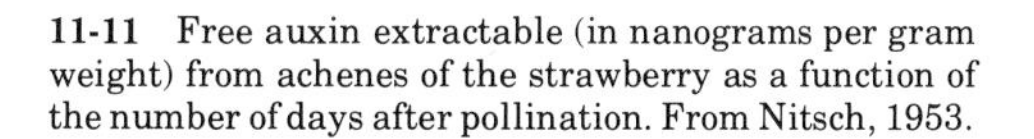
11-11 Free auxin extractable (in nanograms per gram weight) from achenes of the strawberry as a function of the number of days after pollination. From Nitsch, 1953.

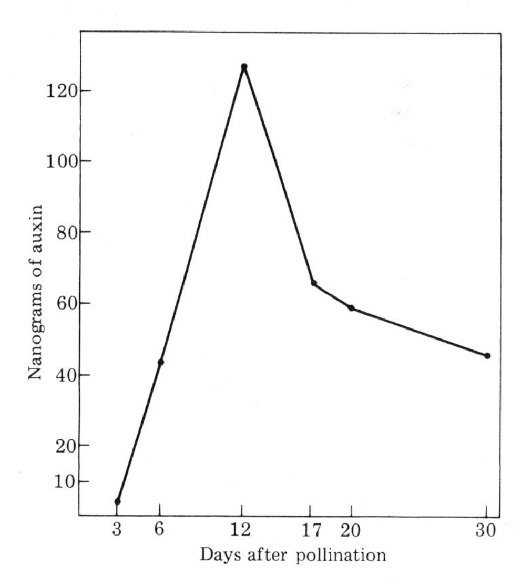

C. FRUIT RIPENING

The idea that a fruit is "ripe" is a somewhat anthropomorphic one; we mean in common parlance that it is ready to eat. Ripening in the edible fruits involves starch hydrolysis and the resulting sugar accumulation, oxidative removal of malic and other acids and of tannins, and softening (probably partial hydrolysis) of the cell walls. Changes in carotenoid and anthocyanin content of the epidermis often occur as well. From the plant's point of view ripening means readiness to release the seed, and in some plants it is almost synonymous with drying, for many "ripe" fruits split open as the ovary walls dry out and scatter the seeds. Thus ripeness is an ill-defined stage in the life of the fruit.

Our knowledge of the control of ripening in edible fruits dates from the days when lemons were picked unripe and shipped to the distant cities in railroad cars heated with kerosene stoves. When more modern convection heaters were installed the temperature was the same or even warmer, but the fruits did not ripen. Evidently the smoke or fumes from the kerosene was playing an important part. Denny in 1924 and Harvey in 1928 identified ethylene as the essential factor in the smoke, and Harvey showed that ethylene hastens the ripening of many types of fruits. Ethylene had been recognized earlier as the factor causing premature leaf fall and loss of geotropism in cases of gas damage to plants. But what was remarkable was the observation that the vapors from other ripe fruits also hastened the ripening. (This is an old observation for any farmer who stores pears; when the first one ripens the others all follow quickly.) Also the vapors from ripe pears inhibited the sprouting of potatoes. Finally in 1934 Gane, working in England, condensed the vapors from ripe apples at low temperature and identified ethylene in them. Thus ethylene is actually *produced by* ripening fruit as well as *causing* ripening; in other words, it is truly a natural fruit-ripening hormone. We have seen in earlier chapters that ethylene exerts many other effects in connection with different aspects of the life of plants, but this one is the most dramatic.

The respiration of detached fruits goes through a fairly repeatable pattern, first followed through with apples by Kidd and West. It can be summarized thus:

Stage	*Respiration rate*
Period of cell division (ca. 4 weeks from fertilization)	Respiration high at first, decreasing rapidly
Period of cell enlargement	Very slow decrease
Period of ripening	Sharp increase ("climacteric")
Period of senescence	Long very slow decrease
Period of morbidity	Small increase, then rapid fall

To some degree the onset and duration of these phases depends on the age at which the fruit was harvested. But it is in the onset of the climacteric increase that ethylene functions. In all fruits studied this sudden rise in respiration closely follows or parallels a similar sudden increase in ethylene production. Since externally applied ethylene can hasten this respiratory rise, it follows that the ethylene production is the cause of it. The reason why ethylene is so especially potent on oranges and lemons is simply that the amounts of ethylene these fruits produce are the lowest. The Burgs have found (table 11-3) that the concentration of ethylene in the intercellular spaces of certain fruits may go up to very high levels, but the concentrations needed to cause ripening are nearly the same in all fruits. The gas withdrawn from the intercellular spaces with a small syringe gives a much more sensitive and reliable criterion than the gas evolved by the whole fruit, which is subject to obstruction by the rather impermeable cuticle and largely depends on the limited number of lenticels. Evidently it takes around 1 part per million of ethylene in the tissue to cause ripening. In view of the varied nature of the fruits this offers a satisfying note of unity. Unfortunately we do not yet understand how the release of ethylene can trigger all the subsequent oxidative and hydrolytic changes that constitute ripening. Sometimes, too, there is evidence that an obscure small respiratory change even precedes the burst of ethylene. There are many unknowns still in this area.

D. THE MODE OF FORMATION OF ETHYLENE

In early studies of this problem it was found that whole or halved fruits produce

Table 11-3

Fruit	Concentration of ethylene produced internally just before climacteric	Minimum concentration of ethylene needed for ripening
Apple (*Malus sylvestris*)	80	ca. 0.2
Avocado (*Persea americana*)	140–180	0.1
Banana (*Musa acuminata*)	6–40	ca. 0.2
Cantaloupe (*Cucumis melo*)	35–75	ca. 0.2
Cherimoya (*Annona cherimoia*)	100–370	1–4
Feijoa (*Feijoa sellowiana*)	20	1–4
Lemon (*Citrus limon*)	—	0.025–0.05
Mango (*Mangifera indica*)	3	0.04–0.4
Orange (*Citrus sinensis*)	0.08–0.6	0.1
Passion fruit (*Passiflora*)	140–220	—
Pear (*Pyrus communis*)	250–500	>0.3
Tomato (*Lycopersicon esculentum*)	4–6	ca. 0.2

From Burg and Burg, 1962; and Biale, 1960.

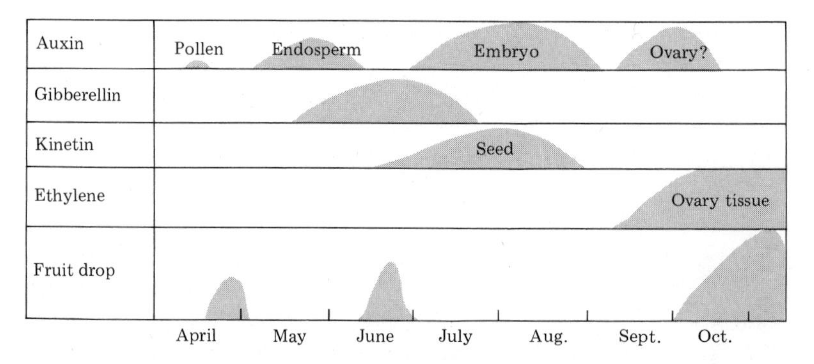

11-12 Schematic diagram (founded mainly on the behavior of pomaceous fruits) of the formation and role of successive waves of hormones during life of the fruit. With acknowledgments to prior concepts of Luckwill and of Van Overbeek.

ethylene in good amounts, and in tomatoes and bananas cutting the fruit into large pieces promotes the formation. Yet ground-up tissues produce little or none. This unfortunate fact militated against unraveling the biosynthetic route for some time. It was observed that oxygen was needed, and that if apples were kept in nitrogen for a time and then transferred to air a burst of ethylene production occurred. Hence an intermediate product, not requiring oxygen for its formation, first accumulates and is then oxidized to yield ethylene. Reagents which inhibit oxidations inhibit ethylene production; and Burg and I found that in fact the concentrations effective on respiration agree closely with those effective on ethylene production, so that the oxidation reaction is certainly an intimate part of the process. Discovery of the need for oxygen stimulated a study of mitochondria with various added substrates, but with little success.

Testing of many labeled substrates with fruit slices only showed that the radioactive label tended to appear in CO_2 with much higher specific activity than in ethylene, indicating that the formation of ethylene from these compounds must be indirect. Then Lieberman and co-workers at the U.S. Dept. of Agriculture found that the amino acid methionine greatly increased ethylene production, and almost simultaneously the Burgs and Clagett in Florida showed that ^{14}C-labeled methionine yielded labeled ethylene of high specific activity. Yang at Davis even found that 80% of the isotope in methionine appeared in the ethylene. Apparently the carboxyl group is released as CO_2, the thiomethyl group is converted to an unidentified material, and the two methylenes give rise to ethylene, thus:

$$\begin{array}{cccc} CH_3\text{—}S & CH_2\text{—}CH_2 & CHNH_2 & COOH \\ \downarrow & \downarrow & \downarrow & \downarrow \\ \text{Nonvolatiles} & C_2H_4 & \text{tissue} & CO_2 \end{array}$$

The ketoacid methional also yields ethylene

in an artificial system but apparently is not an intermediate in this reaction. Ethionine, the analog with a thioethyl group (C_2H_5—S—) inhibits ethylene production in the fruit tissue, confirming the participation of methionine. Yang finds that the first step in the activation of methionine is the introduction, on the S atom, of either a second methyl or an adenosine group.

In studying the effects of other gases the Burgs found that CO_2 inhibits the action of ethylene and drew a parallel between:

$$O{=}C{=}O \quad \text{and} \quad H_2C{=}CH_2$$

Another parallel is with allene, $H_2C{=}C{=}CH_2$, which also inhibits but not so strongly. The action of CO_2, however, explains the older finding that storage of apples in relatively high concentrations of CO_2 (10% or more) greatly retards their senescence. This "gas storage" has been used in fruit warehouses for some time. However, Burg is now developing "hypobaric storage," in which the fruit warehouse is subjected to reduced pressure, whereby the accumulating ethylene is drawn out of the intercellular spaces. Although this procedure unfortunately withdraws other volatiles which contribute to the flavor and aroma of the fruit, it obviously has a promising future.

E. CONCLUSION

The production of auxin by the seeds, the formation of cytokinin in the endosperm, the appearance of gibberellin either from the seeds or perhaps from the conducting system of the whole plant, and finally the liberation of ethylene in the ovary tissue, combine to give us a general picture of the differing roles of hormones in the whole life of the fruit.

Auxin prevents abscission (chap. 13), and therefore the formation of auxin in the young fruits and its transport down the peduncle play a major part in preventing the premature fall of apples and pears early in the season ("June drop"). Only a few percent of the thousands of blossoms on such trees actually set fruit; one estimate is 200 out of 10,000 on a mature tree. In consequence many small ovaries, lacking the auxin supply from fertilized seeds, participate in this June drop.

Figure 11-12, which rests mainly on the studies of rosaceous fruits like the apple, pear, and plum, summarizes the phases and their hormonal control. The chemical nature of the auxins of fruits is still somewhat confused because malic and citric acids, which are almost always present, appear in the ether extracts and tend to promote the elongation of coleoptile and stem segments in the bioassays that are commonly used.

With the ripening of the fruit and the falling of the seed, we have completed the plant's life cycle.

REFERENCES

Biale, J. B. The post-harvest biochemistry of tropical and sub-tropical fruits. *Food Research* 10, 293–354, 1960.

Bünsow, R., and Harder, R. Blütenbildung von *Bryophyllum* durch Gibberellin. *Naturwiss.* 20, 479–480, 1956.

———. Blütenbildung von Adonis und Rudbeckia durch Gibberellin. *Naturwiss.* 16, 453–454, 1957.

Burg, S. P., and Burg, E. A. Role of ethylene in fruit ripening. *Plant Physiol.* 37, 179–189, 1962.

Chailakhian, M. Kh. Substances of plant flowering. *Biol. Plantarum* (Praha) 17, 1–11, 1975.

———, Yanine, L. I., and Frolova, I. A. (Photoperiodic and chemical regulation of flowering of long-short-day species.) *Physiologia Rastenii* (U.S.S.R.) 17, 358–370, 1970.

Chouard, P., and Aghion, D. Modalités de la formation de bourgeons floraux sur les cultures de segments de tige de tabac. *Compt. Rend. Acad. Sci.* 252, 3864, 1961.

Coulter, M. W., and Hamner, K. C. Photoperiodic flowering response of Biloxi soybean in 72-hour cycles. *Plant Physiol.* 39, 848–856, 1964.

Grigorieva, N. Y., Kucherov, V. F., Lozhnikova, V. N., and Chailakhian, M. Kh. Endogenous gibberellins and gibberellin-like substances in long-day and short-day species of tobacco plants; a possible correlation with photoperiodic response. *Phytochem.* 10, 509–517, 1971.

Gustafson, F. G. Probable causes for the difference in facility of producing parthenocarpic fruits in different plants. *Proc. Am. Soc. Hort. Sci.* 38, 479–481, 1940.

Hamner, K. C. Hormones and photoperiodism. *Cold Spring Harbor Symp. Quant. Biol.* 10, 49–59, 1942.

Kidd, F. The respiration of fruits. *Nature* 135, 326–328, 1935.

Melchers, G. The physiology of flower initiation. Lectures given at Imperial College, Univ. of London, 1952.

Nitsch, J. P. Phytohormones et biologie fruitière. 1. *Fruits* 8, 91–97, 1953.

Nitsch, C. Effects of growth substances on the induction of flowering of a short-day plant *in vitro.* In *Biochemistry and physiology of plant growth substances,* ed. F. Wightman and G. Setterfield, pp. 1385–1398. Ottawa: Runge Press, 1968.

Siegelman, H. W., and Hendricks, S. B. Purification and properties of phytochrome, a chromoprotein regulating plant growth. *Fed. Proc.* 24, 863–867, 1965.

Varga, A., and Bruinsma, J. The growth and ripening of tomato fruits at different levels of endogenous cytokinins. *J. Hort. Sci.* 49, 135–142, 1974.

Wardell, W. L., and Skoog, F. Flower formation in excised tobacco stem segments. 1. Methodology and effects of plant hormones. *Plant Physiol.* 44, 1402–1406, 1969.

Hormones and the Senescence of Leaves 12

A. THE GENERAL SYNDROME OF SENESCENCE

We shall be concerned here* strictly with the senescence of leaves as it takes place normally in the growth and development of the whole plant. We have seen that as the axis elongates new leaves are continually formed, and the lowermost leaves gradually senesce. This is seen first by the fading of the chlorophyll and later by a more general loss of leaf constituents. Browning eventually takes place and the leaf wilts and somewhat shrinks. Senescence as a whole is a complex process, some parts of which we know little about. We are especially interested here, however, in what hormones have to do with it and how they function. The prime hormone, which appears virtually to control the senescence process, is cytokinin (cf. chap. 8, part 3D). If we place a droplet of cytokinin on an excised leaf held in darkness and thus about to turn yellow, it prevents the yellowing in the region around the droplet. There results a green zone which can then be measured in various ways, most easily by just extracting the chlorophyll and colorimetering it. In a few plants gibberellic acid, or even a high concentration of an auxin, can have a comparable effect.

The process of senescence is not limited to the disappearance of chlorophyll. In fact, the disappearance of chlorophyll appears to be the second step. The first step is the rapid breakdown of protein, which starts as soon as the leaf is cut off and placed in darkness. If we cut the leaf off and place it in light, the reaction goes somewhat more slowly than in dark, but basically the process is similar in light and in dark. There is a difference of rate, but no evidence of anything qualitative. That is important because, of course, in the whole plant senescence of the lower leaves takes place while they are still in full light; they may be a little shaded by the upper leaves, but they senesce in at least a part of sunlight. It turns out that the average leaf absorbs in the red and blue region, where photosynthesis is effective, from 50% to 60% of the

* In this chapter I am drawing heavily on the work of my own laboratory. Many other workers have contributed to the field, but the points which we have brought out at Santa Cruz exemplify the approach and include most of the major known facts. Indeed, most of the facts are pretty well agreed on, but the problems remain resistant.

incident light. In consequence, the amount of photosynthetically effective red and blue light that gets through is about 40% of the incident light. A second leaf under that one is illuminated with 40% and absorbs again about 60% of that, i.e., 24%, so that only about 16% of the light comes through it. A third leaf under that, again absorbing about 60% of the light it receives, thus lets through only 6.4% of the original intensity. Therefore when there are three levels of leaves in a vegetation or in a crop field the photosynthetic intensity for a fourth leaf is extremely small. In fact if there are more than about three leaves vertically above one another the last one tends to be shaded out. This is not true for trees like pines, with their very thin narrow needles which do not shade one another very much. It is particularly true for broad-leaved plants, whether monocots or dicots. When we analyze the growth of crops in a cornfield, for instance, if there are many more than four leaves one above the other we get little benefit from additional leaves; however, because the sun moves round the sky and because leaves move in the wind, it may take five leaves to produce this degree of shading. The lowermost leaves, therefore, in greatly reduced light do senesce somewhat faster than they would have done in full light. With isolated oat leaves in our experiments, if we transfer from darkness to light, senescence now continues at the light rate. If we transfer from light to darkness the rate of senescence at once changes to the dark rate. So there are two rates, one for light and one for darkness. They show little carryover, and they both are basically the same process, since both involve the breakdown of protein parallel with the loss of chlorophyll and hence with the accumulation of amino acids.

Figure 12-1 will remind you of the general picture. When the leaves are placed for five days in darkness, we see immediately, in the first day, a marked decrease in the protein. It is detectable at six hours, that is, the first quarter of the first day, especially showing as a rise in amino acids. On the next two or three days the process is accelerated, almost as though it were autocatalytic, and there is a rapid pileup of amino acids. By the fifth day most of the protein has been converted to amino acids. These also break down, doubtless by oxidative deamination, yielding some free ammonia. At this stage there are present not only normal amino acids, but some of them have been converted to asparagine, etc., while others have been deaminated. These changes have been followed particularly in tobacco leaves during curing, and one has only to sit next to a cigar smoker in a small room to note that ammonia is indeed present in the leaf.

As far as the chlorophyll is concerned, oat leaves show little change in it in the first day. By the second day it drops off rapidly, following the protein very well. This pattern, with minor variations, seems to be general. In *Rumex* leaves Goldthwaite and Laetsch (1968) did find a slight chlorophyll decrease in the first day or two, but thereafter the rapid drop set in just as in the oat. A mutant of *Festuca* (meadow fescue) has recently been described in which the protein hydrolyzes at the normal rate, but the chlorophyll seems to be stabilized for at least 7 days. Isolated oat chloroplasts be-

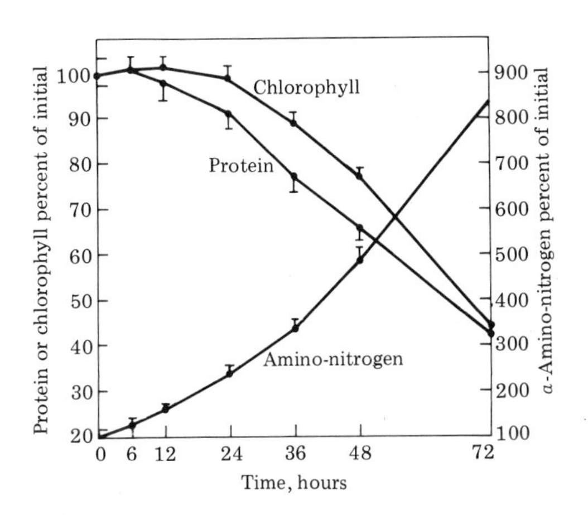

12-1 Changes in chlorophyll, α-amino nitrogen, and protein in 7-day-old *Avena* first leaves, detached and placed in a saturated atmosphere in darkness at 25°C. The vertical lines show ½ the 95% confidence limits. From Martin and Thimann, 1972.

have in a similar way, the chlorophyll showing quite remarkable stability in the dark.

The anatomical effects of senescence are seen mostly in the chloroplasts. The thylakoids become distorted and break down, and lipid globules appear which probably represent the lipids that were in the thylakoids. In the later stages the starch and ribosomes of the plastids disappear. The cytoplasmic contents steadily disappear throughout, but the plasmalemma remains. Barton (1966) who has made perhaps the most detailed study, concludes that the "primary cause" of senescence could be the appearance of a specific enzyme that attacks thylakoids. However, the evidence above favors proteolysis in the cytoplasm as the first step.

B. THE EFFECT OF KINETIN

As we saw in part 3 of chapter 8, kinetin stops the breakdown both of protein and of chlorophyll; the leaves remain green. Furthermore, it prolongs the synthesis of protein which is occurring at the same time. The latter can be measured by applying a labeled amino acid to the leaf and following its conversion to protein. In figure 12-2 the solid line shows the amount of ^{14}C-leucine taken in, and the dashed line shows its conversion to protein. The controls convert to protein from ½ to ⅔ of the amount taken in, but stop doing so after about 40 hours. In the presence of kinetin, intake and protein synthesis both continue longer at the linear rate. Here, then, we have one effect of kinetin at the molecular level—it serves to maintain the intake of amino acid, and to

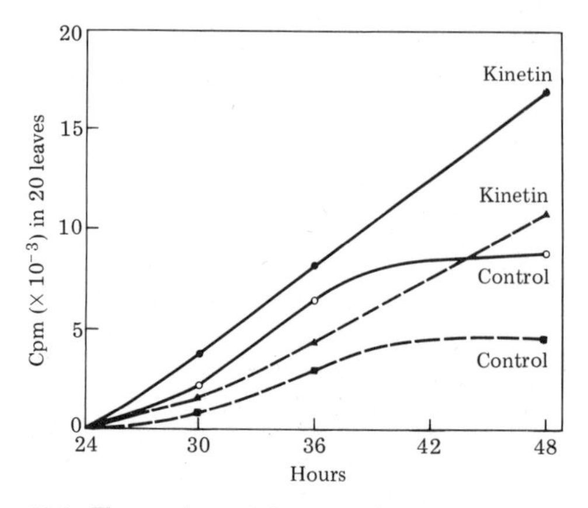

12-2 The uptake and incorporation into protein of ^{14}C-L-leucine with and without kinetin. Solid lines, uptake; dashed lines, incorporation. Vertical lines are the standard deviations. From Shibaoka and Thimann, 1970.

maintain the synthesis of protein from it. Kinetin also prevents most of the increase in RNase.

In looking for a source of cytokinin in roots, Shibaoka and I failed to find much in our pea seedling roots, though Short and Torrey afterwards did find modest amounts, and Weiss and Vaadia had previously reported some in the exudate from sunflower roots. But in making that extraction we came across the fact that the roots contain a substance which *promotes* senescence; this substance turns out to be the common amino acid serine. L-serine antagonizes the whole syndrome; that is, it not only increases the loss of chlorophyll but also accelerates the breakdown of protein, so that serine acts as a kind of anti-cytokinin. D-serine is inactive in this respect.

Figure 12-3 shows the changes of protein, but could almost equally well have shown the chlorophyll change, since after the first 24 hours they go together. The control leaves, after three days, have lost a little more than 50% of their protein. If L-serine is applied, their protein loss is considerably increased. If we protect them by giving them kinetin so that most of their protein is preserved, as we add increasing amounts of serine that effect is steadily antagonized. Finally, at the highest concentration of serine, kinetin exerts almost no effect; in other words, 1/10 M L-serine easily antagonizes 10 ppm kinetin. While this is certainly a fairly high level of serine we must recall that these are tiny droplets on the leaf, from which only a little enters, so that the concentration in the leaf tissue is very much less. We must also recall that serine is one of the most abundant amino acids in most plants; indeed in *Vicia faba* its concentration is second only to that of asparagine.

We made an interesting observation with the original amino acid extract from the roots which contains the serine, namely, that the crude extract does not have the senescence-promoting effect, or has it only very weakly; therefore, the senescence-promoting effect must be balanced by something else in the crude extract. In trying to explain that, we found that the antagonism exerted by L-serine is itself antagonized by L-arginine. There is thus in oat leaves a delicately balanced system of antagonisms. First, in the natural process senescence leads to proteolysis and chlorophyll loss. Then the cytokinins prevent that process. Then L-serine prevents kinetin from preventing that process. And now we find that arginine prevents serine from preventing kinetin from preventing the senescence process! This perhaps gives one an idea of the way in which the amino acids in the leaf maintain a balance with one another. An example is given in figure 12-4. It shows the length of the green zone, which corresponds rather closely with the amount of chlorophyll in the extract. Arginine has no effect over a wide range of concentrations. Serine of course decreases the chlorophyll normally, reducing the length of the green zone; and as arginine is added it brings the effect back, until finally it has completely antagonized the serine. Naturally when proteolysis occurs a lot of serine is liberated, and it may be that this explains the impression that the data give of a sort of autocatalysis. This would require more serine to be produced than arginine, which in

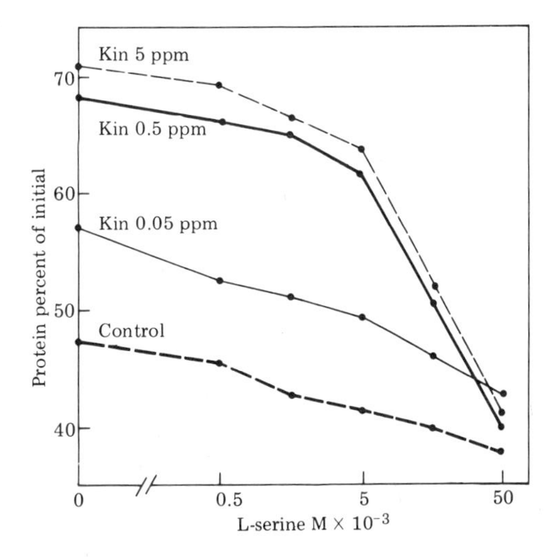

12-3 Effect of L-serine, in presence and absence of kinetin, on the protein contents of detached oat leaves after 3 days in the dark. From Shibaoka and Thimann, 1970.

12-4 Reversal of the serine effect by *L*-arginine. All leaves treated with 0.1 mg/liter kinetin. From Shibaoka and Thimann, 1970.

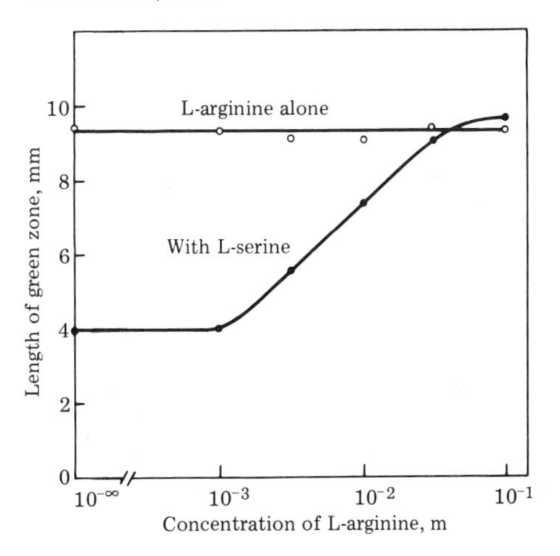

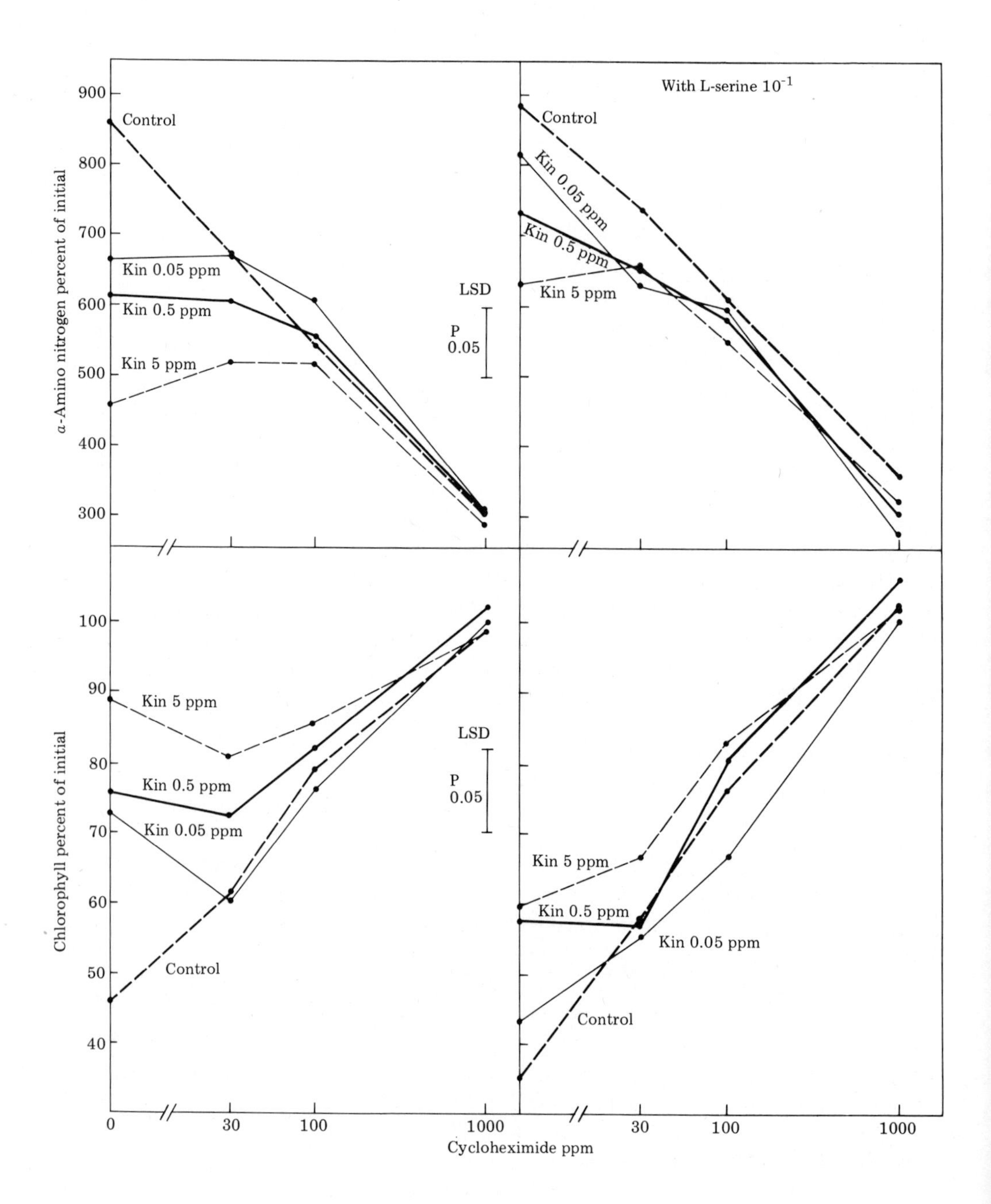
With L-serine 10^{-1}
α-Amino nitrogen percent of initial
900
800
700
600
500
400
300
Control
Kin 0.05 ppm
Kin 0.5 ppm
Kin 5 ppm
LSD
P
0.05
Chlorophyll percent of initial
100
90
80
70
60
50
40
0
30
100
1000
Cycloheximide ppm

some leaves may well happen, for serine is one of the two or three most prevalent amino acids in leaf tissue. No other amino acid acts as strongly as arginine, though several have slight effects. There is an amino acid which does not occur commonly, namely canavanine, which may act as an arginine antagonist; and indeed to a small extent canavanine prevents arginine from preventing serine from preventing kinetin from preventing senescence!

We also found, with oat leaves, that if we prevent the synthesis of protein which takes place in the first days, we prevent senescence. This can be done with only certain inhibitors, of which cycloheximide is the best. The action of cycloheximide is plotted in figure 12-5 against the amino acid nitrogen liberated, and against the protein (it could equally well have been chlorophyll). The free amino acids are brought down almost to their initial low level by cycloheximide, using levels of this inhibitor which inhibit protein synthesis. The chlorophyll is similarly maintained. Suggestively, 6-methylpurine, which is an antagonist for RNA metabolism, works essentially in the same way.

It can be shown directly that cycloheximide at the effective concentrations is inhibiting the incorporation of labeled amino acids into protein. Figure 12-6 (note that this is a log scale) shows that as the concentration of cycloheximide is increased down comes the amount of protein formed. Thus during the observed inhibition of senescence we are in fact inhibiting protein synthesis. We can reasonably deduce, then, that protein synthesis is an essential part of the first stages of this senescence process.

12-5 Action of cycloheximide in preventing senescence, as shown by lowering the α-amino nitrogen (above) and maintaining the chlorophyll (below). The curves on the right show the same response in presence of serine. From Martin and Thimann, 1972.

Since serine and arginine act, one to promote senescence and the other to antagonize that promotion, it is interesting to see what they do to the protease levels in the leaf. The accompanying table gives some pertinent figures. After three days the protein in the leaf has fallen to 50% of its initial value. The specific activity of the protease has increased somewhat after three days, and arginine had only a very slight, if real, effect. Serine has clearly increased the protease activity and if we add arginine to the serine it prevents most of the increase. Serine does not increase the RNase, though it does reverse the effect of kinetin on it.

C. RESPIRATION IN SENESCING LEAVES

So much for the nitrogen metabolism in the leaf. The only thing we see so far about the action of the cytokinin is that it slightly promotes the rate of protein synthesis and it prolongs its duration; it also prolongs the uptake of amino acids, but that last may have little bearing on senescence. A more striking action of cytokinin is to prevent completely the change in respiration which takes place when leaves begin to senesce. Soon after senescence has begun, the oxygen consumption begins to increase, and by about the second or third day it reaches an enormous maximum, some 2½ times the initial value. This great rise is entirely prevented by cytokinin, and at the same concentration as that at which it prevents senescence. This phenomenon poses quite a puzzle, for what could cause respiration to increase by 2 or 3 times? The first possible answer is that because there is protein hydrolysis, and we know also that there is polysaccharide hydrolysis (the leaves contain a fructosan which breaks down to glucose or fructose), respirable substrate is being produced. If the leaf has more substrate it might well have a higher oxidation rate. Apparently this is true to a moderate extent, but it is not the whole phenomenon.

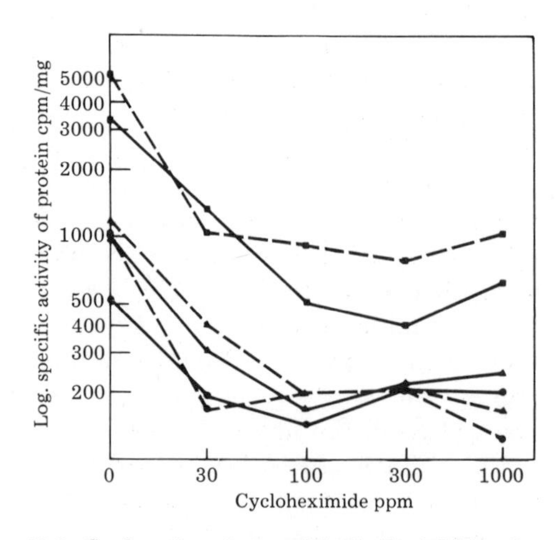

12-6 Confirmation that cycloheximide inhibits the formation of protein, done by measuring the incorporation of ^{14}C-*L*-leucine into protein of leaves. Circles, at 12 hours; triangles, at 24 hours; squares, at 72 hours after application. Solid lines, cycloheximide alone; dashed lines, with kinetin 7.5 mg/liter. The incorporation is reduced by an average factor of 7. From Martin and Thimann, 1972.

Figure 12-7 at upper left shows the oxygen uptake for a given size of leaf segment, placed in the dark. Already in the first day the rate has gone up somewhat, but subsequently there is a large rise followed by a fairly rapid decrease. All during this time kinetin is simply slightly depressing the initial rate of respiration. One might think that since kinetin maintains the leaf proteins and keeps it green, it would be expending energy to do that. On the contrary, it evidently lowers the rate of energy supply via the respiration. Another interesting thing which indicates that substrate is not the main controlling factor is the fact that fully etiolated leaves which have been growing in the dark show exactly the same phenomenon. Figure 12-7 at lower left compares green leaves with etiolated leaves from seedlings that have been growing in a dark room. As soon as they are cut off they show the same phenomenon, and again kinetin prevents that completely, causing only a slight drifting downwards. So the respiratory changes are the same whether

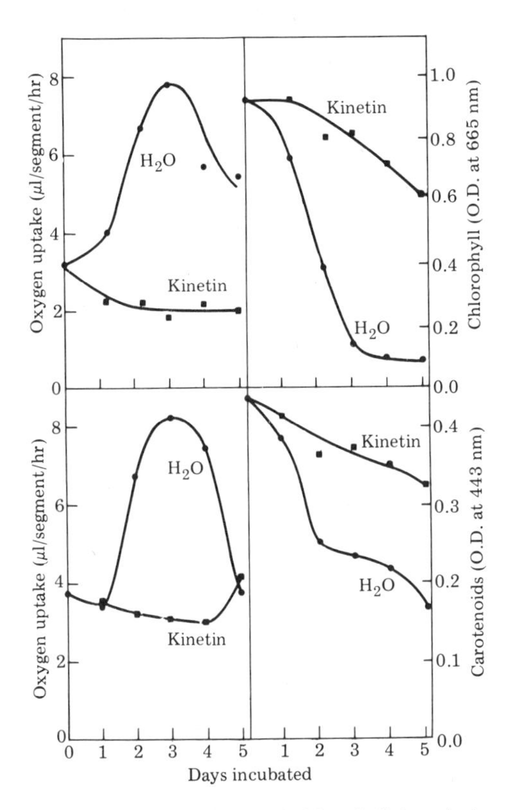

12-7 Above, respiration and chlorophyll loss during senescence in presence and absence of kinetin, in *green* oat leaves. Below, respiration and carotenoid loss, similarly, in *etiolated* oat leaves. From Tetley and Thimann, 1974.

we have a green leaf stuffed with polysaccharide or an etiolated leaf supplied only with the small amount of sugar that comes to it from the seed.

There is a similar parallel with the pigments, as shown in the two figures on the right. The decrease in chlorophyll as usual is very rapid. With these leaves and at this particular concentration, kinetin does not quite hold the greenness 100% but the chlorophyll breakdown is obviously greatly decreased. The fate of the carotenoids in the etiolated leaf is almost parallel; the carotenoids decrease rapidly, and kinetin prevents the decrease. This similarity of the general syndrome certainly makes one doubt that the liberation of sugars can be a prime influence, although it does participate.

Indeed figure 12-8 shows what actually happens to the sugars; the reducing sugar, which is the prime substrate for respiration at day zero, does increase on the second and third day, and kinetin has a moderate lowering effect. But thereafter the reducing sugar does not decrease the way the respiration does; sugar clearly increases with time. Kinetin is unable to prevent that completely, so that by the fourth day the values are the same. As for the sucrose, i.e., the reducing sugar liberated by invertase, the squares, which show the leaves with kinetin, and the circles, which are controls, show little significant difference. Indeed, in the case of kinetin the small difference is in the wrong direction.

When sugar solution is infiltrated into the leaves, it does cause a clear rise in respiration rate—not a 300% rise, but an 80% rise. However, even for sucrose, the effec-

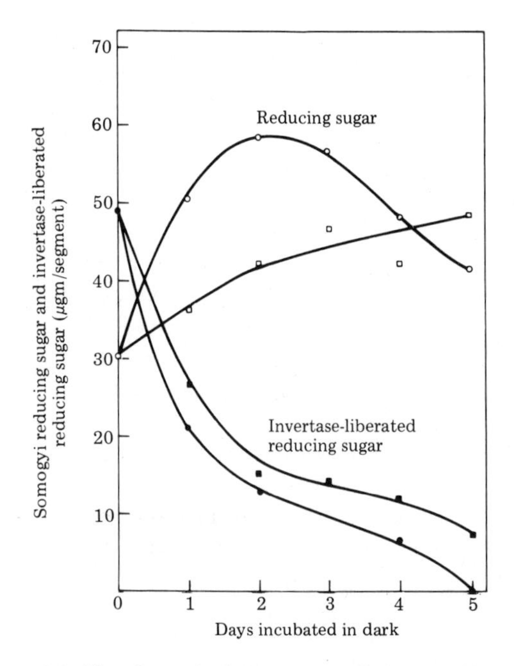

12-8 The release of reducing sugars and of sucrose-like sugar during the senescence of green oat leaves in the dark. Circles, untreated; squares, treated with droplets of kinetin 3 mg/liter. From Tetley and Thimann, 1974.

tive concentration is 1/10 molar. The maximum concentration in the leaves is not nearly so high—only about 30 millimolar. Hence the largest increase in respiration that could be blamed on sugar liberated, at 30mM, might be 40%. The liberation of sugars, therefore, could make only a small contribution to the increase in respiration. There is also a smaller effect due to the liberation of amino acids, of which alanine and some others will increase respiration. But again that effect is not enough to account for the observed respiratory rise.

Another consideration is the respiratory quotient, i.e., the ratio of CO_2 produced to oxygen absorbed. Both for sugars and for alanine this value is 1.0. But the measured RQ of these senescing leaves, especially on the third day, is about 0.7. This low figure would indicate either the oxidation of lipids or the withholding of CO_2 to form organic acids.

This led us to study the respiration of senescing leaves in regard to respiration inhibitors. The result brought up one point of great interest. It has been reported from time to time that the respiration of green leaves and green algae is insensitive to cyanide. It is also insensitive to carbon monoxide, but that is less remarkable because carbon monoxide requires fairly high levels to affect cellular respiration. But cyanide inhibits at least part of the respiration of most animal tissues at around one millimolar or a little above. This is one of the things that people do when they have decided they have had enough of life, namely to drink enough cyanide to reach an internal concentration of one millimolar, only they do not usually calculate quite so coldly.

Figure 12-9, above, shows the sensitivity to cyanide of oat leaves. Etiolated seedling leaves are almost normally sensitive to cyanide; as we increase the cyanide concentration the respiration suddenly drops drastically; at 10 mM it is down to about ⅓ of the control. Thus etiolated leaves behave much like amoebae, tissue slices, crabs, or other experimental animals. But there is a sharp contrast with light-grown green leaves. At a concentration which is nearly enough to kill etiolated leaves completely, the green leaves are uninhibited; in fact their respiration is increased by 30%. Such an increase has been noted before, especially for *Chlorella* by Emerson about 40 years ago; *Chlorella* is not only insensitive to cyanide but its respiration is slightly increased by it. With high cyanide concentrations there is inhibition, but in the range where normal poisoning takes place green tissue is insensitive. An interesting experiment is to take these etiolated leaves and place them in light so that they start to turn green. The right-hand figure shows the slow greening; their chlorophyll level comes up to normal in 60–70 hours. The effect of cyanide on the respiration is measured as the percentage of that in the controls. They start out being very sensitive to cyanide, and the sensitivity steadily decreases until they reach a point where cyanide actually increases their respiration by 22% to 30%. Thus as the leaf turns green it develops the cyanide-insensitive system, and at the same time the net respiration

Table 12-1 Effects of serine and arginine on leaf protease activity

		Protease activity	
	Protein in extract μg/2 ml	Units (nmoles amino acid/hr)	Specific activity (nmoles/hr/μg protein)
Experiment 1			
Initial Value	1663		0.83
Values after 3 days			
Control	829	857	1.04
Arginine	914	870	0.95
Serine	680	1097	1.61
Arginine + serine	709	900	1.27
Experiment 2			
Initial value	1190	763	0.64
Values after 4 days			
Control	699	943	1.36
Arginine	747	908	1.22
Serine	573	1281	2.23
Arginine + serine	667	1088	1.64

From Martin and Thimann, 1972.

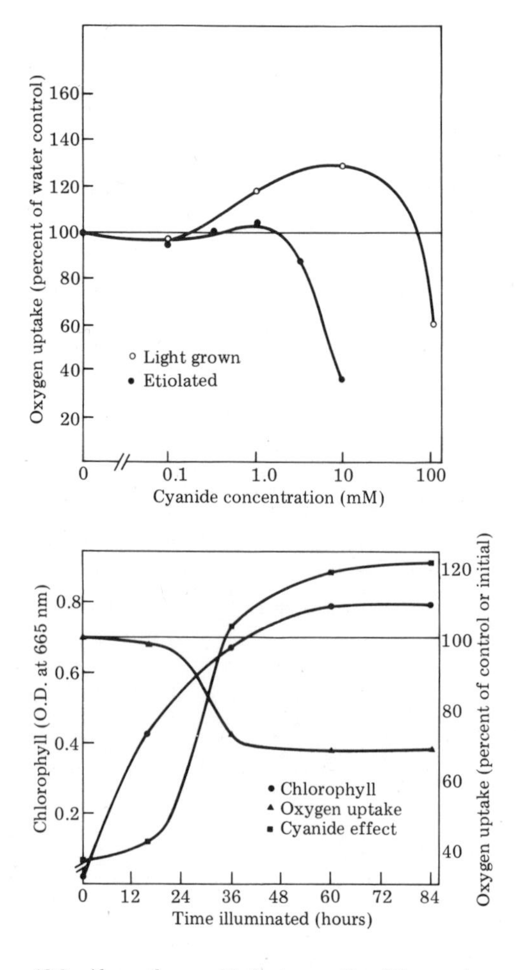

12-9 Above, the sensitivity to cyanide of the respiration of green and etiolated leaves. Below, the change in respiratory rate (triangles), cyanide sensitivity (squares), and chlorophyll content (circles) as etiolated leaves green up in light. From Tetley and Thimann, 1974.

somewhat decreases. The latter point is important because some have suggested that cyanide insensitivity is due to a very large increase in respiration, which is carried on by an alternative oxidase other than the cytochrome system. That happens in some fruits but cannot be the case here because the respiration actually decreases somewhat.

There is a striking parallel between this system and the greening of some algae (fig. 12-10). The experiment was done with *Chlamydomonas*, but it is comparable with *Chlorella*. The algae are raised in darkness and transferred to light; they green up as indicated by the curve of chlorophyll content. As they do so their respiration in the presence of cyanide goes up in beautiful parallel. Their ability to carry out photosynthesis in the light also parallels, but less precisely, since in the early stages of greening the small amount of chlorophyll may not be accompanied by the necessary enzymes, so that it is not very effective in carrying out photosynthesis. The curve of carbon monoxide resistance also goes (in general) parallel. This similarity in the behavior of etiolated leaves and algae needs careful study as a problem in respiration systems.

Another interesting aspect of the respiration of senescing leaves is shown by a study with dinitrophenol, which is the uncoupler *par excellence*; it uncouples phosphorylation from respiration in almost all known tissues. In plant segments it inhibits elongation in response to auxin or gibberellin, as well as inhibiting isodiametric cell enlargement in tissue slices—both types in proportion to log concentration (table 3-1).

The growth inhibition is accompanied by an increase in respiration, which is characteristic of uncoupling. What dinitrophenol does here is that it increases respiration only in leaves whose senescence has been prevented by kinetin. Table 12-2 shows leaves that have been in water in the dark and have lost most of their chlorophyll, i.e. have senesced. Others have been kept in kinetin and have maintained most of their chlorophyll. The leaves that have senesced for three days show very little effect of dinitrophenol—a 3% increase. The leaves that have been preserved by kinetin, on the other hand, show a drastic effect of dinitrophenol—a 50% increase. The second experiment ran for 3½ days instead of 3 days. Again there was a very small increase, 15%, caused by DNP in the senesced leaves, but a 70% increase in the leaves that had been protected by cytokinin. If we follow this phenomenon as a function of time during senescence, we find that the increase in respiration due to dinitrophenol reaches a minimum just at about the time of maximum oxygen uptake (fig. 12-11). At the outset respiration is virtually doubled by dinitrophenol—a 100% raise. But on the third day of senescence dinitrophenol only causes a 10% increase. Thus the respiration becomes resistant not only to cyanide but also to DNP in the senescing leaf. When kinetin keeps it from senescing it remains sensitive to DNP. I believe the explanation is that in the nonsenescent leaf oxygen consumption and ATP formation are tightly coupled. This enables the leaf to use ATP for its normal syntheses. In senescence,

12-10 The rise in resistance to HCN and CO of the respiration during chlorophyll formation and the onset of photosynthesis during germination and greening of the zygospores of *Chlamydomonas*. From Hommersand and Thimann, 1965.

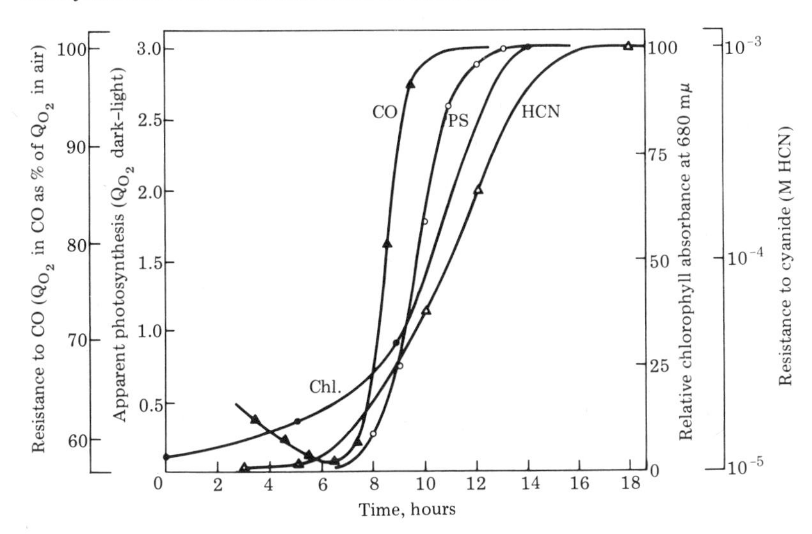

however, these two become uncoupled, and, as a result, the brake which phosphorylation exerts on respiration is released; the respiration soars up to 2.5 times its normal rate, but it does not do the leaf any good, since it allows little energy for synthesis. Hence, if kinetin is present, the most reasonable conclusion one could draw is that it must be restoring or maintaining the initial state, that is, keeping the respiration coupled to the phosphorylation and therefore sensitive to dinitrophenol. This means, then, that we have for the first time the beginnings of an idea of at least one of the modes of action of cytokinin: it acts to keep respiration and ATP synthesis coupled together.

The hypothesis that cytokinin acts to keep respiration coupled to phosphorylation would explain some of the things that cytokinin does. For instance, it greatly promotes the rate of growth of tissue cultures, as we have seen earlier, and growth is a process calling for synthesis and therefore for the effective use of phosphorylating energy. It promotes the growth of organs, in the case of lateral buds, and of course it acts on cell division, for the promotion of which the name cytokinin was originally coined. Both processes require energy. We saw in chapter 8 how kinetin causes amino acids to accumulate at the point where the kinetin is applied (fig. 8-23); and accumulation is well known to be an energy-requiring process dependent on the formation of high-energy phosphates. Skoog and his colleagues found kinetin to cause the phenolic substance scopoletin to be converted to its glucoside scopolin—a synthesis requiring ATP, or some other high-energy phosphate. Then, too, Van't Hof found that DNA synthesis is closely controlled by sucrose, sucrose being a prime respirable substrate yielding ATP to the cells. All these effects, apparently specific, might thus actually be the direct results of a rather nonspecific maintenance of the coupling of respiratory energy to phosphorylation.

Table 12-2 Action of 2,4-Dinitrophenol on respiration of senescent oat leaf segments in presence and absence of kinetin

Experiment	Pretreatment	Chl A_{665}	DNP (0.3 mM, pH 6)	Respiration: O_2 uptake $\mu l\ O_2$/segment·hr	Respiration: Increase %
I. 3 days in dark[1]	water	0.155	0	5.88	3.6
			+	6.09	
	kinetin 3 mg/1	0.783	0	3.59	53.5
			+	5.51	
II. 3.5 days in dark[2]	water		0	6.87	15.4
			+	7.93	
	kinetin 3 mg/1		0	3.62	71.3
			+	6.20	

[1] Values are means of 3 groups of 10 segments each.
[2] Values are means of 2 groups of 10 segments each.
From Tetley and Thimann, 1974.

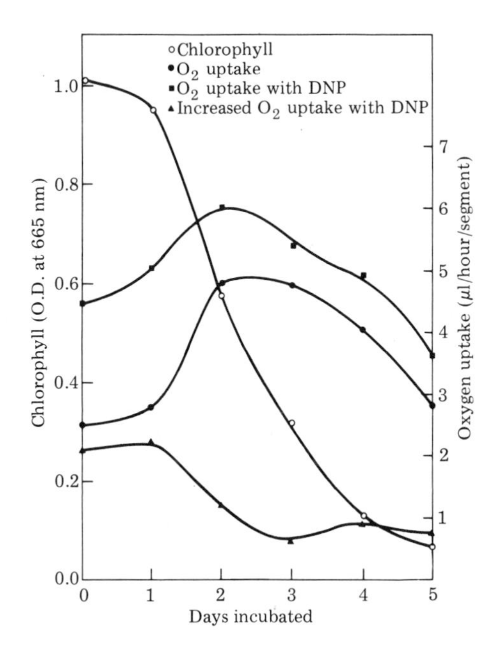

12-11 Showing that the increase in respiration caused by dinitrophenol (triangles) is minimal when the rise in respiration due to senescence is near its peak. Detached green oat leaves in the dark at 25°C. From Tetley and Thimann, 1974.

D. COMPARISON OF CYTOKININ WITH GIBBERELLIN

One difficulty about accepting the above idea is that in a number of plants gibberellin prevents senescence about as well as kinetin. Goldthwaite and Laetsch found this with discs cut from *Rumex* leaves (1968), and figure 12-12a, shows the similar behavior of *Taraxacum* (dandelion) leaves; the chlorophyll and protein are both well maintained. Whyte and Luckwill (1966) and L. Beevers (1966) have made comparable observations. Although this response seems not to be universal, it has been reported already in three plants in different families, namely, *Rumex, Taraxacum,* and *Nasturtium*. In *Rumex*, indeed, figure 12-12b, shows that gibberellic acid is in fact active at lower concentrations than the two cytokinins tested. For the prevention of protein breakdown neither of the hormones tends to be quite as active as on chlorophyll, yet, as we saw above, kinetin can maintain the protein of oat leaves at very close to the initial value. The ribonucleic acid analyses reported in several species also show rather clear-cut maintenance, but the enzyme RNase seems to increase and then decrease during the development of senescence.

There seems to be only one other aspect of plant behavior in which the cytokinins and gibberellins act in the same way, and that is in the germination of light-promoted or light-inhibited seeds, discussed in chapter 2. But in that case, although both types of hormone lead eventually to germination, they almost certainly act on quite different limiting steps of the overall process; and their actions are, at least in part, additive or even synergistic. In senescence they really seem to act in the same way—a great mystery. Although auxins have little effect in oat leaves, or in most other objects studied, yet Sacher has found auxin to delay senescence in bean pods, and some years ago Osborne reported that 2, 4-D delayed senescence in leaves of the cherry laurel; these observations deserve to be revived now that other hormones seem to be predominant. We noted that 6-methylpurine, which is normally an antagonist of purine metabolism, does have at low concentrations a small senescence-promoting effect, and it would be interesting to see its effect in *Rumex, Nasturtium,* or *Taraxacum* leaves. It might be, of course, that both of these hormone types owe their effect on senescence to the ability to keep the tissue supplied with energy; but then this common effect would be in addition to some more specific ways in which they bring about their characteristic and specific effects, such as the formation of buds in the case of cytokinin, or the elongation of leaves and stems in the case of gibberellin.

During senescence one would expect the endogenous hormone levels to decrease. Chin and Beevers certainly found that gibberellins decrease during the senescence of *Nasturtium* leaves (1970), and they also reported an increase in AbA. However, the AbA content of leaves is so sensitive to small changes in the environment that it is hard to be sure that this increase was not due to drying out or other secondary change. Whether cytokinins also decrease in senescing leaves is not clear, but it seems that they must, because their control over senescence is so complete, and it extends to those plants in which gibberellin acts as well.

The apparent similarity of action of cytokinin and gibberellin in senescence, and their quite different actions on other plant processes, brings up a suggestion I made at the Tokyo meeting in 1973, that perhaps growth substances, in general, owe their special importance to having *two* basic functions rather than one. Several pieces of evidence point to the possibility that effectiveness of a hormone means that it has the trick of performing two different actions.

E. SENESCENCE IN WHOLE PLANTS

Back in 1929 Hans Molisch, a pioneer in many fields from plant sugars through purple bacteria to bud dormancy and its release, associated senescence in annual plants with the formation of fruits and seeds. His idea was that these reproductive processes exhausted all the plant's reserves, so that it essentially died of starvation. If the flowers or the young fruits were systematically picked off, the senescence could be almost indefinitely delayed. Many years later, however, Leopold showed that in dioecious plants that formed only male (staminate) flowers, removal of these flowers (which naturally could produce no fruit) still postponed senescence. He deduced that flowers secrete a "senescence factor," but so far this concept, though not unreasonable, has not been supported. There is a comparable phenomenon in some seedlings where, if the adult leaves are systematically removed, the cotyledons will enlarge, stay green, and will not abscise. In mature soy-

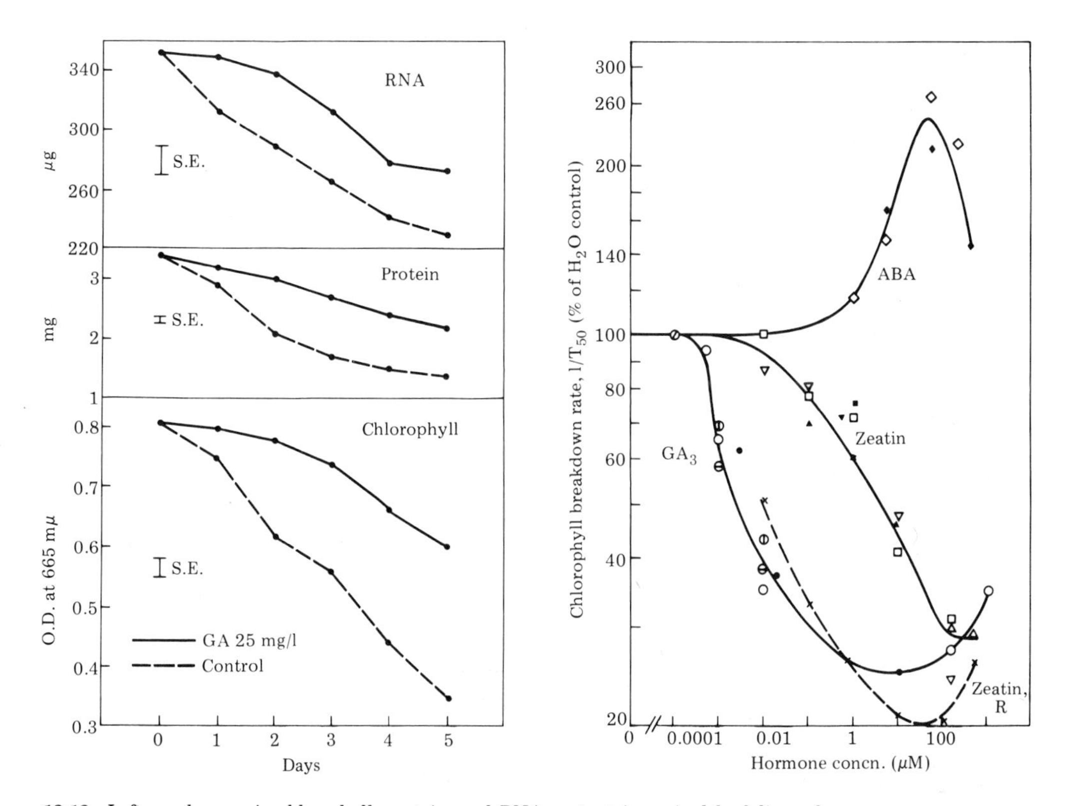

12-12 Left, a: changes in chlorophyll, protein, and RNA content in excised leaf discs of *Taraxacum officinale* in the dark. Solid lines, with gibberellin; dashed lines, controls on water. From Fletcher and Osborne, 1966. Right, b: comparative effectiveness of GA_3, AbA, and zeatin in preventing senescence of *Rumex obtusifolius* leaves in the dark. The data plotted are the rates of chlorophyll breakdown, as percent of that of controls. The different symbols refer to different experiments and show the general repeatability of the data. The dashed line (Zeatin, R) shows the behavior of comparable leaves on which the zeatin was renewed every two days, clearly improving its effectiveness. From Manos and Goldthwaite, 1975; cf. Goldthwaite, 1972.

beans, if the pods are systematically removed, the plants stay green. These relationships are evidently complex, and relationships between leaves have proven somewhat easier to analyze.

In the intact plant the lower leaves tend to senesce while the upper leaves are growing; this happens in virtually all plants. But in deciduous trees in the autumn we do not even see young leaves growing, for all the leaves on the tree senesce at the same time. There are several interesting things about senescence in whole plants. We can approach one of them by using senescence in intact whole leaves as a model. Compare two similar oat leaves; cut one into three segments and follow the rate of loss of protein and gain of amino nitrogen. The changes in the apical and the middle part follow almost identical curves; those in the basal part are very similar, but since it has a little less protein in it, it shows not quite as much gain (fig. 12-13). But basically the curves are parallel; as tissues they all senesce at the same rate. Keep the second leaf intact during senescence and then analyze those same three sections. The middle part exactly follows the curve it would have followed had it been isolated. The basal part, however, contains more amino nitrogen than it would have done if isolated. The excess is accounted for when we see that the apical part begins to accumulate amino acids, but that then their concentration goes down again. Evidently the amino acids which are liberated by proteolysis in the apical part have moved down the leaf and accumulated in the basal part. Thus in the intact leaf there is a polar movement of amino nitrogen, i.e., of amino

12-13 The increase and migration of amino acids within a 7-day-old green oat leaf separated into 3 segments or left intact. From Thimann, Tetley, and Tran Van Thanh, 1974.

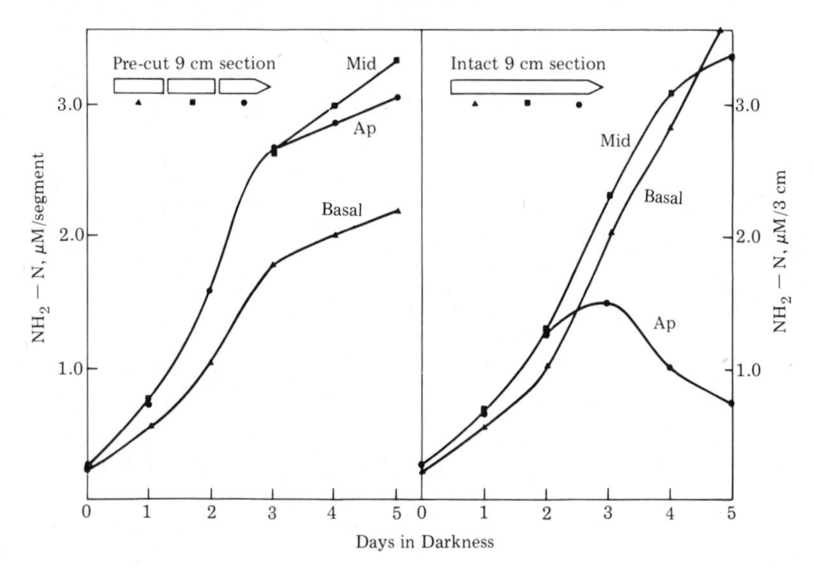

acids, from the apex towards the base. The movement is not rapid compared to the polar movement of auxin, which moves at more than 10 mm an hour. In order to find a clear-cut difference between apical and basal segments in the experiment described, we waited about three days. The amino acids moved some 70 mm in 70 hours, or about 1 mm per hour. The movement is definitely polar; there is no question of any reversal from the base to the apex.

Now we are ready to compare leaves senescing when separated from the rest of the plant and when remaining on the plant, i.e., detached and attached. Figure 12-14 shows the loss of chlorophyll. In the detached leaf, as we have seen, by the fifth day the chlorophyll has come down to about 7% or 8% of the initial value. When the leaves are left attached, there is approximately a day lost, and thereafter they lose chlorophyll at the same rate. Thus the effect of attachment differs from the effect of light, which slows down the rate; attached leaves senesce at about the same rate as detached ones, but are merely a little later starting. So leaving the leaf on the plant does not affect the breakdown of chlorophyll. The behavior of respiration is basically similar. On the first day there is little or no effect; then there is a very large rise in respiration and later a decrease again. When the leaves are on the plant the rise in respiration does not reach so high a maximum level but it takes place at the same time, just after the first day. Thereafter it decreases at the same rate as when the leaves are detached. The most interesting data are given by the amino acids. When the leaves are detached there is a buildup of amino acids, as we have seen, because the proteins are hydrolyzed. When the leaves are attached, however, the amino acids increase and then go down again. The behavior resembles that of the apical segment of the leaf, but this is now the whole leaf attached to the plant. By 2.5 days the amino acids have reached about as high a concentration as they will, and thereafter they decrease, which must mean that the amino acids are going out of the leaf. That in turn means that they are going into the base of the plant, that is, into the crown; and since in a seedling the crown is very slight, they must be moving essentially into the roots.

Table 12-3 shows some data on the roots. The amount of amino acids in the roots at the start of senescence is very small, but after five days the amount has greatly increased. The amount of protein in the same roots has remained the same, so that all this gain in free amino nitrogen must have come in from the leaves. This means that the plant has carefully husbanded its nitrogen, returning it into the roots, where it will be available for the next leaves as they grow up. However, if we put the roots in a cytokinin solution, the plant senesces in essentially the same way but we find much less amino nitrogen and more protein. The total amount of nitrogen is about the same as without the cytokinin, but what has happened is that the cytokinin has made the roots convert much of the free amino nitrogen into protein. I believe that this is actually the first time cytokinin has been shown to promote protein synthesis in roots. The effect is fairly large, about ⅔ of

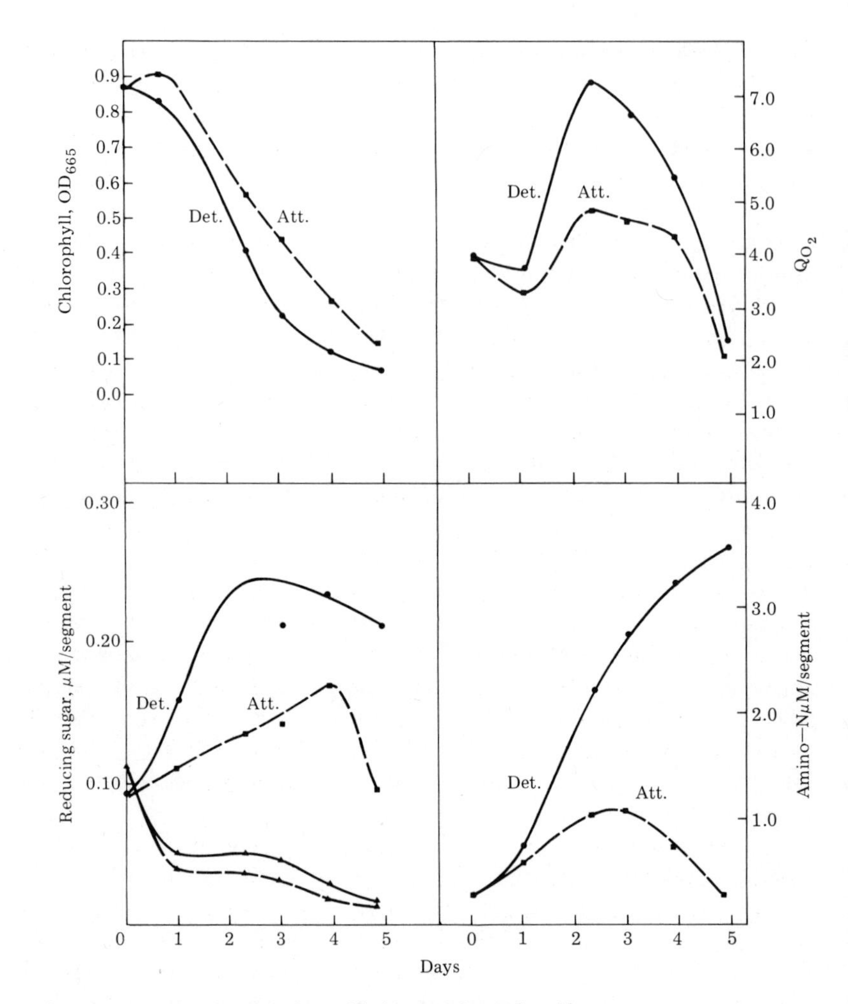

12-14 Chlorophyll, respiratory rate, sugars, and amino acids in senescing 7-day-old oat leaves either detached from the plants (DET) or left on (ATT). Triangles, reducing sugar with kinetin. From Thimann, Tetley, and Tran Van Thanh, 1974.

the added amino nitrogen having been converted to protein in presence of the cytokinin.

Some years ago Drs. Leopold and Kawase did the following experiment: they used young bean plants having only the first and second leaf. They applied cytokinin to the first leaf and then watched the senescence of the second leaf. The senescence of the second leaf was promoted—one of the very few cases where cytokinin actually *promotes* senescence. Figure 12-15 shows that kinetin on the first leaf keeps it dark green, but the second leaf turns yellow (or at least turns yellow faster than it would have done in controls). In the light of what we have just seen in oat seedlings, we can deduce that the cytokinin caused the first leaf not to excrete any of its amino nitrogen, which the second leaf needed; instead the cytokinin made it convert the amino acids into protein, with the result that the second leaf became nitrogen deficient. We can see again how such an influence on synthesis would require energy and therefore would be dependent on keeping up the coupling between respiration and phosphorylation.

The many aspects of the recent revival of interest in senescence would take us too far afield to describe here in detail. For example, to some extent calcium acts like cytokinin. It not only prevents senescence somewhat, but its action augments that of cytokinin (fig. 12-16). Furthermore, in *Rumex* leaves where, as we saw above, gibberellin acts to prevent senescence, the effect of calcium is clearly additive to that of gibberellin. At one-tenth molar, however, calcium has the opposite effect and activates protein hydrolysis. Another metal, most probably iron, helps to initiate senescence. Chelating agents which act on organic iron compounds have a strong effect in preventing or delaying senescence. The meaning of both these sets of observations is not yet understood.

There is a point of some interest in regard to the relative stability of intact leaves and isolated chloroplasts. Since proteolysis in oat leaves precedes the breakdown of chlorophyll (fig. 12-1) it appears as if the chloroplasts are being attacked by enzyme systems originating in the cytoplasm. Indeed, isolated washed chloroplasts in a suitable suspension medium are much more stable than in the leaf, (fig. 12-17) and only after 5 days at 25° do they begin to show signs of senescence. They even main-

Table 12-3 α-Amino nitrogen and protein nitrogen of roots of 10-day-old seedlings after 5 days in benzyladenine in darkness. All values are the means of three parallel experiments, except "initial" which are means of two.

		After 5 days in:	
	Initial	BA (3μg/ml)	H_2O
Amino acids (μg leucine eq/mg fresh wt)	0.83 ± 0.07	2.22 ± 0.14	6.30 ± 0.47
Protein (μg protein/mg fresh wt)	17.24 ± 1.03	21.26 ± 3.76	16.99 ± 2.62
Total	18.07	23.48	23.29

From Thimann, Tetley, and Tran Van Thanh, 1974.

12-15 Senescence of the first trifoliate leaf of *Phaseolus* brought on by treating the two primary leaves with benzyladenine. Photographed after 14 days. From Leopold and Kawase, 1964.

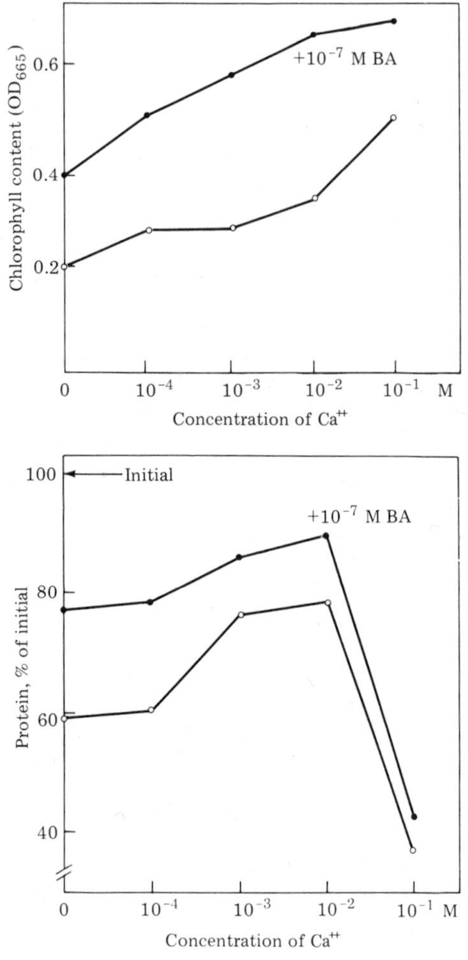

12-16 Effect of calcium chloride on the chlorophyll and protein contents of corn leaf discs, alone and in the presence of 10^{-7}M benzyladenine, after 3-5 days in the dark. From Poovaiah and Leopold, 1973.

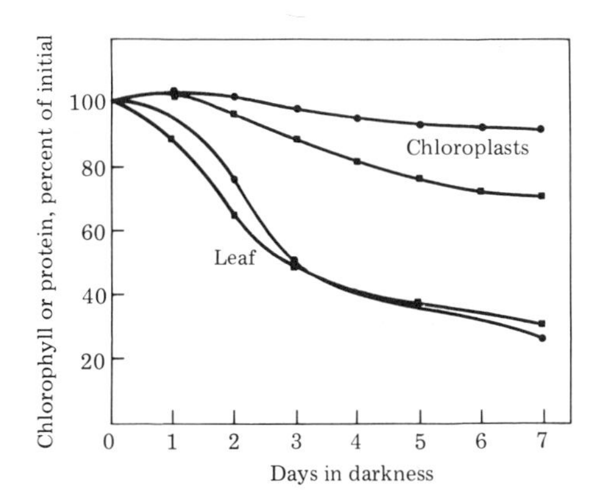

12-17 Comparison of the stability of chlorophyll (circles) and protein (squares) when in isolated chloroplasts and in the leaf. From Choe and Thimann, 1974.

tain some photosynthetic ability for several days, and traces of photosystem I can be detected at up to seven days. In the intact leaf these properties would have decreased rapidly. In order to duplicate the relatively rapid senescence of chloroplasts in the leaf, it was found necessary to add a mixture of enzymes, both proteolytic and lipolytic, and so far the maximum senescence rate attained has not quite equalled the senescence rate of chloroplasts in the intact leaf.

Thus the study of this one phenomenon of senescence brings us into protein metabolism, amino acid transport, the mode of action of the cytokinins, respiration, the inhibitors of respiration, and even photosynthesis; indeed into most of the major aspects of the known physiology of the plant. In the old days senescence was thought of as essentially starvation, and about 10 or 20 years ago papers on senescence would be called "On the Metabolism of Starving Leaves." We are not really dealing with starvation, for senescence is a normal phase of growth and development, first requiring synthesis of hydrolytic enzymes, then requiring the metabolism of the breakdown products, including the husbanding of nitrogenous compounds into other parts. It must be considered as part of normal growth, and correlated with other aspects of the role of hormones in plants.

REFERENCES

Barton, R. Fine structure of mesophyll cells in senescing leaves of *Phaseolus*. *Planta* 71, 314–325, 1966.

Beevers, L. Effect of gibberellic acid on the senescence of leaf discs of Nasturtium *(Tropaeolum majus). Plant Physiol.* 41, 1074–1076, 1966.

Chin, S. T-Y., and Beevers, L. Changes in endogenous growth regulators in Nasturtium leaves during senescence. *Planta* 92, 178–188, 1970.

Fletcher, R. A., and Osborne, D. J. Gibberellin as a regulator of protein and RNA synthesis during senescence in leaf cells of *Taraxacum officinale. Canad. J. Bot.* 44, 739–745, 1966.

Goldthwaite, J. J. Further studies of hormone-regulated senescence in *Rumex* leaf tissue. In *Plant growth substances 1970,* ed. D. J. Carr, pp. 581–588. Berlin: Springer, 1972.

———, and Laetsch, W. M. Control of senescence in *Rumex* leaf discs by gibberellin. *Plant Physiol.* 43, 1855–1858, 1968.

Hommersand, M. H., and Thimann, K. V. Terminal respiration of vegetative cells and zygospores in *Chlamydomonas reinhardi. Plant Physiol.* 40, 1220–1227, 1965.

Leopold, A. C., and Kawase, M. Benzyladenine effects on bean leaf growth and senescence. *Amer. J. Bot.* 51, 294–298, 1964.

Manos, P. J., and Goldthwaite, J. A kinetic analysis of the effects of gibberellic acid, zeatin and abscisic acid on leaf tissue senescence in *Rumex. Plant Physiol.* 55, 192–198, 1975.

Martin, C., and Thimann, K. V. The role of protein synthesis in the senescence of leaves. 1. The formation of protease. *Plant Physiol.* 49, 64–71, 1972.

Poovaiah, B. W., and Leopold, A. C. Deferral of leaf senescence with calcium. *Plant Physiol.* 52, 236–239, 1973.

Shibaoka, H., and Thimann, K. V. Antagonism between kinetin and amino acids: experiments on the mode of action of cytokinins. *Plant Physiol.* 46, 212–220, 1970.

Tetley, R. M., and Thimann, K. V. The metabolism of oat leaves during senescence. 1. Respiration, carbohydrate metabolism and the action of cytokinins. *Plant Physiol.* 54, 294–303, 1974.

Thimann, K. V., Tetley, R. M., and Tran Van Thanh. The metabolism of oat leaves during senescence. 2. Senescence in leaves attached to the plant. *Plant Physiol.* 54, 859–862, 1974.

Thomas, H., and Stoddart, J. L. Separation of chlorophyll degradation from other senescence processes in leaves of a mutant genotype of Meadow Fescue (*Festuca pratensis* L). *Plant Physiol.* 56, 438–444, 1975.

Whyte, P., and Luckwill, L. C. A sensitive bioassay for gibberellins based on retardation of leaf senescence. *Nature* 210, 1360–1361, 1966.

The Abscission of Leaves and Fruits 13

Our plant has been getting steadily older and, as you remember, it flowered the time before last. The leaves are now beginning to pass their maturity and some are turning yellow. We have discussed their senescence and the numerous changes that it involves. Now we come to the last stage in the life of a leaf, namely its falling off (abscission). Fruits and flowers also have to face this ultimate fate, but the leaf occupies center stage in this, the last act.

The career of the leaf of a dicotyledonous plant can be divided into five phases: (1) The formation of the initial primordium is obviously the first, critical creational phase. (2) Second comes the enlargement, and some change of shape usually goes with that. Often the primordia are long and narrow in dimensions but the leaf blade becomes broader as it enlarges; it may become lobed or even compound, so the enlargement and changes in shape go on at the same time. If the enlargement takes place in light (as it usually does) pigmentation, i.e., formation of the photosynthetic pigments and sometimes formation or loss of anthocyanin occurs as well. (3) After these growth changes comes the third phase, a relatively long period of maturity. This is followed by (4) senescence, which we have just discussed, and finally by (5) abscission.

The monocotyledonous leaf shows variations on this theme. In the second phase the elongation is usually far greater and the increase in breadth rather slight. There are a few monocotyledons with lobed leaves, the palms being the most notable example, but we know little about their development and still less about any role of growth hormones in it. Pigmentation and senescence are comparable, but the last phase of abscission is either missing or greatly delayed. The brown leaves of a dried pasture in late summer are familiar reminders of the almost complete lack of abscission in the grass family, and the dried old leaves of the Washingtonia palm hang on around the trunk for years. Ferns are slow to abscise also.

A. ANATOMICAL ASPECTS OF ABSCISSION

Although abscission is basically a phenomenon of dicotyledons, its timing is not the same in all members of the group. Typically, the five phases take place within one growing season, though the leaves of temperate-zone oaks mostly stay on through the winter and abscise in the spring. In "evergreens" the leaves go through a second or even a third growing season before they senesce and abscise. Most pines and spruces keep their leaves for at least three seasons. In *Pinus Cembra* it is claimed that they stay for eight years before they drop off. On the other hand there are a few annual dicots which do not abscise readily. Tobacco abscises very poorly. Rumex hardly abscises at all and the dead leaves hang on, dangling from the base of the stem.

The study of abscission has been carried out partly on leaves and partly on fruits. Very roughly what has been found out for one tends to hold for the other also. The first stages of senescence involve all the cells of the leaf; we have seen how the whole leaf gradually loses chlorophyll, develops proteases, and undergoes other changes. Abscission does not. It involves only a very few cells, usually either the cells at the base of the petiole or at the base of the blade where the blade joins the petiole, or both. It is a narrow zone which is involved in abscission, though it may be multiple, depending on the kind of leaf. In a compound leaf with 10 or 20 leaflets, every one of these can abscise at its base and the whole leaf abscises also at the main base of the rachis. Thus there can be 20 or more *abscission zones*. An abscission zone is popularly thought of as being at the base of the petiole, comprising a set of cells oriented more or less perpendicular to the axis of the petiole, and dividing in that plane. While that does describe the typical abscission zone, many leaves do not have an abscission zone; in *Impatiens*, for example, the cells do not divide at all; when the proper time comes the whole petiole and blade just fall off. Sometimes there is cell division later, on the proximal side after the leaf has fallen off. Very often in old age leaves produce a scarcely visible abscission zone; but, in any case, they become cut off at the point where an abscission zone would normally form. The Boston ivy, which often grows on university buildings (inter alia), has two abscission zones. At the beginning of the fall, the leaf blades fall off at the junction of blade and petiole so that the wall of the building is covered with petioles; one or two weeks later these abscise at their bases. The presence of one or more abscission zones is very characteristic and appears to be species-specific. Figure 13-1 shows the typical phenomenon, as exemplified by *Coleus*, in the work of Wetmore and Jacobs. There are divisions taking place in all the cells across the petiole base, forming a clearly visible abscission zone; and a few days later, the abscission has taken place. On the opposite side, the petiole is just entering into abscission, beginning to break away. The cell division is not essential; it is the separation, the ability of those cells at the base to separate that is most characteristic. The xylem is not involved, but since it is very weak by itself, it breaks off at the first slight shock.* Pods and other dry fruits abscise to scatter the seeds. This process has been little studied, however.

* For those interested in structural aspects, the recent paper of Webster et al. (1976) with 48 references, is listed in the references to this chapter.

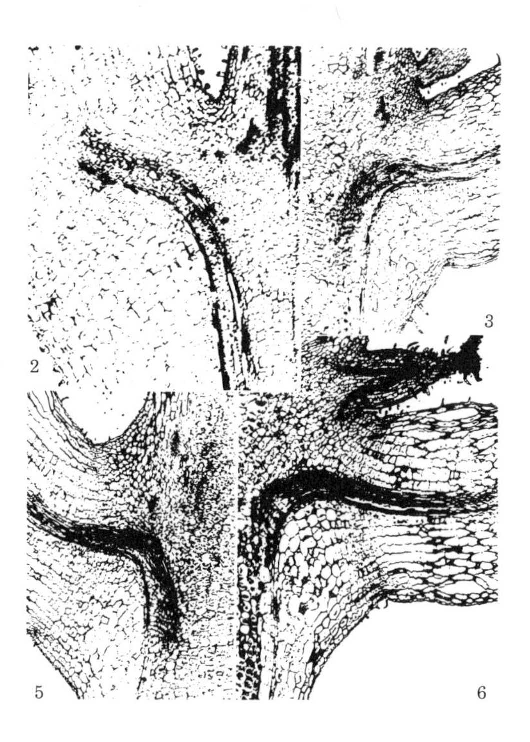

13-1 Photomicrographs of the abscission zones of petioles nos. 2,3,5, and 6 respectively (counting from the apex) of a young *Coleus* plant. The abscission layer is visible already in petiole 3 (upper left) and is well developed in no. 6 (lower right). Abscission generally occurs by the time the leaf has become no. 7 or 8. From Wetmore and Jacobs, 1953.

Now for the analysis. It was long ago shown by Küster (1916), an experimenter with many different interests, that in *Coleus* the blade controls the abscission. If the blade is taken off, the petiole falls very quickly. In his many little experiments, involving no more tools than a pair of scissors, he simply sliced away various parts of the blade, or sliced all the blade away, leaving only the main vein. As a result he showed that it takes only about one square centimeter of leaf tissue at the base to hold it on. After Küster the problem was rather neglected until auxin came into the picture. At that time (1933) there were published the now classical experiments of Laibach and his pupil Mai, who used pollinia of orchids, which are pollen grains joined together with sticky material on the outside which will adhere to the stigma of an orchid. Since these pollinia are extremely rich in auxin, they used them for auxin experiments (at that time auxin was not available as a pure chemical). They showed that if the blade were removed completely and an orchid pollinium applied at the base, then the petiole would stay on. In controls to which a drop of plain lanolin or other control material was applied, the petiole fell off. Thus it was clearly an auxin effect. The very next year (1935), when indoleacetic acid was available, LaRue in Michigan showed that pure indoleacetic acid kept the petiole on. In *Coleus*, *Ricinus* and in many other similar plants the effect of the blade can be completely imitated by auxin. We know that the blade produces auxin, at least when it is young and active, so we deduce that *auxin prevents abscission.* There were some intermediary stages in the argument, one or two pieces of work by Myers and by Jacobs and Wetmore, which showed that the amount of auxin obtainable from a leaf blade by putting it on agar changes with the life of the leaf, decreasing as the leaf gets old. This parallels exactly the ability of the blade to keep the petiole on. Thus it is clearly established that so long as auxin comes from the blade, abscission remains inhibited. That is unfortunately the simplest part of the story.

To follow the preparation for abscission, one cannot wait till the leaf falls; one has to tap it and, to be precise, tap it in a standardized way. This has led to rather elaborate design of basically simple experiments. Some have reduced the procedure to just using the base of the blade, cutting everything else off, including most of the petiole, and also cutting the main stem off and applying materials to one or the other of the cut surfaces.

Figure 13-2 shows the set-up with cotton plants used by Addicott's group at Davis. Luckwill at Bristol, England, has used a similar method with bean plants. Starting with a standard seedling, one takes off the cotyledonary leaves and the apex, leaving a little structure with the two opposite petiole bases and the base of the main axis. Droplets of test solution are applied to these; they are held in a petri dish in a moist atmosphere, and every day they are tapped with a standardized forceps in a standardized way.

Although the simplest statement is that auxin controls abscission by inhibiting, and

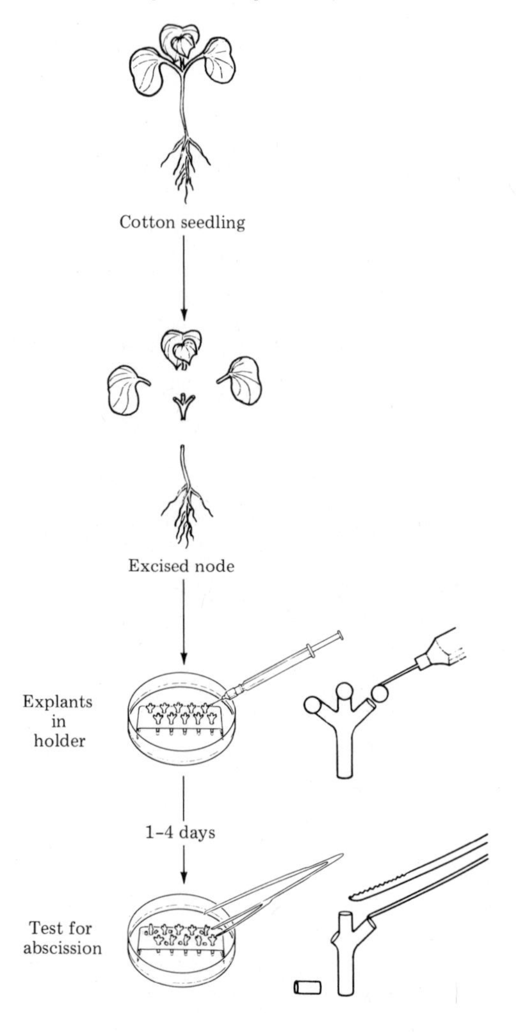

13-2 Bioassay for abscission, using cotton explants. From Addicott et al. 1964.

auxin comes from the young blade, there are some circumstances in which auxin can promote abscission instead of inhibiting it. In some cases auxin promotes at low concentrations and inhibits at high, giving optimum-type curves like those we have seen in regard to growth and bud inhibition. In a few cases auxin promotes abscission if supplied to the stem apex instead of to the petiolar cut surface. More remarkable, perhaps, auxin applied to the base of the petiole, just below what would normally become the abscission zone, will sometimes also promote abscission instead of inhibiting it. Thus we can have an opposite effect just by varying the place at which it is applied. Lastly, an interesting phenomenon discovered by Rubinstein and Leopold is that the time of application of auxin makes the difference. Auxin applied at time zero (fig. 13-3) causes the time required for abscission to be very long, sometimes several weeks. But if we take off the blade and wait perhaps 3 hours and then apply auxin, it now has only a small effect; the petioles perhaps stay on only a few hours longer than the controls. If we wait 6 hours (instead of 3) and apply auxin, now it has the opposite effect—it hastens abscission, so that the time for abscission for the treated plants is less than the time taken by the controls. Thus they have divided the effect of auxin into two stages: stage one, when it inhibits in the classical sense, and stage two, when it promotes.

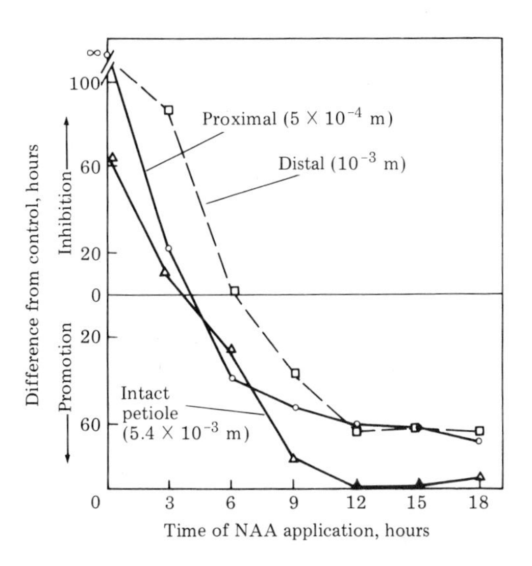

13-3 The change of response from inhibition to promotion of abscission when auxin application is progressively delayed. The circles and squares refer to isolated 10-mm petiole segments with the abscission zone in the center; the triangles are whole hypocotyls with entire petioles attached. From Rubinstein and Leopold, 1963.

Because of these three controlling factors—the proximal effect, the concentration effect, and the time effect—there have been a number of theories about how auxin acts. Most of them are essentially wrong, I believe, for one reason or another. I very much liked Jacobs's idea which was, very simply, that auxin inhibits abscission if it can promote the growth of the petiole. If the petioles are young enough so that they can elongate in response to the auxin, there is inhibition of abscission exerted by the same auxin. It is as though the auxin feels it needs to balance its acts; if it can produce growth in one place, it is happy to inhibit abscission in another place. The parallelism works out perfectly with Jacobs's *Coleus*. However, there are many cases where auxin still inhibits even though the petiole is too old to elongate in response. And 3, 5-D, which is quite inactive as an auxin (chap. 8, part 2), inhibits abscission to a moderate degree.

Secondly, Addicott's idea was that the cells at the base of the petiole like to have their auxin coming in the right direction. They are sensitive to the direction: if the auxin comes from the blade, abscission is inhibited, but if the auxin comes from the stem apex, abscission is promoted. One can imagine these cells nervously weighing their gradient of auxin molecules on either side trying to decide from which direction most of the auxin came and hence whether they should abscise or stay on, as the case may be. Experimentally one can show that there are always cases where time modifies the effect, as it does here. The time effect negates the auxin gradient concept, and there are numerous other cases where low concentrations promote and higher concentrations inhibit.

A third theory was based on pectin methylesterase, the enzyme which takes off the methyl groups from the pectins of the cell wall and by so doing lays the pectins open to hydrolysis by pectinase. Daphne Osborne in England made a particularly strong case for this, proposing that it was the stimulation or the inhibition of the pectin methylesterase in those petiolar cells which determined whether or not abscission would take place. In several cases, however, it was subsequently shown that the changes in this enzyme were either not very marked or were in the wrong direction. For instance, ethylene, which strongly promotes abscission, turns out to inhibit pectin methylesterase, and there were other cases where auxin, in concentrations which affect abscission, had no particular effect on the pectin methylesterase.

The theories are mainly of interest because they show that people were searching for an explanation of these fairly complex phenomena. The situation was made still more complex when it became clear that ethylene has exactly the opposite action to auxin, in that it promotes abscission quite strongly, sometimes at concentrations less than one-hundredth part per million. This realization goes far back in history, dating from the observations of Fahnestock in 1879. Fahnestock's greenhouse in Philadelphia was heated by gas. One day the gas leaked and all the plants, though not killed, were badly damaged. He carefully catalogued all the changes and found that abscission was by far the greatest damage. Some plants had lost all their leaves, while others lost the older leaves and kept a few young ones; depending on the species they all responded strongly. Subsequently the same thing has been seen in street trees where a gas main under the street develops a leak; the trees abscise most of their mature leaves. It was not known at first that the active constituent of the gas was ethylene, but this came through the studies of fruit ripening we have just discussed (chap. 12). Soon it was clear that a major effect of ethylene on plants is to cause abscission. Thus ethylene works in one direction, auxin in the other. Table 13-1 presents a simple example of such action.

The plot began to thicken when it was shown, mainly by Abeles and the Burgs, that auxin, applied in only slightly more than normal levels, causes the production of ethylene. This happens in nodes and petioles too; figure 13-4 shows how bean petiole explants respond immediately to applied NAA by evolving ethylene. Untreated controls produce none. There is a paradox here in that delayed auxin application tends to accelerate abscission rather than to inhibit it (fig. 13-4), but the same delayed application actually decreases the ethylene production. This paradox has yet to be explained.

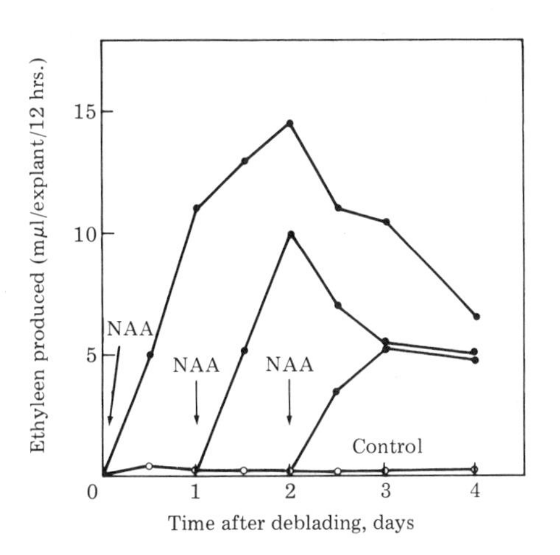

13-4 Production of ethylene when auxin is applied to bean petioles at different times after removing the leaf blades. The ability to make ethylene falls off steadily after deblading. From Rubinstein and Leopold, 1964.

Table 13-1 Effects of ethylene and auxin on the abscission of mature cotton leaves. Groups of about 50 young plants, with the shoots cut off and dipping in water, were treated in an enclosed space and observed for abscission 48 hours later.

Treatment	Percent abscised
Controls	0
Ethrel, 100 ppm, applied	96
Plants first sprayed with sodium salt of naphthalene-acetic acid, then treated with Ethrel as above	6

Data of Hall, 1952.

C. ABSCISIN AND DORMIN

Another aspect of abscission control arose from work on the abscission of cotton fruits. It is important to keep as many cotton fruits on as possible because each forms a boll of cotton; and since many of the young fruits fall off, Dr. Addicott at Davis (with support from the cotton growers) made a study of this. He found that an extract from the young cotton fruits that have fallen off contains something which promotes this abscission. This extract was called abscisin.

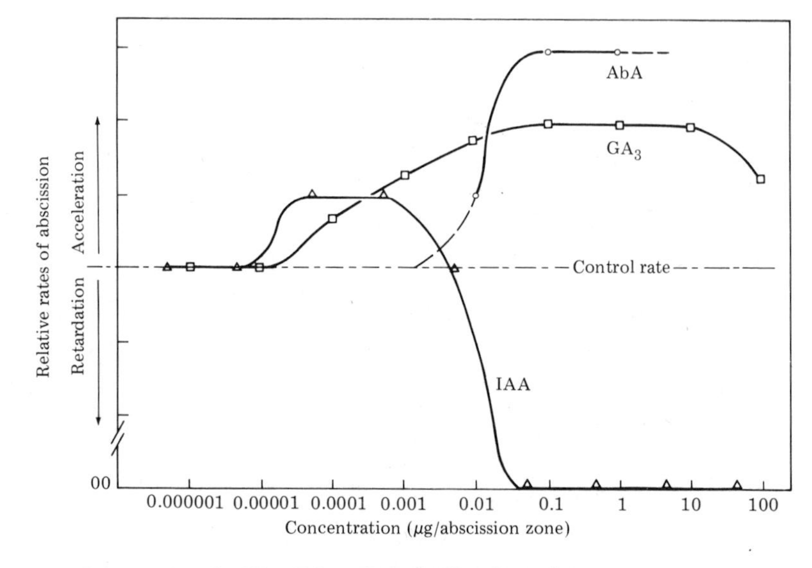

13-5 Diagram of relative rates of abscission produced by AbA, auxin, and gibberellic acid on cotton seedling explants. The actual times needed for 50% abscission were roughly: controls, 4 days; IAA, 3 days; GA_3, 2 days; AbA, 1 day. From Addicott et al. 1964.

Ohkuma, a Japanese worker in Addicott's laboratory, fractionated the abscisin preparation and separated two substances, which they called abscisin I and II. Abscisin II was purified and turned out to be a carotenoid derivative (see formulae figs. 7-9 and 13-6).

Figure 13-5 shows that abscisin II, now called abscisic acid, accelerates the abscission of cotton petioles in the explant test, and works in about the same concentration range as that in which auxin delays the abscission.

At the same time, but 6,000 miles away and quite independently, Wareing in Wales was working on the formation of winter buds in sycamore. In many trees, towards the end of the season, the apex goes on producing leaves, each of which bears an axillary bud. As the days shorten, the time comes when it stops producing leaves, the shoot stops elongating, and the apex now produces only bud scales. These bud scales basically start out as leaf primordia, but instead of enlarging into leaves they form little scales which tightly encase the bud. Wareing found first that this phenomenon, at least in the case of *Acer*, is caused by the short days, but later he found that the leaves that are formed late in the season can yield an extract which, when applied to a growing apex, will cause it to form bud scales. In other words, he had an extract which caused the formation of winter buds. He might have called it scalin, but he resisted the temptation to do that because he realized that he was dealing with passage of the shoot into dormancy for the winter; so he called the material of this extract dormin.

A third piece of research in Wain's laboratory, also in England, stemmed from the fact that when lupines, especially the yellow lupine, *Lupinus luteus* (which Van Steveninck in Australia was also interested in) form mature pods, these promote the abscission of other pods on the same plant. To make a long story short, mature pods produce an extract which promotes abscission of the young pods. This might have been called podin, but fortunately the terminological confusion was avoided because it very quickly turned out, largely through the efforts of Dr. Milborrow at the Shell Research Station in England, that all these three preparations were the same: abscisin, dormin, and the lupine abscission substance were all abscisin II which, because it was an acid, was rechristened abscisic acid. As we saw in chapter 7, the formula looks very much like a breakdown product from half a carotene molecule: it has an important basis for stereoisomerism at the carbon atom next to the ring, which is an asymmetric one, so that there are a D- and an L-form. Furthermore, at carbon atom no. 2 (from the acid group) there is geometrical isomerism, giving rise to an active *cis* form and a much less active *trans* form. The difference in activity between these forms is in the same direction as in the auxin-active cinnamic acids.

At first AbA was thought of as a breakdown product of carotenoids. This idea was greatly stimulated by the remarkable results of Taylor and Burden in England, who exposed the carotenoid violaxanthin to very bright light and showed that it broke down, with partial oxidation, to produce, in good yield, a substance very close to the structure of abscisic acid—so close indeed that it could (at least on paper) be converted to AbA with a further loss of 2H, since it has an aldehyde group where AbA has a carboxyl. Its activity as a growth inhibitor is indeed comparable to that of AbA, especially when the *cis-trans* mixture is used. Because this substance is yellow they called it xanthoxin, a name which also includes the suggestion of an oxidized product. If it were dehydrogenated and the epoxide opened up, one hydrogen transfer would form abscisic aldehyde. In fact an enzyme called epoxygenase can cause the same oxidative breakdown of violaxanthin in the dark. What is more important for physiology, Taylor and Burden (1972) have shown that when tomato shoots take up violaxanthin their AbA content greatly increases.

For a while it seemed likely that this was really the mode by which AbA was formed. Then came the remarkable results of S. T. C. Wright and co-workers, which showed that when leaves wilt the amount of abscisic acid in them goes up very markedly (chap. 7). This would be an important adaptation, since wilted leaves would tend to have their abscission promoted. The rapid appearance of AbA on wilting was at first ascribed to the presence of an inactive, combined form of AbA, tentatively considered to be perhaps a glycoside; but Milborrow, working at the Shell Laboratories in England with careful quantitative assays, could find only slightly more glycoside, perhaps 25 micrograms rather than 20,

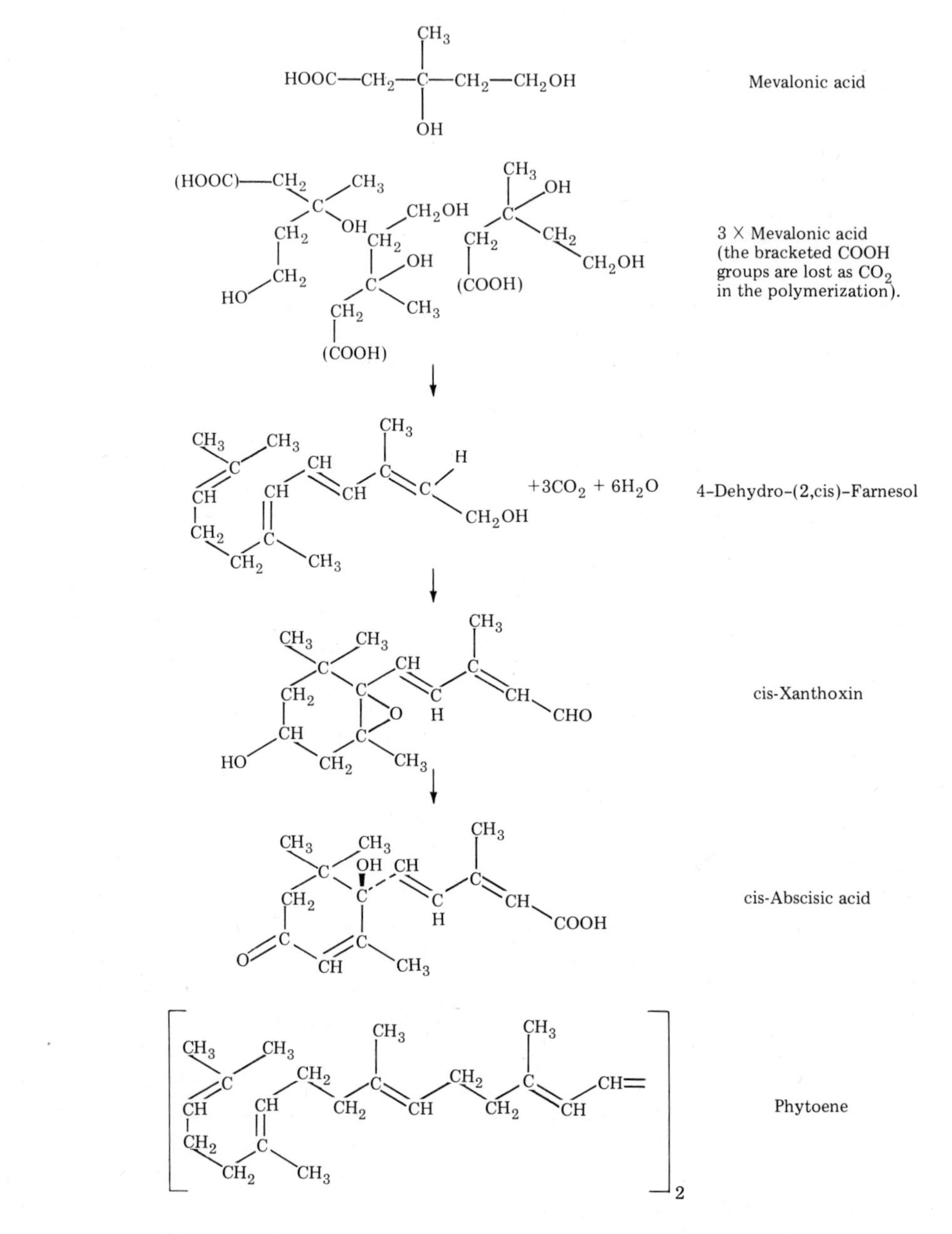

13-6 Probable route from mevalonic acid to *cis*-xanthoxin and to abscisic acid.

than the amount of free abscisic acid, so that it could not account for all the increase. Hence it must be formed anew, and in fairly large amounts—a deduction that would fit with the concept of there being a reservoir of carotenoid present which could rapidly be converted to abscisic acid. On the other hand, the chemical work aimed at applying labeled substances and getting labeled abscisic acid has not supported this idea at all. Again Milborrow and his colleagues were responsible for some very careful work here with labeling. They have, for instance, labeled the substance phytoene, which is a partially unsaturated material (fig. 13-6) in the direct biosynthetic line to carotenoids. ^{14}C-phytoene gave rise to ^{14}C-carotenoids, as had been shown before by many workers, especially Porter and Cole, but unfortunately gave rise to ^{12}C-abscisic acid. On the other hand you will remember from your biochemistry the chain of events that leads to carotenoids, starting with acetate, acetoacetyl Co-A, mevalonate, isopentenyl phosphate and geranyl phosphate, and phytoene. Milborrow therefore used labeled mevalonate, and had no trouble getting the labeled mevalonate to give rise to labeled abscisic acid; in fact the yield was actually somewhat better than the yield from other sources. Thus Milborrow and Cole concluded that carotenoids do not give rise to abscisic acid, but abscisic acid arises in the chain of events from mevalonate which could yield either carotenoids or abscisic acid equally. In other words it lies on a side branch from the carotenoid biosynthetic chain. There was some pretty work in which they labeled a *trans* derivative of farnesol (see formula, fig. 13-6), 2-*trans*-4-dehydrofarnesol, and that gave rise to labeled abscisic acid in a yield 40 times greater than that from mevalonate. On the other hand, 2-*cis*-4-dehydrofarnesol did not give rise to labeled abscisic acid. This shows that farnesol is not a source of general carotenoid-like materials which could be converted subsequently to abscisic acid, but that it can yield abscisic acid directly. This supports the idea of a general route from acetate or mevalonate, branching off later to produce either carotenoids or abscisic acid, so that the two are not directly related. We await final proof and, in any case, no one knows as yet where the branch point occurs in the biosynthetic route. Evidently, however, this is a biogenetic system of fundamental importance to the plant.

Abscisic acid was discovered in three different ways. That means that it must have a number of functions in plants. Although it promotes abscission in cotton fruits, which probably helps to explain why the cotton fruits fall off, that is almost certainly not the whole explanation, since its action in producing abscission is often weak. It produces some abscission in leaves of *Coleus* and cotton, but again its effect is nowhere near as potent as ethylene. It quite strongly promotes (or causes?) the formation of winter buds in the sycamore and several other trees. Similarly it promotes, or perhaps even causes, the formation of turions, resting winter buds, in aquatic plants like *Lemna*. It turns out to inhibit the germination of cereal seedlings, especially wheat and barley seedlings. In these you will remember gibberellin is produced, which causes the aleurone to produce amylase which acts upon the endosperm.

Abscisic acid prevents this secretion of amylase by the aleurone. It also inhibits germination in some other seeds. In those elongation tests in which one simply measures the amount of elongation of a coleoptile or stem segment, it inhibits that too, sometimes fairly strongly, depending on the species. Thus it has a rather general growth-inhibiting effect. For that reason and because its abscission effect is somewhat weak, it would have been better to stick to Dr. Wareing's terminology and call it dormin, in the sense that it promotes dormancy and inhibits the breaking of dormancy; in effect it is a factor working on the side of dormancy against vegetative growth.

Perhaps a more striking effect is that discussed in greater detail in chapter 7, namely the effect on wilting leaves. When leaves wilt the amount of abscisic acid is enormously increased, apparently because certain synthetic processes are stimulated. In 4 hours of wilting Hiron and Wright (1973) found the AbA content of wheat leaves to go up from 23 to 171μg per kg fresh weight. In the first 3 hours of recovery it fell to 116, and the subsequent return to normal was slower. Bean leaves showed even faster responses. The increase in AbA causes the stomata to close, sometimes within 20 to 30 minutes. Thereafter, there is a very interesting phenomenon; if the stomata close (due to dry air) while the roots are still in moist soil or in water they continue taking up water, so that with transpiration prevented the leaf soon regains its water content. Thus a cycle develops: the leaves drop down, the stomata quickly close, the roots go on supplying water, and presently the leaves regain their turgor. Then the stomata reopen, and after a short while the leaves wilt again, and so on. Striking movies have been made of this cyclic recovery, because in spite of the *continuance* of wilting conditions the stomatal closure makes the supply of water available again.

D. THE INTERRELATION OF HORMONES

We saw above that abscission is produced more powerfully by ethylene than by abscisic acid. This response is not limited to leaves. Figure 13-7 shows the effect of ethylene on flower buds. Here the treatment was with ethephon, a sulfate which loses H_2SO_4 quickly and thus releases ethylene locally. The figure presents, not the time to abscise, but the percentage abscission after a given time. In all three clones tested, the effect of the ethephon is almost immediate, while the controls are barely beginning to abscise even at 140 hours.

The effects on abscission are different for each of the hormones. In figure 13-5, auxin concentrations were applied in a logarithmic series on standard cotton explants. There is a zone of low concentrations where auxin causes marked, though moderate, promotion. Above that is a concentration which has no effect, and above that a zone of very drastic inhibition; the abscission actually remains zero for some time. Some fruits show similar behavior (see below). The effects of gibberellin show no such reversal; indeed they are not always as clear as

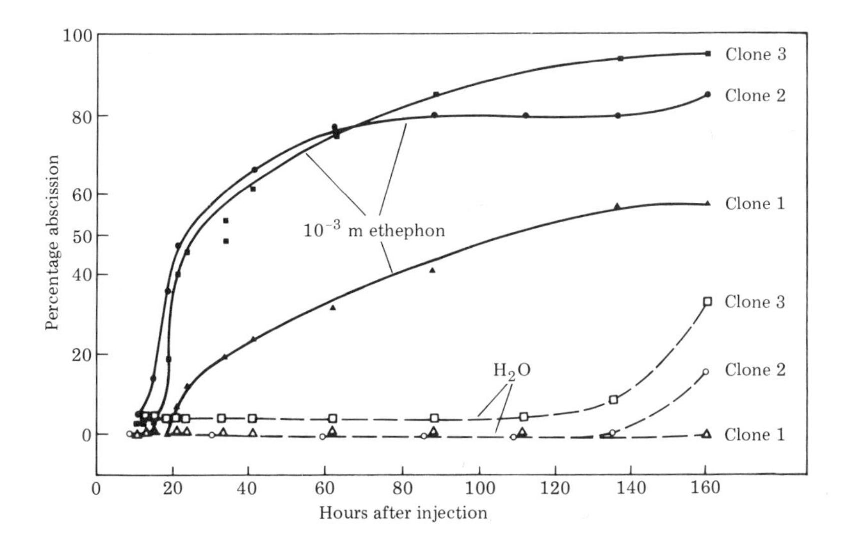

13-7 Effect of a single injection of 2 μl of a 1 mM ethephon solution on the abscission of flower buds in 3 clones of a hybrid *Begonia*. From Ten Cate, and Bruinsma, 1973.

shown here. It promotes abscission over a wide range of concentrations and never really inhibits. Abscisin II, in this experiment with cotton, which is very sensitive, gives 100% abscission but the concentrations needed are on the high side and in the same range as where auxin inhibits. Pretreatment of leaves with kinetin weakly inhibits, but pretreatment with GA strongly promotes abscission in bean leaves (fig. 13-8).

There are some striking and unexplained exceptions. In *Begonia*, the plant with two abscission zones close together, auxin *promotes* abscission. It is not a reaction to low concentrations—all the concentrations promote. If we add to that a substance which we know promotes, namely abscisic acid, it instead reduces the effectiveness of auxin, acting here like an abscission inhibitor. Similarly, with ethylene, which we know promotes and is indeed the most powerful promoter of abscission known, abscisic acid now works in the opposite direction, decreasing the effect of ethylene in causing promotion. AbA also acts to decrease the auxin-induced production of ethylene. Indeed, in another *Coleus* species, both an increase in the production of AbA and a decrease in that of IAA seem to occur together (table 13-2).

Table 13-2 Hormones "diffusing" into an agar receiver (in μg/leaf) from the leaves of *Coleus rehneltianus*

Stage of leaf development	IAA	AbA
Young, enlarging	31	0
Fully grown	3	1
Senescent	2	15

Data of Böttger, 1970.

In *Begonia* flower buds, however, the female flower buds, which stay on, contain nearly 100 times as much extractable auxin as the male buds which abscise, but the Bruinsma group finds that the auxin *content* is not necessarily related to the amount of auxin reaching the abscission zone, and a high abscission rate runs parallel to an apparent blockage of auxin transport out of the bud. Neither the AbA nor the ethylene production correlated with abscission of these buds, and they conclude that access of IAA to the abscission zone "prevented the tissue from becoming sensitive to ethylene" (cf. the implication of table 13-1). According to this view, it is when the tissue becomes low in auxin that the ethylene already present is able to cause abscission.

Thus abscission offers a prime example of a complex hormone interaction. And probably the type of interaction is different in different plants.

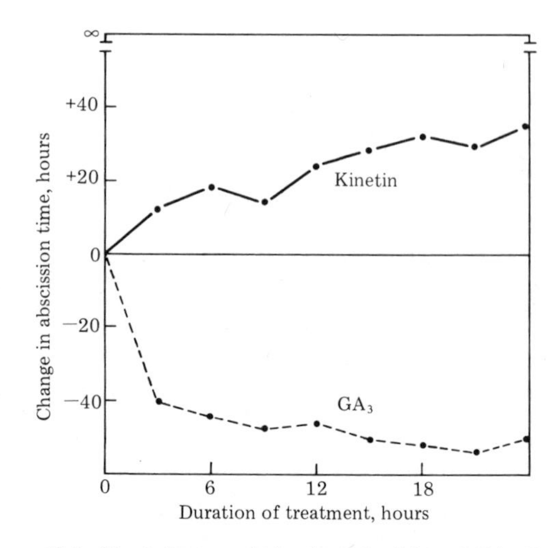

13-8 The influence of kinetin (0.5 mM) and GA_3 (5 mM) on the abscission of bean explants. (Note that acceleration is shown by a *decreased* abscission time.) From Chatterjee and Leopold, 1964.

E. THE ABSCISSION OF FRUITS

There is much uncertainty about the role of abscisic acid in the normal life of the plant. In the abscission of fruits, the findings are very peculiar. Fruits abscise very much like leaves, except that they usually have only one abscission zone, where the pedicel joins the stem. They abscise more or less as they senesce, sometimes a little later. If fruits are abscising too fast, growers apply auxin and inhibit the abscission, as we shall see below. So one might think that the natural abscission, which takes place towards the end of the fruit's life cycle, is due to abscisic acid. Indeed, some abscisic acid is formed as

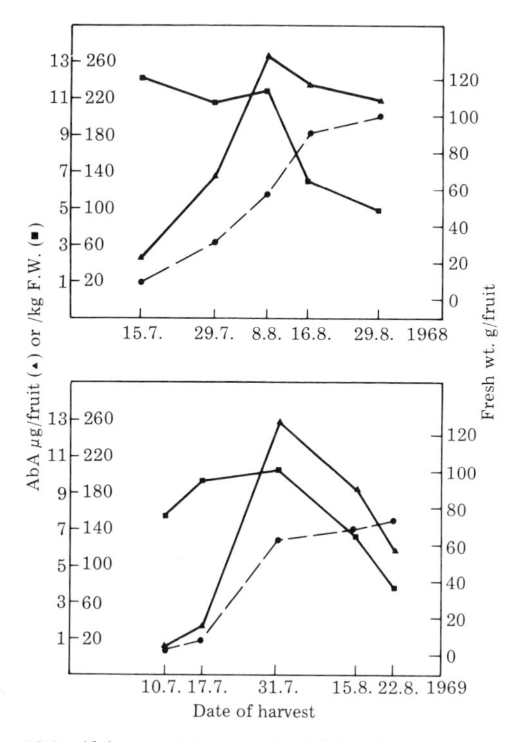

13-9 AbA content in μg per fruit (triangles) or per kg fresh wt. (squares) during maturation and ripening of tomatoes. The dashed lines are fresh wt. Two successive years. From Dörffling, 1970.

fruits mature, but the unfortunate thing for its functional explanation is that generally the amount of abscisic acid decreases as the fruits get ready to abscise. Just at the time when the AbA would be expected to reach its maximum, its level comes down. The abscisic acid is obviously being decomposed, probably by oxidation, to which it is very sensitive. At the time of abscission its amount becomes like the amount in very young fruits which do not abscise at all. Figure 13-9 presents analyses of the AbA in tomatoes during the ripening period, July and August. The abscisic acid level goes up as the fruit is turning ripe but comes down just at the time of abscission. This behavior is repeated in a number of analyses. It inclines one to believe that the chief importance of AbA in the life of plants is to control processes other than abscission.

F. APPLICATIONS OF ABSCISSION CONTROL IN HORTICULTURE

The action of auxin on the abscission of fruits has had extensive use in commercial horticulture. One reason is that the onset of abscission in many types of fruit is rather rapid, so that harvesting must be carefully timed. Also the ripening of apples and late plums and the harvesting of pears commonly occurs at a time of strong winds, at least in North America and western Europe; consequently, farmers can experience considerable losses due to premature fruit drop. As we saw in figure 11-11, the production of auxin by the seeds falls off markedly as they mature and at about the same time ethylene begins to form, so that

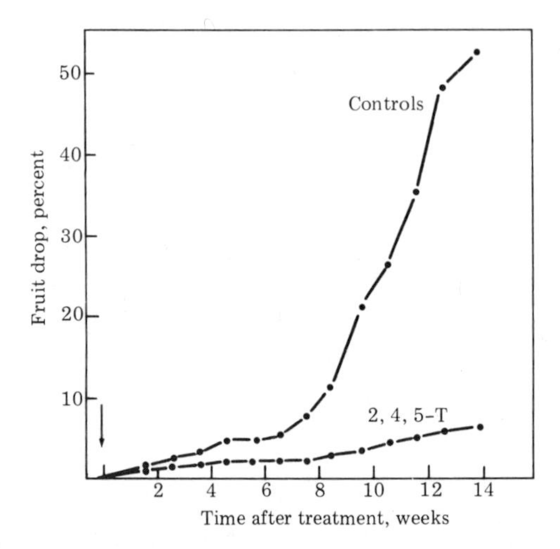

13-10 The inhibition of abscission of apricots by a single spray with 2,4,5-trichlorophenoxyacetic acid (25 ppm). Data of Crane plotted by Leopold, 1958.

two hormonal reactions act in the same direction to hasten abscission. The fall in auxin formation can be readily prevented by spraying the fruit with auxin solution, and this procedure is widely used on apples. Not only does it delay the falling, but it makes it easier for the farmer to plan the harvesting at his convenience rather than the weather's. For this use synthetic auxins are much the best, IAA (probably because of its sensitivity to the oxidase-peroxidase) being relatively ineffective. Naphthaleneacetic acid will keep apples from falling for from 2 to 6 weeks or more, depending on the variety. For apricots, 2, 4-D and 2, 4, 5-T are even more effective, the effect of the latter being to delay the fall more or less indefinitely (fig. 13-10).

Oranges and lemons produce so little ethylene (chap. 11) that they rarely drop prematurely, and indeed the problem for citrus harvesting is the reverse, namely to promote abscission so as to make the picking easier. Spraying with Ethrel or other compounds which release ethylene has been tried, but unfortunately this causes abscission of the leaves as well.

G. ENZYMATIC ACTION

The process of abscission itself, especially in leaves, is very interesting. What we see is this: in the earliest stages, before any abscission can be observed, cavities begin to form in the middle lamella (fig. 13-11). Evidently, then, the first stages involve the production of pectin-hydrolyzing enzymes which attack the middle lamella. Correspondingly, one finds in general that the total content of pectin-hydrolyzing and cellulose-hydrolyzing enzymes does increase at the time of abscission. Pectin methylesterase, which I mentioned at the beginning, was an unfortunate choice because it seems not to have a very close relationship to this phenomenon or, in general, to the gross appearances of abscission. But several pectinases and also Albersheim's pectin transeliminase do increase at about this time. The cellulase increase is quite marked and, what is more significant, it is increased further by ethylene. Here we have a direct connection between ethylene and the abscission-producing enzymes. Daphne Osborne has further shown that the formation of cellulase is depressed by auxin, which again is an effect in the expected direction, though unfortunately of a cellulase rather than of pectinase which appears to be the first reacting system. However, it is satisfactory that ethylene promotes and auxin inhibits.

Where does this enzyme come from, and how does it suddenly begin to attack the cell wall? The situation seems much like that in senescence described earlier. Inhibitors of protein synthesis tend to inhibit abscission. Actinomycin D is very effective in this regard, one of the better inhibitors, but still better is cycloheximide, which, as we saw, inhibits senescence. These two substances, which both prevent the synthesis of proteins, prevent abscission; it seems fair to deduce that the abscission enzymes—the pectinases and cellulases—are mainly formed *de novo*. They are probably synthesized locally by the adjacent cells; enzymes do not travel freely from cell to cell.

I have mentioned that gibberellic acid sometimes has a weak senescence-pro-

13-11 The walls of the cortical cells of a tobacco petiole (near the base) examined after deblading. Arrows point to disintegration beginning in the middle lamella of the wall. The plasmalemma, P, appears intact. From Jensen and Valdovinos, 1968.

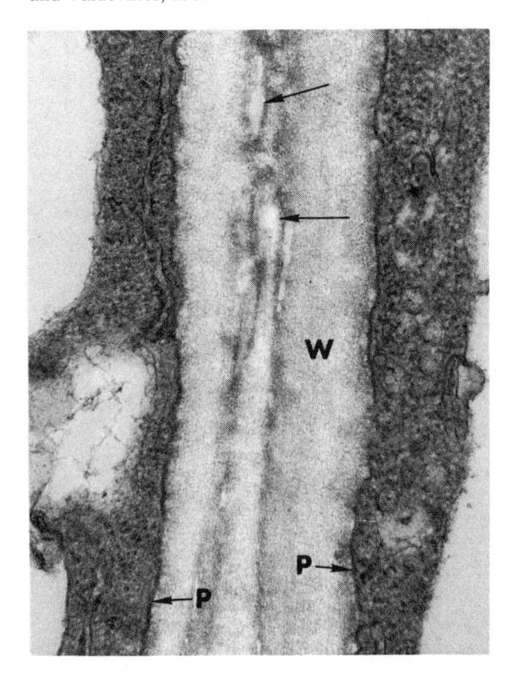

moting effect. Rubinstein and Leopold found a second weak effect, particularly interesting in regard to the relationship with senescence. They found that amino acids promote abscission. Of the amino acids, glycine and alanine were among the most potent. We know that when a leaf is aging and is beginning to senesce, amino acids are liberated, and we saw in chapter 12 that amino acids tended to move polarly out of the leaf, towards the base. That means that the cells in the abscission layer are receiving a stream of amino acids. At the same time they are receiving a lessened stream of auxin. Thus they are subjected to two different influences, both of which act in the same direction. We saw also that in senescence the proteases are, in part at least, formed *de novo*. Thus there is a parallel between the processes in the leaf, where the hydrolytic enzymes are formed *de novo*, and those in the petiole, where the polysaccharidases, cellulase, and pectinase are also formed *de novo*. These phenomena roughly tend to go on at the same time. The first process furthermore supplies amino acids, which promote the second process, abscission.

As far as the auxin effects go, it seems clear that auxin applied to the tissue inhibits abscission by depressing the formation of some of these enzymes. But auxin applied directly on the node promotes abscission, not because it is coming in the wrong direction, for the cells may not mind which way it comes from, but because node tissue, as Burg has subsequently shown, readily produces ethylene when it receives a little more than the physiological levels of auxin. Thus the auxin applied to the node actually produces ethylene, which stimulates abscission. The node therefore operates with a nicely balanced system in which ethylene promotes and auxin inhibits, but in which auxin gives rise to ethylene; this system can work in either direction depending on the exact level of auxin. This happens particularly at nodes, since much less ethylene is formed in internodes or petiolar tissue—something like 30 or 40 times less than in nodes.

The above discussion leaves out the effect of gibberellin, which is not easy to explain. I think it may not be physiologically significant. In many cases it has been shown that gibberellin applied to young plants increases the auxin supply by a factor of two or more, and it might be that the gibberellin is acting via auxin production. Another reason why one is not sure whether it is significant is that, in general, the output of gibberellin from leaves drops off with old age. In young leaves the amount of gibberellin present is clearly largest at the base, so that it would be supplied to the petiole, but in older leaves this is not the case, which means that the gibberellin changes are in the wrong direction. Hence it is questionable whether the gibberellin effect is of importance. It may be of more importance in the abscission of fruits than in that of leaves, but since the effect is generally minor, I fancy that in those cases where it is clear-cut it is acting via auxin.

In any case, with the clear control by auxin, ethylene, and cytokinin, three hormones are involved in the abscission process, providing an example of the general thesis that much of plant behavior is under multiple hormonal control.

Our plant has now completed its life cycle; the fruit has matured and fallen to the ground, and the seeds are dispersed. If it was a deciduous plant, the leaves have senesced and fallen too. We are back at the beginning, with the germination of the new seed.

It only remains to try to elucidate in a little more detail how some of these hormonal actions have been achieved.

REFERENCES

Addicott, F. T., Carns, H. R., Lyon, J. L., Smith, O. E., and McMeans, J. L. On the physiology of abscisins. In *Régulateurs naturels de la croissance végétale,* pp. 687-703. Paris: C.N.R.S., 1964.

Böttger, M. Die hormonale Regulation des Blattfalls bei *Coleus rehneltianus* Berger. II. *Planta* 93, 208-213, 1970.

Chatterjee, S. K., and Leopold, A. C. Kinetin and gibberellin actions on abscission processes. *Plant Physiol.* 39, 334-337, 1964.

Dörffling, K. Quantitative Veränderungen des Abscisinsäuregehaltes während der Fruchtentwicklung von *Solanum lycopersicum* L. *Planta* 93, 233-242, 1970.

Gaur, B. K., and Leopold, A. C. The promotion of abscission by auxin. *Plant Physiol.* 30, 487-490, 1955.

Hall, W. C. Evidence on the auxin-ethylene balance hypothesis of foliar abscission. *Botan. Gaz.* 113, 310-322, 1952.

Hiron, R. W. P., and Wright, S. T. C. The role of endogenous abscisic acid in the response of plants to stress. *J. Exptl. Bot.* 24, 769-781, 1973.

Jensen, T. E., and Valdovinos, J. G. Fine structure of abscission zones. 3. Cytoplasmic changes in abscising pedicels of tobacco and tomato flowers. *Planta* 83, 303-313, 1968.

Laibach, F. Versuche mit Wuchsstoffpaste. *Ber. deut. bot. Ges.* 51, 386-392, 1933.

LaRue, C. D. The role of auxin in the development of intumescences on poplar leaves; in the leaf-fall in *Coleus. Amer. J. Bot.* 22, 908, 1935.

Leopold, A. C. Auxin uses in the control of flowering and fruiting. *Ann. Rev. Plant Physiol.* 9, 281-310, 1958.

Rubinstein, B., and Leopold, A. C. Analysis of the auxin control of bean leaf abscission. *Plant Physiol.* 38, 262-267, 1963.

———. The nature of leaf abscission. *Quart. Rev. Biol.* 39, 356-372, 1964.

Taylor, H. F., and Burden, R. S. Xanthoxin, a new naturally occurring plant growth inhibitor. *Nature* 227, 302-304, 1970.

———. Xanthoxin, a recently discovered plant growth inhibitor. *Proc. Roy. Soc. Ser. B.,* 180, 317-346, 1972.

Ten Cate, C. H. H., Berghoef, J., van den Hoorn, A. M. H., and Bruinsma, J. Hormonal regulation of pedicel abscission in Begonia flower buds. *Physiol. Plant.* 33, 280-284, 1975.

———, and Bruinsma, J. Abscission of flower bud pedicels in Begonia. 1. Effects of plant growth regulating substances on the abscission with intact plants and with explants. *Acta Botan. Neerl.* 22, 666-674, 1973.

Webster, B. D., Dunlap, T. W., and Craig, M. E. Ultrastructural studies of abscission in Phaseolus: localization of peroxidase. *Amer. J. Bot.* 63, 759-770, 1976.

Wetmore, R. H., and Jacobs, W. P. Studies on abscission; the inhibiting effect of auxin. *Amer. J. Bot.* 40, 272-276, 1953.

Concepts of the Mechanism of Action of the Hormones 14

A. PRELIMINARY CONSIDERATIONS

Now that our plant has produced its hormones, reacted to them, and completed its life cycle, it is appropriate to review what knowledge we have about the bases on which the many actions of hormones ultimately rest. The trouble with auxin is that its actions are so numerous and so apparently unrelated. Elongation involves no cell division, requires the uptake of water, sugar and potassium, and is closely linked to respiration. Cambial activation, on the other hand, involves little enlargement but is mainly stimulation of mitoses. Root formation starts with rapid mitoses and then follows with cell elongation. But as soon as root elongation is well launched and the root initial is outside the stem, auxin inhibits its elongation. And apical dominance is clearly an inhibition of elongation.

Much of the work has focused upon elongation, but here also there are two complications: (a) Different cells or tissues respond differently, as for instance the outer and inner layers of the stem in the slit pea-stem curvatures (even *Avena* coleoptiles, with only 6 cell layers, show the reaction beautifully); (b) The growth always shows optimum curves, generally giving the best response at around 2–10 mg IAA per liter or $1\text{–}5 \times 10^{-5}$M. At first this seemed to be a true function of the mode of action, but later studies by Abeles and by the Burgs have given these optima a quite different complexion. For auxin in superoptimal concentrations causes nearly all plant tissues to form ethylene, which is an inhibitor of elongation as well as of some other types of growth. The concentration that triggers ethylene production is just about the same as that which starts the decrease in growth, as shown in figure 14-1. There is a lag of 2 hours or so before ethylene begins to appear, so that growth may start vigorously in high auxin concentrations and then quickly level off or even stop. Not only is this important in explaining a very common optimum response, but it raises a strange question: how can auxin both stimulate growth and cause ethylene production, for these appear to be such unrelated effects? As yet the question is unanswerable. Leaving aside these two complications, we will take up a series of concepts of the basis of auxin action, one by one, and finally see what can be said about the mechanisms of action of the other hor-

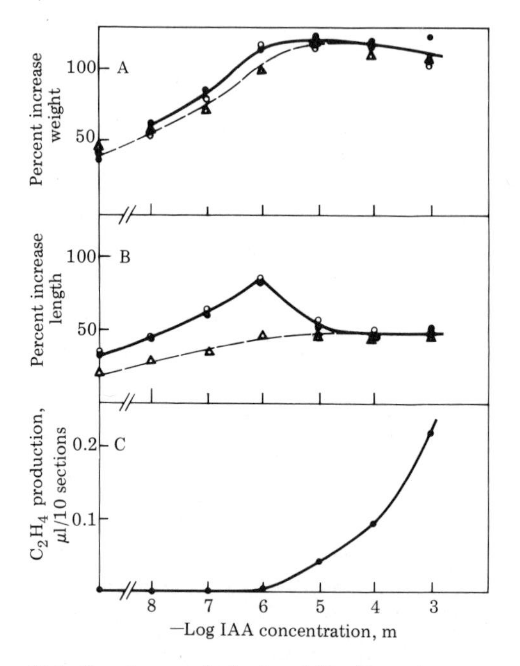

14-1 Top: increase in fresh weight of pea stem segments in IAA (circles) or IAA plus ethylene (triangles). Middle: increase in length of the same. Bottom: evolution of ethylene by the segments. All as function of IAA concentration. From Burg and Burg, 1968.

mones. One must bear in mind that in nature several hormones are normally present at once, so that when one applies a single hormone exogenously its effect may be modified by, or even dependent on, endogenous hormones already present in the tissues.

B. AUXIN AND CYTOPLASMIC STREAMING

One of the first effects to be observed when auxin is applied to responsive tissue is an acceleration of the rate of streaming, or cyclosis, in the cells. It starts almost at once (fig. 14-2), certainly within 2 minutes in the *Avena* coleoptile (at 25°C), while the soonest that an acceleration of growth rate can be detected is, under normal conditions, about 10 minutes. (By raising the temperature and increasing the auxin concentration Zenk, in Germany, has reduced this time lag for growth acceleration almost to zero, but it is not certain whether his growth effect may be the result of acidity in the auxin solution.) The acceleration of streaming, which occurs at physiologically low auxin concentrations such as 0.2 mg per liter, soon slows down again, but it can be maintained by the addition of sugars—glucose, fructose, or sucrose (fig. 14-2). Since elongation of coleoptile segments also requires sugar for its continuation (chap. 3) this is a very suggestive parallel. It is annulled, as is the respiration, by cyanide, and is clearly dependent on the access of oxygen, as is growth. In roots whose growth is accelerated only by much lower auxin concentrations, the streaming also is accelerated by very low concentrations, even showing an effect at 10^{-10}M.

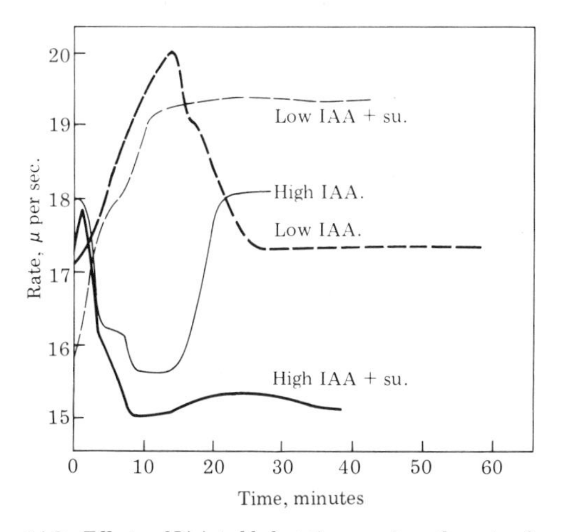

14-2 Effects of IAA (added at time zero) on the rate of cytoplasmic streaming in epidermal cells of the oat coleoptile. Low IAA concentration causes a transient acceleration; high concentration, a transient deceleration. In both cases sucrose greatly prolongs the effect. From data of Sweeney and Thimann, 1938.

The occurrence of a similar acceleration was seen in that classical object, the stamen hairs of *Tradescantia*, by Kelso and Turner in Australia. Further, in the long cells of pine cambium, also, Kaufman observed similar streaming acceleration and found it sensitized by addition of fructose and/or malate. Synthetic auxins act about as effectively as indoleacetic acid, while non-auxins have not been found to act in this way, so that it appears to be a true auxin effect. How an increase in streaming might promote growth is not at all clear, but one could speculate that it might bring increased supplies of carbohydrate to the cell wall, etc.

However, as a *necessary* step in the action there is some evidence against it. The drug Cytochalasin B stops the streaming of cytoplasm in *slime molds*, and Ray's group at Stanford have tested this compound against the streaming in the *Avena* coleoptile—it stops it. This is of interest in itself since it offers prima facie evidence that the muscle-like actomyosin whose contraction and relaxations power the streaming in the slime mold is also in control of the process in plant cells. But for our present inquiry the question is, how does it affect growth? Figure 14-3 shows that the largest part of the growth promotion by auxin is in fact prevented. Indeed, after 2 hours the treated sections grow at exactly the same rate as controls. But because there is a small growth promotion by auxin while the streaming has stopped, the Stanford group have concluded that the acceleration of streaming is not a *necessary* first phase of the auxin effect. To accept this, one would want to be quite sure that a minimal amount of streaming was not still continuing, which is difficult. And one must ask, if the effect on streaming is not important, why is growth so drastically affected? As yet, it is hard to draw a firm conclusion either way.

C. AUXIN AND PERMEABILITY OF THE CELL MEMBRANE

One of the standard class experiments in plant physiology is the study of plasmolysis and deplasmolysis in the cells of *Rhoeo discolor*. These cells, because of their bright purple vacuoles, make most convenient experimental material. So, when von Guttenberg reported that pretreatment with physiological concentrations of IAA delayed the deplasmolysis, i.e., that it increased the outflow rate of water or decreased its inflow, there was much interest. Subsequently Sacher reported that in the endocarp cells of bean pods both plasmolysis *and* deplasmolysis were delayed by IAA, and he concluded that auxin tightens and constricts the pores of the plasmalemma. One can follow water movement by using D_2O and making density measurements, or by using T_2O and making radioactivity measurements. Neither method has confirmed an effect of auxin on water inflow or outflow. Ordin and Bonner measured the inflow of D_2O in coleoptile segments and found no change. Samuel and I measured the outflow of previously accumulated T_2O in potato slices and found that it was actually accelerated by auxin, indicating that the pores of the membrane were somewhat opened. Figure 14-4 is a record of the actual outflow of the radioac-

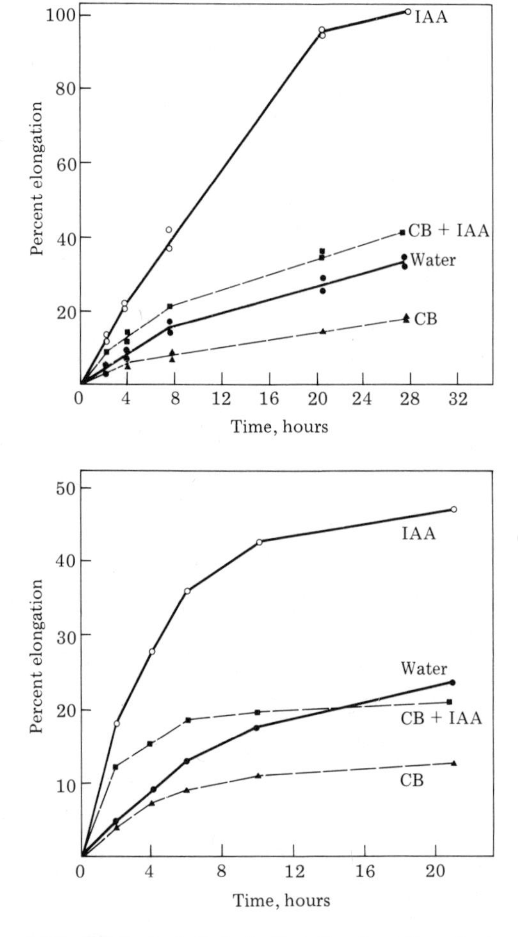

14-3 Effect of cytochalasin B on the elongation of oat coleoptiles (above) and of pea stem segments (below) with and without IAA. Pretreated for 2 hours with cytochalasin B; IAA added at time zero. From Cande et al. 1973.

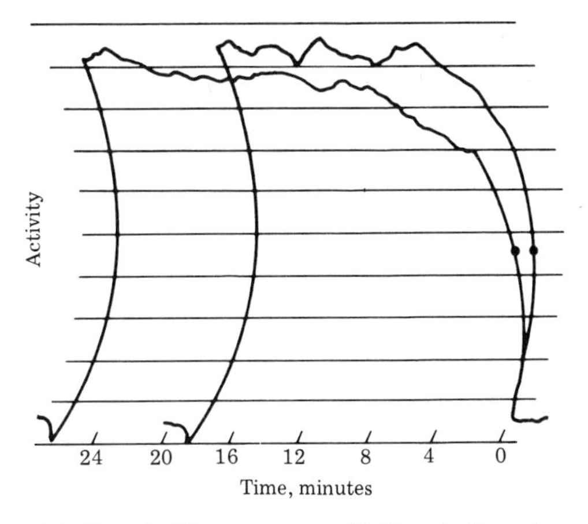

14-4 Record of the appearance of tritium in the outer solution from 1-mm potato discs pre-exposed to NAA and T_2O. Lower curve, discs kept in water ($H_2O + T_2O$); upper curve, discs kept in NAA. The times to reach half-maximum value are from 1 to 2 minutes. From Thimann and Samuel, 1955.

14-5 Times to reach half-maximum value of the radioactivity of the outer solution by efflux from potato discs of different thicknesses. Lower curve, discs kept in NAA; upper curve, discs kept in water. From Thimann and Samuel, 1955.

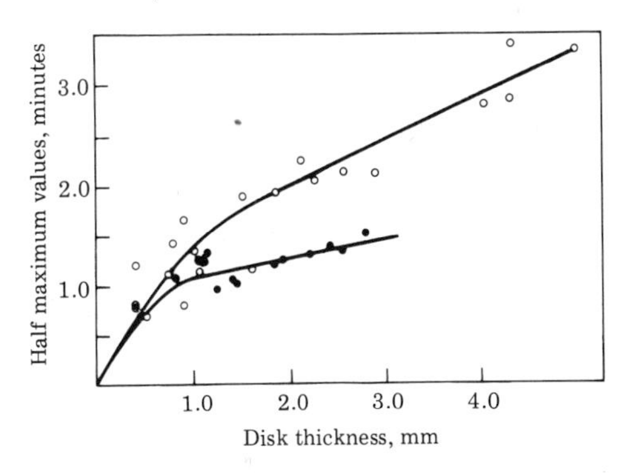

tivity, and figure 14-5 shows how the rate varies with the thickness of the slice, since the thicker the slice the more cells would be participating. The apparent increase in efflux is probably due to the fact that the slices had been exposed to the auxin (naphthaleneacetic acid) for 48 hours and thus may have begun their growth response. If so, this would give a useful new pointer to events in growth, namely, that in the first stages the cell membrane (the plasmalemma) becomes extended faster than new material can be incorporated into it. Recently Kang and Burg reported an *immediate* effect of auxin in increasing the rate of water efflux, but the findings are disputed.

The absolute rate of water outflow as shown in figure 14-4 is surprisingly high, so high that a minor change in it, even if established, could not account for the difference in growth rate caused by auxin. In the figure, equilibrium was shown to be reached in 4 minutes, while growth, of course, continues for hours or days. The concept that the prime control of cell enlargement is *via* changes in permeability to water must therefore be abandoned, for in so permeable a membrane small permeability changes could hardly be of critical importance.

D. THE ROLE OF OXIDATIVE METABOLISM

The approach to understanding the role of hormones in growth through the study of respiration and oxidative metabolism was described in part 1 of chapter 3. The main results can be summarized in a few sentences: (1) Oxygen is needed for the response to auxin in the tissues studied. (2) All reagents that inhibit oxidative steps inhibit the auxin response, though similar evidence for other hormones has not been clearly presented. The list of inhibitors includes numerous sulfhydryl reagents, carbon monoxide, fluoride, fluoroacetate, and dinitrophenol. (3) The auxin response is uniformly more sensitive to such inhibitors than is the overall oxygen consumption. (4) When growth ensues there is a distinct increase in the rate of oxygen consumption, and the auxin concentrations effective in these two responses agree closely. Again, as with inhibition, such increases are proportionately smaller than the increase in growth rate. Even with the prize example of artichoke slices, whose respiratory rate increases 2.5 times, the growth *rate* in optimal auxin concentration increases 10 times.

The most reasonable interpretation of these facts is that cell enlargement is not simply dependent on the presence of oxidative metabolism, but is an oxygen-consuming process *in its own right.*

One additional fact to be brought in is that growth can be inhibited by removing the osmotic gradient, especially by use of mannitol. This sugar alcohol does not enter the cells appreciably. However, tuber tissue when thus subjected to an external osmoticum, and therefore to a decrease in its turgor, proceeds to increase its internal osmotic pressure, probably by hydrolyzing polymers. Because of the parallel with the thermostatic mechanism in animals, I have suggested this process be called *manostasis*. For our analysis the mechanism is less important than the fact that when mannitol prevents the growth response it prevents the respiratory increase too. Thus respiration and the growth response are certainly linked in some indirect way, perhaps via the stimulated synthesis of certain specific cellular materials, among which cell wall and membrane constituents must be prominent.

E. AUXIN AND THE SYNTHESIS OF ENZYMES AND PROTEINS

Putting together the findings of numerous workers, it seems that a total of 13 specific enzymes have been recorded as showing increases due to auxin treatment. Among these a few show very large increases. In tobacco pith tissue, *pectin methylesterase* and *ascorbic acid oxidase* show the biggest increases, the latter reaching 4 times the control value. In chicory root and artichoke tuber, *invertase* is increased by a factor of almost 20 times, especially with 2,4-D as auxin. More suggestively linked to likely requirements for growth is the finding of MacLachlan, at Montreal, that in pea stems cellulase (β-1, 4-glucanase) is increased by about 12 times. Most of the increase could be traced to the microsomes (fig. 14-6), in which the factor can be 30 times! Cellulose is certainly the most rigid fraction of the cell wall, so that its hydrolysis would seem a strong candidate for a key role in growth. Unfortunately, however, the increase is somewhat slow, reaching its full value in 4 days, by which time growth of the treated region is likely to have stopped. Also gibberellic acid causes good growth in the pea seedling, yet no increase in cellulase could be detected with GA. *Avena* coleoptile seg-

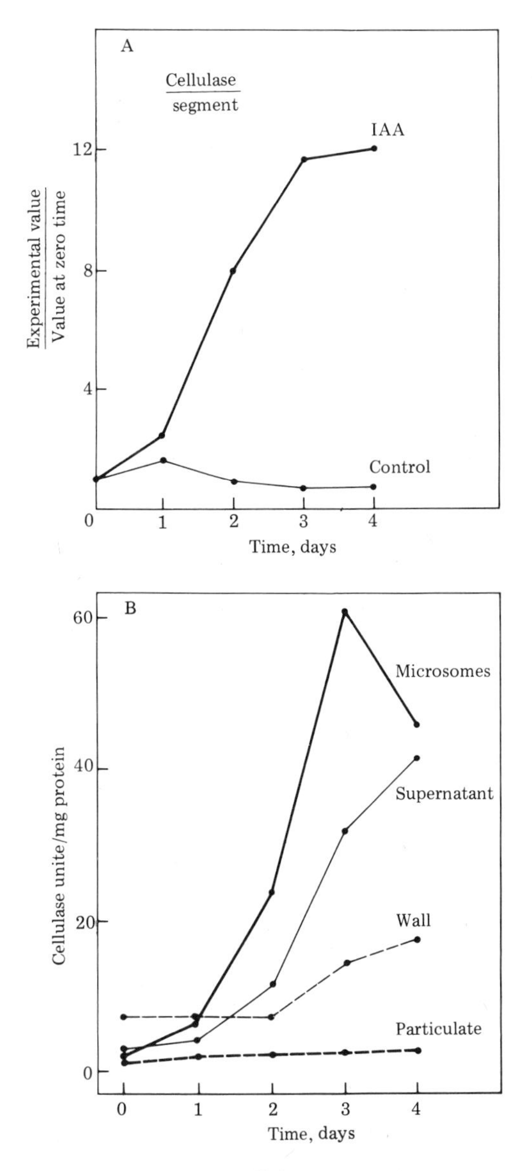

14-6 Development of cellulase activity in pea stem segments after IAA has been applied on day zero. Above, enzyme per segment. Below, enzyme (expressed as per mg protein) in individual subcellular fractions. From Davies and MacLachlan, 1968.

ments show no growth response when treated with purified (and highly active) cellulase. Since the treated regions of the pea stems swelled laterally, it may well be that ethylene, produced in response to the auxin, was responsible for the cellulase increase.

All the enzyme increases took place over relatively long time intervals, so that they could well represent the results of auxin action rather than its primary seat. Furthermore, an increase in protein synthesis would require an increase, or a stimulation, of messenger ribonucleic acids (m-RNAs), so that one must look further back than the synthesized enzyme when looking for the basic mode of action. As a matter of fact, auxin does increase the total RNA of a variety of plants. This has been investigated in recent work mainly by measuring the rate of incorporation of labeled single bases—uracil, uridine, or orotic acid. The increases are large, ranging from 70% to 300% in different tissues. Even isolated pea nuclei, which can still make RNA when suspended in a mixture of all four triphosphates, show an increase in such labelling, when treated with 2, 4-D, of some 90%.

As to the types of RNA that are thus influenced by auxin, the evidence is less clear. Auxin-treated soybean hypocotyls show a change towards forming the larger polymers. However, the labeled RNA from artichoke tissue and pea stem segments shows increases in all fractions, the ribosomal RNA showing the largest effects.

Quite apart from these effects on RNA, some of which remain to be clarified, and the effects on specific enzymes summarized above, there is plenty of evidence that synthesis of proteins of *some* kind is essential to growth. The evidence for this statement has been given by numerous inhibitor studies, using the following compounds with known routes of action: actinomycin D, which attaches to DNA and prevents it from being translated into messenger-RNA; chloramphenicol, which binds to the ribosomes and prevents them from acting; puromycin, which structurally resembles one end of a nucleic acid and thus inhibits amino acids from attaching to the ribosomes; 5-fluorouridine and 5-fluorodesoxyuridine, which compete with RNA and DNA, respectively; 4-fluorophenylalanine, which is competitive with both phenylalanine and tyrosine in the actual synthesis of proteins.

All of these inhibit growth, some drastically. Actinomycin D, providing it is given some time before the auxin, is particularly powerful, causing 50% growth inhibition at concentrations around 2×10^{-6}M (fig. 14-7). If applied within an hour or so of the IAA, it is apt to have little or no effect, presumably because the messenger for growth has already been formed, or perhaps also because, being a large peptide, it is only slowly taken up. Chloramphenicol also has a powerful action, though higher concentrations are needed, and the parallelism between the two effects, namely on growth and on protein synthesis, is striking (fig. 14-8). It has a small effect on oxygen uptake as well, unfortunately, but comparison of its respiratory effect with that of the much more powerful sulfhydryl reagents described in chapter 3 makes it most unlikely that its growth inhibition could be ascribed to an effect on respiration.

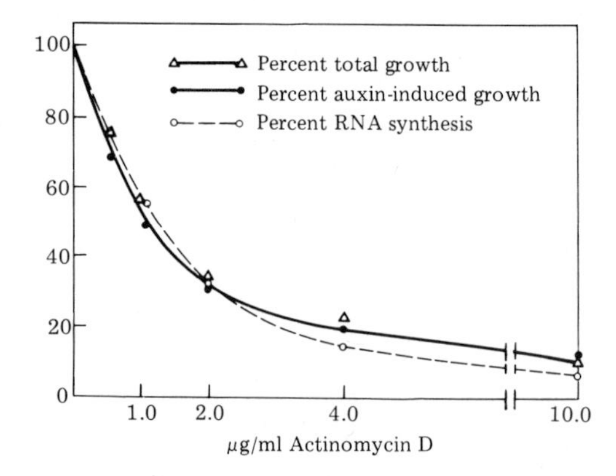

14-7 Parallel between the inhibiting effects of actinomycin D on growth in controls, growth in auxin, and the synthesis of an RNA fraction, probably m-RNA. All measurements after 4 hours. From Key and Ingle, 1968.

14-8 Comparison of the percentage inhibition by chloramphenicol of elongation (open circles) and of incorporation of ^{14}C-leucine into protein (filled circles), using oat coleoptile segments in presence of IAA. From Noodén and Thimann, 1965.

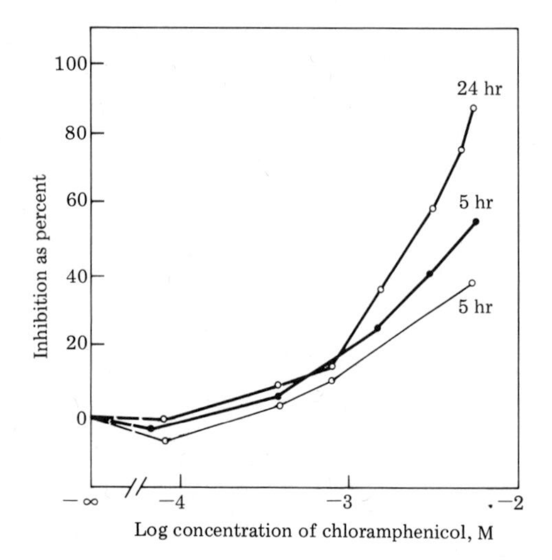

Puromycin causes 50% inhibition of growth at 3×10^{-5}M. Although 4-fluorophenylalanine is somewhat less potent in actual concentrations needed, its inhibition is clearly located, because it can be partially reversed by adding phenylalanine.

These actions on elongation are paralleled by actions on the plasticity, measured by bending or stretching techniques (chap. 3). Chloramphenicol and cycloheximide both inhibit the auxin-induced increase in plasticity, and so does actinomycin D if supplied 3 hours before the auxin. However, the degree of inhibition of plasticity does not always check with the degree of inhibition of growth; thus when the growth inhibitor is mannitol, the IAA effect on plasticity is still detectable.

The deduction from inhibitor studies is borne out by direct measurements of protein synthesis from the rate of incorporation of labeled amino acids, as influenced by auxin. These also show general increases, sometimes as great as 50%, though the parallelism with effects on growth is not always so clear. However, there is really no reason to expect that growth would be dependent on an increase in *total* protein. On the contrary, Cleland, in Seattle, has postulated a growth-limiting protein (GLP), specific for the enlargement process; it may or may not be an enzyme. Earlier, a protein rich in hydroxyproline which was detected as a constituent of cell walls was thought to be such a growth-controlling protein, but it has now become clear that this material accumulates as growth comes to an end; and indeed it may help to prevent further extension in much the same way as, on an intercellular scale, the scalariform lignification of metaxylem is thought to prevent elongation.

If growth is indeed directly linked to synthesis of one or more special types of protein one could see why oxidative metabolism would be involved, for it is only in this way that nearly all of the needed ATP is produced in nonphotosynthetic cells.* The growth that can go on in partially anaerobic conditions, as in flooded rice plants, is perhaps made possible by oxygen diffusion through intercellular spaces. In more general terms, the strong effect of high nitrogen fertilization on vegetative growth may have its explanation along these lines.

Some workers are reluctant to accept the main conclusion brought out here, for the reason that auxin (though not gibberellin) shows very rapid effects. Under the microscope one can detect growth acceleration beginning about 15 minutes after auxin application, and the increase in streaming rate begins within 2 minutes. The length of the lag period before growth starts has therefore been carefully studied; examples are shown in figure 14-9. With high auxin concentrations it can be very short. In contrast, the long reaction series from DNA to messenger RNA to transfer RNA to aminoacyl-RNA to protein synthesis on the ribosomes takes, at least in the bacteria where it is most fully understood, not less than 20 minutes. The observed increases in RNA in plants take place over periods of an hour or more. This reservation is reason-

* Assuming 19 molecules of ATP are needed to synthesize one molecule of amino acid and convert it into a peptide, S. S. Chen (1972) has calculated that as much as 50% of the respiration of a growing tissue culture could be used in this way to form protein.

able, although the transfer of data from bacteria to higher plants is not necessarily safe. But, since all this evidence cannot be lightly set aside, another explanation is possible, namely that *two* effects are involved, one which initiates growth and one which sustains it. The former may well be located at the plasmalemma, while the latter would be in some way connected with the nucleus or its direct products. This concept will be taken up again below.

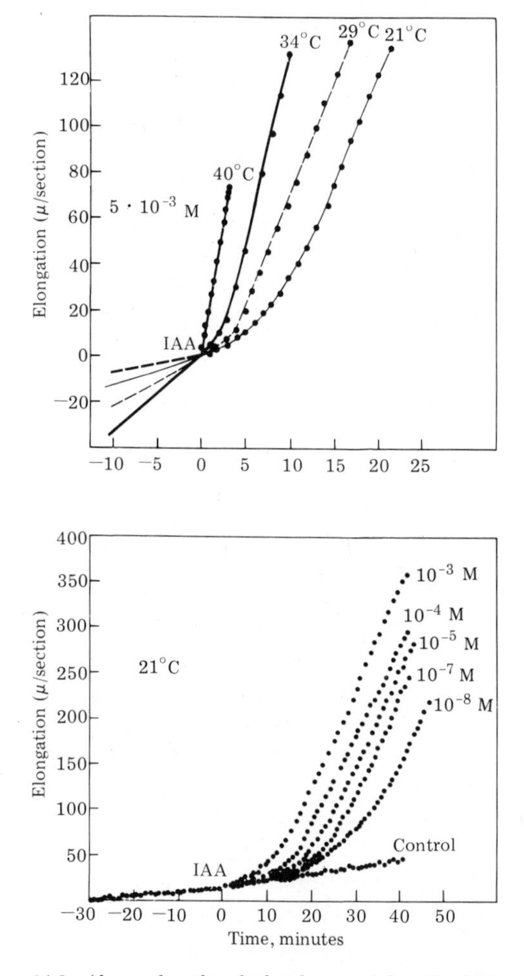

14-9 Above, length of the lag period with IAA 5×10^{-3}M at four different temperatures. Below, length of the lag period (at 21°C and pH 4.7) with five concentrations of IAA. From Nissl and Zenk, 1969.

F. HORMONES AND THE PROPERTIES OF THE CELL WALL

In addition to the above areas in which modes of action have been sought, a major one in regard to enlargement is in the physiology and components of the cell wall. The process of cell enlargement means that new wall, and new membrane, must be constantly laid down, at least to keep pace with, and perhaps even to control, whatever is going on in the cell contents.

As we saw in chapter 3, work on the changes in cell walls during growth has gone on actively ever since the first studies by Heyn and Söding in the 1930s. Stretching and bending experiments, especially with plasmolyzed plant material, have led to the general conclusion that the effect of auxin is mainly to increase the plastic extensibility of those tissues that respond. This evidence was presented in chapter 3. Evidently this mechanism does not hold for gibberellin, however; Yoda and Ashida in Kyoto have shown that gibberellin stimulates elongation without apparently causing any such increase, and there may even be some decrease in plasticity in certain

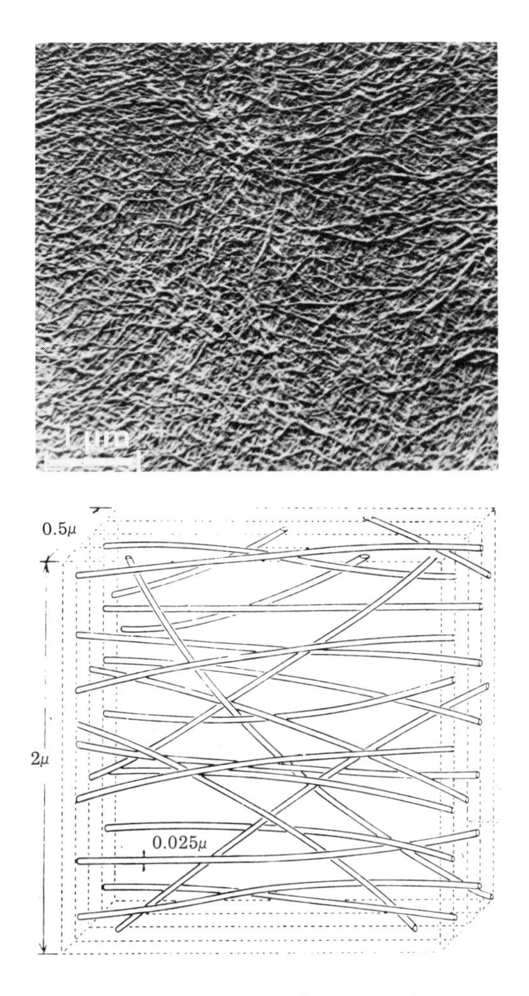

14-10 Above, primary wall of a corn seedling (root meristem); for microscopy the wall has necessarily been dried down. Below, diagram of the fibrillar structure of a primary wall; when fully hydrated as shown, the microfibrils occupy about 2.5% of the volume. From Frey-Wyssling, 1976.

tissues after gibberellin treatment. But in any case the conclusion that cell wall properties may control enlargement was especially important in that it directed attention to the cell wall as a possible seat of the growth process. In view of the equal importance of the cell membrane (plasmalemma), however, it is curious that so far little attention has been paid to it as another possible seat of the growth-controlling changes.

Electron micrographs of a cell wall show that it consists of fibrils of polysaccharide more or less buried in an isotropic material (fig. 14-10). There are clearly two predominating angles to the main axis, and in young growing material these angles are large. After physical extension the angles will become smaller; the slope will be steeper. And new fibrils have to be interpolated in order to maintain the amount of material per unit length. Under the influence of auxin the new fibrils are evidently again set at a large angle to the axis, for the *average* orientation remains about constant, as was shown long ago. The optically active fibrils are embedded in a nonfibrillar substance, and the whole represents a group of different polysaccharides that are laid together to make one structure. Ray's elaborate analyses show that this complex can be considered roughly as divided into four classes: The cellulose, which in the coleoptile represents about 40% of the total wall material, is the most fibrillar material and the most rigid; the polyuronide, mainly pectin and pectic acids, comprises part of the nonfibrillar stuff, and the middle lamella between the cells consists almost wholly of polyuronide; the two other major materials are arabogalactan, an anhydride of arabinose and galactose, and xyloglucan, an anhydride of xylose and glucose. This fourth material, xyloglucan, is thought to represent an important constituent of the binding between the polyuronide material and the cellulose. It is not proven that this is so, but there is some evidence that xyloglucan has a function in growth, as we shall see; and the general impression is that it affects the binding, so that it is now beginning to occupy an interesting place in the concept.

It was an early observation too (following the work of Heyn), made by James Bonner, that when the coleoptile segments are grown in auxin and sugar solution, the weight of cell wall per unit length remains about the same—that is to say, the cell is able to keep up with whatever stretching takes place by laying in additional polysaccharides. If, however, the segments are grown in auxin alone without any supply of sugar, the weight of cell wall per unit length *decreases*. Thus there is some stretching which is not followed by filling in of materials. Or one may conclude that there is temporary hydrolysis of one or other of these materials, which allows stretching. And that suggests that there is probably some hydrolysis going on *during the whole process of wall growth.*

Several hydrolytic enzymes acting on cell wall materials are known. We saw earlier that the connection between cellulase itself and elongation is a bit tenuous, that the changes in cellulase are very slow, though they are indeed large. Cellulase can hardly, therefore, be a controlling factor in growth, and there are other reasons to doubt that the stretch or hydrolysis of its substrate, the cellulose, can be a major factor in growth either. Besides, we know that the enzyme is present in most growing tissues. Ruesink has studied the effect of adding a powerful cellulase to coleoptile segments, but it did not result in any increased elongation.

Along with the cellulose structure, which consists of β,1-4 linkages, there are also some β,1-3 linkages, making a glucan which is rather like cellulose, though it does not confer as much tensile strength as the 1-4 linkage.

A few years ago, at the time of the 1969 Botanical Congress, Heyn, returning to the field after 30 years, found that the 1-3 glucanase in coleoptiles increased after treatment with auxin. Correspondingly Masuda, in Japan, has prepared a β,1-3 glucanase from some fungi and simply added it to coleoptile segments, with the result that it does cause some elongation, though not nearly the growth that auxin can cause. However, the result is of some interest because the fibrillar structure of the β,1-3 glucan probably contributes to the tensile strength of the cell wall, and if that is weakened it may make it possible for the tissue to elongate somewhat. But since the elongation is so much smaller than can be brought about by auxin, the role of this enzyme, or of the β,1-3 linkage, as a controlling factor in growth is still open to doubt.

In all this we have assumed that extension is brought about by the osmotic force, which induces water to enter; the cytoplasm does have a higher osmotic potential than the outer solution, for even if sugar is added its molarity is very low. The optimum sugar concentration is 2% sucrose, and that is 0.06 M, so this concept is perfectly acceptable. However, what is not so acceptable is the idea that the cell wall is a relatively inert material which, once deposited, does not further enter into metabolism. We noted above that the weight of the cell wall per unit length, and the orientation of the fibrils, roughly remain constant during growth. Therefore hydrolysis and redeposition are going on simultaneously, or at least side by side, and the redeposition requires, of course, the action of *synthetic* systems. Thus some attention has been given to the possible action of synthetases in keeping the cell wall supplied.

Hassid's work on cellulose synthesis has shown that the reaction is brought about by way of a guanidine diphosphate-glucose compound, and, with an enzyme preparation and this GDP glucose, cellulose can indeed be formed *in vitro*. However, the work on this and other synthetase enzymes has shed little light on their possible role in growth until recently. Now Ray at Stanford has brought out some interesting facts on the role of glucan synthetase in cell wall formation. The amount of this enzyme does increase in the presence of auxin (fig. 14-11). It may not increase quite fast enough to be a primary seat of action since it is at least 20 minutes before any increase is detectable, whereas an increase in growth can be detected within 10–20 minutes of adding auxin, and an increase in streaming, as we saw above, is seen within two minutes. Nevertheless this is one of the first demonstrations of a cell wall enzyme which gives a rapid and quantitative response to auxin.

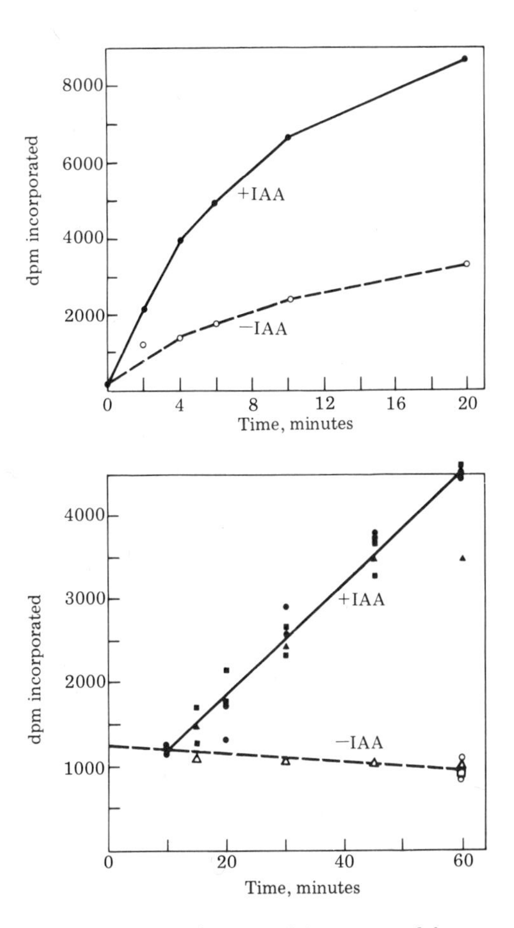

14-11 Glucan synthetase activity, measured by extracting the particles from pea stem segments and incubating with ^{14}C-UDP-glucose. Above, time course; segments first held for 3 hours to allow the synthetic activity to decrease; then 1 hour in sucrose; then IAA added at time zero. Below, concentration plot. All at 35°C . From Ray, 1973.

Figure 14-11 (top) shows the time curve for the change in the glucan synthetase activity in pea stems. It is measured by supplying ^{14}C-glucose and measuring the amount of incorporation of C^{14} into the insoluble glucan. Auxin is supplied at time zero; at ten minutes the data show no change; at 15 minutes the points are scattered; at 20 minutes there is no doubt of a strong increase in the incorporation. And probably at fifteen minutes the increase may also be real.

A plot of the quantitative effect against the concentration of auxin shows a linear relationship (fig. 14-11, bottom); 10^{-6} and 10^{-5}M auxin concentrations represent the normal physiological range. We must note, however, that there is plenty of the enzyme present without auxin treatment. And the activity after auxin is supplied is no more than doubled. The increase is thus not dramatic, but it is quite clear. Also the isolated enzyme turns out to be present in particles and the enzyme activity in these particles increases after auxin treatment, as it does in the tissue. At 17×10^{-6}M IAA, which is a good physiological concentration, the effect is clear. Thus we have here a wall-synthesizing enzyme which apparently responds nicely to auxin.

The process of deposition of the cell wall is very complicated. Generally it is believed that vesicles, representing sizable carbohydrate content, are excreted from the Golgi bodies and make their way to the wall and are somehow polymerized *in situ*. There is one oat variety in which the electron microscope shows that the wall is made up of long objects like compressed spheres (these were discovered by O'Brien and are sometimes referred to irreverently as O'Brien's bananas); but it is only in the one variety that they are clearly to be seen. It is tempting to think that these are the actual Golgi vesicles which have made their way into the wall and have been incorporated, but that is merely speculation.

In any case most people are agreed that this method is the way in which the wall is laid down. And, if so, we must ask where the glucan-synthesizing enzyme is located; must it be located in the cell wall itself? And if so, does it simply take ordinary sugar molecules and link them together end to end, presumably with a GDP or ADP attached to make them reactive, or is the whole thing more complex? We cannot really answer these questions.

One interesting thing that has come up in the work of Labavitch and Ray is that in the course of cell wall deposition there is also some breakdown. Apparently a xyloglucan residue, with perhaps a half-dozen sugar molecules in it, is released and goes into soluble form. It comes out of the wall and can be isolated in small amounts, and the amount liberated is nearly proportional to the elongation (fig. 14-12). The explanation of this could be that after it comes off the cell wall it then goes back on again. In that case it might be a wall-property-changing factor. If an oligosaccharide comes off, is moved a little way, and then is put back on again, obviously this might have a big effect on the properties of the cell wall. Interesting also in connection with the need for one or more oxidative reactions, described earlier, is the fact that this release of a small oligosaccharide seems to be inhibited by inhibitors of the oxidative

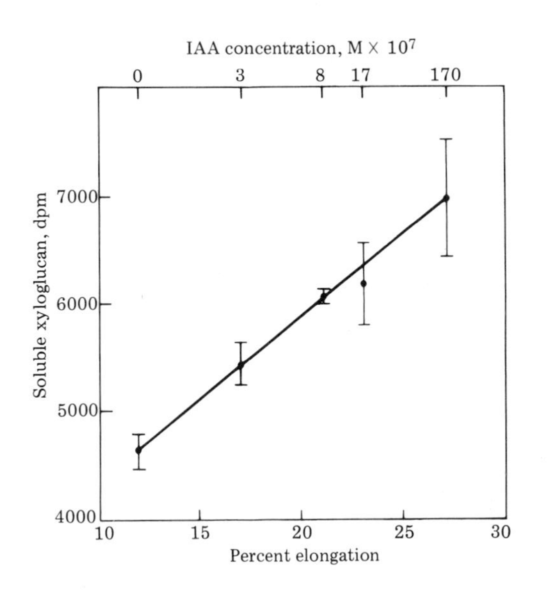

14-12 Proportionality between elongation and the water-soluble xyloglucan fraction in pea stem segments pre-incubated in ^{14}C-glucose, then in ^{12}C-glucose plus serial concentrations of IAA for 7 hours and extracted. From Labavitch and Ray, 1974.

processes. We saw above that arsenite, iodoacetate, mercurials, and fluoride have strong inhibiting effects on growth. And the release of a soluble xyloglucan is inhibited strongly by arsenite and fluoride. This shows that the release requires energy, perhaps for the formation of an ADP-type sugar derivative.

Furthermore, when growth is inhibited by adding mannitol, so that enlargement does not take place, the small xyloglucan is still released. That is significant because one could always imagine that as a result of wall extension and laying down of polysaccharides some fragments get left over; these would not have to do with the action of the hormone but would be the *result* of extension. However, since sufficient mannitol to inhibit growth does not prevent the release of the xyloglucan, such a reservation does not apply.

The one clear impression from these beginning attempts is that the cell wall is not as we have thought of it in the past. In Frey-Wyssling's book and other treatments of cell wall material, it is commonly presented as being an excretion from the cell—material which is excreted and laid down there to form the rather rigid cell wall, which then has to be expanded by special means. The newer accumulating data indicates, as mentioned above, that the cell wall is just as much a seat of metabolism as is the cytoplasm. There are enzymes there which cause hydrolysis, the release of fragments which are then reattached and so on, with both hydrolytic and synthetic enzymes acting to modify the properties and make it extensible under osmotic force.

Among other constituents of the cell wall is a peculiar protein. Any cell wall preparations, however carefully washed, contain nitrogen—usually not a great deal, about 2%—but since protein is 1/6 nitrogen, that means they contain 12% protein. And this is an odd protein in that it is very high in hydroxyproline. Much has been made of this, especially by Lamport, who first noted that it contained hydroxyproline, and drew a parallel between this and the collagen of animals because collagen, which is an important part of muscle, is also very rich in hydroxyproline. At first it was thought that this hydroxy linkage, being a possible seat of a connection between protein and carbohydrate, might be the seat of the extension process; indeed, the protein was even called *extensin*. But it now appears we should not use that name, because when hydroxyproline appears in the wall in large amounts, extension has ceased. It is usally an end-product of growth, not a product which mediates extension. When the protein is synthesized, it is proline, not hydroxyproline, which is incorporated into the protein chain—and only after it is in the chain is it hydroxylated. (The same process takes place in collagen formation.) Hydroxyproline, if added to growing stem or coleoptile segments, not only does not help, but actually slows down, growth, perhaps by competing with the proline. The hydroxylation process requires a special enzyme which is dependent on its iron content and can be inhibited by dipyridyl, which strongly chelates iron and thus acts as an iron action inhibitor.

There may be other interesting components in the cell wall that have not yet been uncovered, and as polysaccharide chemistry becomes more sophisticated we can certainly expect that some new details will come up. But whether they will shed any light on elongation, I do not know. Basically the known materials suffice to satisfy the overall chemical analysis.

G. GROWTH AND HYDROGEN ION BALANCE

Now we come to the last approach, which is also a very old one, revived now, and in some respects it is the one that is hottest at the moment. It stems from an old observation made by Siegfried Strugger, a very careful German cell biologist. It was essentially this: if one places growing plant tissues like coleoptiles or etiolated seedling stems in acid solutions, they often show a slight swelling. And if they are peeled on one side, then they curve so that the peeled side is on the outside, on the convex side. These are what Strugger referred to as acid curvatures in 1934.

One might think that, since that is the wounded side, it would be able to respond less than the other side, and there are other conditions under which tissues will curve the opposite way because of the wound. (Curvature towards a wound is called traumatotropism.) But "acid growth" means that the part without its cuticle has absorbed the acid medium faster than the intact side and has grown more as a result. The reaction is truly growth; it is not reversible by placing in an osmotic solution, so it is truly irreversible elongation. Even without externally added auxins, if one places young coleoptile segments in an acid solution, they will grow quite rapidly. They begin to grow even sooner then they do in auxin. The growth is not as great as in auxin but is maintained for quite a long while. Bonner at Cal Tech (1934) and Miss van Santen in Utrecht (1940) studied this phenomenon, and both related acid growth to the presence of auxin. Old tissue which no longer yields diffusible auxin does not give any acid growth. Both workers concluded, therefore, that there was auxin in responsive tissue in some inactive form and that the addition of hydrogen ions activated it. Bonner was even able to show that part of the response followed the pH dissociation curve of an acid of the strength of auxin (pK = 4.75). It was as though auxin in the salt form has no activity, auxin in the undissociated acid form is what causes growth, and the acid pH converts one to the other.

The second interesting fact is that if, to young stem or coleoptile segments in an unbuffered medium, one adds auxin, the medium *becomes acid*. Thus acid causes growth, and the tissues produce acid. One is reminded of the puzzle of ethylene, namely that ethylene causes fruits to ripen and ripening produces ethylene. Coleoptiles in an unbuffered medium can bring the pH to about 3.8, pea stems somewhat less, but they act clearly in the same direction. It takes about half an hour for a detectable increase in the external hydrogen ion concentration, which is again slow for a primary seat of action, but the connection between pH and growth is evidently very close.

The top of figure 14-13 shows the time course of the pH of an auxin solution in which segments are growing. The solution

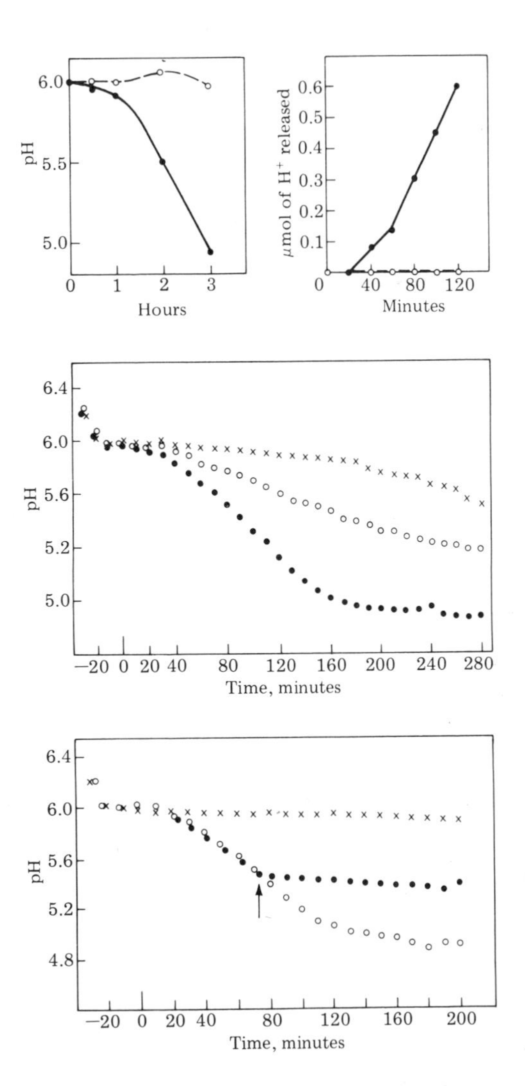

14-13 Top: time course of the pH change and the accumulating acidity from oat coleoptile segments in pH 6 buffer. IAA (filled circles) added at time zero: open circles, controls. From Cleland, 1973. Middle: as above, but IAA 10^{-9}M (circles), 10^{-5}M (dots), and controls (crosses). From Rayle, 1973. Bottom: as above, but IAA 10^{-5}M (open circles), with cycloheximide (10 ppm) added at arrow (filled circles). Crosses, controls. From Rayle, 1973.

was brought precisely to pH 6.0. The first measurement, at half an hour, shows a barely significant increase in hydrogen ions, but since it falls smoothly on the curve the small increment may be significant. Had the segments been peeled the change would have been faster. The second figure shows the proportionality between IAA concentration and the pH change resulting. The third figure shows that cycloheximide stops all pH change.

These data have led to a fresh concept of how auxin acts. It is proposed that auxin causes essentially *acid growth*. It makes the cell (via a newly formed protein) produce hydrogen ions which then go into the cell wall, acidifying the wall materials; this somehow makes it enlarge.

Now how would this happen? Because of the enzymes in the cell wall, which are evidently important for growth? The excretion of H^+ ions might make the pH of the wall favorable to their activity. Or because of some inherent property of the cell wall structure?

These questions have been made more pressing by a recent discovery of Dr. Marré in Italy, a discovery which was breathtaking at the time. Some of Marré's colleagues in plant pathology isolated a fungal toxin, produced by the parasite *Fusicoccum amygdalae*, a disease of almond trees in which the tissues swell and become very abnormal. This fungal toxin, fusicoccin, as we saw in chapter 8, is a glucoside, not a free acid and, unlike auxin, contains no indole (see formula in chap. 8). It also contains several hydroxyl groups, which in all biochemical studies of auxins have been shown to decrease auxin activity. It has none of the chemical features associated with the auxins that we know, yet nevertheless it acts in most ways like an auxin.* In fact, on pea stem segments it causes more elongation than IAA does. The concentration needed for maximum elongation is 10 to 20 times higher than that of IAA, but the absolute elongation resulting is clearly greater than that due to optimal auxin. Now it turns out that this substance, which although unlike the auxins nevertheless causes elongation, also causes acidification of the medium just as IAA does.

Figure 14-14 compares the elongations caused by IAA and by fusicoccin. The fusicoccin causes even greater elongation than the auxin. These elongations are paralleled by the excretion of hydrogen ions. What is more, the amount of hydrogen ions excreted is proportional to the amount of elongation, whether caused by auxin, by fusicoccin, or by sundry other synthetic substances (fig. 14-15). Marré has been too cautious to draw a straight line through the points, but it is clear that the auxins and fusicoccin fall on the same curve.

Mannitol does not inhibit the excretion of hydrogen ions, even when it inhibits the elongation, so that again we conclude it is not the *result* of growth. In figure 14-16 we compare the elongation and the hydrogen ion excretion. As the mannitol concentration increases, the elongation decreases, as expected, whether the growth is promoted by fusicoccin or by auxin. But the excretion of hydrogen ions is not inhibited at all; in fact it even tends to increase somewhat.

* Unlike auxin, however, fusicoccin does not stimulate ethylene production.

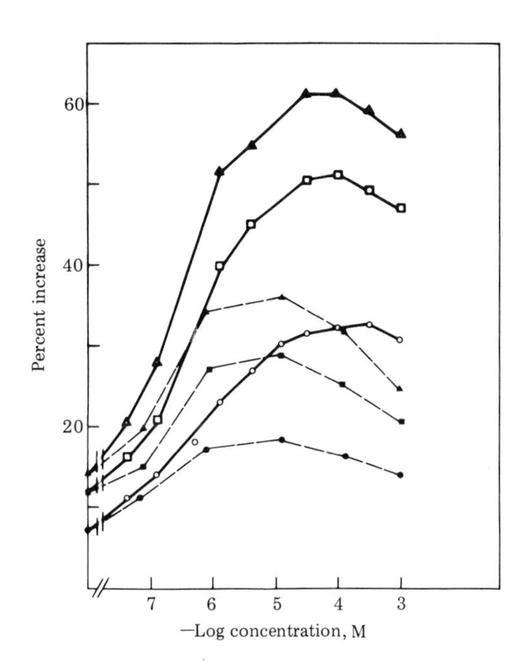

14-14 Growth, measured as % increase in fresh weight, of pea stem segments in IAA (dashed lines) or in fusicoccin (solid lines). Measured after 2 hours (circles), 4 hours (squares), or 6 hours (triangles) at 25°C. From Lado et al. 1973.

I speculated above that the excretion of hydrogen ions might activate some enzymes in the cell wall, which are there at a pH unfavorable for their activity; when they are activated they would hydrolyze off fragments which would change the properties of the wall material and thus allow growth. An attractive thought was that the xyloglucan fragment which is released could be the causal agent, its release and reattachment being a function of pH. Acid pH does indeed release it, but Dr. Albersheim in Colorado has shown that its binding, or reattachment, is not a function of pH. So that concept is not so simple. Indeed, enzymes may not be involved at all, for Masuda's group finds isolated, frozen and thawed pea stem epidermis to show extension responses both to auxin and to acid pH. Some preparations of *Nitella* cell wall seem to behave similarly. Some inherent property of the cell wall material itself is apparently involved.

It is evident that, to achieve the proton excretion, energy is required, so there must be respiration and also, apparently, some protein synthesis. Since both phosphorylation inhibitors and protein synthesis inhibitors inhibit growth, that picture has excellent coherence. Indeed, table 14-1 shows that the general degree of inhibition by several reagents is about the same with fusicoccin as with auxin.

An important observation that supports the whole concept in another way was reported at the Tokyo meeting by Kasamo and Yamaki. They found that labeled auxin administered to bean hypocotyls becomes bound to a centrifuge fraction that consists largely of plasma membranes. This fraction

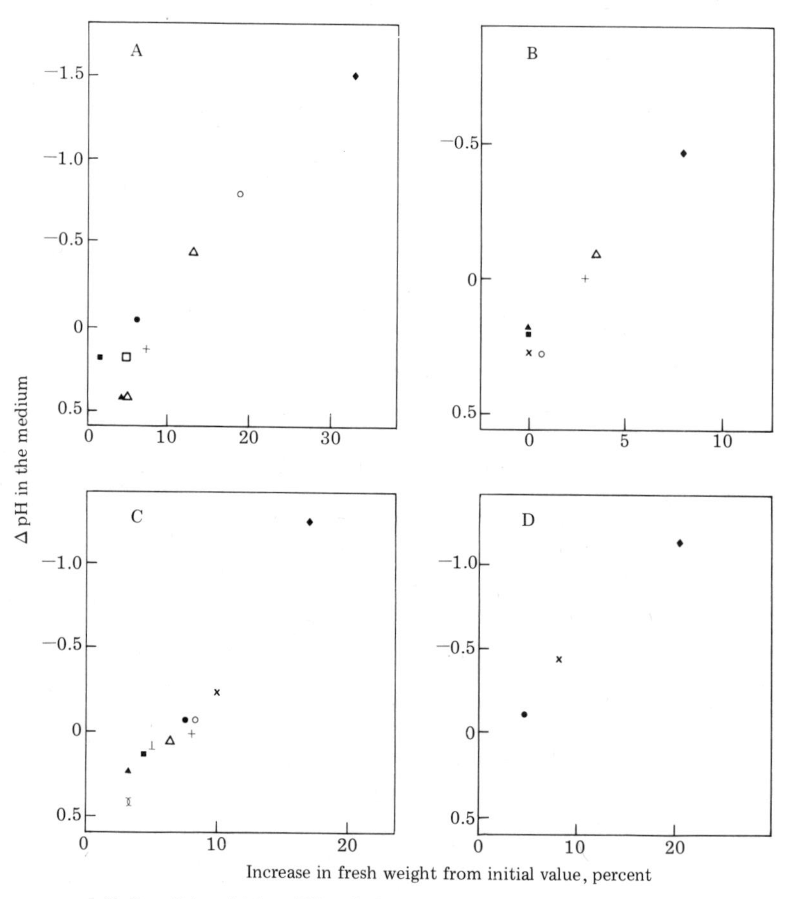

14-15 Growth plotted against pH change for four different plant materials. Top left, pea stems; top right, tomato leaf segments; below left, *Cucurbita* cotyledons; below right, radish cotyledons. Black circles, controls; cross, benzyladenine; open circles, IAA; diamond shapes, fusicoccin. The other symbols show the action of inhibitors; CH, cycloheximide; DNP, dinitrophenol. From Marré et al. 1974a.

contains an ATPase, stimulated by magnesium. The ATPase might well make the fraction able to supply energy for proton transport.

Now of course one cannot excrete positive ions from a neutral cytoplasm, even if an energy source is available, without either excreting negative ions at the same time or taking in positive ions from outside. Apparently the latter is what occurs. It is an old observation that potassium ions play a large part in growth; we have seen in chapter 3 how often growth is promoted by potassium ions, especially in longer term experiments. And Marré now has evidence that when auxin causes an excretion of hydrogen ions into the solution, it soon afterwards causes a rather rapid intake of potassium ions. In the first two hours K^+ ions may drift out, probably being liberated from the free space, but thereafter they are rapidly absorbed. If the tissue is growing in an unbuffered medium without potassium ions present, it probably withdraws potassium ions from the cell wall into the cytoplasm.

This observation is most interesting for another reason. We have seen that abscisic acid in general acts as the opposite of a growth hormone. It tends to slow down growth and to inhibit many different processes. One thing that AbA does very strikingly is to cause the rapid closure of stomata, as we saw in two earlier chapters. Now, what happens when stomata open and close? The process appears to be controlled by the movement of potassium ions. When the stomata open, they do so because the guard cells take in potassium from the adjacent cells, and when they close they excrete potassium into the surrounding medium. AbA has been shown by Mansfield and Jones (1971) to cause potassium to leave the guard cells. Conversely, fusicoccin causes the stomata to open in light, and its action is potassium-dependent. Thus, it is possible that AbA is acting on fundamentally the same system as is auxin (or fusicoccin), only AbA is working in the opposite direction. And recently Reed and Bonner (1974) have found AbA to inhibit the uptake of potassium into coleoptile segments; this powerful inhibition is shown in figure 14-17, and strengthens the above suggestion.

In any case, Marré's data show that

Table 14-1 The effects of inhibitors of metabolism and of RNA and protein synthesis on the increase in fresh weight induced by IAA or fusicoccin in segments of etiolated pea internodes

Solution	Percent increase in fresh weight	Percentage inhibition caused by:			
		Fluoride	Dinitro-phenol	Actinomycin D	Cyclo-heximide
Water	7.7	21	69	45	83
IAA	22.5	30	67	50	91
Fusicoccin	33.5	28.5	75	34	72

Concentrations: IAA, 5×10^{-5}M; FC, 1.5×10^{-5}M; NaF, 4×10^{-3}M; 2,4-DNP, 10^{-4}M; Act. D, 80 mg/liter; CYC., 100 mg/liter. From Lado et al. 1973.

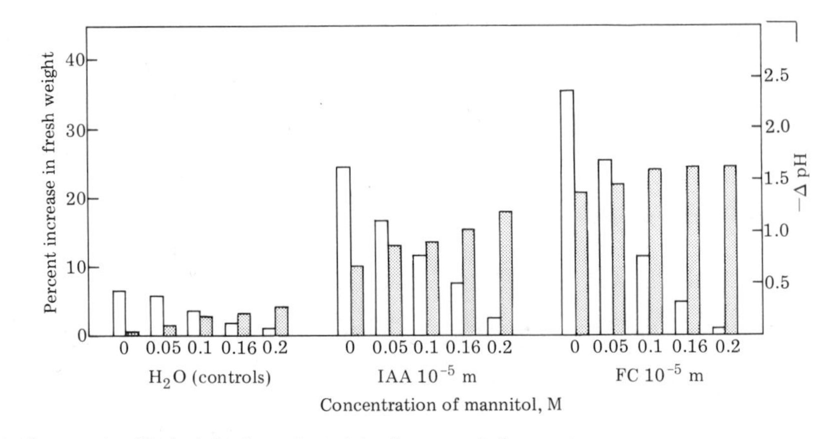

14-16 Showing that the decrease in pH (shaded columns) persists when mannitol prevents elongation (open columns), both with IAA and with fusicoccin. From Marré et al. 1973.

potassium uptake and movement are closely intertwined with proton extrusion, if not actually a specific part of the process. In the first hour or so potassium comes out into the medium, as protons do also, but thereafter potassium is strongly reabsorbed. Striking evidence for the promotion of the proton extrusion by potassium, and not by sodium, is shown in table 14-2. It appears that localisation of the ions plays a crucial part in the processes leading to elongation; potassium moves out of the wall (or the vacuole) at first, then crosses the membrane as the hydrogen ions cross in the other direction. Something of that kind seems to occur. Ultimately the vacuole, which contains many salts, would have to be the source of potassium for the prolonged growth which auxin can cause in an unbuffered medium. For without adding any potassium salt many tissues can continue to elongate in auxin for at least 16 to 18 hours.

There is even a third aspect of plant behavior which may turn out to have its basis in this hydrogen-potassium ion exchange. Bünning, who has led much of the exploration of the diurnal or "circadian" movements of leaves, has found that the antibiotic Valinomycin causes a shift in the phase of these movements in *Phaseolus*. Now Valinomycin owes its antibiotic potency to its ability to inhibit the transport of potassium ions across membranes. Since it is only intact rooted (and therefore fertilized) plants that are able to maintain prolonged circadian rhythms, a supply of potassium ions might well be the limiting factor over long periods. Since it is changes in turgor that underlie the leaf movements, it may be that the immediate mechanism of these changes is about the same as that of stomatal movements, and only the *initiating* stimulus (light in one case, the diurnal clock in the other) is different. Thus, although many small points will have to be cleared up, it is an exciting idea, even though frankly speculative at present, that

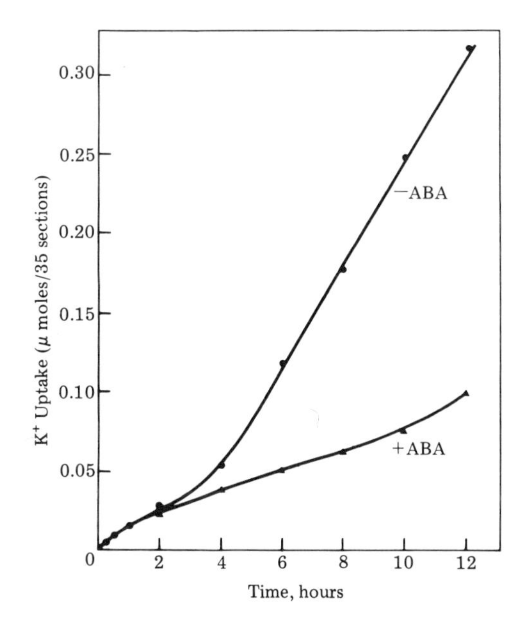

14-17 The uptake of potassium ions by freshly cut oat coleoptile segments in presence and absence of AbA (10 ppm). Both solutions contained IAA. From Reed and Bonner, 1974.

Table 14-2 Change in acidity of the medium after 2-mm pea stem segments have grown in Fusicoccin. The figures are mμ equivalents of hydrogen ion in the solution.

Experiment No.	No salt	NaCl added	KCl added
In 4 hours, after 7 hours in Fusicoccin			
I	40	—	270
II	24	30	208
In 3 hours, after 2.5 hours in pH 5.0			
III	10	−20	408

Data of Marré et al. 1974 b.

there is a possible basic mechanism here for elongation, stomatal movement, and diurnal rhythms.

Before one can accept the idea that auxin action on elongation is *only* due to hydrogen ion excretion, it must be noted that there are some important reservations to be made. In the first place, growth caused by acid media is commonly transient, lasting only an hour or two, while growth caused by auxin is long lasting. In the case of potato or artichoke tuber slices it may continue for 4 to 6 days. Still more serious is the fact that if tissues are placed in an acid medium and allowed to grow until they stop growing, then rinsed and placed in an auxin solution, they will renew growth. Yet if the growth were due only to the hydrogen ion balance and this has come to an end, it should not be possible for auxin to renew it. Those experiments in the past have been of a long-term nature, and more careful work over short growth periods will be needed before this can be cleared up.

It is hard to see, also, how the effect of auxin at the organ level can be brought into the ion-extrusion picture. Root formation, bud inhibition, and ethylene production seem so unrelated both to cell enlargement and to each other, that one is forced to accept the idea that, at least for such processes, an action either on the nucleus or at the m-RNA level must occur. The extensive evidence summarized in section E above cannot be lightly set aside.

H. THE MODES OF ACTION OF OTHER HORMONES

As far as gibberellin is concerned, we know a great deal more about the chemistry of this group than about the physiology. We do know that its action on stem elongation is different from that of auxin, since when it produces elongation it does not increase the plastic extensibility of the tissue. Also in the work on cellulase described above there was no effect whatever of gibberellic acid. With isolated pea stem segments there is good reason to accept the deduction that the actions of gibberellin and auxin are additive. Pea stem segments in auxin show the usual response with, of course, an optimum curve (probably due to the production of ethylene at high auxin concentrations). In gibberellin they show a comparable response, though with a much less marked optimum. In auxin plus gibberellin they show either an approximate summation of two responses, or else, in some special cases, a marked synergism. These facts all suggest that they have two separate and independent sites of action.

As to its action at the molecular level, we have a hint from the work of Kessler's group in Israel, who find that (roughly) the more DNA a plant has in its nuclei, the less sensitive it is to GA. Most of the plants in which applied GA causes vigorous elongation contain less than about 50 picograms ($= 50 \times 10^{-12}$ grams) of DNA per nucleus, and those with more than 50 picograms show little or no growth response. Also when CCC, a gibberellin antagonist, inhibits the gibberellin-induced elongation in cucumber stems, part of the endogenous GA is shifted from the free to the bound form. These workers therefore believe that GA somehow binds to DNA (see Snir & Kessler, 1975).

The mode of action of ethylene we also know little about. Burg believes that the reaction involves the absorption of ethylene on to a metal atom, perhaps in a metallic enzyme. A number of metals form ethylenic complexes (especially with ethylenediamine), and the action of inhibitors might be to displace ethylene from such a metalloenzyme. The action of CO_2 is certainly competitive.

Only in the case of cytokinins do we have an obvious clue to the action, because these are adenine derivatives and thus may act through the presence of adenine in one or several of its functions. The most suggestive example here is the group of cytokinins that are found in transfer-RNA. For most of the t-RNAs contain not only the usual adenine, guanine, cytosine, and thymine bases but also contain substituted adenines. Several contain 6-methyladenine, and now several have been found to contain larger substituent groups which confer cytokinin activity. Especially notable are the 6-isopentenyl derivative (2iP) and its 2-methylthio derivative (ms-2iP). Both of these are found, and in more than one t-RNA. They occur in t-RNAs for several different amino acids, including those for serine, leucine, phenylalanine, and tyrosine. All those contain 2-methylthio, 6-isopentenyl adenine as part of the molecule, and in those studies where the structure of the whole molecule has been worked out, the cytokinin-active substituted adenine is usually adjacent to the anti-codon, where the attached amino acid would be located and hence where the specificity would be sharpest.

The cytokinin of one t-RNA of tobacco leaves and of wheat germ is zeatin; curiously enough it is in the *cis*-configuration

(4-hydroxy, 3-methyl, *cis*-2-butenylaminopurine) which is only 1% to 2% as active as the *trans* form. There is also some zeatin in the t-RNA from pea stems, and this is a mixture of *cis*- and *trans*- forms. Cytokins also occur free in plants, perhaps *via* breakdown of RNAs, but there is evidence that they can also be formed independently (see Chen *et al.*, 1976).

The Wisconsin group has expended a lot of effort in determining whether applied synthetic cytokinins are incorporated into t-RNAs. Skoog and Leonard and their co-workers have applied synthetic benzylaminopurine, labeled with C^{14}, to actively growing tissue cultures and then isolated the t-RNA and looked for the label. The C^{14} is indeed there, but the experiments are not quite conclusive because it represents only a small fraction of the t-RNA. Much of the t-RNA contains other cytokinins which would normally have to be synthesized anew by the tissue. However, it is satisfactory that a little of the benzylaminopurine does find its way there.

But even if we know that that is the route of cytokinin's entry we must still ask, why does it have so many effects? For it is characteristic of cytokinin that, among other things, it produces buds on tissue cultures, it releases the inhibition of inhibited axillary lateral buds, it promotes the synthesis of protein, and it antagonizes senescence. One could always take refuge in the generality that a compound which is to become a component of a nucleic acid, the centre of activity, may influence a number of different phenomena. But in any case there is no clear idea of how this should be.

And now I come back to what we noted at the outset of this chapter, namely, that whenever one studies the effects of these hormones, one is struck by the fact that they seem to control a number of processes, but that they can be processes which have no obvious relationship to one another. Thus auxin produces elongation and an increase in respiration and has many varied morphogenetic effects. Cytokinin certainly has more than one effect. Gibberellin causes enormous elongation in rosette plants and it activates the amylase in barley seeds; these seem to be quite unrelated phenomena. Of course they may have a single basis; but it is also possible that they do not have a single basis. Perhaps we have too determinedly thought of *the* mode of action.

It is a principle in evolution, and certainly in biochemistry, that compounds become of great importance in biology if they do more than one thing; if they bring about two apparently different results, then they are much more crucial and are retained in evolution by the organism. An obvious example is, of course, DNA. DNA does two things: first, it multiplies itself, a very unusual phenomenon; no other substance does this, for it is the result of its double helical structure; second, it can be transcribed into RNA. These are two functions which apparently are unrelated, although they basically arise from its characteristic structure. If it had only one of those two functions, it would not be nearly so important. The other striking example is ATP itself. ATP is not merely a phosphate-transferring agent which allows biochemical synthesis; it actually enters into the composition of such substances as Coenzyme A, which contains a whole ATP molecule. And ATP becomes attached to sugars to make those ADP-glucose, UDP-glucose types of compounds which are then activated sugars, able to polymerize and to isomerize.

Thus, several compounds are known which owe their special biological importance to their ability to carry out *more than one* function, and I feel that the same thing may be true at the organ level. The ear is a characteristic example; the ear is not only an organ of hearing but—what seems like a totally unrelated function—an organ of balance. The animal ear has these two functions, which must be somehow related because they always go together throughout evolution. At a subtler level we can draw a parallel with the eye pigments, which have two functions, one for dark and light and one to tell colors. And we saw in an earlier chapter how the phototropism of seedlings is dual, too.

It is worth considering that this may be a general phenomenon in biology. The dual function may be of critical importance whether at the organ level, the organelle level, or the chemical level.

REFERENCES

Abeles, F. B., and Rubinstein, B. Regulation of ethylene evolution and leaf abscission by auxin. *Plant Physiol.* 39, 963-969, 1964.

Bonner, J. The relation of hydrogen ions to the growth of the *Avena* coleoptile. *Protoplasma* 21, 406-423, 1934.

Burg, S. P., and Burg, E. A. Molecular requirements for the biological activity of ethylene. *Plant Physiol.* 42, 144-152, 1967.

———. Auxin-stimulated ethylene formation; its relationship to auxin-inhibited growth, root geotropism and other plant processes. In *Biochemistry and physiology of plant growth substances*, ed. F. Wightman and G. Setterfield, pp. 1275-1294. Ottawa: Runge Press, 1968.

Cande, W. Z., Goldsmith, M. H. M., and Ray, P. M. Polar auxin transport and auxin-induced elongation in the absence of cytoplasmic streaming. *Planta* 111, 279-296, 1973.

Chen, S. S. Energy budgeted for protein synthesis in cultured plant cells. *Taiwania* 18, 5-8, 1973.

Chen, C.-M., Eckert, R. L., and McChesney, J. D. Evidence for the biosynthesis of transfer-RNA-free cytokinin. *F.E.B.S. Letters* 64, 429-434, 1976.

Cleland, R. Auxin-induced hydrogen ion secretion from *Avena* coleoptiles. *Proc. Nat. Acad. Sci.* 70, 3092-3093, 1973.

Davies, E., and MacLachlan, G. A. Effects of IAA on intracellular distribution of β-glucanase activities in the pea epicotyl. *Arch. Biochem. Biophys.* 128, 595-600, 1968.

Frey-Wyssling, A. *The Plant Cell Wall.* Berlin: Borntraeger, 1976.

Hértel, R., and Flory, R. Auxin movement in corn coleoptiles. *Planta* 82, 123-144, 1968.

Heyn, A. N. J. Glucanase activity in coleoptiles of *Avena. Arch. Biochem. Biophys.* 132, 442-449, 1969.

Kaufman, D., and Thimann, K. V. Cytoplasmic streaming in the cambium of white pine. In *The physiology of forest trees,* ed. K. V. Thimann, pp. 479-492. New York: Ronald Press, 1958.

Kelso, J. M., and Turner, J. S. Protoplasmic streaming in *Tradescantia*. 1. The effects of IAA and other growth-promoting substances on streaming. *Australian J. Biol. Sci.* 8, 19-35, 1955.

Key, J. L., and Ingle, J. RNA metabolism in response to auxin. In *Biochemistry and physiology of plant growth substances,* ed. F. Wightman and G. Setterfield, pp. 711-722. Ottawa: Runge Press, 1968.

Labavitch, J. M., and Ray, P. M. Relationship between promotion of xyloglucan metabolism and induction of elongation by IAA. *Plant Physiol.* 54, 499-502, 1974.

Lado, P., Rasi-Caldogno, F., Pennachioni, A., and Marré, E. Mechanism of the growth-promoting action of Fusicoccin. *Planta* 110, 311-320, 1973.

Lamport, D. T. A. Hydroxyproline-D-glycosidic linkage of the plant cell wall glycoprotein Extensin. *Nature* 216, 1322-1324, 1967.

Mansfield, T. A., and Jones, R. J. Effects of abscisic acid on potassium uptake and starch content of stomatal guard cells. *Planta* 101, 147-158, 1971; Squire, G. R. and Mansfield, T. A. The action of fusicoccin on stomatal guard cells and subsidiary cells. *New Phytol.* 73, 433-440, 1974.

Marré, E., Colombo, R., Lado, P. and Rasi-Caldogno, F. Correlation between proton extrusion and stimulation of cell enlargement. Effects of Fusicoccin and of cytokinins on leaf fragments and isolated cotyledons. *Plant Sci. Letters* 2, 139-150, 1974 a.

———, Lado, P., Rasi-Caldogno, F., and Colombo, R. Correlation between cell enlargement in pea internode segments and decrease in the pH of the medium of incubation. 1. Effects of Fusicoccin, natural and synthetic auxins and mannitol. *Plant Sci. Letters* 1, 179-184, 1973.

———, Lado, P., Rasi-Caldogno, F., Colombo, R., and De Michelis, M. I. Evidence for the coupling of proton extrusion to potassium uptake in pea internode segments treated with Fusicoccin or auxin. *Plant Sci. Letters* 3, 365-379, 1974 *b*.

Masuda, Y., Oi, S., and Satomura, Y. Further studies on the role of cell wall-degrading enzymes in cell wall loosening in oat coleoptiles. *Plant and Cell Physiol.* 11, 631-638, 1970.

———, and Yamamoto, R. Effect of auxin on 1,3-glucanase activity in *Avena* coleoptiles, *Development, Growth and Differentiation* 11, 287-296, 1970.

Nissl, D., and Zenk, M. H. Evidence against induction of protein synthesis during auxin-induced elongation of *Avena* coleoptiles. *Planta* 89, 323-341, 1969.

Noodén, L. D., and Thimann, K. V. Evidence for a requirement for protein synthesis for auxin-induced cell enlargment. *Proc. Nat. Acad. Sci.* 50, 194-200, 1963.

———. Inhibition of protein synthesis and of auxin-induced growth by chloramphenicol. *Plant Physiol.* 40, 193-201, 1965.

O'Brien, T. P. Note on an unusual structure in the outer epidermal wall of the *Avena* coleoptile. *Protoplasma* 60, 136-140, 1965.

Ordin, L., Applewhite, T. H., and Bonner, J.

Auxin-induced water uptake by *Avena* coleoptile sections. *Plant Physiol.* 31, 44-53, 1956.

Ray, P. M. Regulation of β-glucan synthetase activity by auxin in pea stem tissue. 1. Kinetic aspects. 2. Metabolic requirements. *Plant Physiol.* 51, 601-608, 609-614, 1973.

Rayle, D. L. Auxin-induced hydrogen ion secretion in *Avena* coleoptiles and its implications. *Planta* 114, 63-73, 1973.

Reed, N-M. M., and Bonner, B. A. The effect of abscisic acid on the uptake of potassium and chloride into *Avena* coleoptile sections. *Planta* 116, 173-185, 1974.

Sacher, J. A. Studies on auxin-membrane permeability relations in fruit and leaf tissues. *Plant Physiol.* 34, 365-372, 1959.

Snir, I., and Kessler, B. Relationships between the cellular content of DNA in plants and their sensitivity to gibberellin. *Plant Sci. Letters* 5, 163-170 (1975).

Strugger, S. Beiträge zur Physiologie des Wachstums. 1. Zur protoplasma-physiologischen Kausalanalyse des Streckungswachstums. *Jahrb. Wiss. Bot.* 79, 406-471, 1934.

Sweeney, B. M., and Thimann, K. V. The effect of auxins on protoplasmic streaming. 2. *J. Gen. Physiol.* 21, 439-461, 1938.

Thimann, K. V., and Samuel, E. W. The permeability of potato tissue to water. *Proc. Nat. Acad. Sci.* 41, 1029-1033, 1955.

Valent, B. S., and Albersheim, P. The structure of plant cell walls. 5. On the binding of xyloglucan to cellulose fibers. *Plant Physiol.* 54, 105-108, 1974.

Van Santen, A. M. A. Groei, groeistof en pH. Dissertation, Utrecht, 1940.

Walker, G. C., Leonard, N. J., Armstrong, D. J., Murai, N., and Skoog, F. The mode of incorporation of 6-benzyl-aminopurine into tobacco callus t-ribonucleic acid. A double labeling determination. *Plant Physiol.* 54, 737-743, 1974.

Yoda, S., and Ashida, J. Effect of gibberellin on the extensibility of the pea stem. *Nature* 182, 879-880, 1958; *Plant and Cell Physiol.* 1, 99-105, 1960.

Supplementary Reading

GENERAL

Audus, L. J. 1972. *Plant Growth Substances.* Vol. 1, *Chemistry and Physiology*. London: Leonard Hill, 533 pp.

Galston, A. W., and Davies, P. J. 1971. *Control Mechanisms in Plant Development.* Englewood Cliffs, New Jersey: Prentice Hall, 184 pp.

Leopold, A. C., and Kriedemann, P. E. 1975. *Plant Growth and Development,* Second edition. New York: McGraw-Hill, 545 pp.

Phillips, I. D. J. 1971. *Introduction to the Biochemistry and Physiology of Plant Growth Hormones.* New York and London: McGraw Hill, 173 pp.

Steward, F. C., and Krikorian, A. D. 1972. *Plants, Chemicals and Growth.* New York: Academic Press, 232 pp.

Thimann, K. V., with sections by L. G. Paleg and C. A. West, and by F. Skoog and R. Schmitz. 1972. *Plant Physiology: A Treatise,* ed. F. C. Steward. Vol. VI B, *The Natural Plant Hormones.* New York: Academic Press, 365 pp.

Wareing, P. F., and Phillips, I. D. J. 1970. *The Control of Growth and Differentiation in Plants.* Oxford and New York: Pergamon Press, 303 pp.

Went, F. W. 1974. Reflections and speculations. In *Annual Revs. of Plant Physiol.* 25: 1–25.

Wilkins, M. B., ed. 1969. *Physiology of Plant Growth and Development.* London: McGraw Hill, 695 pp. (18 chapters).

SYMPOSIA

Biochemistry and Physiology of Plant Growth Substances, ed. F. Wightman and G. Setterfield, 1968. Ottawa: Runge Press, 1642 pp. (112 papers).

Plant Growth Substances 1970, ed. D. J. Carr, 1972. Berlin, Heidelberg and New York: Springer Verlag, 837 pp. (108 papers).

Hormonal Regulation in Plant Growth and Development, ed. H. Kaldewey and Y. Vardar, 1972. Weinheim: Verlag Chemie, 524 pp. (40 papers and discussion).

Plant Growth Substances 1973. 1974, Tokyo: Hirokawa Publishing Co. 1242 pp. (148 papers).

SPECIAL TOPICS: BOOKS

Khan, A. A., ed. 1977. *The Physiology and Biochemistry of Seed Dormancy and Germination.* A.S.P. Amsterdam: Biology and Medicine Press (in press).

Kozlowski, T., ed. 1973. *The Shedding of Plant Parts.* New York: Academic Press, 560 pp.

Krishnamoorthy, H. N., ed. 1976. *Gibberellins and Plant Growth.* New Delhi: Wiley Eastern, 356 pp.

Smith, H. *Phytochrome and Photomorphogenesis.* 1975. New York: McGraw-Hill, 242 pp.

Vince-Prue, D. *Photoperiodism in Plants.* 1975. New York: McGraw-Hill, 424 pp.

SPECIAL TOPICS: REVIEW ARTICLES

With the exception of the last entry, all of the following are in *Annual Review of Plant Physiology,* Annual Reviews, Inc., Palo Alto, California:

Abeles, F. B. 1972. Biosynthesis and mechanism of action of ethylene. 23:259–292.

Dure, L. S. III. 1975. Seed formation. 26:259–278.

Evans, M. L. 1974. Rapid responses to plant hormones. 25:195–223.

Hall, R. H. 1973. Cytokinins as a probe of developmental processes. 24:415–444.

Jones, R. L. 1973. Gibberellins: their physiological role. 24:571–598.

Juniper, B. 1976. Geotropism. 27:385–406.

Mayer, A. M. and Shain, Y. 1974. Control of seed germination. 25:167–193.

Milborrow, B. V. 1974. The chemistry and physiology of Abscisic acid. 25:259–307.

Phillips, I. D. J. 1975. Apical dominance. 26:341–367.

Sacher, J. A. 1973. Senescence and postharvest physiology. 24:197–224.

Scott, T. K. 1972. Auxins and roots. 23:235–258.

Sequeira, L. 1973. Hormone metabolism in diseased plants. 24:353–380.

Torrey, J. G. 1976. Root hormones and plant growth. 27:435–459.

Wain, R. L. 1975. Some developments in research on plant growth inhibitors. *Proc. Roy. Soc.,* ser. B. 191:335–352.

Author Index

Figures in italics refer to the bibliographies at the ends of the chapters

Subject Index